Emergent Nature

Patterns, Growth and Scaling in the Sciences

Miroslav M. Novak
School of Mathematics, Kingston University, UK
Editor

Emergent Nature

Patterns, Growth and Scaling in the Sciences

World Scientific
New Jersey • London • Singapore • Hong Kong

Published by

World Scientific Publishing Co. Pte. Ltd.

P O Box 128, Farrer Road, Singapore 912805

USA office: Suite 1B, 1060 Main Street, River Edge, NJ 07661

UK office: 57 Shelton Street, Covent Garden, London WC2H 9HE

British Library Cataloguing-in-Publication Data
A catalogue record for this book is available from the British Library.

EMERGENT NATURE

ISBN 981-02-4910-1

Printed in Singapore by Mainland Press

to Doreen, Karl and Erika

Preface

This book, *Emergent Nature*, provides something of interest to every reader in the general field of nonlinear dynamics. The abundance of papers from numerous disciplines makes it exciting reading and provides a unifying thread through the topics such as

ray tracing	ecology	soil electrical conductivity
foetal heartbeat	metabolic cycle of plants	solar magnetic fields
rotating DLA-clusters	modelling fractal surfaces	monkey cortical neurons
tree morphology	random wavelets	occurrence of earthquakes
structure of peptides	cancer growth	structures in architecture

The tools of nonlinear dynamics frequently succeed in classifying and correctly describing these diverse fields and provide a common link. This modern approach has gained universal approval in the last decade or two.

Papers in this book are based on presentations at the 7th international conference, *Fractal 2002*, exploring the above-mentioned issues. The conferences are now regular and well established among the nonlinear series of conferences and provide a unique and genial atmosphere to foster exchange and incubation of ideas. This travelling conference series, organized in different geographical regions, is to encourage international collaborations. Among the many distinguishing features of this series is its multidisciplinary nature, which has been growing steadily.

There are three papers, based on the invited talks by the eminent authorities in their respective fields, J.-P. Bouchaud (France), M. G. Velarde (Spain) and B. J. West (USA).

The *Fractal 2002* conference was partially supported by the Department of the Navy Grant, issued by the Office of Naval Research International Field Office.

The conference was made possible through the generous help of the following members of the programme committee (in alphabetical order):

F. T. Arecchi (*Italy*), Y. Bar-Yam (*USA*), A. Coniglio (*Italy*), M. Daoud (*France*), K. Falconer (*UK*), J.-F. Gouyet (*France*), A. Holden (*UK*), A. Hubler (*USA*), R. Kapral (*Canada*), M. S. Keane (*The Netherlands*), C. M. Kim (*Korea*), J. Klafter (*Israel*), J. Levy-Vehel (*France*), R. S. MacKay (*UK*), J. Marro (*Spain*), E. Mosekilde (*Denmark*), M. M. Novak (*UK – Chair*), T. Puu (*Sweden*), G. Radons (*Germany*), D. Saupe (*Germany*), P. F. Stadler (*Austria*), H. E. Stanley (*USA*), W. H. Steeb (*South Africa*), B. J. West (*USA*), J. Z. Zhang (*P. R. of China*), and Y.-C. Zhang (*Switzerland*).

Details on the next conference of this series will be posted on the following website http://www.kingston.ac.uk/fractal/.

M. M. Novak
Kingston-upon-Thames, UK

Contents

MODELING CEREBELLAR DYNAMICS

MANUEL G. VELARDE AND VALERI A. MAKAROV
Instituto Pluridisciplinar, UCM, Paseo Juan XXIII, 1, Madrid 28040, Spain
E-mail: velarde@fluidos.pluri.ucm.es and makarov@fluidos.pluri.ucm.es

RODOLFO R. LLINAS
Department of Physiology and Neuroscience, Medical Center, New York University, New York 10016, U.S.A.

Modeling various functional circuits in the Central Nervous System (CNS) is crucial for the understanding of its specific and global functions. In addition to understanding brain function, such modeling is essential in designing autonomous artificial systems mimicking particular CNS functions in a new effort towards developing neuro-based robots. The intrinsic rhythmic activity of the CNS is known to be essential to its functional organization. Such rhythmicity is supported by the electrical activity of single neurons and by the existence of well-defined feedback and feedforward neuronal circuit loops. These circuits allow selection and control of particular global rhythms as a resonance, synchronous or synergetic property in such neuronal clusters. Because the properties of such neurons, and the loops they generate, can be directly investigated information from neuroscience research gives important insight into network functions -and their relevance to particular CNS functions. An example is the olivo-cerebellar circuit responsible for fine-tuning of motor performance and control of movements. It involves the Inferior Olive (IO), a cell cluster at the lower brainstem, whose neurons project excitatory signals through their axons (the climbing fibers) into Purkinje cells (PC) in the cerebellar cortex and by collaterals of such axons to the Cerebellar Nuclei (CN). In turn the PCs send inhibitory messages to the CN. Because CN neurons generate inhibitory effect on to the IO neurons such circuit can be viewed as a self organizing neuronal clock system. Here a model is proposed to account for such features. We also give evidence that noise that is unavoidable in biophysical systems can be taken to advantage by this neurobiological clock.

1 Introduction. Cerebellum lore, underlying basic concepts and explanation of terms

There are about 10^{10}–10^{11} neurons in the brain with some 10^3–10^5 synapses or connections per neuron, hence about 10^{15} synapses. Synapses, the junctions between neurons, are chemical or electrical. The former are unidirectional, excitatory or inhibitory, exhibit a delay in transmission of about one millisecond, help amplifying action potential signaling between neurons (like a TV or FM repeat station) and possess plasticity. Electrical synapses are either unidirectional or bidirectional (depending on channel conductance due to the molecular structure of the junction channels), preserve depolarization or hyperpolarization upon transmission, exhibit no delay in response and due to dissipation losses lower the intensity of inputs. Gap junctions (the resultant "electrotonic coupling") support passive electrical communication (no gain increase from the original signal; Bennett, 1997; numbered references are given below). Incidentally, it was for long time known that synapses seem to obey Darwinian evolution in the early stage of development after birth (Changeux, 1983), and, moreover, appear and disappear in the course of time but it is only recently that it has been established that the (human) brain does produce new neurons in adulthood.

The functional characteristics of a neuron are the outcome of a complex interplay between its intrinsic membrane properties and its synaptic-interactions (Llinas, 1988). Signals, action potentials, travel along neuronal axons at speeds in the range 10^{-1}m/s–10^2m/s (a maximum speed of about 200km/h) depending on axon diameter. Generally, axons operate over long distances without loss. The speed is limited by the internal longitudinal resistance of the core conductor and by capacitative losses. Roughly, the electrical resistance of an axon of, say, length one meter is about that of 10^{10} kilometers of (0.7 mm) standard copper wire. The signals that travel along these axons are all or none electrical events continuously boosted, at every step, by inward current, somewhat similar to that in *active* lattices (Nekorkin and Velarde, 2001). In myelinated axons the boosting currents are introduced at particular sites along the nerve (nodes of Ranvier) and so such conduction is known as saltatory (Huxley and Stampfli, 1949). Action potentials themselves are transmembrane voltage pulses/spikes/(dissipative) solitons of about 10^2mV amplitude. In view of the thickness (10^2Å) of axonal membranes this voltage corresponds to a 10^2kV/cm field which is about the limit of dielectric rupture (at the edge of membrane instability). Neurons (composed of dendrites, soma or metabolic/biochemical center, and the axon) have a resting membrane potential in the range of -60 to -80mV. This resting potential is a DC driving force (e.m.f.) *actively* generated across the membrane that supports the gated flux of ions responsible for action potential production. The resting potential results from the different concentrations of Potassium (K) ions inside and outside the cell membrane. When the membrane potential becomes positive to the resting potential the cell is said to be depolarized; when the membrane potential becomes negative to the resting potential it is said to be hyperpolarized. It is possible to depolarize or hyperpolarize a neuron by piercing the cell's membrane with a microelectrode and passing a positive or negative current through the electrode. Exceeding a certain value (threshold voltage about -40 mV), generates a regenerative (avalanche) opening voltage-gated (Sodium, Na) channels that becomes self-regenerative (like a chain reaction). Because the concentration of Na ions outside the axon is about ten times the corresponding value inside and the resting potential is negative (opposite charge to the Na ion) this electrochemical driving force produces inward Na movement across the membrane. The resulting fast inward Na-current accounts for the rapid upstroke of the action potential (up to +55mV). The inward Na-current is followed by a slower outward K-current which repolarizes the membrane (the concentration of K ions inside the axon is about five times the corresponding value outside; Hodgkin and Huxley, 1952). We shall recall later a more complex behavior exhibited by neurons in which Calcium (Ca) channels play a significant role. Following the action potential, there is a refractory period during which no action potential can be induced, which is due primarily to Na-channel inactivation; like in most reaction–diffusion systems, e.g., forest fires this refractoriness ensures one-sided propagation. Thus large, high, brief increases in conductances to Na and K are the key electrochemical processes underlying action potentials in neurons. The increase in K-conductance exhibits a long tail beyond the time required to bring the potential to resting level.

This tail contributes to an after-hyperpolarization and, moreover, contributes to decreased excitability as the K-conductance represents a resistive shunt that serves as a leak to any synaptic excitatory current.

Any stimulus large enough to generate an action potential in the axon produces the same amplitude action potential, regardless of the stimulus strength (amplitude). However, other action potential features depend on the strength of the stimulus. For instance, the time delay (latency) from the initial time of the stimulus to the peak of the action potential is shorter for a stronger stimulus, this is called "utilization time". Furthermore, if a sustained depolarizing stimulus above threshold leads to action potential generation at a certain frequency, the stronger is the depolarizing signal the higher is the frequency of action potential firing as the response of the neuron within a limit given by the Na-channel inactivation dynamics.

Generally, neurons have endogenous electrical activity and hence they are not mere input-output devices. Some neurons do not fire spontaneously at all (*silent* neurons), others may beat (*pacing* neurons) or burst in a regular manner (Llinas, 1988). *Bursting* neurons tend to exhibit rhythmic behaviors generating regular bursts of action potentials that are separated by hyperpolarization of their membranes. Although external stimulation can change the firing rate of a neuron, or inhibit it altogether, the mechanisms that drive repetitive firing are generally intrinsic to the neuron itself and do not require continual synaptic activation or other external stimuli.

1.1 The Cerebellum

Figure 1 is a drawing of the lateral view of the human brain showing the cerebellum, details about its mid-sagittal section, and a folium, showing the three layers of cerebellar cortex and the white matter. The cerebellum lies at the back of the skull behind the brain stem and under the great hemispheres of the *cerebrum.* Its name is a Latin diminutive of *cerebrum* and means simply "lesser brain". Superficially that is an adequate description of the cerebellum: it is much smaller than the cerebrum but shares certain morphological features with it. Also it has more neurons than the rest of the brain put together. As in the cerebrum, the highest functions in the cerebellum in part relate to the thin layer of gray matter that makes up the cortex and, as in the cerebrum, this layer is elaborately folded and wrinkled to increase its area. The folds are in fact much deeper and more closely spaced than those of the cerebral cortex. Both the structure and the function of the cerebellum have been known, at least in terms of broad principles, since the end of the 19th century. The fundamentals of cerebellar anatomy were established in 1888 by Ramon y Cajal [1,2]. The first reliable clue to the function of the cerebellum was provided by L. Luciani, who discovered that experimental animals deprived of cerebellum suffer disturbances of coordination and equilibrium. The physiology of the cerebellar circuit was originally described in the middle of last century (Eccles, Ito and Szentagothai, 1967; Llinas, 1969). Below we say more about this question.

Figure 2 corresponds to a Purkinje cell (PC) of the human cerebellum as depicted by Ramon y Cajal (1897). The dendrites of a PC in humans receive about

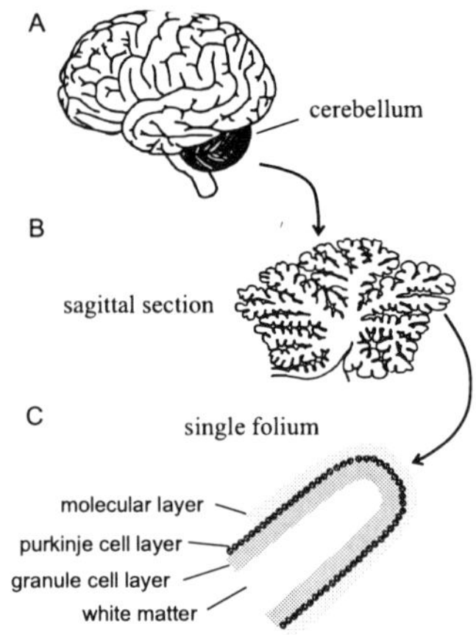

Figure 1. Brain and cerebellum (taken from Llinas and Walton, 1988).

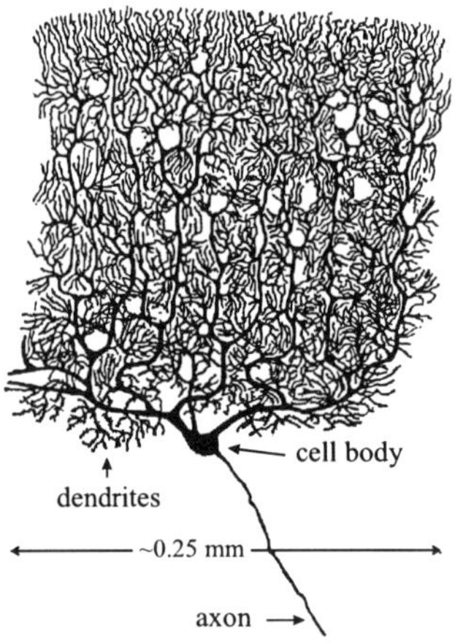

Figure 2. Purkinje cell (redrawn after Ramon y Cajal, 1897).

two hundred thousand synaptic inputs from the granule cell layer and only one climbing fiber from the IO. A PC may contact as many as thirty NC, but most contacts are made with a few PC. In addition to this divergence, there is convergence since there are about twenty or more NC for each PC. There are almost a thousand PC axons for each NC. On the other hand "the arrangement of the PC dendrites provides a clue to its functional organization" (Llinas and Walton, 1988). The entire mass of repeatedly bifurcating branches is confined to a single plane. Moreover, the planes of all PC dendrites in a given region are parallel to each other, so that the dendritic arrays of the cells stack up in neat ranks; adjacent cells in a single plane form equally neat, but overlapping files. Thus parallel fibers (from the granule layer) running perpendicular to the plane of the dendrites intersect a

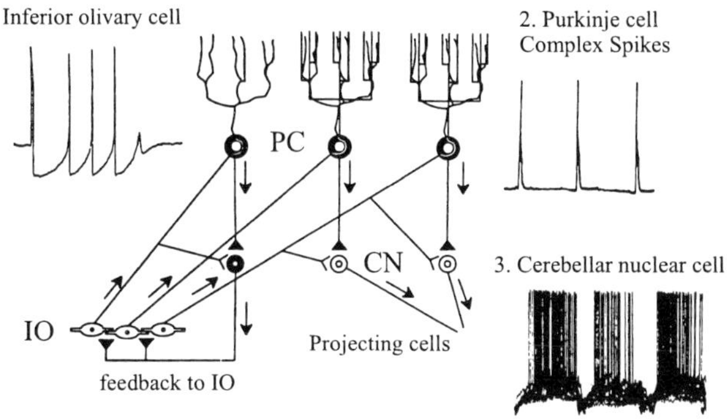

Figure 3. Sketch of the olivo-cerebellar system with typical spiking and bursting produced by the various neurons involved in the dynamics (taken from Llinas and Walton, 1988).

great many PC, by the very manner in which these elements are organized.

Figure 3 sketches the organization of the olivo-cerebellar system [Inferior Olive (IO)–(PC)–Nuclear Cerebellar (NC) neurons]. A Superior Olive exists which was linked by Ramon y Cajal to audition.

The *cerebellum* has undergone enormous elaboration throughout evolution. Its own evolution suggests that its function has become more important during the span of vertebrate history. In *homo sapiens* its size seems to have increased four-fold in the past million years while the entire brain has increased three-fold. In man the surface is about 5×10^4 cm^2 (for a frog is about 4×10^3 cm^2). The cerebellum has one third of brain mass and occupies about ten percent of overall volume. It contains more than half of the total number of neurons in the brain. Incidentally, some electric fish (*mormiridae*) have a gigantic cerebellum making up to some 70% of the total brain weight. The *cerebellum* is highly regular and largely repetitive with units/neurons of (relatively) simple structure. All areas of the cerebellum seem to perform similar function but each area performs its function on a different set of inputs. It has been said, that of all parts of the brain the *cerebellum* might be most readily likened to a (standard electronic) computer.

Although the cerebellum appears involved in coordination of interrelated muscle activities, fine control of reflexes, tuning of motor events and sensorimotor integration, it is a regulatory organ (clock function) that supervises rather than command movements. It does not initiate movement. It seems to compensate for errors in movement by comparing intention with performance. Thus, it has been said that the cerebellum may be a virtual reality center creating simulations of our movements to help control the real thing. Indeed, its disfunction does not produce alteration of sensations or paralysis as it does not play a *primary* role in either sensory or motor function. However, lesions of the cerebellum produce well defined and often devastating changes in the ability of the rest of the nervous system to generate simple and the elegant motor sequences that normal animals utilize to attain motor goals (and motor coordination). Excellence in motor coordination is obviously an adaptative change, and it is enough of an advantage to sustain the development of a specialized brain center committed primarily to that purpose.

1.2 The Inferior Olive

The IO acts like a relay station for a collection of pathways that convey sensory information from the limbs. In the rat has some $10^4 - 10^5$ neurons while in humans the figure is 10^6 in an apparent 3D architecture. Actually, when unfolded the IO is topologically composed of 2D layers. The PC (cerebellar cortex) is rather planar (2D tissue, Fig. 2) with $10^4 - 10^5$ units for rats, 10^6 for cats and $10^6 - 10^7$ for humans. The number of CN cells in the cat is in the range $10^4 - 10^5$ with a ratio PC/CN less than 30/1 while in humans this ratio is a bit below 40/1 and shows an almost constant value throughout mammalians.

In adults each PC receives synaptic input from only a single climbing fiber coming from the Inferior Olive (spontaneous firing rate is about once per second). Each IO neuron generates up to ten climbing fibers but each climbing fiber contacts only one PC with some three hundred synaptic contacts (all excitatory), quite a secured transmission. Indeed, climbing fibers seem to inform PC of errors caused by their misperformance, and hence the synaptic connection between climbing cells and PC is one of the most powerful in the entire nervous system. Single action potentials originated at low rates in the IO elicit very large Ca spikes by their excitatory postsynaptic potentials in both the soma and dendrites of the PC (hence complex spikes, about one per second) that in turn trigger large spikes by a high frequency burst of smaller Na action potentials. IO neurons affect some ten PC that are rather far from each other. In turn one PC (inhibitorily) connects to some thirty to forty CN cells. Complex activity at the PC level provides a direct picture of IO activity.

1.3 The issue of rhythmicity and temporal discontinuity

In conclusion of this introductory Section let us emphasize the importance of the rhythmic time–setting properties of the olivo-cerebellar system and its role in the proper execution of active movements in animals with cerebellar control (Llinas and Sasaki, 1989). Evidence exists that motor coordination is organized in a non–continuous manner and in such a way that movements are generated on carrier rhythm due to oscillation and resonance between central and peripheral clocks. In fact it has been known for years that the onset of movement in humans whether voluntary or involuntary occurs in phase with physiological tremor (whose essential feature is that of motion which is sustained and regular). Long ago, Sherrington (1910) proposed that motricity occurs on a background of 10–12 Hz tremor. Incidentally, the *Scherzo* of Schuman's Quartet with piano (op. 47 in E flat major; 1842) requires rhythmic movement of the hand eight times per second (in its *Finale. Vivace* the *Finale. Allegro assai* of piano concerto in A major, No. 23, K488, 1786, by Mozart is evoked). Jazz saxophonist Charlie Parker was capable of fingering notes at a rate of 11 per second (e.g. *Bird gets the worm*). Presumably there was no time for him to hear and react to one note before he played the next and hence his fingers, even when improvising, were responding to a preprogrammed pattern of neural activity accounting for groups of notes. A simple syllable (la) can be

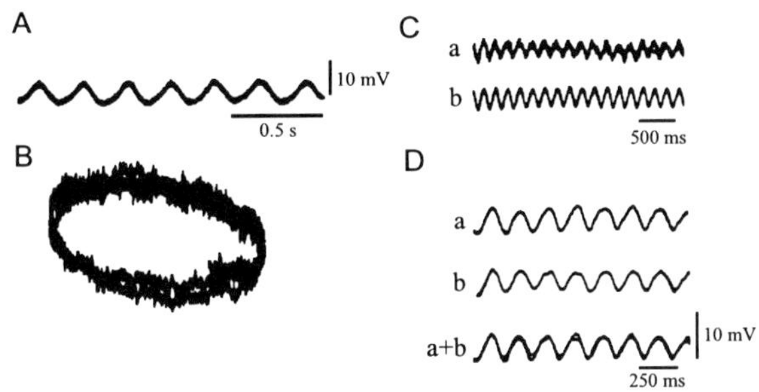

Figure 4. IO intrinsic properties. Spontaneous oscillations of the membrane potential and its synchronicity in two IO neurons simultaneously recorded intracellularly. A) spontaneous oscillatory property of the membrane potential. B) Lissajous figure to illustrate the regularity of the spontaneous oscillation. The x axis of the oscilloscope was derived by a sinusoidal wave form of 4Hz. C) superimposed traces of spontaneous membrane potential oscillations recorded simultaneously from two olivary neurons (a, b). D) an average record of six traces recorded during the same time interval from the two cells in B and superimposed in a and b. Note that the membrane potential of both neurons oscillated in phase and with similar amplitude (taken from Llinas and Yarom, 1986).

repeated about 11 times per second. For comparison note that the throb rate of a deep organ pipe is about 13 per second. More and more evidence accumulates indicating that the maximal frequency of voluntary movement never exceeds that of physiological tremor (i.e. no one can move faster that they can tremble). Leaving aside possible wind bias, reaction time to departure (sound) signal and rescaling by stride length (elite sprinters may react in 10^2 ms and may take 200 ms/stride) this sets an absolute bound (minimum) for world record in, say, the 100m race. Ultimately, then it is not surprising that the olivo-cerebellar system, which controls such rhythmicity, should be intimately intertwined with the actual control of coordination, that is, with the function of the cerebellum as a whole.

For background material and supplementary reading on neurophysiology, neurodynamics, dissipative structures and synergetics, and nonlinear theory of (active) lattices or neural networks see Refs. [1-28].

2 From structure to single cell function

Before embarking in the description of the model cerebellar loop let us recall the salient functional features of neurons in the IO, PC and CN [29-45]. Figures 4–7 illustrate spontaneous firing of IO neurons, synchronicity in behavior, response to depolarizing and hyperpolarizing signals and to various other stimuli. In pharmacologically untreated slices, intracellular recordings of cell pairs revealed the presence of synchronous oscillations in the IO (Figs. 4 and 5). Elsewhere [46] these typical features and further experimental data have recently been reproduced using a composite model-unit made of Fitz-Hugh-Nagumo-Schlögl excitatory elements [20] and a Van der Pol (robust) oscillator and we shall not delve on this matter here. Needless to say, a given (nonlinear) model may display a rich variety of behaviors when several parameters are involved and, furthermore, with a broad range of val-

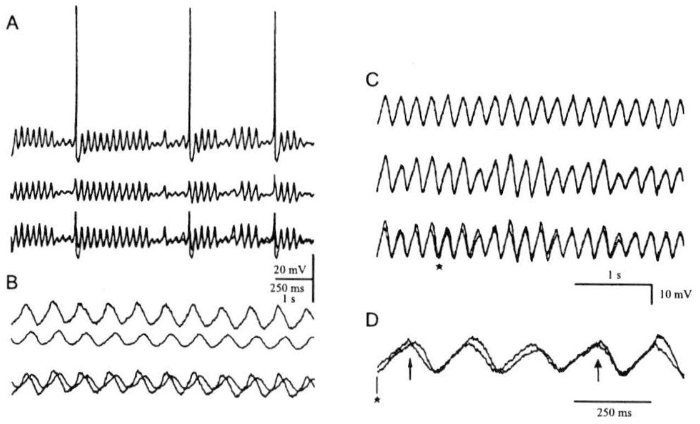

Figure 5. IO intrinsic properties. In-phase synchronous spontaneous oscillations of two nearby IO neurons for varying distance between them. A. 150 μm apart, B. 350 μm apart, C. 50 μm apart. D. Enlarged view of C from the star point (third recording) to show details of the synchronicity (arrows emphasize it) of two cells in C (taken from Benardo and Foster, 1986).

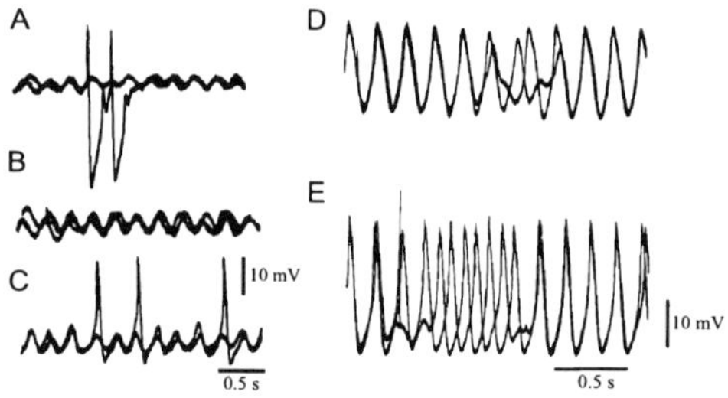

Figure 6. IO intrinsic properties. Separation of single cell excitability from the subthreshold oscillatory property. A-C) superimposed traces of membrane potential at rest (B), at 15mV depolarization (A) and at 24mV hyperpolarization (C). Note that while the spontaneous oscillations may trigger dendritic (A) and somatic (C) Ca^{2+} spikes, the frequency of the oscillations was not modified by the change in membrane potential. D) two superimposed traces of spontaneous oscillations at two levels of hyperpolarization. Upper panel at −5mV and lower at −10mV from rest level. Each wave of oscillation triggers a somatic Ca^{2+} response. In the upper trace three spikes occur out of step with respect to the oscillations. In the lower trace the somatic spike is larger, and out of step firing occur for four cycles. Records taken with TTX in the bath (taken from Llinas and Yarom, 1986).

ues. Thus a model–unit/neuron (or a cluster) may behave in many dynamic modes (subthreshold oscillations, spiking, bursting, etc.) depending on (physiological) conditions or stimuli.

Figure 8 neatly depicts the two firing levels when an IO neuron is activated by a double ramp current pulse [35]. It is worth emphasizing what these authors have demonstrated about such complex response of IO cells to direct stimulation (e.g. a depolarizing stimulus of 0.5nA during 70ms). Indeed, IO neurons exhibit somatic Na-spikes followed by a prolonged depolarization generated by a high-threshold dendritic Ca-current. The latter in turn triggers a K-conductance that will lead

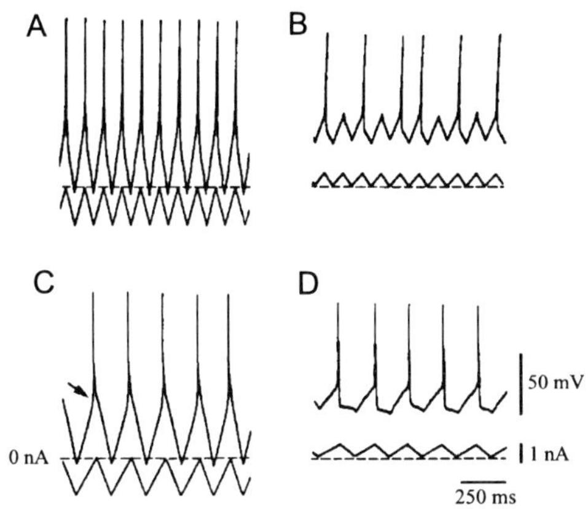

Figure 7. IO intrinsic properties. Intracellular recording from an olivary neuron during cell activation by sets of double-ramp current injection at two different membrane potential levels. The frequencies of the current ramps were 9Hz (A, B) and 5Hz (C, D), and the membrane potentials −78mV (A, C) and −48mV (B, D). Note that in A the cell fired at each peak but not in B. Also in C and D the neuron responded to the current injection at both levels of membrane potentials; however, at hyperpolarizing level (C), the generation of action potentials preceded the peak current (arrow). Null current level is indicated for all records (0 nA) (taken from Llinas and Yarom, 1986).

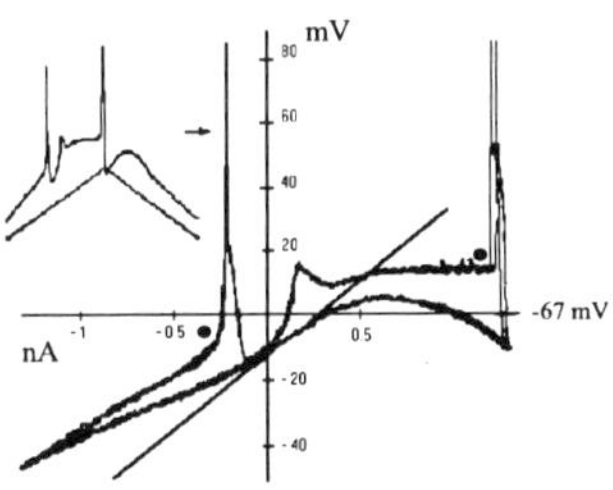

Figure 8. IO intrinsic properties. Two firing levels (dots) demonstrated when an IO neuron is activated by a double ramp (400ms, 6.3nA/s) current pulse [inset shows the current injection (lower trace) and the voltage response of the cell (upper trace)]. Resting potential, −67mV (taken from Yarom and Llinas, 1987).

from the hyperpolarized state to a low-threshold (rebound) somatic Ca-spike that may be strong enough to yield the condition for another Na spike and hence periodic neuron firing.

Figure 9 illustrates the intrinsic properties of mammalian PC recorded in vitro when subject to prolonged, threshold current pulses injected in their soma [30,31].

Figure 10 provides the corresponding features of CN neurons when subject to various depolarizing and hyperpolarizing currents.

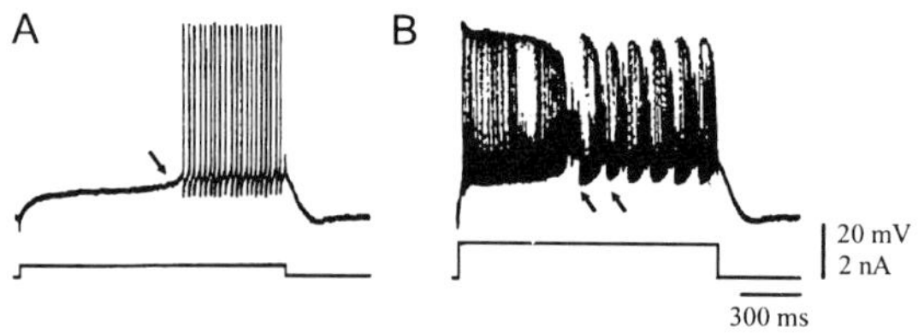

Figure 9. Intrinsic properties of mammalian PC in vitro. Repetitive firing by direct stimulation. A) A threshold current stimulus produces a repetitive activation after an initial local response (arrow). B) Increases in current injection amplitude produce high frequency firing and an oscillatory behavior marked with arrows (taken from Llinas and Sugimori, 1981).

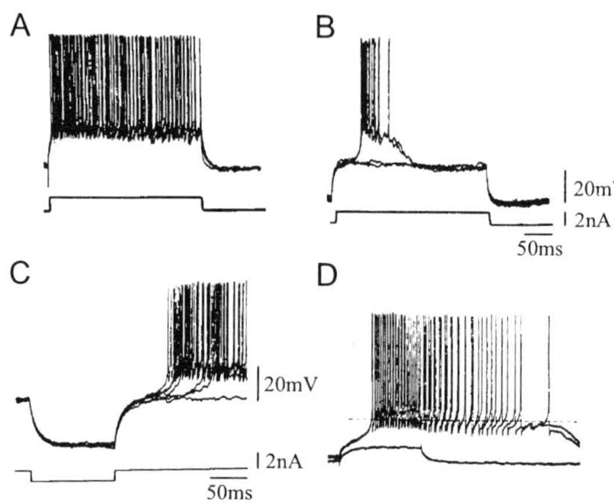

Figure 10. Intrinsic properties of CN neurons. A) A depolarizing current injection from resting potential elicits tonic firing. B) When the same strength current pulse is delivered from a hyperpolarized membrane level, an all-or-none burst response is elicited. C) Hyperpolarizing current injection from the resting potential elicits a strong rebound burst of action potentials from a slow depolarization. D) Response to current injection from a hyperpolarized level (resting potential marked by broken line) (taken from Llinas and Muhlethaler, 1988).

3 A model for the olivo-cerebellar loop and clock

Multi-electrode experiments with PC in the rodent cerebellar cortex have shown that the number of cells producing isochronous spike clusters is relatively small for spontaneous activity and increases with neuropharmacological intervention with drugs such as harmaline (loosely speaking further hyperpolarizes IO neurons) or picrotoxin (prevents decoupling gap junctions between IO neurons). In the latter case almost all neurons are grouped into one cluster and fire together. This reentry provides a means for an external stimulus to control the sensitivity of the loops IO-PC-CN and IO-CN. Such modulations allow the formation of well-organized patterns of global activity, which are of significance in motor coordination. Figure 11 summarizes data obtained from 48 Purkinje cells arranged from lateral (R, top) to medial (M, bottom), as a function of time [47]. Multichannel data involv-

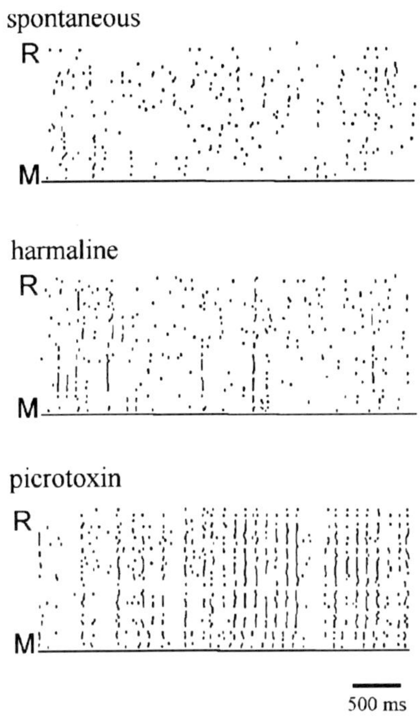

Figure 11. Raster display of complex spike activity in PC recorded from 44 cells arranged in the mediolateral direction: under control condition (spontaneous), after harmaline injection and picrotoxin injection. Bars correspond to spiking in a given unit (taken from Yamamoto, Fukuda and Llinas, 2001).

ing simultaneous changes in the space and time domains are demonstrated using a raster display allowing visual inspection of the timing of complex spike activity. Synchronous firing is enhanced in the presence of drugs. After injection of harmaline the mean complex spike firing frequency increased from 1.44 Hz recorded under controlled conditions (spontaneous activity) to 2.75 Hz in the presence of harmaline. Picrotoxin injection further enhanced synchronous complex spike firing. The complex spike activity after picrotoxin injection is characterized by two phases: oscillatory firing at 120 ms intervals followed by a low- frequency period lasting several seconds (Fig. 11, picrotoxin).

In accordance to the functional organization of the olivo-cerebellar system (Fig. 12a) a schematic diagram (Fig. 12b) suffices for our purpose [49]. It involves three interconnected 2-dimensional lattices. The bottom layer mimics an array of IO neurons. For simplicity, each IO neuron is electrically coupled only to its nearest-neighbors. When a given IO neuron reaches threshold an action potential is generated in the axon and transmitted to the corresponding CN. We assume that signals propagate rather fast (time delay is less than 3 % of the base oscillation period), and hence the time intervals of transferring a spike from the IO to the CN are the same for all axons and negligible. To mimic excitatory properties of

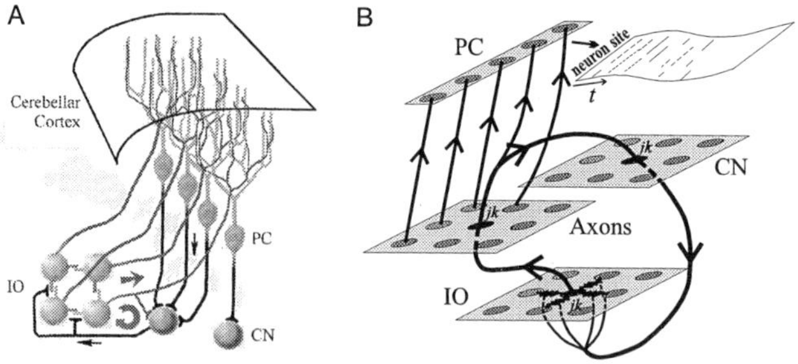

Figure 12. A) Organization of the Olivo-Cerebellar system. B) Schematic diagram of A reduced here to intralayer couplings in the IO lattice and interlayer couplings. The bottom, middle and top lattices correspond, respectively, to IO, Axons and CN (part A is taken from Llinas and Welsh, 1993; part B is from Velarde, Nekorkin, Makarov, Makarenko and Llinas, 2001).

the axons we use an intermediate or middle lattice of uncoupled units (Fig. 12b, Axons). In addition, the activity of this layer reflects the spiking behavior of the PC since the same axons that give collaterals to CN neurons terminate as climbing fibers in PCs. Finally, the top lattice involved in the loop corresponds to CN and collects and feedbacks to the first lattice decreasing the Inhibitory Postsynaptic Potential (IPSP) that corresponds to temporary disruption of couplings between IO neurons in some vicinity of a given neuron. To visualize spiking activity of the system we introduce one extra linear array that can be associated with PCs. It gets input from all or some closed part of the axons. Then, similar to the experimental procedure we illustrate the spatio-temporal evolution of the system by means of "raster displays", plotting a small vertical bar for each spiking event in the PC neurons (Fig. 12b, Neuron Site).

Since neurons have numerous different ion channels with complex and often noisy behavior, we take IO neurons as harmonic oscillators subjected to noise. Thus for the IO layer we have a 2-dimensional $n \times n$ lattice

$$\frac{dz_{jk}}{dt} = (i\omega_0 - \gamma)\, z_{jk} + \sum_{lm \in L} d_{jk}^{lm}(t)\,(z_{lm} - z_{jk}) + i\sqrt{2D}\xi_{jk}(t), \tag{1}$$

where the pair jk denotes a site in the lattice; z is a complex variable characterizing the dynamics of a neuron; γ is the damping constant; ω_0 is the angular oscillation frequency, $\omega_0 = 2\pi \times 10$ Hz; $\xi_{jk}(t)$ is a zero mean delta correlated noise sources; d_{jk}^{lm} accounts for the coefficient of the electrical coupling or bond between neurons at sites jk and lm (as already said its actual value depends on the dynamics of neurons in the top CN lattice). The sum in the right hand side of (1) is taken over neighboring neurons. Clearly, a harmonic oscillator is not only an oversimplified model of an IO neuron but, moreover, it does not account for the robust, albeit quasiharmonic, IO neuron oscillations (Figs. 4–6). As earlier mentioned we have elsewhere [46] provided a more realistic nonlinear model. Note, however, that a harmonic oscillator with delay does provide robust (limit cycle) oscillations. As we here would like to focus on the salient functional features of the olivo-cerebellar loop the harmonic oscillator suffices for our purpose [50-52].

The lattice (1) produces oscillations with a frequency band peaked around ω_0. These oscillations are the external input for the second $n \times n$ lattice (Fig. 12b, Axons) consisting of excitable, FitzHugh-Nagumo-Schlögl (see, e.g. Ref. [20]) units

$$\varepsilon \frac{du_{jk}}{dt} = f(u_{jk}) - v_{jk}, \quad \frac{dv_{jk}}{dt} = u_{jk} - I_{jk}(t), \tag{2}$$

where ε is a smallness parameter, u_{jk} and v_{jk} mimic, respectively, voltage and recovery variables in the corresponding site of the axon layer and nonlinear function is $f(u) = \alpha\left(-u^5/5 + 2u^3/3 - 2u^2\right)$. The parameter α allows us to tune the duration of a pulse that set to $T_{\text{sp}} = 4$ ms. Axons (and PC) are taken uncoupled according to experiment and get excitation from IO neurons via the activation current I_{jk} which depends on the corresponding variable in the IO layer

$$I_{jk} = x_{jk} - I_0, \quad \text{with } x_{jk} = \text{Re} z_{jk}.$$

I_0 is a positive constant controlling the level of hyperpolarization.

The CNs get excitatory signals from the corresponding axons and then inhibit couplings in the IO lattice by decreasing IPSP for time intervals about 30 ms. Accordingly, for the CN (upper) lattice we take

$$\tau \frac{dw_{jk}}{dt} = -w_{jk} + \Theta(u_{jk}), \quad \text{with } \Theta(u) = \frac{\exp\left[10(u + I_0 - 0.6)\right]}{1 + \exp\left[10(u + I_0 - 0.6)\right]}, \tag{3}$$

where w is the variable mimicking the CN response and τ is the time scale of decay. To have a pulse lasting for approximately 30 ms we put $\tau = 0.08$. The pulse formed by the variable w reduces the electrical coupling in the vicinity of a given neuron in the IO lattice. Thus for the coupling coefficients, $d_{jk}^{lm}(t)$, in the system (1) we can write

$$d_{jk}^{lm} = d \frac{1}{1 + \Gamma \sum w_{pq}}, \tag{4}$$

where d is the coupling coefficient in the absence of the feedback loop; Γ reflects the decoupling strength and summation is taken over neurons which can destroy the coupling (lm)-(jk).

Let us assume now that the electrical coupling strength between neurons in the IO lattice is constant and does not depend on the dynamics of the CN lattice ($\Gamma = 0$). Such mode can be achieved in experiments by blocking the GABA decoupling with picrotoxin or lesions of the CN. Experimental results show a strong enhancement of synchronicity in complex spike firing of PC (Fig. 11, picrotoxin). In such mode the noise will excite oscillations in the units in the master lattice which are not correlated for $d = 0$, and becomes more and more correlated with the increase of the coupling coefficient. Consequently, due to the inter-lattice interaction we get in PC spike trains with high level of coherence. However the synchronization cannot be perfect due to the noisy origin of the oscillations. Numerical integrations with different values of the coupling coefficient d using 15×15 lattices have been carried to explore the validity of the proposed model. Figure 13 shows the results of two computations for both vanishing coupling strength ($d = 0$) and strong coupling ($d = 200$). Clearly, the spiking trains have no coherence in the uncoupled case while they are highly synchronous in the case of strong coupling. In the last case, like in

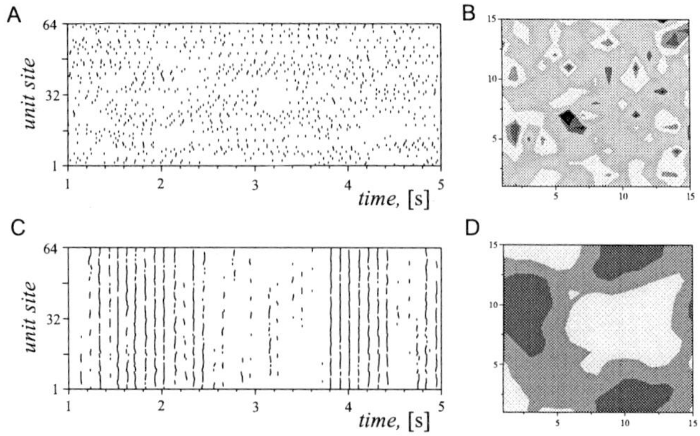

Figure 13. Picrotoxin-like simulations. A,C) raster displays of spiking activity in Axon layer (only an (8×8) square part of the 15×15 lattice is shown); B,D) snapshots of oscillations in the IO-layer (color corresponds to the value of x). A,B) vanishing coupling ($d = 0$); C,D) strong coupling ($d = 200$) (taken from Velarde, Nekorkin, Makarov, Makarenko and Llinas, 2001).

experiments with picrotoxin injection (Fig. 11, picrotoxin) spiking activity has two typical phases: a) high frequency sequence of spikes (with about 100 ms interspike intervals) and b) silent period lasting up to a few seconds (Fig. 13c). Figures 13b and 13c show snapshots of oscillations in the IO lattice. For the uncoupled case the distribution of variable $x(t)$ over the IO lattice is random. In contrast, for strong coupling big clusters with practically the same value of x are observed.

For the closed feedback loop ($\Gamma \neq 0$) the coupling coefficients between IO neurons are not constant, but functions of time. Figure 14 shows schematically the loop operation. Each coupling linking two neurons in the IO lattice can be altered by the signals activated by these neurons via a feedback loop. The signals pass via Axons to CN and finally come back in the form of pulses lasting 30 ms, which reduce the coupling coefficient for the time of their period due to (4). If two pulses are not synchronous then they superpose and strongly decrease the coupling (twice relative to the synchronous case). Thus the average coupling between neurons is less than for vanishing Γ. Consequently, the coherence level of oscillations in the IO layer and the spike trains in the Axon layer will be smaller than in picrotoxin-like experiments for the same parameter values. Moreover, superposition of decoupling events prevents unlimited cluster growth. If there are two clusters of neurons firing non-synchronously, then the mean intra-coupling will be higher than the coupling between clusters. This helps small clusters to survive, which indeed is important for fine-tuning in motor coordination.

Besides decoupling the value of the hyperpolarizing current, I_0, should significantly influence on the dynamics of the system. In experiments, injecting harmaline, which increases the hyperpolarization of IO neurons, leads to a decreasing spiking activity (the average frequency of spike events) of PC (and axons), but spike trains become more coherent (Fig. 11, harmaline). To test our model for its ability to satisfy this experimental fact we have made several computations with different values of I_0. Figure 15 shows two examples of spike trains in the Axon layer on the

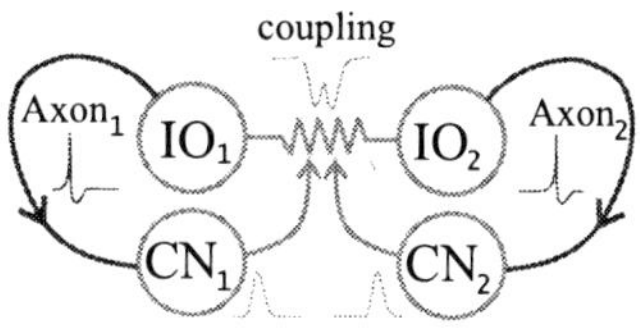

Figure 14. Decoupling mechanism between two IO cells. Two pathways convey feedback signals to the (electric) coupling linking IO neurons.

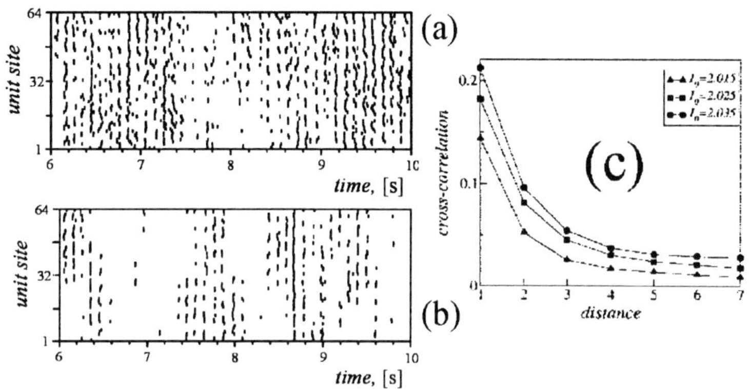

Figure 15. Influence of hyperpolarizing current, I_0, on the coherence of spike trains in the Axon lattice with switched-on feedback loop. (a) $I_0 = 2.025$; (b) $I_0 = 2.035$; (c) Cross-correlation function for different values of I_0 (taken from Velarde, Nekorkin, Makarov, Makarenko and Llinas, 2001).

raster display and cross-correlation functions for different values of I_0. As expected, the spiking activity for low level of hyperpolarization ($I_0 = 2.025$ on Fig. 15a) is higher than for a high level ($I_0 = 2.035$ on Fig. 15b) and correlation diminishes between units in the first case (Fig. 15c).

4 Further details about the role of noise

Neurons have numerous different ion channels with often noisy behavior and hence fluctuations in parameter values and other quantities are unavoidable. Note that, for instance, in preparations of IO neurons, the (subthreshold) rhythmic activity is observed only in a fraction of them. In fact, their loose electrotonic coupling helps increasing the percentage of IO neurons exhibiting spontaneous sustained oscillations. Taken axons separately, if a single one is electrically stimulated with a current intensity close to threshold, an action potential is generated in a fraction of the trials. Such experimental result is due to fluctuations in thresholds and not uncontrolled variability in the stimulus parameters. Conductance fluctuations and variability in the current-voltage (I-V) relations act as an intrinsic noise source. In vivo, as part of a living organism, membranes are constantly being repaired and do

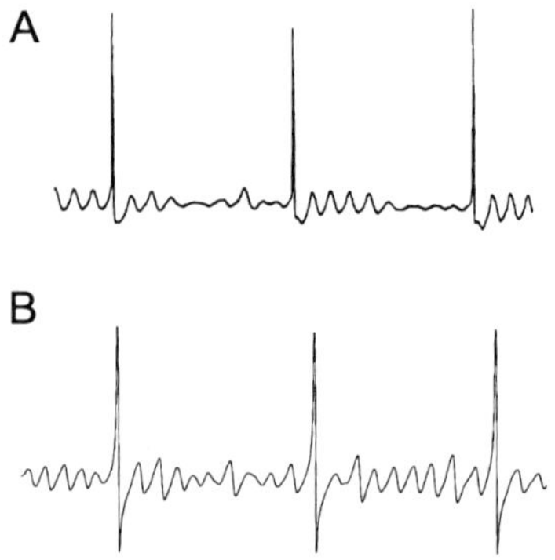

Figure 16. Real IO neuron spiking (A) versus model output (5) (B) (part A is taken from Benardo and Foster, 1986).

not necessarily always have the same response pattern.

Without embarking in a serious study of the dynamics of fluctuations or noise (additive or multiplicative, white or colored, finite correlated like a delay) let us comment on a significant feature of the interplay between (nonlinear, complex) dynamics and the simplest possible noise [53]. Take an IO neuron as an oscillator subjected to noise which in the simplest case is

$$\varepsilon \frac{du}{dt} = u(u-a)(1-u) - v, \quad \frac{dv}{dt} = g(u-b) + \sqrt{2D}\xi, \tag{5}$$

Equation (5) is a modified FitzHugh-Nagumo-Schlögl neuron model [20] with noise (term $\xi(t)$). As earlier, u and v loosely account for voltage and recovery variables. Function g is a monotonically increasing function introduced here to get flexibility for tuning independently the time scales of subthreshold oscillations and spikes. We consider the system (5) operating slightly beyond the supercritical Andronov-Hopf bifurcation. The behavior of the time series obtained from the model (5) shows fairly good agreement with experimental traces of oscillations in the IO neuron (Fig. 16). Furthermore, it turns out that Eq. (5) possesses features of stochastic resonance behavior. To illustrate this we compute the characteristic correlation time [54],

$$\tau_c = \int_0^\infty C^2(\tau)\,\mathrm{d}t, \tag{6}$$

where $C(\tau)$ is the normalized autocorrelation function. Figure 17a shows τ_c as a function of noise intensity, which has a rather well-pronounced maximum at moderate noise level. This indicates the existence of coherence resonance in the system, i.e., the spike train is "more-ordered" for intermediate noise intensity than for low or high noise levels. The relevance of stochastic resonance to neuron dynamics stems from the fact that in the vicinity of a bifurcation a neuron behaves much like a bistable system. Stochastic resonance mechanisms have been suggested to lie at the basis of multimodal interspike intervals and enhanced signal to noise ratio.

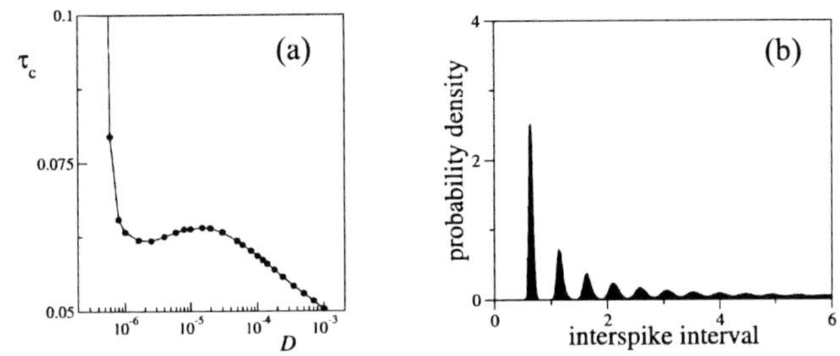

Figure 17. Resonance phenomena in model (5). (a) Correlation time vs noise intensity D. Maximum at moderate noise intensity indicates stochastic-resonance behavior. (b) Interspike interval histogram. Equidistant peaks shows regularization of spike events due to intrinsic dynamics of the model, with no need of external signal, and noise (taken from Makarov, Nekorkin and Velarde, 2001).

The coherence of oscillations produced by the system has a maximum at moderate noise intensity. Accordingly, noise can regularize the dynamics of a neuron (5). Besides, the system shows imperfect phase locking between interspike intervals and low amplitude quasiharmonic oscillations (Fig. 16 and Fig. 17b). The voltage-like variable has maximum probability to fire when the system passes the maxima of subthreshold oscillations. Consequently, the interspike intervals histogram (Fig. 17b) has clear equidistant peaks. The distance between peaks corresponds to the oscillation period of subthreshold oscillations. Then for the imperfect phase locking no need exists of an external signal as it follows from the FitzHugh-Nagumo-Schlögl *intrinsic* dynamics. This reinforces the idea that neurons may be, generally, operating, close to bifurcation points, at the edge of (dynamic) instability (earlier we noted the huge fields involved in axon spike transmission) and small perturbations, for example due to a noisy environment, can give rise to significant firing activity that in turn, and this is the crucial item, can be regularized by their intrinsic dynamics. Elsewhere [53] we have discussed this question in further details.

Acknowledgments

The authors acknowledge fruitful discussions with Dr. P. Arena, Dr. M. Argentina, Prof. J. Ayers, Prof. H. Cruse, Dr. O. Ekeberg, Prof. A. Fdez. de Molina, Prof. L. Fortuna, Dr. A. Giaquinta, Prof. S. Grillner, Prof. H. Haken, Dr. V. B. Kazantsev, Dr. E. Lang, Dr. V. I. Makarenko, Prof. V. I. Nekorkin, Prof. G. Nicolis, and Dr. G. Pratt. Research supported by the Spanish Ministry of Science and Technology under Grant PB 96-599.

References

1. S. Ramon y Cajal, *La Textura del Sistema Nervioso del Hombre y los Vertebrados*, 3 vols., Moya, Madrid (1897-1904). French translation: *Histologie du systeme nerveux de l'homme et des vertebres*, 2 vols., Maloine, Paris (1910,

1911); reprinted CSIC, Madrid (1972). English translation: *Histology of the Nervous System of Man and Vertebrates*, 2 vols., Univ. Press, Oxford (1995). See also *Histology* (10th Ed.) Wood, Baltimore (1933).
2. S. Ramon y Cajal, Rev. Trimestr. Histol. Normal. Patol. **1**, 305-315 (1888).
3. E. R. Kandel, J. H. Schwartz and T. M. Jessel, *Principles of Neural Science* (4th Ed.) Mc Graw-Hill, N.Y. (2000).
4. B. Hille, *Ionic Channels of Excitable Membranes* (2nd Ed.), Sinauer Ass. Inc., Sunderland, Mass. (1992).
5. I. B. Levitan and L. K. Kaczmarek, *The Neuron Cell and Molecular Biology* (2nd Ed.), Univ. Press, Oxford (1997).
6. R. R. Llinas (Ed.), *The Biology of the Brain. From Neurons to Networks* (Readings from *Scientific American* Magazine), Freeman, N.Y. (1988).
7. R. R. Llinas, *The Squid Giant Synapse. A Model for Chemical Transmission*, Univ. Press, Oxford (1999).
8. R. R. Llinas and K. D. Walton, *Cerebellum*, in *The Synaptic Organization of the Brain* (4th Ed.), ed. G. M. Shepherd (Univ. Press, Oxford, 1988), ch. 7, pp. 255-288.
9. R. R. Llinas, *Electrophysiology of the Cerebelar Networks*, in *Handbook of Physiology*. Sect. 1. *The Nervous System*, vol. II, ed. V. B. Brooks (Am. Physiol. Soc., Bethesda, MD, 1981), ch. 17, pp. 831-876.
10. R. R. Llinas and J. I. Simpson, *Cerebellar Control of Movement*, in *Handbook of Behavioral Neurobiology*, vol. 5, eds. A. L. Towe and E. S. Luschei (Plenum, N.Y., 1981), ch. 5. pp. 231-302.
11. G. M. Shepherd, *Foundations of the Neuron Doctrine*, Univ. Press, Oxford (1991).
12. X.-J. Wang and J. Rinzel, *Oscillatory and bursting properties of neurons*, in *The Handbook of Brain Theory and Neural Networks*, ed. M. A. Arbib (M.I.T. Press, Boston, 1995), pp. 686-691.
13. J. P. Changeux, *L'homme neuronal*, Fayard, Paris (1983).
14. *Trends in Neurosciences, Special issue: Cerebellum*, **21**(9), 367-419 (1998); See also the companion volume of the sister journal *Trends in Cognitive Sciences.*
15. H. Haken, *Principles of Brain Functioning*, Springer-Verlag, Berlin (1996).
16. H. Haken, *Synergetics. An Introduction* (3rd Ed.), Springer-Verlag, Berlin (1983).
17. H. Haken, *Advanced Synergetics*, Springer-Verlag, Berlin (1983).
18. G. Nicolis and I. Prigogine, *Self-Organization in Non-equilibrium Systems*, Wiley, N.Y. (1977).
19. G. Nicolis, *Introduction to Nonlinear Science*, Univ. Press, Cambridge (1995).
20. V. I. Nekorkin and M. G. Velarde, *Synergetics of Active Lattices (Patterns, Waves, Solitons, Chaos)*, Springer-Verlag, Berlin (2001) to appear.
21. M. V. L. Bennet, J. Neurocytol. **26**, 349-366 (2000).
22. A. F. Huxley and R. Stampfli, J. Physiol. **108**, 315-339 (1949).
23. A. L. Hodgkin and A. F. Huxley, J. Physiol. **117**, 500-544 (1952).
24. J. C. Eccles, M. Ito and J. Szentagothai, *The Cerebellum as a Neuronal Machine*, Springer-Verlag, Berlin (1967).
25. E. M. Izhikevich, Int. J. Bifurcation Chaos **10**, 1171-1266 (2000).

26. J. M. Kowalski, G. L. Albert, B. K. Rhoades and G. W. Gross, Neural Netw. **5**, 805-822 (1992).
27. Y. Manor, J. Rinzel, I. Segev and Y. Yarom, J. Neurophysiol. **77**, 2736-2752 (1997).
28. N. Schweighofer, K. Doya and M. Kawato, J. Neurophysiol. **82**, 804-817 (1999).
29. R. R. Llinas (Ed.), *Neurobiology of Cerebellar Evolution and Development*, American Medical Association, Chicago (1969).
30. R. R. Llinas and M. Sugimori, J. Physiol. **305**, 171-195 (1980a).
31. R. R. Llinas and M. Sugimori, J. Physiol. **305**, 197-213 (1980b).
32. R. R. Llinas and Y. Yarom, J. Physiol. **315**, 549-567 (1981a).
33. R. R. Llinas and Y. Yarom, J. Physiol. **315**, 569-584 (1981b).
34. R. R. Llinas and Y. Yarom, J. Physiol. **376**, 163-182 (1986).
35. Y. Yarom and R. R. Llinas, J. Neurosci. **7**, 116-1176 (1987).
36. R. R. Llinas and M. Muhlethaler, J. Physiol. **404**, 215-240 (1988).
37. R. R. Llinas, Science **242**, 1654-1664 (1988).
38. K. Sasaki, J. M. Bower and R. R. Llinas, Eur. J. Neurosci. **1**, 572-586 (1989).
39. R. R. Llinas and K. Sasaki, Eur. J. Neurosci. **1**, 587-602 (1989).
40. E. J. Lang, I. Sugihara and R. R. Llinas, J. Neurophysiol. **76**, 255-275 (1996).
41. R. R. Llinas and J. P. Welsh, Curr. Opin. Neurobiol. **3**, 958-965 (1993).
42. J. P. Welsh and R. R. Llinas, in *The Cerebellum: From Structure to Control*, eds. C. I. de Zeeuw, P. Strata and J. Voogd, *Progr. Brain Research* **114** (1997), ch. 26, pp. 449-461.
43. J. P. Welsh, E. Lang, I. Sugihara and R. R. Llinas, Nature (London) **374**, 453-457 (1995).
44. Y. Yarom, Oscillatory behavior of olivary neurons, in Experimental Brain Res. Series **17**, 209-220 (1989)
45. L. S. Benardo and R. E. Foster, Brain Res. Bulletin **17**, 773-784 (1986).
46. M. G. Velarde, V. I. Nekorkin, V. B. Kazantsev, V. I. Makarenko and R. R. Llinas, Neural Netw. (2001) to appear.
47. T. Yamamoto, M. Fukuda and R. R. Llinas, Eur. J. Neurosci. **13**, 327-339 (2001).
48. M. Fukuda, T. Yamamoto and R. R. Llinas, Eur. J. Neurosci. **13**, 315-326 (2001).
49. M. G. Velarde, V. I. Nekorkin, V. A. Makarov, V. I. Makarenko and R. R. Llinas, Neural Netw. (2001) submitted.
50. V. I. Nekorkin, V. B. Kazantsev and M. G. Velarde, Eur. Phys. J. B **16**, 147-155 (2000).
51. A. Giaquinta, M. Argentina and M. G. Velarde, J. Stat. Phys. **101**, 665-678 (2000).
52. V. A. Makarov, V. I. Nekorkin and M. G. Velarde, Int. J. Bifurcation Chaos **11**, 109-122 (2001).
53. V. A. Makarov, V. I. Nekorkin and M. G. Velarde, Phys. Rev. Lett. **86**, 3431-3434 (2001).
54. A. S. Pikovsky and J. Kurths, Phys. Rev. Lett. **78**, 775-778 (1997).

TWO AND THREE DIMENSIONAL COMPUTER SIMULATION OF CANCER GROWTH

SABÁS FLORES ASCENCIO
HÉCTOR PÉREZ MEANA
MARIKO NAKANO MIYATAKE
Graduate School of the National Polytechnic Institute of Mexico. Av. Santa Ana #1000, Col. San Francisco Culhuacán, Delegación Coyoacán, México. D.F, C.P. 04430.
E-mail sfa@servidor.unam.mx

The qualitative and quantitative comparison of simulated growth patterns with histological patterns of primary tumors may provide additional information about the morphology and the functional properties of cancer. In this paper we show a simple two and three dimensional model to simulate the growth of carcinoma. Our simulation includes cell proliferation, reciprocal influence among cells, cell division and cell death. The results show fractals structures and its gyration radius, number of cells on tumor periphery and fractal dimension, characterizes every simulated pattern growth.

1 Introduction

We believe that histological patterns of primary tumors provide additional information about the qualitative and quantitative comparison of simulated growth patterns with morphology and the functional properties of cancer. If we consider that in humans cancer denotes a collection of more than 100 forms of diseases involving cells, which divide when they should not [1]. This excessive, autonomous and non-homeostatic cell growth occurs in multi-cellular organisms, like animals and plants. Cancer cells accumulate some genetic and phenotypic changes [1,3]. Cancer cells are endowed with the same resistance and survival abilities exhibited by unicellular organisms under stress through geological times and catastrophes. Several mechanisms of cancer, such as various physical, chemical and biological carcinogen agents, produce an initial lesion characterized by an unusual proliferation of transformed cells. This clonal adaptation represents a defense response of the organism, which normally disappears via a programmed pathway of differentiation. With the advance of tumor progression a sequence of events occurs. Aberrant growth in a subset of cells within the initial lesion may generate a dysplasia that contains many morphologically atypical cells similar to some cells appearing in invasive, primary cancers. Later, the tumor progression, the so-called intermediated lesions or carcinoma, all cells of the lesion are atypical and grow in contiguous array. This phenomenal growth is temporally unrestricted, connected to the tissue of origin or is only micro-invasive and does not occur in mesenchyme (conjunctive or connective tissue) of primary site. In later lesions, a *nidus of cells*(a group of cells with same characteristics), exhibits most properties of overt cancer cells including the ability to grow in the mesenchyme of the tissue site, metastatic competence and tumor cell heterogeneity acquired by reciprocal interaction with the extra cellular matrix. Late in the evolution of primary tumors, cells with metastatic potential may overcome the primary cancer. It is important to note that the sequential lesions leading to a primary cancer are the result of local, not global changes within a lesion. This scheme of tumor progression is universal [4]. It is well known [5,10] that stressed bacterial colonies can develop complex patterns, irregular fractal shapes with details on many length scales. Indeed, biological growth almost invariably leads to the formation of complex forms and the fractal design is ubiquitous in living systems [11,12]. Thus, there is current interest in the search for basic principles of growth in living organisms, which are the most complex and challenging self-organized systems. Specifically, we are interested in two aspects: the patterns generated by primary tumors in a normal tissue and the way to simulate them by computer.

For the simulation, the diffusion controlled cluster formation, is a non-equilibrium irreversible growth process that gives rise to low-density random fractal objects. A wide variety of phenomena in nature and human activities lead to formation of this type of aggregates: metal particle aggregation, crystal growth governed by heat diffusion, solidification of alloys and more [13,16]. Some models of the growth of such aggregates have been proposed and studied so far. The simplest one is the lattice model of Eden [17] in which particles are added at random, one by one, to the nearest neighbors of already occupied sites. The Eden model leads to a relatively compact cluster while a large number of experimental observations of the DLA growth phenomena indicate structures whose density correlation falls off as a fractional power of distance. The DLA model that reproduces experimentally observed structures was proposed by Witten and Sander [18]. In Witten and Sander's model, one seed particle forms the initial aggregate. Other particle appears long distance from the aggregate and performs a Brownian motion, call it simple random walk, on a lattice until they come in contact with the structure and attach to it, or *die* having wandered far away from the cluster.

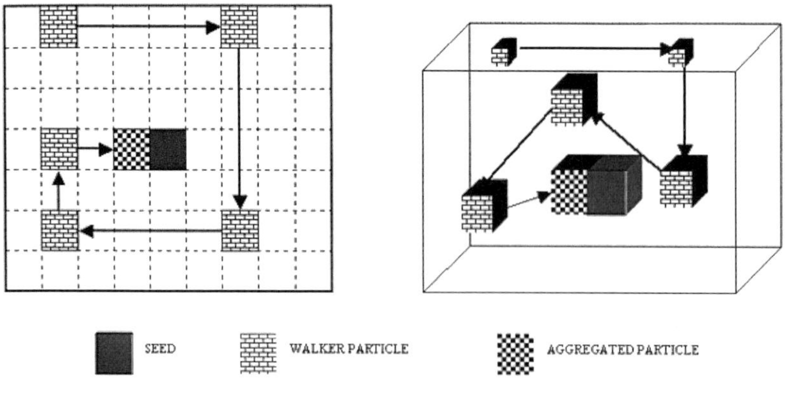

Figure 1. Monte Carlo simulation controlled cluster formation using a two and three lattice model

The relationship among an Eden type growth and a DLA type growth was studied by Williams [27]. In that work both models were related by means of a single parameter of growth.

2 Simulation

The basic biological principles included in this model are cells proliferation, motility and death, as well as the reciprocal influence among cells mediated by autocrine and paracrine growth factors, motility factors [19,21]. Such stimulatory proteins, secreted and released by each cancer cell, move through the extra cellular space and bind to specific receptors on the surface of the cells. Usually, these proteins act on nearby cancer cells (paracrine factors) but they may also turn back and stimulate the same cells that just secreted them (autocrine factors). All these growth factors, influence the microenvironment of each cell and hence the resulting morphological patterns. In our model, the tissue is represented by a cube lattice containing up to 1000 x 1000 x 1000 sites, and a square lattice containing up 1000 x 1000 sites, for three and two dimension is respectively, both with fixed boundary conditions. Although a tumor mass is composed of different cell subpopulations [5], we shall consider binary variables $C_{i,\ j}$ (i, j=1,2,..L), where Ci,j = 0 represents a normal cell and Ci,j = 1, represents a cancer cell in two dimensions and $C_{i,\ j,k}$(i,j,k=1,2,3,4,5...L) where $C_{i,j,k}$ =0 represents a normal cell and Ci,j,k = 1, represents a cancer cell in the case of three dimensions.
To the computer simulation, we identify a cell in the three dimensional space with a cube, and left low vertex coordinates are *(x,y,z)* lower case. See figure 2.

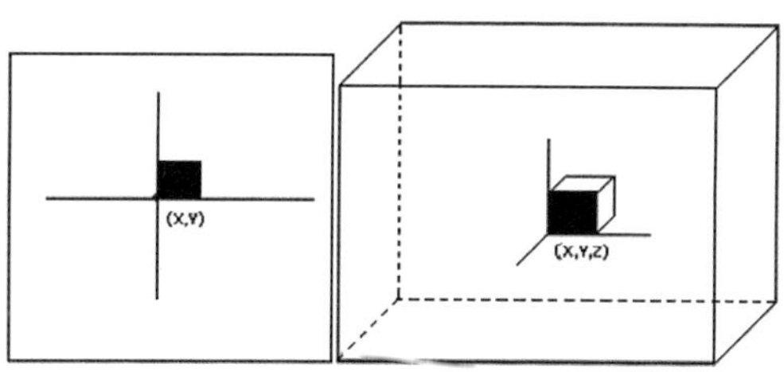

Figure 2. For the simulation, be in two or in three dimensions, each cell is represented and it is perfectly defined by the coordinates of the left inferior vertex.

As initial *seed,* a single cancer cell, defined by its coordinate in the center of the lattice, is chosen in agreement with the theory of clonal origin of cancer [22]. As exemplified by the tumor progression of melanocytic neoplasia[5], the net growth of the lesion occurs at its periphery and the cancers slowly enlarge along the radii of an imperfect circle. So, each cell on the tumor periphery can be selected at Brownian motion on the lattice, with equal probability and carry out one of three actions[23]: cell division, with probability $P_d(div)$, cell motility, with probability $P_m(mov)$, cell death, with probability $P_{dl}(del)$. These parameters, representing the behavior of isolated cells, are fixed and satisfy the normalization condition:

$$P_d(div)+P_m(mov)+P_{dl}(del)=1 \tag{1}$$

After placing a *seed* in the centre of the lattice, a long distance from this initial seed, at a random point on a circumference S of radius R a new particle is added, with the probability described before. This particle then performs a Brownian motion on a lattice, if it reaches any site of the lattice located at distance $R_K >> R$, it is eliminated and the process begins again. If the particle reaches the point adjacent to the original seed, it becomes a part of the growing cluster. The procedure is repeated until cluster of sufficient size is formed. Our model includes as special limits the Eden[17] and the Williams and Bjerkness[24] models in which, $P_m(mov) = P_{dl}(del) = 0$ and $P_m(mov) = 0$, respectively, are fixed probabilities for all cells. However, the cell-cell interactions mediated by growth factors introduce the possibility that distinct cells show different local probabilities for a given action[23]. Thus, choosing the tentative action always according to its intrinsic probability $P_{act}(act)$, each selected cell on tumor periphery will implement or not this tentative action in response to its local microenvironment. The local probability for a particular action that will be carried out on the selected peripherical cell is given by: [23]

$$P_{(act)} = \begin{Bmatrix} Po_{(act)} \ldots\ldots\ldots\ldots if \langle C_{i,j} \rangle \le C^*_{act} \\ Po_{(act)} + [1 - Po_{(act)}] \dfrac{\langle C_{i,j} \rangle - C_{act}}{A_{act} + < C_{i,j} > - C^*_{act}} \end{Bmatrix} \tag{2}$$

which the average growth factors sensed by the selected cell at the case of two dimensions is[23]:

$$\langle C_{i,j} \rangle = \frac{1}{5} \sum_{\langle i,j \rangle} C_{i,j} \tag{3}$$

The sum extends over the Von Neumann neighborhood of the site (i,j). C^*_{act} is a threshold value of the growth factor concentration above which the intrinsic probability P(act) is influenced by the micro-environment of the cell. A_{act} is a parameter, which control this influence. The growth factors concentration obeys the diffusion equation:

$$\frac{\partial c(r,t)}{\partial t} = D\nabla^2 c(r,t) + \sum_{cells} \delta(r - r_i)c_o - \Gamma c(r,t) \tag{4}$$

which includes the natural degradation of growth factors and the cells acting as their sources. The boundary conditions satisfied by the concentration field is c(r,t) = 0 on a large circle of radius R centered on the initial seed. The cell motility and division changes from pure random walk to a movement and growth along the opposed direction of the gradient of the concentration field. Thus, when cell division occurs, the daughter cell will occupy the nearest-neighbor normal site of the selected cell with the smaller concentration. The same holds in the case of cancer cell migration by a moving distance of one lattice constant. In the case of cell death, a normal one replaces the cancer cell. Above the threshold C^*_{act}, Eq. (2) represents a function which increases monotonically as the growth factor concentration increases and becomes saturated at high $<C_{i,j}>$ values. Thus, as the cancer growth progress, an increasing number of abnormal cells leave the replicative pool by death, differentiation or lack nutrients, which means that cell death probability increases as a natural consequence of an increase in cell numbers and consequently, in the growth factors concentration. The same holds true for cell division, which is widely stimulated by autocrine and other growth factors, secreted by cancer cells. The cell motility against a concentration gradient seems to be a chemotactic, (gradient-sensitive) response to cell-derived cytokines. Such as the autocrine motility factor, secreted by a variety of tumor cells and also to the progressive destruction of the extra-cellular matrix near the border of the tumor by a series of enzymes that includes collagenases, metalloproteinases, among others, released by cancer cells. In the model, the probabilities *Po(div)*, *Po(del)* and *Po(mov)* take into account the intrinsic properties of cancer cells, while the parameters c^*_{act} , A_{act} , *D*, *Co* and Γ characterize the cellular response and the diffusion of growth factors responsible by the chemotactic interactions among cancer cells. For the computational simulation, the growth model simulations were implemented according to the following procedure: at each time step, Eq. (4) is solved numerically to calculate the growth factors concentration field at any lattice site, and the actual number of cells in the tumor periphery, S is counted. Then, periphery cancer cells are sequentially selected at random with equal probability. For each one of them, a tentative action (division, death or movement) is chosen and implemented or not according to the corresponding local probability determined by Eq. (2). If the selected cancer cell divides then, its nearest-neighbor normal site with the smaller concentration of growth factors will be occupied by its daughter cell. The same rule holds true if the cancer cell moves and in addition, its site of origin will be occupied by a normal cell. Finally, if the selected cancer cell dies, then a normal cell will occupy its site. Each time that a particular cell implements some action, the table of peripherical cells is updated and also, this cell is prohibited to be selected again along the sequence. At the end of this sequence of S, a new time step begins and the entire procedure (solution of the diffusion equation and application of the growth rules) is iterated. If the chemical interactions among cancer cells are neglected, then the new occupied sites for both, cell division and motility, is chosen at random from the normal nearest neighbors of the selected cancer cell. In our computational experiments, we utilized the procedure of the particle generation proposed in [25]. We utilized a Hash function to increase speed in the calculation[26]. The procedure is repeated until the cluster is large enough.

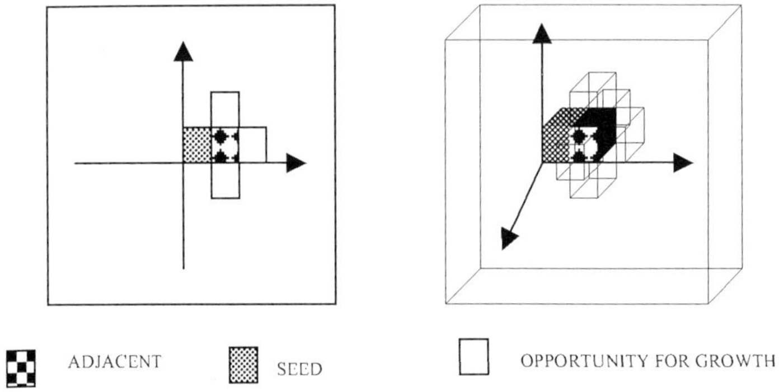

Figure 3. The Hash function applied to each cube that represents a cell and their possibilities of growing.

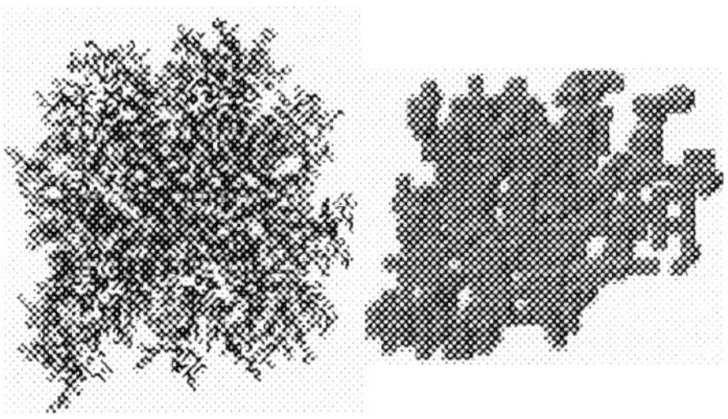

Figure 4. Clusters in two and three dimensions with 4000 and 3000 cells respectively, obtained using the described procedure and the Hash function proposed in the text.

3 Simulation results

Cell patterns obtained by the simulations of the model are shown in Figures 5 and 6.

Figure 5. Typical two dimensional cluster of 1000, 3000 y 4000 cells obtained using the procedure

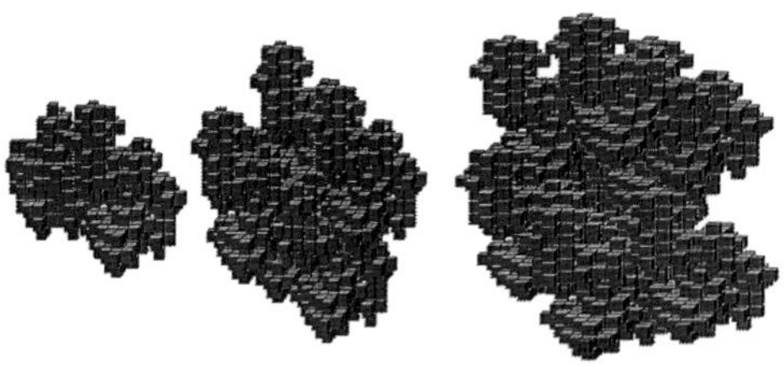

Figure 6. Typical two dimensional cluster of 1000, 5000 y 40000 cells obtained using the procedure described. We use the Hash function proposed in the text for three dimensional model.

Consistent with clinical observations, all the patterns are roughly *imperfect circles* bounded by an irregular perimeter. Therefore, the gyration radius, the fractal dimension, and the number of cells on tumor periphery can quantitatively describe the geometry of the simulated tumors. The growth patterns without communication among cells are compact and their surface complexity increases with increasing cell motility. The number of cells on the tumor periphery S, gyration radius R_g, and surface roughness W are defined as[23]:

$$R_g = \left(\frac{i}{N} \sum_{cells} r_i^2 \right)^{1/2} \tag{5}$$

$$W = \left(\frac{1}{S} \sum_{periphery} (r_i - \langle r_i \rangle)^2 \right)^{1/2} \tag{6}$$

$$\begin{aligned} R_g &\sim N^{\nu} \\ S &\sim N^{\delta} \\ W &\sim N^{\delta} \end{aligned} \tag{7}$$

In which r_i is the distance of the cancer cell i from the center of the lattice, N the total number of cancer cells, S the number of peripherical cells and $\langle r_i \rangle$ the average tumor surface radius. The exponents of N, (δ and ν), approach the Eden value 0.5 as N tends to infinity even in the case of high cell motility. The roughness exponent σ seems to increase very slowly towards the asymptotic value 0.5 of the Eden model. For growth patterns containing 10^5 cancer cells, the exponent δ assumes the values in models of 0.14 (without motility), 0.12 (low motility) and 0.08 (high motility). For 10^6 particles these values increase to 0.16, 0.14 and 0.11, respectively. In comparison, our results for the Eden limit are δ =0.18 for clusters containing 10^5 particles, consistent with Plischke and Rácz simulations [26], and δ =0.27 for patterns with 10^6 particles. The growth patterns for the model including chemotactic interactions among cancer cells, exhibit a morphological transition, which depends on the cell motility probability. For low cellular motility, 0<=*Po(mov)*<=0.4, the patterns are compact from the beginning with an increasing surface complexity, as evidenced by the initial presence of increasing internal holes. In contrast, for high cell motility, *Po(mov)*>=0.50, the growth patterns are constituted by single cells or small cell clusters highly dispersed in the tissue. All the cells are on tumor periphery and the initial invaded tissue is larger than that for low motility case. This rapid and invasive growth is afterwards slowed down while the tumor pattern became progressively more compact. See the figure 7 where the fractal dimensions of the patterns are plotted against the total number of cancer cells.

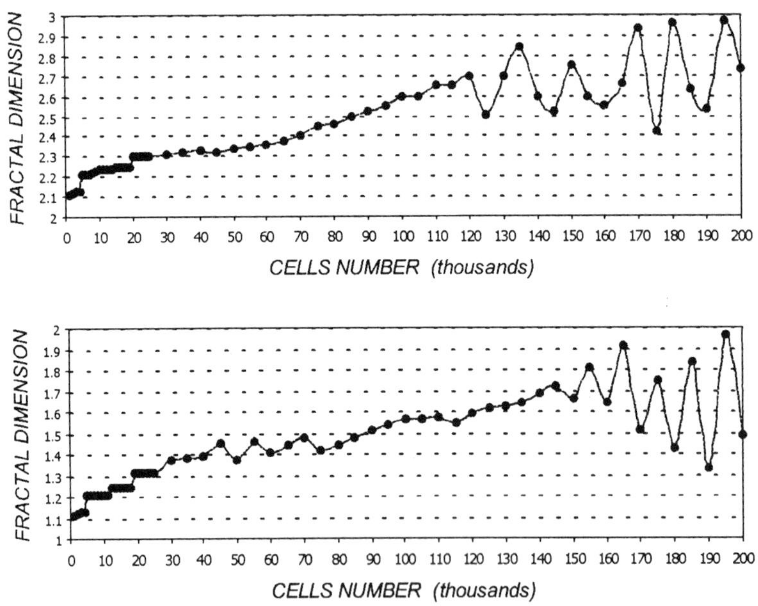

Figure 7. Number of cells in the growth vs. their fractal dimension. In three and two dimensions respectively. The graphics show instability for heaps with great quantity of cells.

The patterns demonstrate that the diffusion of growth factors affects the initial growth of the tumor. Indeed, at the beginning, a separating force acts among cancer cells since the growth factors are initially concentrated around the center of the cluster and the cell movement is against the gradient of these chemicals. Cells tend to move isotropically, which favors more compact growth patterns. This effect is shown in figure 8.

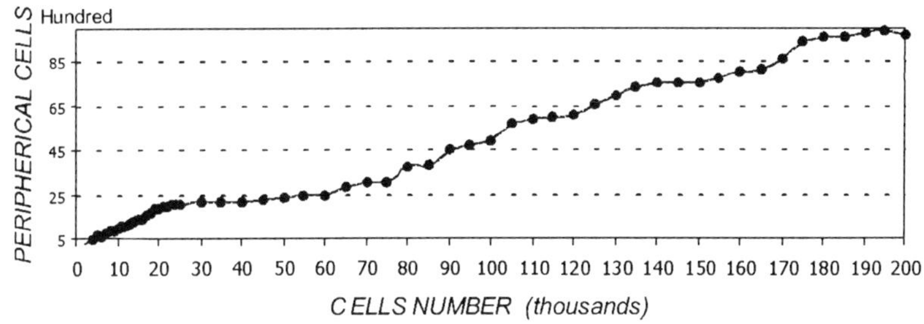

Figure 8. Number of peripherical cancer cells S vs total number of cancer cells N.

The number of peripherical cancer cells increases with the cell motility. Again, the exponents δ and σ approach the Eden value 0.5 asymptotically with N, as in the case without chemotactic interactions among cancer cells. So, the asymptotic patterns are always compact as should be expected for cancer growth, since, if the disease did not kill its host, cancer cells would completely invade the normal tissue, given sufficient time. It is important to emphasize what is the mechanism that leads to a growth process driven by diffusion of growth factors to generate asymptotically large compact structures. The cancer cells themselves are the main sources of growth factors and continuously increase the local concentration of these proteins that then propagate to neighboring sites. As a consequence, the concentration of growth factors throughout the growth perimeter increases, thus *masking* the diffusion field and generating more uniform gradients and growth. Therefore, as previously mentioned, it is the local amplification of growth factor levels by the autocrine release of growth factors that causes the simulated patterns to become progressively more compact as the number of cancer cells increases. Finally, we shall compare the simulated patterns with some real histological tumor patterns. A section of a primary human gastric carcinoma and skin carcinoma are shown in figure 9.

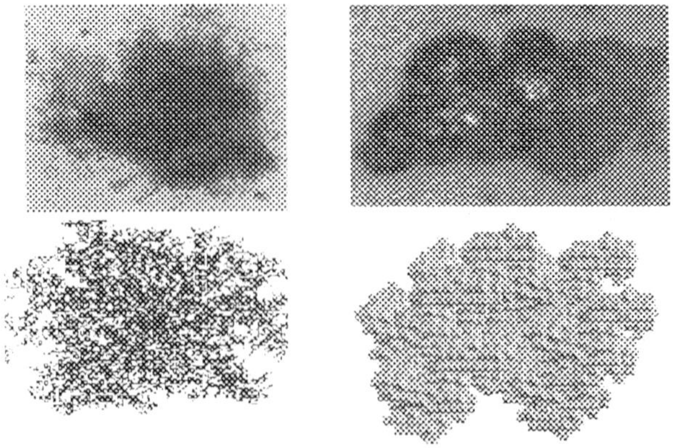

Figure 9. Human gastric carcinoma vs. simulation in two dimensions and skin carcinoma vs. simulation in three dimensions

These photographs are very similar to the patterns exhibited in our simulation in two and three dimensions. In figure 9, an explant from a human gastric carcinoma is shown. This compact cluster with a complex surface is similar to the patterns simulated, corresponding to intermediate cell motility. At the beginning, we assume that similarities between these real and simulated patterns suggest that some of the functional properties of cancer cells are similar to those built into our model. But, since distinct growth models can probably generate morphologically related patterns, a quantitative analysis of human and animal primary cancer assumes a central role.

4 Conclusions

A clinically detectable primary cancer contains at least 10^9 cells and thus, initial growth patterns appear to be inaccessible to the experiments. On the other hand, histological specimens of primary cancer only provide a static view of the dynamical evolution of tumor growth and, in addition, in a live observation method with resolution at the cellular level able to follow the tumor progress is a difficult problem to overcome. We believe that the analysis of medical images using fractals can give a diagnosis element, although this element is not definitive. Therefore, we consider that at the moment a comparison between the computer simulations and the real cancerous growths is a first stage of further studies.

References

1. R.S. Cotran, V. Kumar, S.L. Robbins (Eds.), Robbins *Pathologic Basis of Disease*, 5th ed., W.B.Saunders Co., London, 1994.

2. J. O'D. McGee, P.G. Isaacson, N.A. Wright (Eds.), Oxford *Textbook of Pathology*, Oxford University Press, Oxford, 1992.
3. J.A. Anderson (Ed.), *Muir's Textbook of Pathology*, 12th ed., Edward Arnold, London, 1987. W.H. Clark, J. Cancer 64 (1991) 631.
4. E. Ben Jacob, H. Shmueli, O. Shochet, A. Tenenbaum, Physica A 187 (1992) 378.
5. E. Ben Jacob, A. Tenenbaum, O. Shochet, O. Avidan, Physica A 202 (1994).
6. E. Ben Jacob, O. Shochet, A. Tenenbaum, I. Cohen, A. Czirók, T. Vicsek, *"Generic modelling of cooperative growth patterns in bacterial colonies"*, Nature 368 (1994) 46.
7. E. Ben Jacob, I. Cohen, O. Shochet, I. Aranson, H. Levine, L. Tsimiring, Nature 373 (1995) 566.
8. E. Ben Jacob, I. Cohen, O. Shochet, A. Czirók, T. Vicsek, , "Cooperative Formation of Chiral Patterns during Growth of Bacterial Colonies", Phys. Rev. Lett. 75 (1995) 2899.
9. I. Cohen, A. Czirók, E. Ben Jacob, Physica A 233 (1996) 678.
10. T.F. Nonnenmacher, G.A. Losa, E.R. Weibel (Eds.), *Fractals in Biology and Medicine*, Birkh. Auser Verlag, Basel, 1994.
11. J.B. Bassingthwaighte, L.S. Liebovitch, B.J. West, *Fractal Physiology*, Oxford University Press, New York, 1994.
12. M.J. Vilela, M.L. Martins, S.R. Boschetti, J. Pathol. 177 (1995) 103.
13. L. Pietronero, E Tosatti (Eds) *"Fractal in Phisical"*, Elsevier. Amsterdam (1986)
14. R. Jullien, R. Botet, *"Aggregation and Fractal Aggregates"* , Word Scientific. Singapure
15. T. Vicsek, *"Fractal growth phenomena"*, 2nd edition, World Scientific. Singapore (1987).
16. M. Eden, in Proc. *"Proc. of the Forth Berkeley Symposium on Math. Static and Probability"*, IV (1961) 223.
17. T.A. Witten, L.A Sander, *"Diffusion-limited Aggregation, a kinetic critical phenomenon"* Physical Review, Letter 41 (1981) pp. 1400.
18. R.A. Weinberg, *"How cancer arises"*, Scientific American (1996) 32.
19. C.G. Schirren, K. Scharffetter, R. Hein, O. Braun-Falco, T. Krieg, J. *"Tumor necrosis factor alpha induces invasiveness of human skin fibroblasts in vitro"*, Invest. Dermatol. 94 (1990) 706
20. L.A. Liotta, R. Mandler, G. Murano, D.A. Katz, R.K. Gordon, P.K. Chiang, E. Schimann, *Proc. Natl. Acad. Sci.* USA 83 (1986) 3302.
21. P.C. Nowell, Science 194 (1976) 23.
22. S.C. Ferreira Junior, M.L. Martins , M.J. Vilela, *"Growth model for primary cancer"*, Elsevier, Physica A 261 (1998) 569-580
23. T. Williams, R. Bjerkness, Nature 236 (1972) 19.
24. P. Meakin, *"Diffusion-controlled cluster formation in 2-6 dimensional space"*, Phys. Rev. A27(1983) pp. 1495

25. M. Plischke, Z. Rcz, "*Active Zone of Growing Clusters: Diffusion-Limited Aggregation and the Eden Model*", Phys. Rev. Lett. 53 (1984) 415.
26. S. Flores Ascencio, H. Pérez Meana, M. Nakano Miyatake, "*A Three Dimensional Growth Model for Primary Cancer*", Telecommunications and Radio Engineering, Edit. Bagell House, Inc. N.Y, Approved.
27. H.T. Williams et al. "Two-dimensional growth models", Phys. Lett. A 250 (1998) 105.

STRUCTURAL AND DYNAMIC COMPLEXITY OF AN ELASTIN-RELATED PEPTIDE

ANTONIO BONELLI
CISIT, Universita' della Basilicata, via Nazario Sauro 85
85100 Potenza, Italy
E-mail: bonelli@unibas.it

LUCIANO D'ALESSIO
Dipartimento di Chimica, Universita' della Basilicata, via Nazario Sauro 85
85100 Potenza, Italy
E-mail: dalessio@unibas.it

STEFANO RUFFO
Dipartimento di Energetica, Universita' di Firenze, via Santa Marta 3
50139 Firenze, Italy
E-mail: ruffo@avanzi.de.unifi.it

ANTONIO MARIO TAMBURRO
Dipartimento di Chimica, Universita' della Basilicata, via Nazario Sauro 85
85100 Potenza, Italy
E-mail: tamburro@unibas.it

The supramolecular organization and the conformational properties of the elastin-related octapeptide ALGGGALG, and its polymer, are studied by electron microscopy and molecular dynamics simulations. Our results evidence the complex nature of the single molecule and its aggregates, both from the structural and the dynamic points of view, indicating that a low-level complex behaviour exists in the building blocks of the elastin molecule. In particular a remarkable molecular mobility is observed around a central hinge zone. This confirms the role of the phase transition fractal-to-Euclidean in the framework of the well-known entropic mechanism of elastin elasticity.

1 Introduction

The carbon atom has a unique ability to build-up complex structures at all geometric dimensions ranging from 0 to 3. In fact, the classic elemental carbon solids, graphite and diamond, are essentially two- and three-dimensional respectively. The recently discovered fullerene buckyballs [1], when observed from a sufficiently distant point of view (interstellar dust), can be considered as point-like zero-dimensional objects. As a consequence, a fullerene nano-powder can be presumably a Cantor-like system, with dimension between 0 and 3, depending on its lacunarity.

On the other hand, the fullerene nanotubes [2] are essentially one-dimensional, and this kind of space filling has been confirmed experimentally through the measure of the temperature-dependent specific heat [3]. A random coil of nanotubes can obviously reach higher dimensions. Finally, if one considers the complex structures and dynamics of the organic compounds, including biomolecules and biopolymers, together with the macroscopic shapes of all living organisms, then the fractal and dimensional possibilities are practically unlimited. In particular, this seems true for elastin which is the protein responsible for the elasticity of many vertebrate tissues including the skin, lungs, ligaments and large arteries [4]. It is also involved in pathological processes like atheromatosis and arteriosclerosis [5], emphysema [6] and some skin diseases [4]. The elastin native form consists of polypeptide chains cross-linked by several covalent bonds [7]: from the topological point of view it is constituted by a three-dimensional network of polymeric chains, joined together at a number of connection sites, that can be considered when hydrated as a polymer gel [8]. The most interesting property of the protein and its derivatives is the supramolecular organization, that we have investigated in previous works [9, 10], which shows a variety of fractal patterns, going from leaf-like aggregates to compact-fibrils structures, and from percolation-cluster shapes to twisted-rope hierarchy. Our results are on-line with the well known fractal structure of many protein molecules [11, 12].

The main property of elastin is its entropy-driven elasticity: indeed it is well established [13] that the restoring force responsible for the elastic behaviour of the molecule is of entropic nature, i.e. it originates from the disorder increase associated with the transition from the stretched to the relaxed state. In other words the recovering force can be expressed as:

$$F = -T\left(\frac{\partial S}{\partial x}\right)_{T,V}$$

where S = k ln W is the entropy (disorder) and W is the statistical weight, namely the number of available states or the volume of the phase space spanning the system. In this respect it should be noted that the disorder of a molecular system can arise not only from its geometry, but also from the characteristics of its motion. Therefore one can recognize two main sources of molecular disorder, namely the configurational and the dynamic entropy; both are involved in the elasticity mechanism of an elastomer.

Recently [14, 15] we have proposed new model for the elastin elasticity based on the concept of the transition-to-chaos in the motion of suitable recurring peptide sequences, as the source of the entropy increase accompanying the transition from the stretched to the relaxed state.

In the present work we have studied the dynamics, the supramolecular and polymeric organization of the octapeptide ALGGGALG, which is a fragment of elastin containing the rather flexible sequence of three consecutive glycine amino acidic residues, with the purpose to evaluate its role in the elastic behaviour of the protein.

2 Experimental

The elastin-related octapeptide was synthesized by the Solid-Phase Peptide Synthesis method using the Fmoc/DCC/HOBt chemistry [16]. Poly(ALGGGALG) was obtained by the DMSO/DPPA/TEA chemistry [17]. The polymer molar weight distribution, determined from poly-acrilamide gel electrophoresis, was in the range 3000-14000 a.m.u. To obtain suitable samples for TEM observations, the octapeptide was dissolved in distilled water at a concentration of 20 mg/ml, whereas the polymer was equilibrated in water at 4 °C in order to obtain a saturated solution. A drop of the solutions was placed onto a carbon coated copper grid (400 mesh), rinsed with HPLC grade water and stained with a 2% uranyl acetate solution.

The micro graphs were submitted for a digital image processing for fractal dimension calculation. They were imported via scanner in a Macintosh system as 256-grey-tone images, with a resolution of 300 dpi (dots per inch), and stored as PICT files. The pictures are then opened from an application for image analysis [18] and converted in a binary (black and white) version by selecting a suitable threshold value in order to capture the main morphological feature of the photos. The binary pictures are then loaded into an application [19] for the calculation of the fractal dimension by the box-counting algorithm [20]. The fractal (box) dimension is obtained as the slope of the log(N(s)) vs log(1/s) graph, where N(s) is the number of square box of side s containing at least one black pixel of the binary image.

3 Theoretical Model

The molecule submitted to molecular dynamic simulations was the peptide ACE-ALGGGALG-NME, where an acetyl group was added at the N terminal of the peptide chain and an N-methyl was attached at the C terminal, in order to obtain uncharged ends. Calculations were performed with the AMBER 4.1 program [21] on a DEC Alpha 2100 under the UNIX 4.0 operating system. The molecular force field was derived from a potential energy model including stretching, bending, torsion, non-bonded and H-bonded terms [22]; the external forces are determined by the interaction with an external heat bath [23] at the constant temperature of 300 K. The atomic positions and velocities as functions of time are obtained by numerical integration of the classical equations of motion through the Verlet leap-frog algorithm [24].

In our simulation the integration step was 2 fs and data were stored every 500 steps; the total length of the dynamics was 1300 ps. Before the dynamics, the molecule was submitted to a Newton-Raphson energy optimization, starting from a fully extended chain conformation, to find a local potential energy minimum. Afterward the molecule was placed in a box of 300 uniformly distributed random water molecules, with periodic boundary conditions, and the dynamics was started. The motion of the octapeptide was followed by the calculation of the kinetic, potential and total energies, inter-chain dihedral angles ϕ (CNCC), ψ (NCCN) and ω (CCNC), and the end-to-end distance at every time step.

The results were analysed by standard signal analysis techniques (Fourier power spectra, Lyapounov exponents, Poincarè delay map) in order to extract information about the system geometry and mobility.

4 Results and discussion

4.1 Electron Microscopy Investigations

TEM observations of the octapeptide show a typical fractal morphology: a DLA-like cluster (Fig. 1) characteirized by the box dimension 1.72. On the other hand the morphology of the polymer can be described, depending on the magnification, as a fibril-tree structure (Fig. 2) whose dimension is 1.27, or a twisted-rope hierarchy (Fig. 3) with a dimension of 1.55. The reported fractal dimensions appear reasonable when compared with literature values [25] and are consistent with the picture of a diffusion driven aggregation process of elongated molecules.

These findings are very common in the microscopic structure of elastin and its derivatives, indicating that the protein and its molecular fragments have the peculiar characteristic to build-up self-similar fractal aggregates. In particular, the twisted-rope morphology was observed over four orders of magnitude, from tens of nanometers to hundreds of microns [26] Our results indicate that the fractality, or statistical scale invariance, of the elastin molecule is a property existing at a fundamental level of the structural organization, because it is present also in its smaller pieces. Therefore, the microscopic chaos, responsible of the conformation entropy, is nearly scale-insensitive and this is probably a consequence of the intrinsic internal homothety (repetition) of the

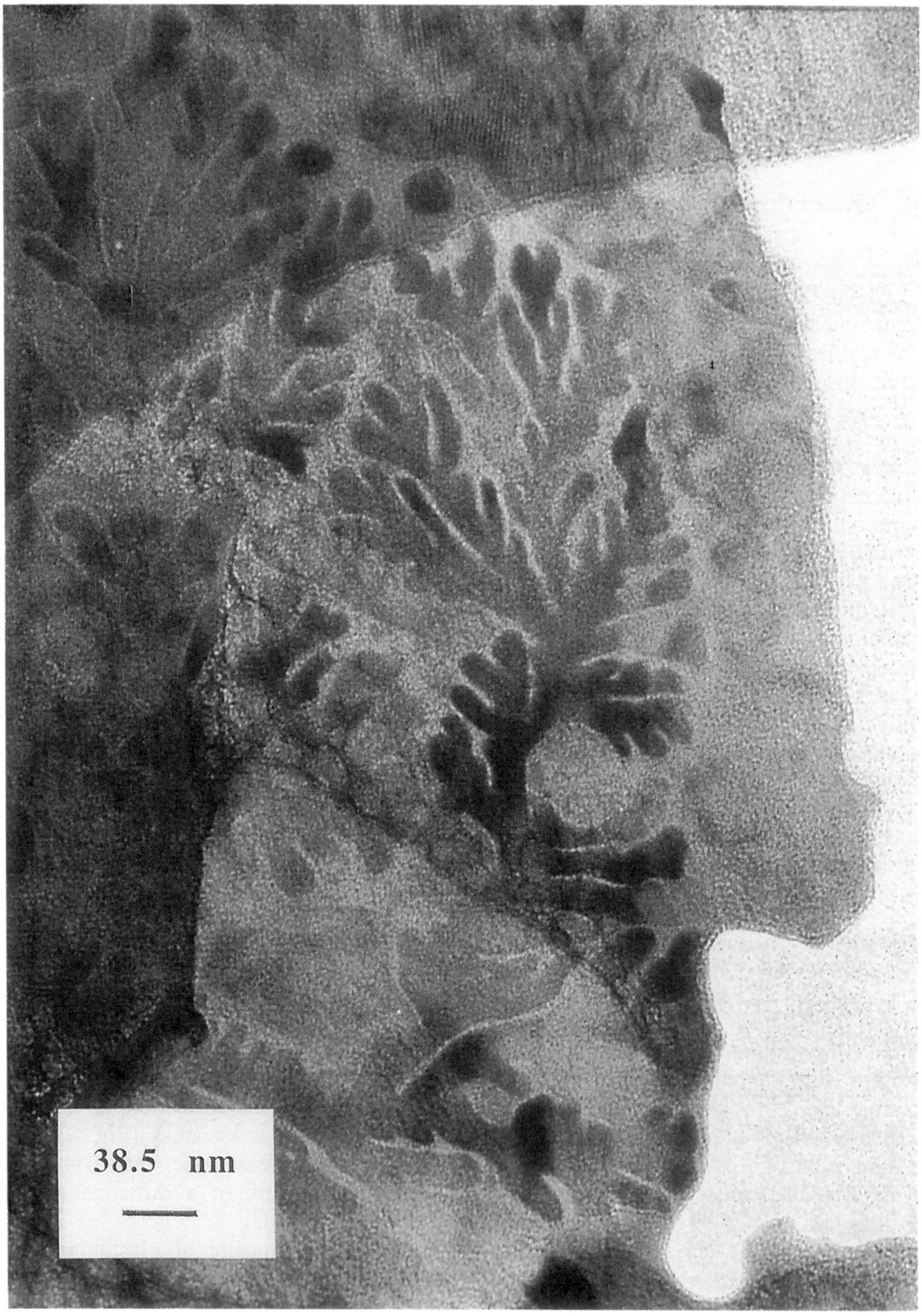

Figure 1. TEM micro-graph of the peptide ALGGGALG showing a DLA-like structure.

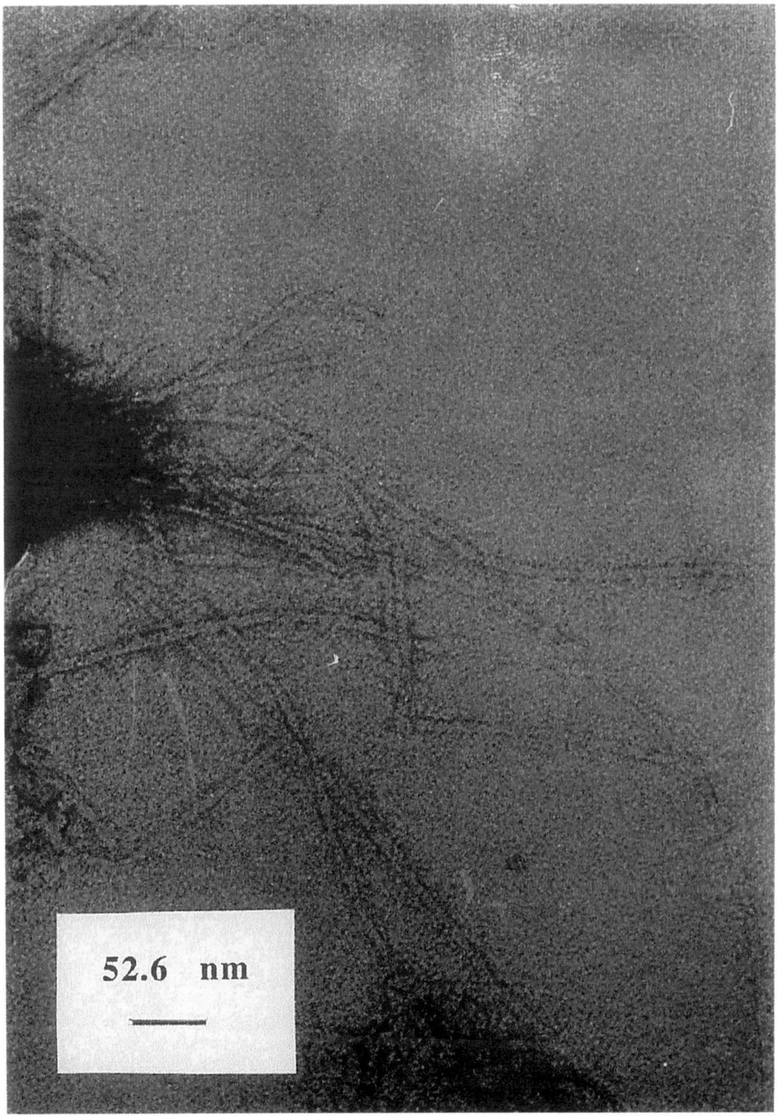

Figure 2. TEM micro-graph of the poly(ALGGGALG) with a fibril-tree organization.

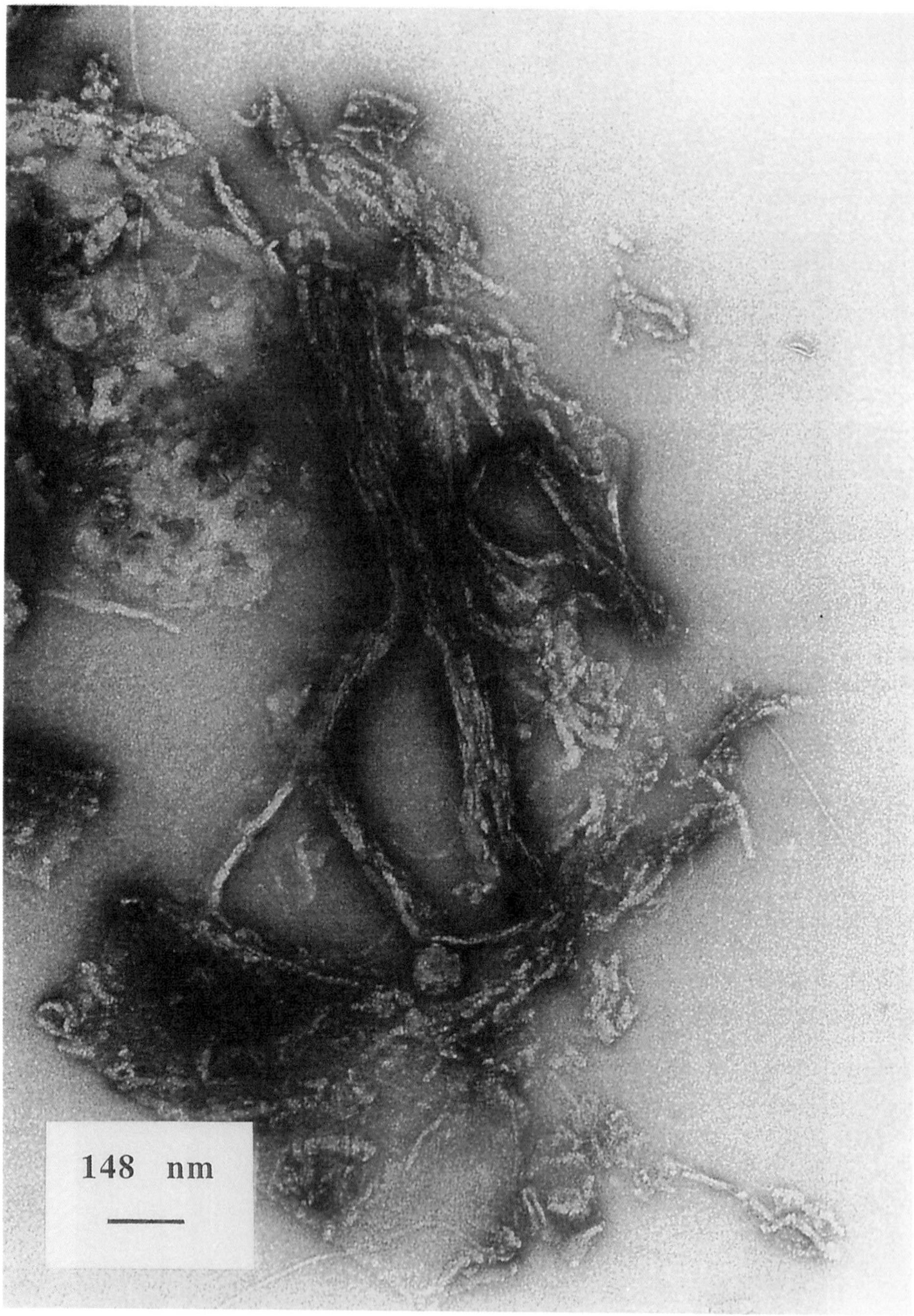

Figure 3. TEM micro-graph of the poly(ALGGGALG) with a twisted-rope hierarchy.

amino acid sequence [27].

4.2 Molecular Dynamics Simulations

The analysis of time evolution of the molecular kinetic and potential energies shows that the peptide molecule thermalization occurs after 600 ps of the dynamics. In Table 1 the values of the dihedral angles ϕ, ψ and ω (numbered from the N to C ends of the peptide) are reported for the conformations corresponding both to the start and the end of the dynamics; the corresponding molecular shapes are reported in fig. 4. From this, one observes that as the dynamics goes on, the molecule assume a bent conformation without H bonds.

Table 1. Molecular dihedral angles.

Angle	Initial conformation	After optimization	End of dynamics
ϕ (1)	-144.54	-116.02	-80.39
ψ (1)	172.12	168.27	161.02
ω (1)	178.87	169.95	173.85
ϕ (2)	-144.95	-84.39	-68.61
ψ (2)	152.62	153.67	141.93
ω (2)	177.86	177.95	-176.09
ϕ (3)	-167.70	-162.12	174.80
ψ (3)	-179.25	-171.00	171.31
ω (3)	-179.85	174.49	-175.63
ϕ (4)	179.82	94.43	84.29
ψ (4)	-179.98	102.18	1.77
ω (4)	180.00	-179.68	169.93
ϕ (5)	-179.98	-140.90	-55.03
ψ (5)	-179.95	172.54	168.51
ω (5)	179.41	178.57	-168.07
ϕ (6)	-144.63	-90.88	-74.96
ψ (6)	172.12	151.82	178.37
ω (6)	178.89	-176.70	-169.10
ϕ (7)	-144.97	-113.22	-136.54
ψ (7)	152.56	145.22	117.48
ω (7)	177.94	175.59	175.22
ϕ (8)	-168.15	-139.00	-75.81
ψ (8)	-179.17	-179.42	165.51
ω (8)	-179.87	179.86	-174.49

The analysis of the $\psi(4)$ and $\phi(5)$ dihedral angles, during the dynamics (Fig. 5), indicates that a so-called phase transition occurs near 600 ps, which is the fingerprint of a strong conformational change involving the central part of the molecule. This is also confirmed by a decrease of the end-to-end distance between the first an the last C atom (Fig. 6), which is an expression of the folding degree of the molecule.

The characteristics of molecular motion in the equilibrium state are well described by the delay map (Poincarè section) of the end-to-end distance in a phase space reconstruction: we have observed a diffusion-like trajectory characteristic of a fractional Brownian motion. This finding is also confirmed by a low frequency dominated Fourier power spectrum, and a value of the maximum Lyapounov exponent very close to zero (0.015).

In an attempt to evaluate the dependence of the dynamics on the initial conditions, we have carried out a second simulation starting from a different molecular conformation. The new starting point was a partially-extended structure (Polyproline II [28]) similar to the experimentally found conformation [29] of our octapeptide in aqueous solution. The final structure obtained in this case is not far from the initial one, even though a certain flexibility is found, as in the previous case. A comparison between the two dynamics indicates that, in the final state, the dihedral angles near the ends of the chain are quite the same, and this suggests that the Polyproline II conformation probably corresponds to a local energy minimum.

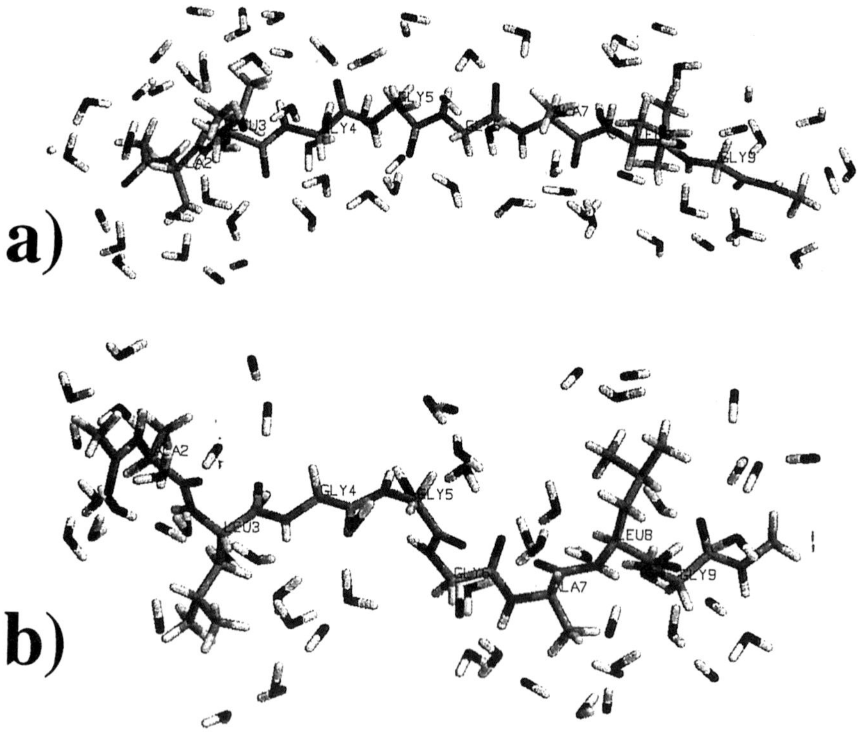

Figure 4. Snapshots of the ACE-ALGGGALG-NME molecule conformation (a) after optimization and (b) at the end of the dynamics.

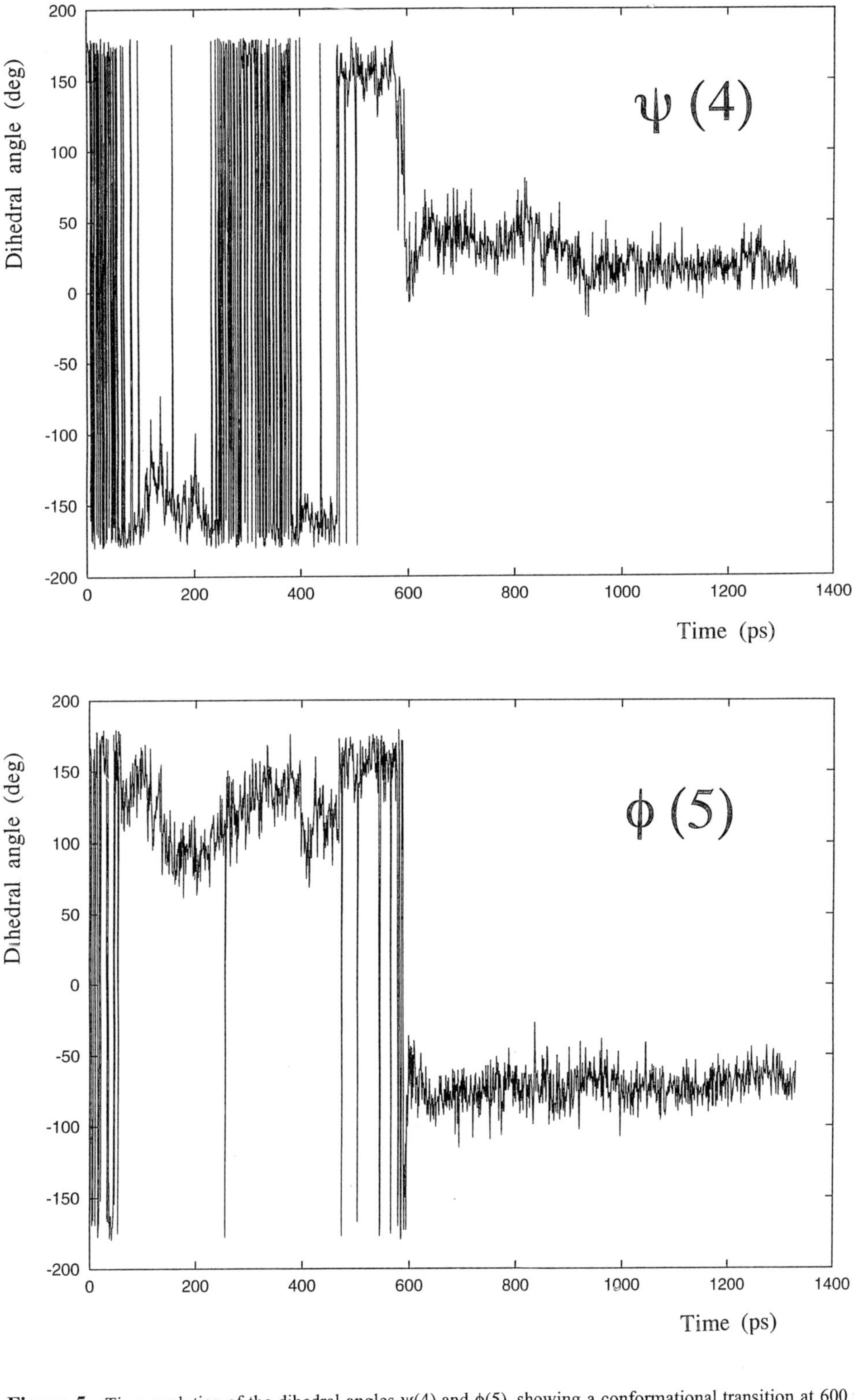

Figure 5. Time evolution of the dihedral angles ψ(4) and ϕ(5) showing a conformational transition at 600 ps.

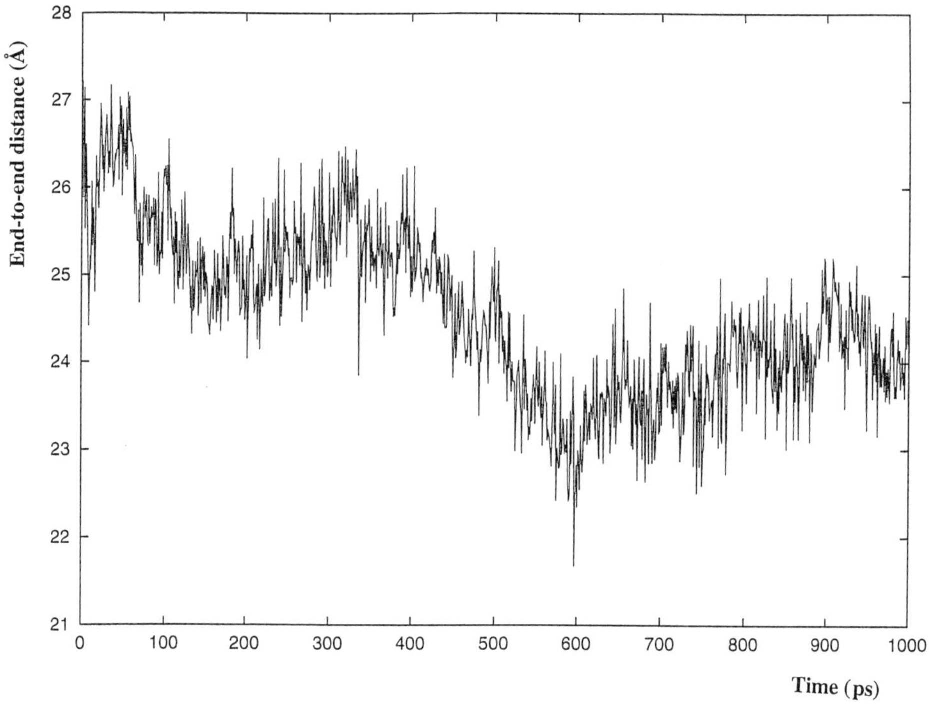

Figure 6. Time evolution of the end-to-end distance showing a molecular shrinkage at 600 ps.

5 Conclusions

The main conclusion that can be obtained from our molecular dynamics simulations is the great mobility of the octapeptide chain in correspondence of the GGG sequence. In fact the molecule is able to change its conformation, starting from a rather rigid extended geometry, through a rotation around a central pivotal point (that cannot be evidenced by the sole energy optimization). This flexibility plays a key role in elastin elastic properties, because it gives to the poly-peptide chain a further degree of freedom increasing the entropy of the relaxed state, with respect to the stretched one. The remarkable flexibility of the chain is also confirmed by the large amplitude Brownian motion of the end-to-end distance, in contrast with a small amplitude quasi-periodic motion expected for a more rigid structure.

The chaotic motion of the molecule, as evidenced by the global dynamics, can be interpreted in terms of temporal fractality [30], i.e. a time evolution independent from the level of description. This, together with the space fractality observed in the electronic micro-graphs, leads to the idea of the "space-time fractality" as a scale-independent link between structure and dynamics. It is also a new paradigm for the complexity of the ALGGGALG peptide and of the elastin molecule. In this framework the elastin elasticity entropic mechanism arises in a straightforward way from the internal homothety of the macromolecule, as previously suggested [31].

References

1. H. W. Kroto, D. R. M. Walton (Eds.), *The fullerenes. New horizons for the chemistry, physics and astrophysics of carbon* (Cambridge University Press, 1993).
2. B. I. Yakobson, R. E. Smalley, *Am. Sci.* **85** (1997) p. 324.
3. J. Hone, B. Batlogg, Z. Benes, A. T. Johnson, J. E. Fischer, *Science* **289**(2000) p. 1730.
4. A. M. Tamburro, J. M. Davidson (Eds.), *Elastin: chemical and biological aspects,* Proceedings of the International Congress (Maratea, Italy, October 10-13,1988, Congedo Editore, 1990).
5. P. M. Royce, B. Steinmann (Eds.), *Connective tissue and its heritable disorders* (Wiley-Liss, 1993) p. 167.
6. B. Carp, A. Janoff, *Exp. Lung Res.* **1** (1980) p. 237.
7. L. B. Sandberg, W. R. Gray, C. Franzblau (Eds.), *Elastin and elastic tissue* (Plenum, 1977).
8. M. E. Cates, M. R. Evans (Eds.), *Soft and fragile matter* (SUSSP Publications and Institute of Physics Publishing, 2000) p. 168.
9. A. M. Tamburro, A. De Stradis, L. D'Alessio, J. *Biomol. Structure & Dynamics* **12** (1995) p. 1161.
10. L. D'Alessio, A. M. Tamburro, A. De Stradis, Observation of fractal structures in the supramolecular organization of protein molecules, in INRIA, *L'ingenieur et les fractales - Fractals in engineering* (Delft, The Netherlands, June 14-16, 1999) p. 130.
11. D. Avnir (Ed.), *The fractal approach to heterogeneous chemistry* (John Wiley & Sons, 1989) p. 407.
12. W. J. Rothschild, *Fractals in chemistry* (John Wiley & Sons, 1998) p. 139.
13. J. Flory, *Principles of polymer chemistry* (Cornell University Press, 1953) p. 464.
14. V. Villani, L. D'Alessio, A. M. Tamburro, Contribution of GLY-X-GLY-GLY sequences to elastin elasticity. Development of the transition to chaos mechanism of elasticity, in *Elastin and elastic tissue*, Proceeding of an International Conference (Maratea, Italy, 17-20 October, 1996).
15. V. Villani, L. D'Alessio, A. M. Tamburro, *J. Chem. Soc. Perkin Trans.* **2** (1997) p. 2375.
16. F. Bisaccia, M. A. Castiglione Morelli, M. De Biasi, S. Traniello, S. Spisani, A. M. Tamburro, *Int. J. Peptide Protein Res.* **44** (1994) p. 332.
17. N. Nishi, K. Hagiwara, S. Tokura, *I. Peptide Protein Res.* **30** (1987) p. 275.
18. W. Rasband, *Image*, Version 1.38 (1991).
19. P. Bourke, *Fractal Dimension Calculator*, Version 1.0 (1992).
20. A. Harrison, *Fractals in chemistry* (Oxford University Press, 1995) p. 16.
21. D. A. Pearlman, D. A. Case, J. W. Caldwell, W. S. Ross, T. E. Cheatham III, D. M. Ferguson, G. L. Seibel, U. Chandra Singh, P. K. Weiner, P. A. Kollman, *AMBER 4.1* (University of California, 1995).
22. S. J. Weiner, P. A. Kollman, D. T. Nguyen, D. A. Case, *J. Comp. Chem.* 7(1986) p. 230.
23. H. J. C. Berendsen, J. P. M. Postma, W. F. van Gunsteren, A. Di Nola, J. R. Haak, *J. Chem. Phys.* **81** (1984) p. 3684.
24. L. Verlet, *Phys. Rev.* **159** (1967) p. 98.
25. S. Havlin, D. Ben-Avraham, *Adv. Phys.* **36** (1987) p. 695.
26. A. M. Tamburro, D. Daga Gordini, V. Guantieri, A. De Stradis, in N. Russo (Ed.), *Properties and chemistry of biomolecular systems* (Kluwer Academic Publishers, 1994) p. 389.

27. G. Cicila, M. May, N. Ornstein-Goldstein, Z. Indik, S. Morrow, H. S. Yeh, J.C. Rosembloom, C. Boyd, J. Rosembloom, K. Yoon, *Biochemistry* **24** (1985) p. 3075.
28. S. S. Zimmerman, M. S. Pottle, G. Nemethy, H. A. Scheraga, *Macromolecules* **10** (1977) p. 19.
29. M. Martino, A. Bavoso, V. Guantieri, A. Coviello, A. M. Tamburro, *J. Mol. Struct.* **519** (2000) p. 173.
30. S. Vrobel, *Fractal Time* (The Institute for Advanced Interdisciplinary Research, 1998) p. 11.
31. A. M. Tamburro, The supramolecular structures of elastin and related synthetic polypeptides: scale invariant weaving, in G. Savelli (Ed.), *Macrocyclic and supramolecular chemistry in Italy* (Centro Stampa Università di Perugia, Italy, 1995) p. 265.

CUMULATIVE EFFECTIVE HÖLDER EXPONENT BASED INDICATOR FOR REAL-TIME FETAL HEARTBEAT ANALYSIS DURING LABOUR

ZBIGNIEW R. STRUZIK
Centrum voor Wiskunde en Informatica (CWI)
Postbus 94079, NL-1090 GB, Amsterdam, The Netherlands
email: Zbigniew.Struzik@cwi.nl

WILLEM J. VAN WIJNGAARDEN
Afdeling Verloskunde en Gynaecologie, H4 - 205,
Academisch Medisch Centrum,
Postbus 22660, 1100 DD Amsterdam, The Netherlands,
email: W.J.vanWijngaarden@amc.uva.nl

We introduce a special purpose cumulative indicator, capturing in real time the cumulative deviation from the reference level of the exponent h (local roughness, Hölder exponent) of the fetal heartbeat during labour. We verify that the indicator applied to the variability component of the heartbeat coincides with the fetal outcome as determined by blood samples. The variability component is obtained from running real time decomposition of fetal heartbeat into independent components using an adaptation of an oversampled Haar wavelet transform. The particular filters used and resolutions applied are motivated by obstetricial insight/practice. The methodology described has the potential for real-time monitoring of the fetus during labour and for the prediction of the fetal outcome, allerting the attending staff in the case of (threatening) hypoxia.

1 Introduction

Methods of wavelet transform modulus maxima (WTMM) based multifractal analysis (MF) and detrended fluctuation analysis (DFA) have been demonstrated to be suitable for capturing scaling and correlation characteristics of the fluctuations in human heartbeat intervals[1,2,3,4]. These characteristics, obtained under a variety of conditions, have also been shown to reflect deviations in the heartbeat due to a variety of malfunctions[5,6,4,7] and physiological behaviour[8,9,10]. Obviously, these results can be considered for clinical applications.

Unfortunately, such standard methods of statistical analysis of heartbeat signals are not directly applicable to the practical problem of evaluating (the characteristics of) the fetal heartbeat in real-time during labour.

There are several reasons for this. One is that statistical techniques use long stretches of data to provide estimates of global measures (entities), like correlation exponents or multifractal spectra. A typical measurement requires over 30,000 samples (2^{15}) to provide reliable estimates of scaling for the extraction of exponents and reliable transformation from scaling exponents to the domain of multifractal spectra.

The typical heartrate of a fetus is about 130 beats per minute. The required time stretch for acquiring a sufficiently long data set (2^{15}) is thus about 250 minutes. By this time, the baby is often already born. Decisions about an intervention (such as a Caesarean section) have to be taken on the basis of 10 – 60 minutes long observations of the heartbeat.[a]

In addition to this, correlation exponents and multifractal spectra derived from them are rather sensitive to spikes, missing data, noise bursts and boundary effects.[b] Such erroneous data can cause a dramatic alteration of the results, which is particularly bad due to the lack of indication of where the errors come from[c] - a problem inherent to global statistical techniques. Fetal heartbeats during labour are recorded in difficult circumstances and subject to frequent data fall out and a high level of noise. Missing data in fetal heartbeat records can amount to about 20 – 40%. This does not include spikes and other noise.

Another serious problem for standard methods of analysis is the presence of the so-called decelerations in fetal heartbeat. These are sudden drops in the heartrate of the fetus, which can be caused by a number of events such as compression of the umbilical cord, increased intracranial pressure of the fetus during contractions or (temporary) hypoxia. The deficiency of oxygen caused by the contractions is compensated for by slowing down the heartrate (the so-called whale effect). Thus the presence of these 20 – 200 beats long drops in the heartrate does not necessarily imply a severe condition of hypoxia. However, they severely distort any standard method of heartrate analysis - the amplitude of the decelerations can be about one degree of magnitude larger than that of the residual fluctuations of the heartbeat.

Lastly, but importantly, global statistical techniques are usually computationally expensive. However, the most unsuitable for real-time applications is the lack of update mechanism on new samples. Taken as they are, such techniques would require recalculation of all the coefficients of the time series on each new sample - an unacceptable approach even in the days of cheap computational power.

Although these constraints practically disqualify standard methods from being used in medical practice, adaptations of standard approaches are possible. We have, therefore, considered the above-mentioned problems and designed a methodology which is capable of providing the required characteristics of fetal heartbeat in real-time.

The steps taken were thus the design of a running, incremental real-time decomposition scheme[13], which is capable of separating meaningful components of the fetal heartbeat. The criterium for the decomposition used here is not arbitrary (or driven by properties of the decomposition) as in standard wavelet decomposition, but designed to capture the unique features of the fetal heartbeat. These features are meant to reflect the components of the heartbeat which are analysed by obstetricians in standard clinical practice. The effect of such a decomposition is that the high frequency variability component can be analysed separately from the deceleration component. In this way, the influence of decelerations on the spectrum of exponents can be minimized. Additionally, this way of decomposing the signal gives the possibility of providing the obstetrician with estimates reflecting

[a]It has to be noted that fetal heartbeat is the only indicator of the state of the fetus. Therefore the characteristics determined from it are crucial in taking the decision to carry out an operative delivery in case of suspected hypoxia.

[b]This can particularly affect higher moment calculations, resulting in serious errors in the MF spectrum.[11,12]

[c]despite the fact that techniques like WTMM provide localised information

traditionally observed entities.

Instead of the standard correlation analysis on long stretches of data, a simplified (effective) Hölder exponent estimate h has been used[14]. It has been proven to provide useful roughness characteristics in the context of heartbeat or financial analysis[10,15]. In this work, we use it to analyse the local roughness of the variability component. This is a novel approach in obstetrics, since the standard way of estimating variability level refers to amplitude sensitive standard deviation rather than scale-free characteristics like the local roughness exponent.

By introducing a special cumulative indicator, we have been able to verify that the increase of the exponent h (local roughness exponent) of the variability component of the fetal heartbeat relates to adverse blood gas values of the fetus (hypoxia condition). It has, therefore, a potential for monitoring the fetus and for the prediction of the fetal outcome. This observation confirms the reported observations on adults, where the increase in the (global) correlation exponent corresponding to the loss of anti-correlation has been attributed to a number of malfunctions in the cardiac system.[d]

The contents of this paper are divided as follows. In section 2, we introduce a running time decomposition of the fetal heartrate (FHR). Most of the technology used here is self-contained, but the reader may want to check Ref.[13] for technical details of real-time, Haar-type, wavelet decomposition. In section 3, basic operators are introduced, which act on the decomposition coefficients. The operators reveal and enhance collective behaviour in the fluctuations of the coefficients. Finally, in section 4, a cumulative operator is defined, which is next applied to the effective Hölder exponent of the variability component of the fetal heartbeat. Section 5 closes the paper with conclusions and future plans.

2 Separating Meaningful Components of Fetal Heartbeat

Fetal heartbeat is monitored during labour, as it is the only indicator of the well-being of the fetus. It is, therefore, used to alert the attending staff of possible hypoxia requiring direct intervention - a Caesarean section or pH and Base Excess estimation from a fetal scalp blood sample taken during labour. The heartbeat is usually analysed from the point of view of its three main characteristics:

- the level of the baseline - the low frequency outline of the heartbeat without spikes, high frequency variability and without decelerations and accelerations
- the presence and frequency of decelerations and accelerations
- the level of variability - the high frequency 'noise' of the heartbeat.

Using these characteristics, obstetricians can predict a good outcome very well. However, in cases of 'bad' fetal heartrate patterns, half the time the fetal outcome is good and operative intervention may have been carried out unnecessarily. It turns out that we can conveniently approximate these characteristics in an appropriate decomposition.

[d]It should be noted here that most malfunctions correspond with an insufficient supply of oxygen.

Unlike in the more traditional multiresolutions schemes with dyadic resolution separation,[16,17] we apply only two resolutions (scales) of approximation at the large separation of one decade (i.e. resolutions $a_1 = 200$ and $a_2 = 20$). The total decomposition plane is, therefore, partitioned into three sections.

The resolution $a_1 = 200$ beats is intended to approach the so-called baseline of the heartbeat - the low frequency 'backbone' of the fluctuations. This is where the adaptive filter proves necessary. The resolution of $a_1 = 200$ beats is suitable for approximating the baseline, but not in the presence of decelerations (or accelerations, sudden jumps of the heartbeat). The moving average filter performing approximation of the heartbeat at $a_1 = 200$ beats is thus equipped with a threshold mechanism which does not accept jumps or drops larger than one standard deviation from the (historic N samples) mean value of the signal.

$$\mathrm{MAA}_{\mathrm{a}_1}(f_i) = \frac{1}{a_1} \sum_{i=-a_1/2}^{i=a_1/2} (f_i)\,\delta(i) \tag{1}$$

where

$$\delta(i) = \begin{cases} 1 & \text{for } \; (f_i - mean(f_i, i = i - N, \ldots, i)) < stdev(f_i, i = i - N, \ldots, i) \\ 0 & \text{otherwise}\,. \end{cases}$$

The second approximation level, centred at $a_2 = 20$ beats, uses a simple block smoothing function. It is intended to separate the highest frequencies, which we attribute to the so-called variability component.

$$\mathrm{MA}_{\mathrm{a}_2}(f_i) = \frac{1}{a_2} \sum_{i=-a_2/2}^{i=a_2/2} (f_i) \tag{2}$$

Just like in the standard wavelet decomposition, the multiresolution bands of decomposition are obtained from subtracting the multiresolution approximations at subsequent resolutions. As we have used only two levels of approximation, three bands of resolution are obtained:

1. The first band of our decomposition is the entire low frequency component separated by the adaptive MA filter at the $a_1 = 200$ resolution level. It captures features of resolution less than $a_1 = 200$ and amplitude less than one standard deviation from the (historic) mean value of the signal. The filter used to define the baseline approximation level, Eqn. 1, is a low-pass filter, therefore the entire resolution band of low frequencies - the baseline, simply becomes:

$$B_i = \mathrm{MAA}_{200}(f_i)\,. \tag{3}$$

2. The second band is obtained from subtracting the baseline approximation level from the variability approximation level. This contains the middle range of frequencies, capturing combined accelerations and decelerations.

$$AD_i = \mathrm{MA}_{20}(f_i) - \mathrm{MAA}_{200}(f_i) \tag{4}$$

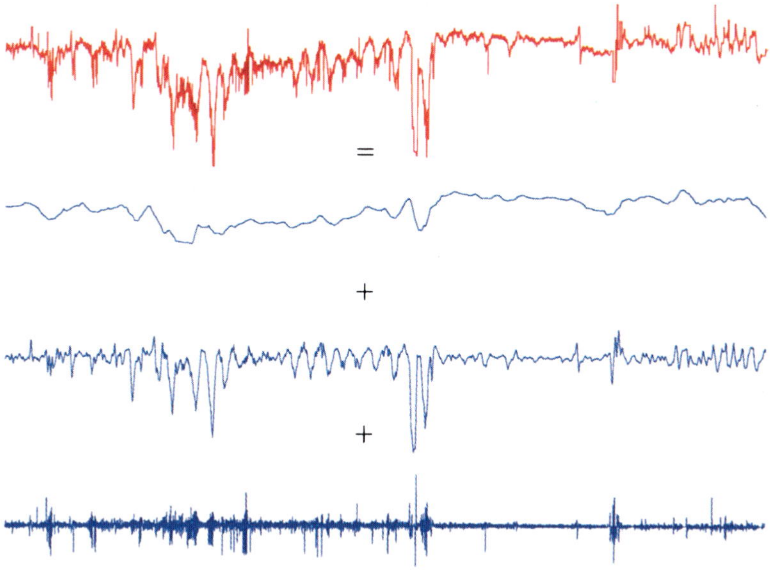

Figure 1. The decomposition of the fetal heartbeat time series. From top to bottom, the original time series, the low frequency baseline, the combined acceleration-deceleration line, the variability residue. All three components are orthogonal and can be summed up to give the original.

3. Finally, the variability range of resolutions is obtained by subtracting the original time series from the approximation at the resolution level $a_2 = 20$. This variability is thus the signal, less baseline fluctuations, and less decelerations and accelerations component.

$$V_i = f_i - \mathrm{MA}_{20}(f_i) \tag{5}$$

The result of applying such a procedure is a complete orthogonal decomposition into three components of the time series: the baseline, the combined line of accelerations and decelerations, plus the residual variability component. Of course, the original time series can be restored by simply summing up all three components, see figure 1.

3 Constructing Meaningful Real-time Indicators

The components of the fetal heartbeat thus obtained carry information, which is (visually) analysed by the obstetrician. In order to mimic a visual evaluation of the trends and collective behaviour of the fluctuations and features in the decomposition, we apply running MA_{1000} filters.

The local average effective acceleration/deceleration obtained in this way, see figure 2, indicates the effective level of deceleration which can be monitored and evaluated on-line. We will, however, not further pursue here the discussion of the relevance of this characteristic, leaving it to a separate communication. Rather, we intend to focus exclusively on the diagnostic capabilities of variability, or rather its roughness exponent.

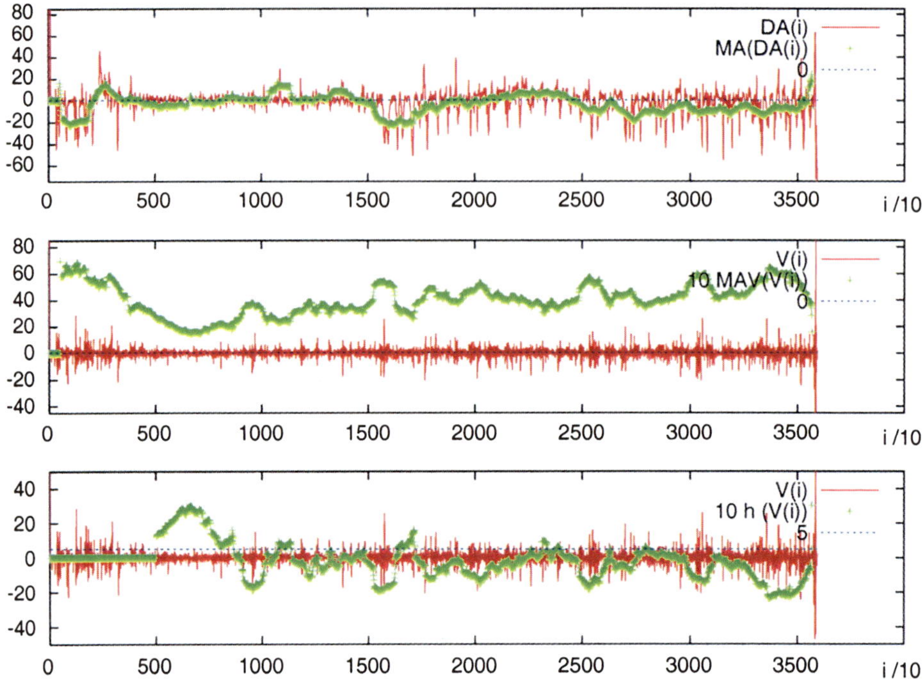

Figure 2. Real-time indicators, from top to bottom: local average dec-/acceleration $MA_{1000}(DA_i)$, local average variability $MAV_{1000}(V_i)$, local 'Hölder' exponent $h_{eff}(V_i)$. Both the $MAV_{1000}(V_i)$ and $h_{eff}(V_i)$ have been rescaled by a factor 10.

One could be tempted to apply a similar moving average filter to the variability component. Of course, this time it would have to be applied to the square amplitude, as the variability is usually perceived as the (mean) amplitude of high frequency deviations:

$$\mathrm{MAV_a}(V_i) = \sqrt{\frac{1}{a}\sum_{i=-a/2}^{i=a/2} {V_i}^2}\,. \tag{6}$$

Such a fixed resolution indicator (for example at $a = 1000$) would perhaps be considered an adequate measure of variability. However, it is known that our visual system picks scale-free characteristics when evaluating measures of roughness or variability. Therefore, even from the point of view of a visual evaluation, a fixed scale measure is not necessarily the best indicator.

It has been demonstrated [1-10] that scale-free measures of variability may be more suitable for diagnostic purposes.[e] There are a number of methods providing scale-free characteristics, most of them are, however, unsuitable for real-time local analysis of non-stationary time series. Therefore, for this study we use an adaptation of the effective Hölder exponent measure of local roughness[14,11]. The original concept has already been used in the study of adult heartbeat, using the WTMM representation[10].

The local effective roughness of the variability component V is defined as the logarithmic increase of the (standard deviation) of variability across scales/resolutions:

$$h_{eff}(V_i) = \frac{\log(\mathrm{MAV}_{a_l}(V_i)) - \log(\mathrm{MAV}_{a_h}(V_i))}{\log(a_l) - \log(a_h)} \,. \tag{7}$$

Two relatively rough resolutions ($a_h = 1000$ and $a_l = 10000$) have been used for the evaluation of the local h.[f] Also, we have used a simplification of the local effective exponent concept from Ref.[14,11], in that a simple block function (moving average filter) is used instead of the wavelet. Using the moving average filters makes possible incorporating the local h evaluation in the real-time incremental decomposition framework, as described above. Of course the local h analysis with smoothing block kernels is only possible due to the fact that the variability component V (as defined in Eqn. 5) is effectively free of any trends.

4 Cumulative Hölder Exponent Based Real-time Indicator

There is no reason why the local Hölder exponent of the variability $h(V_i)$ should be stationary. It reflects dynamic changes in the condition of the fetus and the degree of stress to which it is subjected. Despite the fact that stress has a rapid effect on the heartbeat, the effects on the state of the fetus can be long term. This is why short dynamic changes in the heartbeat characteristic may not be relevant and not representative to the state of the fetus.

Rather than using a long observation window which would capture trend behaviour, we use a cumulative indicator, which still has the resolution of the mean local h used in the previous section. The cumulative h is defined from the beginning of the observation and with respect to some normal reference level h_{ref}:

$$h_{cum}(V_i) = -\sum_{l=1}^{i}(h_{eff}(V_l) - h_{ref}) \,. \tag{8}$$

The minus sign is introduced to give the h_{cum} indicator increasing direction when the level of local correlations is lower than h_{ref}. This corresponds with a healthy condition. The case of higher correlations is associated with problems and, therefore, the accumulation of positive difference $(h_{eff}(V_l) - h_{ref})$ will lead to decreasing cumulative h.

[e] Of course, this does not exclude using amplitude based measures of variability in parallel.

[f] Note that the actual (temporal) resolution of the local h so obtained is defined by the higher (finer) resolution level - the coarse reference resolution can be accumulated from past historic samples.

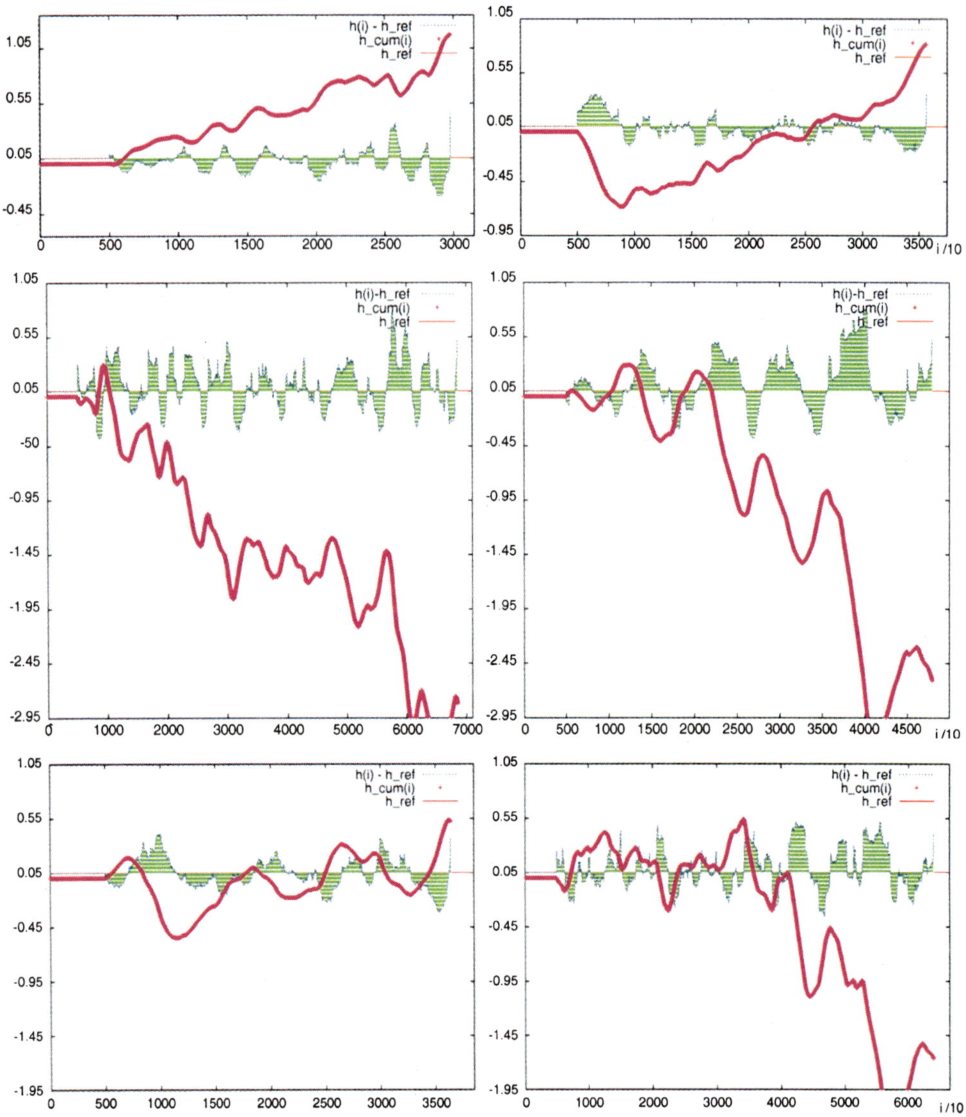

Figure 3. Six different time series analysed using the cumulative Hölder exponent based real-time indicator. The time series correspond with three good outcomes (healthy) and three bad outcomes (hypoxia). The good outcomes can be identified by the indicator oscillating near zero or steadily increasing. In the case of hypoxia, the indicator plunges (down towards negative values). Problems can occur at any moment during labour, even after a stable condition, as is visible in the last of the six plots (lowest row, right). The first part of the plot is flat since this period is required for initialisation - acquiring a reference for the upper value for the Hölder exponent evaluation. Both the cumulative Hölder exponent $h_{cum}(V_i)$ (red line) and the deviation of the Hölder from the reference value $h_{ref} = 0.05$ (blue filled curve) are plotted. The $h_{cum}(V_i)$ has been rescaled by a factor 0.01.

We have tested several examples of fetal heartbeats and found a good correlation with the fetal outcome, as determined by the blood tests. In figure 3, we plot six cases where three represent good outcomes and three bad outcomes. The cumulative indicator steadily increasing or remaining within some margin of fluctuations indicates no problems and a good prediction. When the indicator plunges down, it calls for intervention. This can, of course, happen at any moment during labour. The nature of this process is dramatically non-stationary, and a period of positive evaluation can be interrupted at any stage (for example by the occlusion of the umbilical cord due to movement). One of our examples (figure 3 lower right), shows the cumulative indicator plunging after a prolonged homeostasis.

5 Concluding Remarks

Fetal heartbeat during labour is a highly non-stationary process which needs to be monitored in real-time.

We have presented a methodology of decomposition of fetal heartbeat into meaningful components suitable for real-time monitoring of the fetus. We have also introduced a real-time cumulative indicator, based on the effective Hölder exponent of the variability component of fetal hearbeat. The indicator has been demonstrated to provide insight into the highly non-stationary nature of the variability properites of the fetal heartbeat. It has also shown correlation with the blood samples, motivating its use as a monitoring and a predictive tool.

The exact parameters of the decomposition are subject to tuning. Also, other local measures of roughness of the variability component may provide better and more detailed insight. Combining the observations derived from (all) the components of the heartbeat (and possibly external knowledge) into a predictive tool may result in a substantially better predictive power.

References

1. M. Kobayashi, T. Musha, 1/f fluctuation of heartbeat period, *IEEE Trans Biomed. Eng.*, **29**, 456-457 (1981).
2. C.-K. Peng, J. Mietus, J.M. Hausdorff, S. Havlin, H.E. Stanley and A.L. Goldberger Long-Range Anticorrelations and Non-Gaussian Bahavior of the Heartbeat *Phys. Rev. Lett.*, **70**, 1343-1346 (1993).
3. J. B. Bassingthwaighte, L. S. Liebovitch and B. J. West. *Fractal Physiology*, (Oxford University Press, 1994).
4. P.Ch. Ivanov, M.G. Rosenblum, L.A. Nunes Amaral, Z.R. Struzik, S. Havlin, A.L. Goldberger and H.E. Stanley, Multifractality in Human Heartbeat Dynamics, *Nature* **399**, pp. 461–465, (1999).
5. R.G. Turccot, M.C. Teich, Fractal Character of the Electrocardiogram: Distinguishing Heart-failure and Normal Patients, *Ann. Biomed. Eng.* **24**, 269-293 (1996).
6. M. Meyer, Scaling Properties of Heartbeat Interval Fluctuations in Health and Disease, in *Fractals and Beyond*, Ed. M.M. Novak, (World Scientific, 1998), pp. 33-42.

7. P.Ch. Ivanov, L.A. Nunes Amaral, A.L. Goldberger, S. Havlin, M.G. Rosenblum, H.E. Stanley and Z.R. Struzik, From 1/f Noise to Multifractal Cascades in Heartbeat Dynamics, *Chaos* **11** Special Issue on Unsolved Problems of Noise, Ed. Derek Abbott, to appear, (2001).
8. P. Ch. Ivanov, A. Bunde, L. A. N. Amaral, J. Fritsch-Yelle, R. M. Baevsky, S. Havlin, H. E. Stanley, and A. L. Goldberger, Sleep-wake Differences in Scaling Behavior of the Human Heartbeat: Analysis of Terrestrial and Long-term Space Flight Data, *Europhys. Lett.* **48**, 594-600 (1999).
9. A. Bunde, S. Havlin, J.W. Kantelhardt, T. Penzel, J.-H. Peter and K. Voigt, Correlated and Uncorrelated Regions in Heart-rate Fluctuations During Sleep, *Phys. Rev. Lett.* **85**, 3736 (2000).
10. Z. R. Struzik. Revealing Local Variablity Properties of Human Heartbeat Intervals with the Local Effective Hoelder Exponent. Fractals **9**, No 1, (2001).
11. Z.R. Struzik, Determining Local Singularity Strengths and their Spectra with the Wavelet Transform, Fractals **8**, No 2, (2000).
12. Z. R. Struzik, A. P. J. M. Siebes. Outlier Detection and Localisation with Wavelet Based Multifractal Formalism. Technical Report INS-R0008, CWI, Amsterdam, The Netherlands, (2000). Available from `www.cwi.nl/~zbyszek`
13. Z. R. Struzik, Oversampling the Haar Wavelet Transform, CWI report INS-R0102, (2001). Available from `www.cwi.nl/~zbyszek`
14. Z. R. Struzik, Local Effective Hölder Exponent Estimation on the Wavelet Transform Maxima Tree, in *Fractals: Theory and Applications in Engineering*, Eds: M. Dekking, J. Lévy Véhel, E. Lutton, C. Tricot, Springer Verlag, pp. 93–112, (1999).
15. Z. R. Struzik, Wavelet Methods in (Financial) Time-series Processing, *Physica A*, **296** (1-2), 307-319, (2001).
16. S. Mallat, A Theory for Multiresolution Signal Decomposition: The Wavelet Representation, *IEEE Trans. PAMI*, **11**, pp. 674-693, (1989).
17. I. Daubechies, *Ten Lectures on Wavelets*. S.I.A.M., (1992).

FRACTAL FEATURES IN THE NONSTATIONARITY OF PHYSIOLOGICAL TIME SERIES

PLAMEN CH. IVANOV[1,3], PEDRO BERNAOLA-GALVÁN[1,2], LUÍS A. NUNES AMARAL[1,3], AND H. EUGENE STANLEY[1]

[1] *Center for Polymer Studies and Department of Physics, Boston University, Boston, MA 02215*
Email: plamen@argento.bu.edu

[2] *Departamento de Física Aplicada II, Universidad de Málaga E-29071, Spain*
Email: rick@uma.es

[3] *Cardiovascular Division, Harvard Medical School, Beth Israel Deaconess Medical Center, Boston, MA 02215*

We present a new method to probe the nonstationarity of a signal by partitioning it into segments with different mean values. We find that the lengths of these segments follow a power-law distribution for a nonstationary time series representative of a complex dynamics, namely the human heartbeat. This scale-invariant structure cannot be explained by the presence of correlations in the data. We find also a common functional form describing the differences in mean heart rates between consecutive segments, but with different parameters for healthy individuals and for patients with heart failure. These findings may provide information into the way heart rate variability is reduced with disease. The approach we persent may be used on a wide range of physiologic signals.

1 Introduction

A time series is stationary if the mean, standard deviation and all higher moments, as well as the correlation functions, are invariant under time translation [1]. Signals that do not obey these conditions are nonstationary. Nonstationarity is a prominent feature of biological variability that can be associated with regimes (segments) of different statistical properties. The borders between different segments can be gradual or abrupt (Fig. 1).

A major problem in contemporary physiology is the nonstationarity of time series generated under free-running conditions [2]. Physiological signals obtained under widely-varying conditions raise serious challenges to both technical and fundamental aspects of time series analysis. By filtering out effects of nonstationarity, much work has focused on "intrinsic properties" of physiological signals [3]. This approach is based on the implicit assumption that the nonstationarity arises simply from changes in environmental conditions — e.g., different daily activities — so environmental "noise" could be treated as a "trend" and distinguished from the more subtle fluctuations that may reveal intrinsic correlation properties of the dynamics. Indeed, important scale-invariant features in physiological processes were recently revealed after filtering out masking effects of nonstationarity [4]. However, nonstationarity itself is also an important feature of physiological time series and is known to change from healthy to pathological conditions [5], suggesting more than only environmental conditions are reflected in the phenomena. Thus one would expect that there is perhaps a non-trivial structure associated with the nonstationarity in physiological signals, which may change with disease. To test this hypothesis we fo-

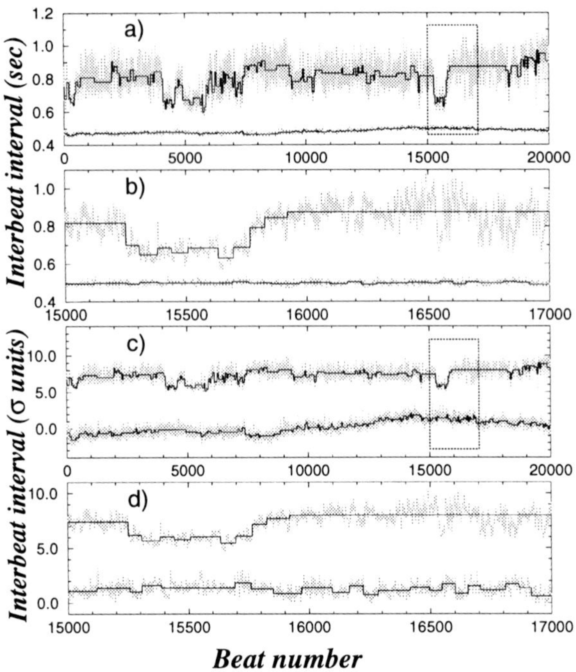

Figure 1. (a) Plot of 20,000 interbeat intervals ($\approx$ 6 hours) for a healthy subject (upper curve) and a subject with heart failure (bottom curve). Note the larger variability and patchiness for the healthy record. (b) Magnification of a small fraction (2000 beats) of the signals in (a). (c) Same signals as displayed in (a) after subtracting the global average and dividing by the global standard deviation; after this normalization both signals appear very similar. (d) Magnification of a small fraction (2000 beats) of the signals in (c).

cus on one statistical property, the mean heart rate, which is related to physiologic responses and is commonly used for medical evaluation.

2 Segmentation Algorithm

The problem we address is the partition of a nonstationary time series, which is composed of many segments with different mean value, in such a way as to maximize the difference in the mean values between adjacent segments. We apply the following procedure: we move a sliding pointer from left to right along the signal. At each position of the pointer, we compute the mean of the subset of the signal to the left of the pointer ($\mu_{\rm left}$) and to the right ($\mu_{\rm right}$). To measure the difference between $\mu_{\rm left}$ and $\mu_{\rm right}$, we compute the t-statistic [6]

$$t \equiv \left| \frac{\mu_{\rm left} - \mu_{\rm right}}{s_D} \right| , \tag{1}$$

where $s_D = \left(\frac{s^2_{\rm left} + s^2_{\rm right}}{N_{\rm left} + N_{\rm right} - 2} \right)^{1/2} \left(\frac{1}{N_{\rm left}} + \frac{1}{N_{\rm right}} \right)^{1/2}$ is the pooled variance [6] and $s_{\rm left}$ and $s_{\rm right}$ are the standard deviations of the data to the left and to the right of the pointer respectively, and $N_{\rm left}$ and $N_{\rm right}$ are the number of points to the left and to the right of the pointer respectively.

We next determine the position of the pointer for which t reaches its maximum value, $t_{\rm max}$, and compute the statistical significance of $t_{\rm max}$. The significance level $\mathcal{P}(\tau)$ of a possible cutting point with $t_{\rm max} = \tau$ is defined as the probability of obtaining the value τ or lower values within a random sequence: $\mathcal{P}(\tau) = \text{Prob}\{t_{\rm max} \leq \tau\}$. Thus, a series of N random numbers of fixed mean would remain unsegmented with probability $\mathcal{P}(\tau)$. This statement remains true for a series of random numbers of any length. We have carried out the following experiments: we generate 100,000 random series of given length N and fixed mean, and we segment them at significance level $\mathcal{P}_0$. In all experiments the ratio between the number of series which remain undivided and the total number of series is very close to $\mathcal{P}_0$, independently of N. Note that the probability $\mathcal{P}(\tau)$ is not the same as the used for the standard Student's test.

As we could not obtain $\mathcal{P}(\tau)$ in a closed analytical form, we have developed an suitable approximation by means of Monte Carlo simulations. $\mathcal{P}(\tau) \approx \left[1 - \text{I}_{\frac{\nu}{\nu+\tau^2}}(\delta\nu, \delta)\right]^{\gamma}$, where $\gamma = 4.19 \ln N - 11.54$, $\delta = 0.40$, N is the size of the sequence or subsequence to be split, $\nu = N - 2$ is the degrees of freedom, and $I_x(a, b)$ is the incomplete beta function [6].

We check if this significance exceeds a selected threshold $\mathcal{P}_0$, usually taken to be 95%. If so, then the signal is cut at this point into two subsequences; otherwise the signal remains undivided. If the sequence is cut, the procedure continues recursively for each of the two resulting subsequences created by each cut. Before a new cut is accepted, we also compute t between the right-hand new segment and its right neighbor (obtained by a previous cut) and the t between the left-hand new segment and its left neighbor (also obtained by a previous cut) and check if both values of t have a statistical significance exceeding $\mathcal{P}_0$. If so, we proceed with the new cut; otherwise we do not cut. This ensures that all resulting segments have a statistically significant difference in their means. The process stops when none of the possible cutting points has a significance exceeding $\mathcal{P}_0$, and we say that the signal has been segmented at the "significance level $\mathcal{P}_0$" (Fig. 2).

Our method leads to partitioning of a time series into segments with well-defined means, each significantly different from the mean of the adjacent segments (Fig. 1). This allows us to probe the nonstationarity in a signal through the statistical analysis of the properties of the segments.

3 Application to Human Heart Rate Data

Here we consider 47 datasets from 18 healthy subjects, 17 records of cosmonauts during orbital flight and 12 patients with congestive heart failure [7]. We separately analyze 6–hour long subsets of each dataset, corresponding to the periods when the subject is awake or sleeping. Figure 1 shows a representative dataset of a healthy subject, and a subject with heart failure. Superposed on the interbeat interval series, we also plot the segments obtained by means of our segmentation algorithm.

To quantify the nonstationarity in heart rate variability, we study the statistical properties of the segments corresponding to parts of the signal with significantly

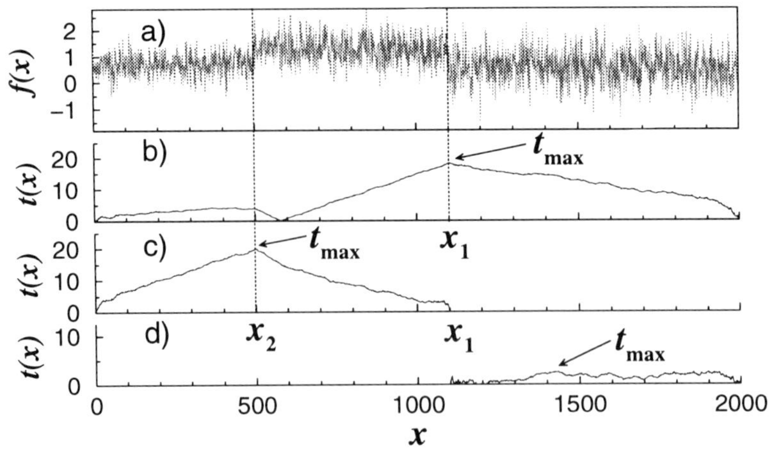

Figure 2. (a) An artificial time series $f(x)$ composed of three segments with different mean values. (b) Values of the statistic t, defined in Eq. (1), obtained by moving the pointer along the time series. Note that $t_{\max}$ is reached at x_1. We find that if $\mathcal{P}(t_{\max}) \geq \mathcal{P}_0 = 95\%$, and so we cut the series at x_1. (c) We iterate the procedure with the segment $[0, x_1]$. We find that $\mathcal{P}(t_{\max}) \geq 95\%$ and we also find that the significance of t computed between $[x_2, x_1]$ and $[x_2, 2000]$ is greater than 95%, so the series is cut at x_2. (d) We iterate the procedure with the segment $[x_1, 2000]$. Now, $\mathcal{P}(t_{\max}) \leq 95\%$, so this segment is not cut. Our procedure has a limitation for the extreme case of a long segment with a given mean, followed by a short segment with a different mean, which again is followed by a long segment with a mean identical to the mean of the first segment. First, we note that this is a very unlikely event in real data. However, even in this extreme case the algorithm could provide a good segmentation, if we lower sufficiently the significance level, $\mathcal{P}_0$. In fact, more often one can find in real data a situation when the second long segment has a mean value very close but not identical to the mean value of the first long segment; in such a case the procedure works accurately.

different mean values. To characterize the segments, we analyze two quantities: (i) the length of the segments and (ii) the absolute values of the differences between the mean values of consecutive segments, which we call jumps.

3.1 Distribution of segment lengths

Healthy subjects typically exhibit nonstationary behavior associated with large variability, trends, and segments with large differences in their mean values, while data from heart failure subjects are characterized by reduced variability and appear to be more homogeneous (Fig. 1) [5]. Thus, one might expect that signals from healthy subjects will be characterized by a large number of segments, while signals from heart failure subjects will exhibit a smaller number of segments (i.e., the average length of the segments for healthy subjects could be expected to be smaller than for heart failure subjects).

We find that the distribution of segment lengths for the healthy subjects is well-described by a power law with similar exponents, indicating absence of a characteristic length for the segments. Surprisingly, we also find that this power law

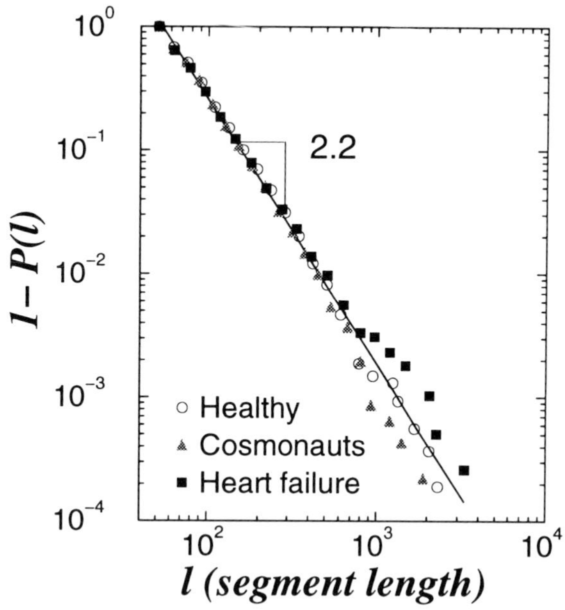

Figure 3. Probability of finding segments with a length ℓ larger than a given value for the segments obtained from all subjects in the healthy, cosmonauts and heart failure groups during during daily activity. The significance level is fixed to $\mathcal{P}_0 = 95\%$, and the imposed minimum length of the segments is $\ell_0 = 50$ beats. For all three groups we find a power law in the distribution of segment lengths with exponent $\beta \approx 2.2$.

remains unchanged for records obtained from cosmonauts during orbital flight (under conditions of micro-gravity) and for patients with heart failure (Fig. 3). A similar common type of behavior is also observed from 6–hour records during sleep for all three groups [8].

To verify the results of the segmentation procedure, we perform several tests. First, we check the validity of the observed power law in the distribution of segment lengths. We generate a surrogate signal formed by joining segments of white noise with standard deviation $\sigma = 0.5$, and mean values chosen randomly from the interval $[0, 1]$. We choose the lengths of these segments from a power-law distribution with a given exponent. Even when the difference between the mean values of adjacent segments is smaller than the standard deviation of the noise inside the segments, we find that our procedure partitions the surrogate signal into segments with lengths that reproduce the original power-law distribution [Fig. 4(a)]. This test shows that the distributions obtained after segmenting surrogate data with similar values of their exponents, appear clearly different from each other, making more plausible that the distributions obtained for the lengths of the segments for the healthy, cosmonauts and congestive heart failure subjects (Fig.3) follow indeed an identical distribution.

Second, we test if the observed power-law distribution for the segment lengths is simply due to the known presence of long-range correlations in the heartbeat interval series [9]. We generate correlated linear noise [10] with the same correlation exponent as the heartbeat data and find that the distribution of segment lengths obtained for the linear noise differs from the distribution obtained for the heartbeat data [Fig. 4(b)]. For the noise, the distribution decays faster, which means that these signals are more segmented than the heart data. In fact, for different linear noises with a broad range of correlation exponents, we do not find power-law behavior in the distribution of the segments. Thus we conclude that the linear correlations are not sufficient to explain the power-law distribution of segment lengths in the heartbeat data.

3.2 Differences between the mean values of consecutive segments (jumps)

Different healthy records can be characterized by different overall variance, depending on the activity and the individual characteristics of the subjects. Moreover, subjects with heart failure exhibit interbeat intervals with lower mean and reduced beat-to-beat variability (lower standard deviation). Thus one can trivially assume that these larger jumps in healthy records are due only to the fact that their average standard deviation is larger [Fig. 1(a)(b)]. In order to systematically compare the statistical properties of the jumps between different individuals and different groups, we normalize each time series by subtracting the global average (over 6 hours) and dividing by the global standard deviation. In this way, all individual time series have zero mean and unit standard deviation [Fig. 1(c)(d)]. Such a normalization does not affect the results of our segmentation procedure.

We find that both the healthy subjects and the cosmonauts follow identical distributions, but the distribution of the jumps obtained from the heart failure group are markedly different — centered around lower values — indicating that, even after normalization, there is a higher probability for smaller jumps compared to the healthy subjects [Fig. 5(a)]. Note that the distributions for all groups appear to follow an identical homogeneous functional form, so we can collapse these distributions on top of each other by means of a homogeneous transformation [Fig. 5(b)]. The ratio between the scaling parameters used in this transformation gives us a factor by which this feature of the heart rate variability is reduced for the subjects with heart failure as compared to the healthy subjects. This finding indicates that, although the heart rate variability is reduced with disease, there may be a common structure to this variability, reflected in the identical functional form. These observations agree with previously reported results for the distribution of heartbeat fluctuations obtained by means of wavelet and Hilbert transforms [11].

4 Summary

We present a new method, conceptually simple and computationally efficient, to partition a nonstationary signal into segments with different mean values. The method accurately recognizes segments with different mean values even in the presence of noise with large amplitude, and can be applied without restriction to various physiological nonstationary signals. We raise the hypothesis that a non-trivial

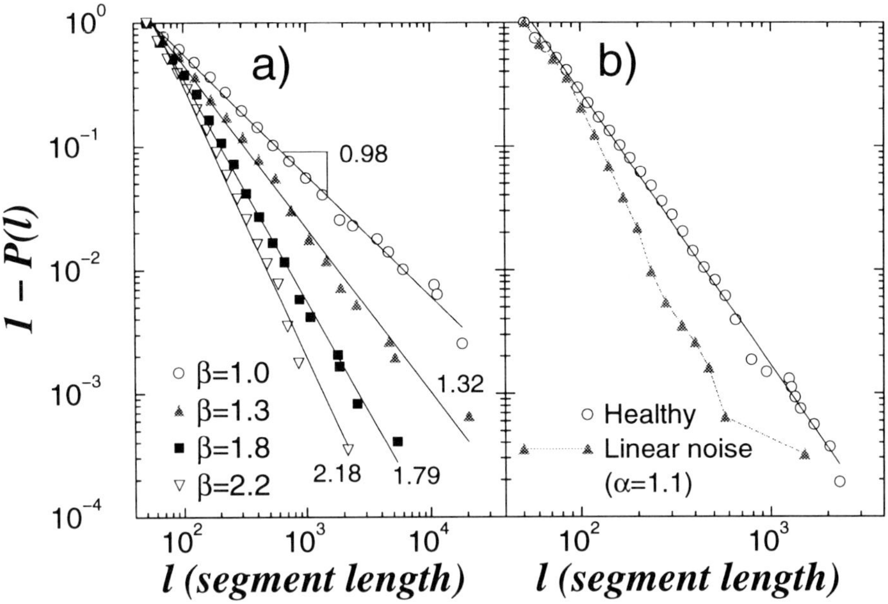

Figure 4. (a) Testing the validity of the observed power-law behavior in the distribution of segment lengths. We generate a surrogate signal formed by joining segments of white noise with standard deviation $\sigma = 0.5$ and average values chosen randomly from the interval $[0, 1]$. We chose the lengths of these segments from a power-law distribution with a given exponent β. The test shows that the distributions obtained after segmenting the surrogate data generated from power-law distributions with nearby values of their exponents appear clearly separated. This suggests that the distributions for the healthy, cosmonauts and congestive heart failure subjects in Fig. 3 are indeed identical. (b) Testing the effect of correlations in the heartbeat fluctuations on the segmentation. We generate 10 realizations, each with length of 26,000 points, of a linear Gaussian-distributed correlated noise with an exponent $\alpha = 1.1$ [10]. This exponent is calculated using the detrended fluctuation analysis method and is identical to the exponent α observed for the heartbeat data [9]. The distribution of segment lengths for this correlated noise does not follow the power law found for the heartbeat data. This test suggests that the observed scale-invariant behavior in the distributions of segment lengths in the heartbeat is not simply due to the correlations. According to the results in (a), the differences found between heartbeat data and correlated noise are significant. To verify that the curvature found in the distribution of segments for the noise is not due to finite size effects, we also repeated the test with longer realizations of the noise (1,000,000 data-points).

structure may be associated with the nonstationarity in physiological signals. We test our hypothesis on records from healthy individuals, subjects with congestive heart failure and cosmonauts during orbital flight.

For heart interbeat interval time series we surprisingly find that there is a scale-invariant structure, a power law, associated with the lengths of segments with different means. This power law is characterized by the same scaling exponent for all three groups in our database. Moreover, this power law cannot be explained by the presence of correlations in the signal. We also find that the differences in mean heart rates between consecutive segments, which we call jumps, display a

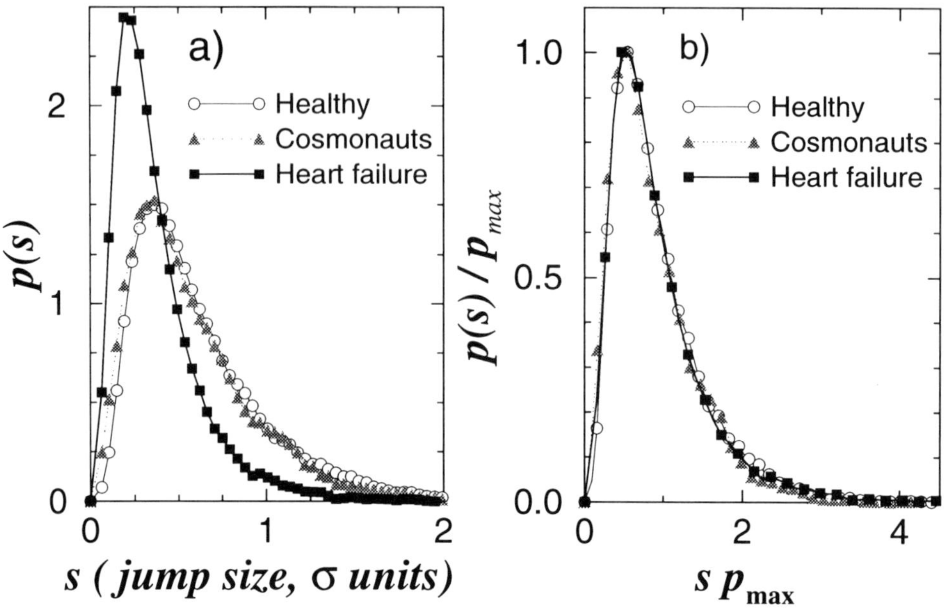

Figure 5. (a) Probability distribution of the absolute value of the difference between the mean values ('jumps') of consecutive segments. Both healthy and cosmonaut subjects follow an identical distribution while the heart-failure subjects follow a quite different distribution with higher probability for small jumps consistent with reports of smaller variability in heart failure subjects [5]. All distributions are normalized to unit area. Note that, the distributions are plotted in units of standard deviation, and that the results present a *striking* difference between the healthy and heart failure group, which otherwise cannot be seen (by eye) from the raw data after normalization [Fig. 1(d)]. (b) Same probability distributions as in (a), after rescaling $P(s)$ by $P_{\max}$, and s by $1/P_{\max}$. This homogeneous transformation preserves the normalization to unit area. The data points collapse onto a single curve.

common functional form, but with different parameters for healthy individuals and for patients with heart failure.

An important question raised by our results regards the physiologic meaning of the finding of identical distributions of segment length for both disease and healthy subjects. This finding is very unexpected because these two groups have radically distinct levels of physical activity and of neuroautonomic control of the heart rate [12], and there is no clear explanation to it. In fact, it raises a totally new scientific question, namely, what is the origin of the average heart rate nonstationarity? Our results would suggest the possibility that there may be a very basic physiological mechanism accounting for this scaling property. The fact that we find identical distributions even for data from cosmonauts during orbital flight (conditions of microgravity) indicates that the statistics of change of heart rate (i.e. the statistics of lengths of segments with different mean) exhibits universal properties. On the other hand, the reduced variability observed in the records from failure subjects (i.e. higher probability for smaller jumps) is perhaps related to reduced responsiveness to stimuli which can lead to a change in the mean heart rate.

Acknowledgments

We thank NIH/National Center for Research Resources (P41RR13622), The Mathers Charitable Foundation, and the Spanish Government grants BIO99-0651-CO2-01 for support.

References

1. R.L. Stratonovich, *Topics in the Theory of Random Noise*, vol.1 (Gordon and Breach, New York, 1981).
2. R.I. Kitney and O. Rompelman, *The Study of Heart Rate Variability* (Oxford Univ. Press, London, 1980); J.B. Bassingthwaighte, L.S. Liebovitch and B.J. West, *Fractal Physiology* (Oxford Univ. Press, New York, 1994); B.J. West, *Fractal Physiology and Chaos in Medicine* (World Scientific, Singapore, 1990).
3. H. Kantz and T. Schreiber, *Nonlinear Time Series Analysis.* (Cambridge Univ. Press, Cambridge, 1997); T. Schreiber, Phys. Rev. Lett. **78**, 843 (1997); A. Witt, J. Kurths and A. Pikovsky, Phys. Rev. E. **58**, 1800 (1998); G. Mayer-Kress, Integ. Physiol. Behav. Sci. **29**, 205 (1994); R. Hegger, H. Kantz, and L. Matassini, Phys. Rev. Lett **84**, 3197 (2000).
4. M. Kobayashi and T. Musha, IEEE Trans Biomed Eng. **29**, 456 (1982); J.M. Hausdorff *et al.*, J. Appl. Physiol. **80**, 1448 (1996); M.F. Shlesinger, Ann. NY Acad. Sci. **504**, 214 (1987); L.S. Liebovitch, Biophys. J. **55**, 373 (1989); A. Arneodo *et al.*, Physica D **96**, 291 (1996); P. Ch. Ivanov *et al.*, Physica A, **249**, 587 (1998); P. Ch. Ivanov *et al.*, Europhys. Lett., **48**, 594 (1999).
5. M.M. Wolf *et al.*, Med. J. Aust. **2**, 52 (1978); C. Guilleminault *et al.*, Lancet **1**, 126 (1984); A.L. Goldberger *et al.*, Experientia **44**, 983 (1988).
6. W.H. Press *et al.*, *Numerical Recipes in FORTRAN* (Cambridge University Press, Cambridge, 1994).
7. A.L. Goldberger *et al.*, Circulation **101**, e215-e220 (2000). The data used in this study were provided, without cost, by PhysioNet (http://www.physionet.org/), a public service of the Research Resource for Complex Physiologic Signals, under a grant from the NIH/National Center for Research Resources (P41 RR13622).
8. However, for the records during sleep, the distribution exhibits a crossover at a characteristic segment length of 700 beats, which might be related to the presence of sleep phases. This crossover indicates a smaller number of segments with short length.
9. C.-K. Peng *et al.*, Chaos **5**, 82 (1995).
10. H.A. Makse *et al.*, Phys. Rev. E **53**, 5445 (1996)
11. P.Ch. Ivanov *et al.*, Nature **383**, 323 (1996); M. Meyer *et al.*, Integ. Physiol. and Behav. Sci. **33**, 344 (1998).
12. P.Ch. Ivanov *et al.*, Nature **399**, 461 (1999).

FRACTAL ANALYSIS OF AGGREGATES OF NON-UNIFORMLY SIZED PARTICLES: AN APPLICATION TO MACAQUE MONKEY CORTICAL PYRAMIDAL NEURONS

B.I. HENRY

Department of Applied Mathematics, School of Mathematics, University of New South Wales, Sydney NSW 2052, Australia

E-mail: B.Henry@unsw.edu.au

P.R. HOF[A,B,C], P. ROTHNIE[D,C] AND S.L. WEARNE[D,B,C]

[A] *Kastor Neurobiology of Aging Laboratories,*

[B] *Fishberg Research Center for Neurobiology,*

[C] *Computational Neurobiology and Imaging Center,*

[D] *Department of Biomathematical Sciences,*

Mount Sinai School of Medicine, New York, New York, 10029-6574

E-mail: patrick.hof@mssm.edu, paul@camelot.mssm.edu, susan@camelot.mssm.edu

A variant of the cumulative mass method is developed for measuring the multifractal dimension spectrum of three-dimensional aggregates composed of particles of different sizes. The method is applied to measuring the mass fractal dimensions of pyramidal neurons of the prefrontal cortex of macaque monkeys, digitized with standard 3-dimensional tracing software. Fractal dimension estimates obtained from our approach are found to be useful for distinguishing two functionally different neuronal types which are visually similar.

Short title: *Fractal Analysis of Pyramidal Neurons*

PACS numbers: 05.40.+j, 68.70.+w

1 Introduction

Measurements of fractal dimensions for geometric or mass multifractals can be obtained from the box counting method or the cumulative mass method[1]. The cumulative mass method is particularly popular in applications because it can provide reliable estimates of fractal dimensions for smaller cluster sizes than the box-counting dimension. The cumulative mass method has been used extensively in the fractal analysis of clusters grown using computer growth models such as diffusion-limited aggregation[1,2,3,4], in which the growing aggregates are composed of identical sized particles. The cumulative mass method has also been employed in the fractal analysis of neuronal morphology[5,6,7,8]. In these studies, neuronal morphology has been represented as digitized camera lucida images which can again be regarded as aggregates comprising identical sized particles (pixels). Computer-assisted digitization is increasingly becoming the preferred method of capturing neuronal morphology for geometric and morphometric analysis. Standard 3-D digitization software represents dendritic branches as consecutive cylinders of varying length and diameter, which at high resolution suffer information loss if converted to pixel images in 2-D, and are computationally too demanding to convert to high resolution voxel images in 3-D.

In Section 2 of this paper we introduce a variant of the cumulative mass method that can be employed in the fractal analysis of aggregates composed of particles of different sizes. In Section 3 we explore an application of this method to the fractal analysis of two different types of macaque monkey cortical pyramidal neurons represented as sequential cylinder segments. The neurons investigated comprise two functional types: those furnishing long corticocortical pathways and those projecting locally, forming short-range networks[9]. The former typically support the transmission of information along hierarchies of organized cortical regions, linking distant and functionally different areas of the cerebral cortex, whereas the latter subserve lattices of connections within a given cortical region. Long corticocortically projecting neurons thus support the transfer of increasingly complex sensory or motor information while short projecting neurons may enable local binding of the converging information within a given cortical domain[10]. Although the neurons that constitute these functionally distinct pathways share a general pyramidal morphology, subtle differences in the complexity of their dendritic arbors are yet to be elucidated. Morphometric factors such as mass, branching structure, dendritic taper and rate of change of mass and surface area with distance from the soma are known to affect dendritic integration[11,12,13,14] and the efficacy of action potential backpropagation[15], which is crucial in synaptic plasticity. Because standard 3-D digitization procedures cannot capture fine dendritic varicosities and spine morphology, which are basic determinants of perimeter complexity, in this study we have concentrated on mass measures of fractal dimension. The application of the cumulative mass method to the fractal analysis of these neurons in this paper is an important first step. The results of our analysis are summarized in Section 4 where we report that mass fractal dimension estimates from the cumulative mass method provide a useful discriminant between short projecting neurons and long projecting neurons. The paper concludes with a discussion in Section 5.

2 Cumulative Mass Method for Non-Uniformly Sized Particles

The cumulative mass method was originally introduced[1] to measure the multifractal dimension spectrum for aggregates consisting of uniform sized particles. Consider a cluster of overall size L comprising N particles each of size a. Cover the cluster with boxes of size l, where $a \ll l \ll L$. Let $M_i(l)$ be the mass of particles in the ith box of size l, and M_0 be the total mass of the cluster. The generalized dimensions are defined by the scaling relation

$$\sum_i \left(\frac{M_i(l)}{M_0}\right)^q \sim \left(\frac{l}{L}\right)^{(q-1)D_q}, \tag{1}$$

where the sum over i is a sum over all boxes that contain particles. The finite particle size a and the finite cluster size L provide lower and upper cut-off length scales for the scaling relation. By considering the quantity

$$p_i(l) = \frac{M_i(l)}{M_0}$$

as a probability distribution, Vicsek and colleagues[1] rewrote the left hand side of Eq.(1) as

$$\sum_i \left(\frac{M_i(l)}{M_0}\right)^q = \frac{\sum_i \left(\frac{M_i(l)}{M_0}\right)^{q-1}\left(\frac{M_i(l)}{M_0}\right)}{\sum_i \left(\frac{M_i(l)}{M_0}\right)} = \left\langle \left(\frac{M_i(l)}{M_0}\right)^{q-1} \right\rangle .$$

The above ensemble average is taken with respect to the normalized probability distribution $p_i(l) = \frac{M_i(l)}{M_0}$, the probability that a randomly chosen particle in the cluster is inside the ith box of size l. Since an average over the lattice boxes with respect to the probability distribution $\frac{M_i(l)}{M_0}$ is equivalent to an average over randomly selected centres, the scaling relation

$$\left\langle \left(\frac{M_i(l)}{M_0}\right)^{q-1} \right\rangle \sim \left(\frac{l}{L}\right)^{(q-1)D_q}$$

will hold when the averaging is made with respect to a uniform probability distribution over the fractal. In the application to growing aggregates this scaling relation is implemented as

$$\left\langle \left(\frac{M(R)}{M_0}\right)^{q-1} \right\rangle \sim \left(\frac{R}{L}\right)^{(q-1)D_q} \tag{2}$$

where $M(R)$ is the mass contained in a ball of size R surrounding a randomly selected particle of the aggregate within the radius of gyration.

Reliable estimates for the mass fractal dimension can be obtained using this approach based on about 10% of the aggregate particles as centres[2]. As an aside we note that the application of this formalism to neuronal morphology in reference[8] incorrectly uses the exponent q on the left hand side of Eq.(2) rather than the exponent $(q-1)$.

An average over randomly selected particles is inappropriate for aggregates comprising non-uniformly sized particles. Instead it is necessary to sum over all possible particles taking the different masses of the particles into account. First note that if all particles do have equal mass, then an average over randomly selected particles is equivalent to an average over all particles, so that we can write

$$\frac{1}{N}\sum_{j=1}^{N} \left(\frac{M_j(R)}{M_0}\right)^{q-1} \sim \left(\frac{R}{L}\right)^{(q-1)D_q} . \tag{3}$$

Now taking the mass of each particle into account (and restricting the sum to particles within the radius of gyration where the scaling laws should persist) we have

$$\sum_{k=1}^{\hat{N}_G} \left(\frac{\hat{M}_k}{M_G}\right)\left(\frac{M_k(R)}{M_0}\right)^{q-1} \sim \left(\frac{R}{R_G}\right)^{(q-1)D_q} . \tag{4}$$

In this equation R_G is the radius of gyration, $\hat{N}_G$ is the number of particles inside R_G, $\hat{M}_k$ is the mass of the kth such particle, M_G is the total mass inside R_G and $M_k(R)$ is the total mass inside a ball of radius R centred on the centre of the kth particle. The dimensions D_q can be obtained from the straight line slopes in a log-log plot of the scaling relation, Eq.(4), using balls of size R in the range $r < R < R_G$ where r is the size of the largest particle in the cluster. Following[1] we define the cumulative mass dimension of order q for non-uniformly sized particle aggregates as

$$D_q = \frac{1}{q-1} \frac{\log\left[\sum_{k=1}^{\hat{N}_G} \left(\frac{\hat{M}_k}{M_G}\right) \left(\frac{M_k(R)}{M_0}\right)\right]}{\log\left(\frac{R}{R_G}\right)} \tag{5}$$

for R in the range $r < R < R_G$. The mass fractal dimension, D_m, for non-uniformly size particle aggregates is then defined by Eq.(5), with $q = 2$, i.e., $D_m = D_2$. In practical applications the dimensions D_q defined by this equation may not be constant over the full range $r < R < R_G$ and so estimates of D_q from the slopes of straight line portions in $\log - \log$ plots are employed. An obvious shortcoming in applying Eq.(4) directly is that it is an approximate scaling relation which assumes among other things that: i) any particle with its centre inside a ball of radius R is fully contained within that ball, and ii) the mass scaling in balls centred on the centres of particles is the same as the mass scaling in balls centred on other points of the same particle. If each particle is homogeneous, these two shortcomings can be reduced by replacing each large particle by a cluster of smaller particles with equivalent total mass and occupying approximately the same volume. In the next section we demonstrate this procedure in replacing cylinder segments by equivalent chains of spheres.

3 Fractal Analysis of Pyramidal Neurons

3.1 Data Acquisition

Materials from four adult male long-tailed macaque monkeys (*Macaca fascicularis*, 10-12 years old) were used in the present study. All experimental protocols were conducted within the NIH guidlines for animal research and were approved by the Institutional Animal Care and Use Committee (IACUC) at Mount Sinai School of Medicine. These animals received intracortical injections of the retrograde tracer Fast Blue (Molecular Probes, Eugene, OR; 4% aqueous solution) in area 46 of the prefrontal cortex to identify projection neurons as previously described[9]. The animals were then perfused transcardially under deep anaesthesia[9], with cold 1% paraformaldehyde in phosphate-buffered saline (PBS) and then for 14 minutes with cold 4% paraformaldehyde in PBS. Following perfusion, 4 mm-thick blocks were dissected out of area 46 and the superior temporal cortex, postfixed for 2 hours in 4% paraformaldehyde, and cut at 400 μm on a Vibratome. For intracellular injection of corticocortically-projecting neurons, these sections were immersed in PBS. Fast Blue-containing neurons were identified under epifluorescence with a UV filter, impaled, and loaded with 5% Lucifer Yellow (Molecular Probes). Neurons were subsequently traced and reconstructed three-dimensionally at 100x magnification using a computer-assisted morphometry system consisting of a Zeiss Axiophot photomicroscope equipped with a Zeiss MSP65 computer-controlled motorized stage (Zeiss, Oberkochen, Germany), a Zeiss ZVS-47E video camera system (Zeiss, Thornwood,

NY), a Macintosh 840 AV microcomputer, and custom designed morphometry software (NeuroZoom[16], NeuroGL [Computational Neurobiology and Imaging Center, Mount Sinai School of Medicine, New York]).

3.2 Cylinders to Chains of Spheres

The digitized morphologic data described above are available as a set of cylindrical segments of specified diameter and location. The fractal analysis of these data sets can be found by employing the scaling relation, Eq.(4), with each cylinder segment considered as a separate particle. As discussed above, however, while this scaling relation properly accounts for the different masses of the cylinder segments it does not take into account their different volumes. To address this, we replaced each cylinder segment by a chain of uniformly sized spheres, and then used the spheres as individual particles in the scaling analysis.

Consider a given cylinder segment of length L, radius r and end points $\vec{x}_a$ and $\vec{x}_b$. We wish to replace each cylinder by a chain of n spheres of uniform radius R so that the chain of spheres is the best single line packing approximation to the cylinder, i.e., we wish to satisfy the dual conditions $\pi r^2 L = n\frac{4}{3}\pi R^3$ and $L = 2nR$. It is a simple matter to solve these simultaneous equations from which we deduce the following two cases:

1. $L \leq \sqrt{6}r$:
   The cylinder is replaced by a single sphere with centre
   $$\vec{x}_c = \frac{\vec{x}_a + \vec{x}_b}{2}$$
   and radius
   $$R = \left(\frac{3r^2L}{4}\right)^{\frac{1}{3}} .$$

2. $L > \sqrt{6}r$:
   The cylinder is replaced by $n = [\frac{L}{\sqrt{6}r}]$ spheres (where $[x]$ denotes the greatest integer not exceeding x), with centres
   $$\vec{x}_j = \vec{x}_a + \left(\frac{2j-1}{2n}\right)(\vec{x}_b - \vec{x}_a) \qquad j = 1, \ldots, n$$
   and radii
   $$R = \left(\frac{3r^2L}{4n}\right)^{\frac{1}{3}} .$$

Fig. 1 shows a portion of a neuron represented by cylinder segments and the same portion represented by a chain of spheres using the above approach.

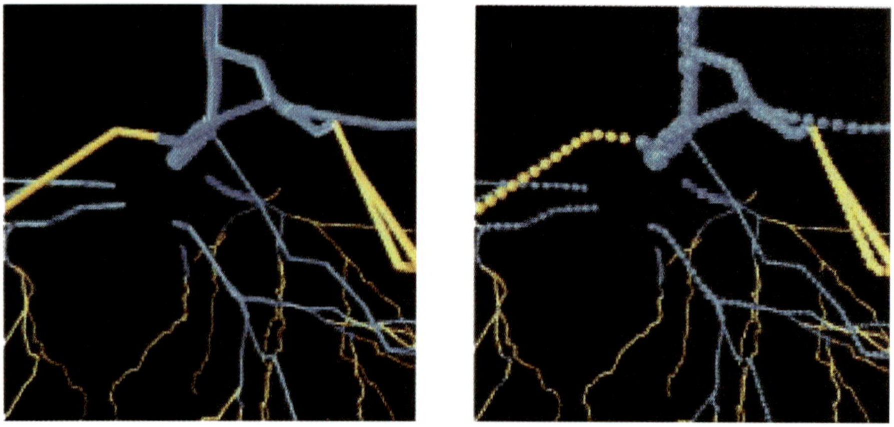

Figure 1. Representation of the same portion of a neuron using cylinder segments (left), and chains of spheres (right).

3.3 Mass Fractal Dimensions

The mass fractal dimensions, D_m, of the neurons are obtained by using the scaling relation, Eq.(4), with $q = 2$, and each cylinder segment replaced by a chain of spheres. The scaling relation is used taking all particle spheres within the radius of gyration as centres for fifty balls of size R uniformly spaced on a log scale in the range $[R_G/8, R_G/2]$. The restriction to the central portion of the neuron within its radius of gyration avoids artefacts associated with the sparsely branching distal regions of the tree. Fig. 2 shows a representative pyramidal neuron and a ball of size R_G centred on the centre of mass of this neuron. It is clear that most of the complexity is contained within this ball. The morphology of the neuron shown in figure 2 and the other pyramidal neurons in this study is characterized by a pyramid-shaped soma, and an overall triangular-shaped dendritic arbor.

The lower ball radius $R_G/8$ used in the scaling analysis was always found to be above the radius of the largest aggregate particle sphere. The slope of the best fit straight line portion in the log-log plots was obtained by identifying the plateau portion in a plot of successive slopes from ten point moving averages. The plateau portion was found to occur towards the upper end of the ball size, $R \approx R_G/2$. Fig. 3(a) shows a log-log plot for a typical neuron; a plot of successive slopes for this neuron using ten point moving averages is shown in Fig. 3(b).

4 Results

Mathematical and statistical analyses were performed using Matlab (The MathWorks, Natick, MA), MAPLE, Fortran90 and C/C^{++}. Neurons were reconstructed

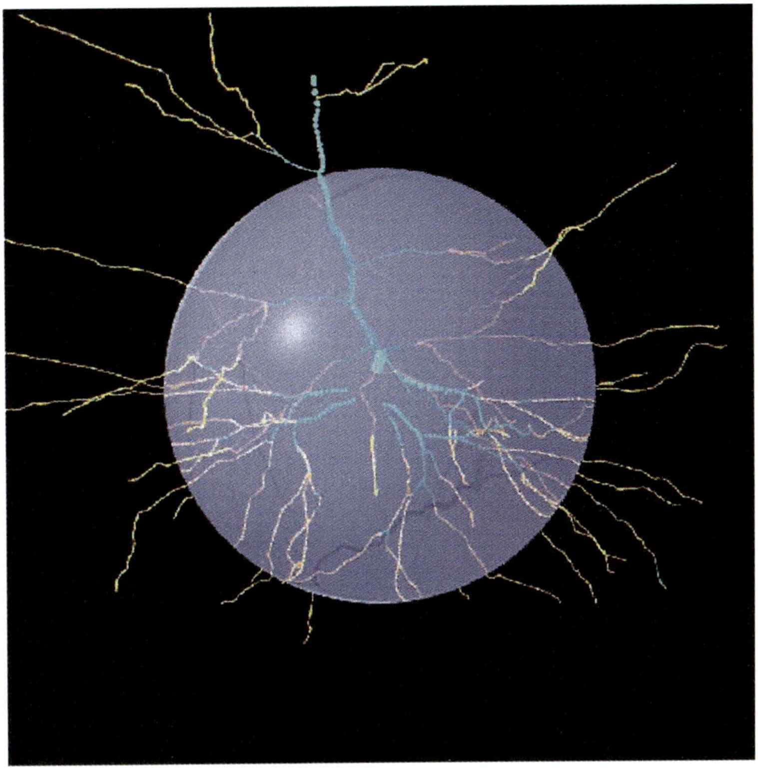

Figure 2. A representative pyramidal neuron shown with a ball of radius equal to the radius of gyration of the neuron which is centred on the centre of mass of the neuron.

and surface rendered in 3-D using custom designed software (NeuroGL). Inspection of relative frequency histograms of D_m revealed that the two neuron classes were approximately normally distributed, and differences between mean values were assessed with t-tests for independent samples. A total of 35 pyramidal cells, comprising 16 long projecting and 19 short projecting neurons from four macaque monkeys were analyzed.

Despite the visual similarity of branching patterns in the two neuron types (compare Fig. 4A and Fig. 4B), the mass fractal dimension D_m was significantly different for the two classes. Short projecting neurons had significantly higher mass dimensions than long (mean D_m for short $= 1.62 \pm 0.21$ s.d.; mean D_m for long $=$

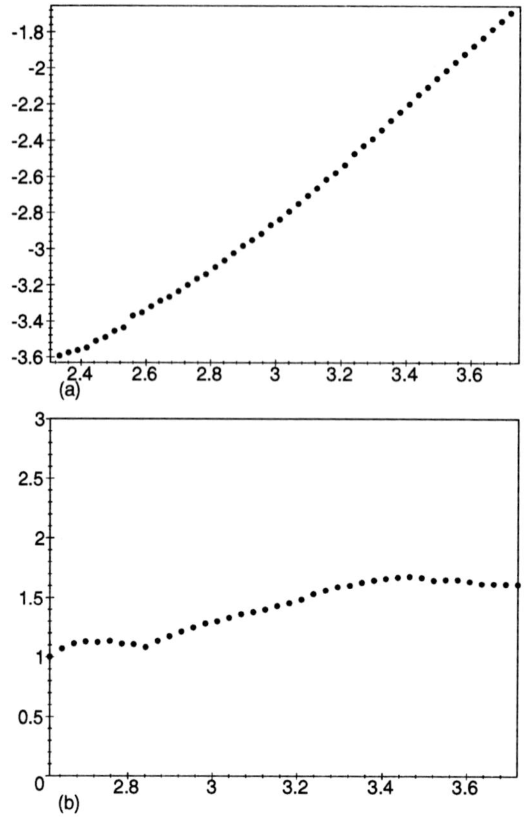

Figure 3. Mass dimension scaling data for a representative pyramidal neuron. The plots show: a) $\log\left[\sum_{k=1}^{\hat{N}_G}\left(\frac{\hat{M}_k}{M_G}\right)\left(\frac{M_k(R)}{M_0}\right)\right]$ versus $\log\left[\left(\frac{R}{R_G}\right)\right]$ and b) successive slopes, D_m, from a ten point moving average of the straight line of best fit in this log-log plot.

1.44 ± 0.19; $p = 0.0155$), (Table 1). These values for D_m are somewhat higher than values calculated using the 2-D dilation method for pyramidal neurons in macaque monkey[17]. However these higher values for D_m are consistent with the generally higher fractal dimensions found using the cumulative mass method as opposed to box-counting or coastline methods[6,7].

The calculated mass dimensions associated with the neurons shown in Fig. 4 at first may appear counterintuitive. The long projecting neuron **A** appears slightly more complex than its short projecting counterpart **B**. To elucidate this result, we compared the number of branches originating from the soma, along with the total number of branchpoints, in long and short projecting neurons. Long projecting neurons had significantly more branches (mean 6.25 ± 0.86) than short projecting neurons (mean 5.21 ± 1.08, $p = 0.0039$). Long neurons also had more branch points (mean 92.13 ± 15.34) than short neurons (mean 72.58 ± 17.71, $p = 0.0015$). This greater number of branches and branch points compensates visually for the lower

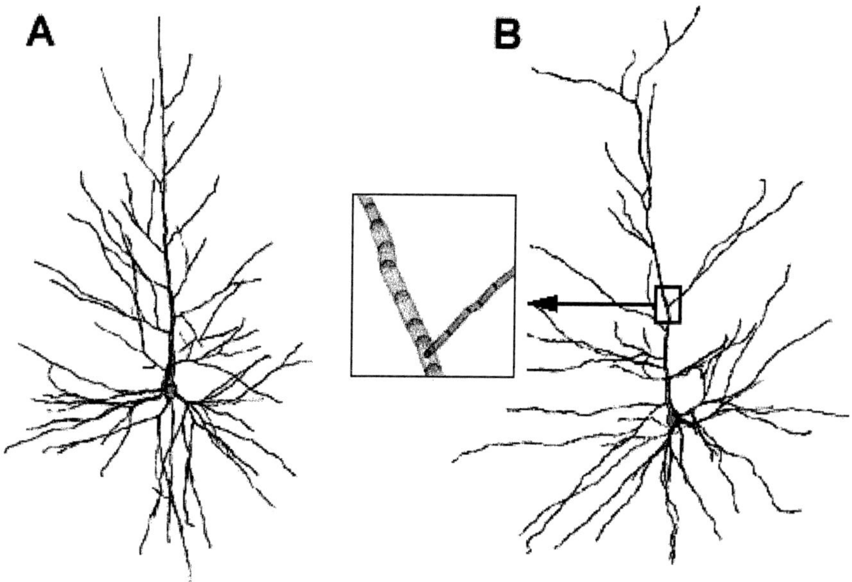

Figure 4. Morphology of two functionally distinct neurons; **A**: long projection neuron and **B**: short projection neuron. The insert shows successive cylinder segments from which the dendritic tree is reconstructed.

rate of increase in mass measured by D_m, increasing the apparent complexity of long projecting neurons relative to short.

Neuron	N	D_m	**S.D.**	No. Branches	**S.D.**	No. Branchpoints	**S.D.**
Long	16	1.44	0.19	6.25	0.86	93.13	15.34
Short	19	1.62	0.21	5.21	1.08	72.58	17.71

Table 1. Summary of statistics for long and short projecting neurons. The values in columns 3,5 and 7 represent means for D_m, number of branches, and number of branchpoints, respectively.

5 Discussion

An overall goal of our research is to develop measures of dendritic branching geometry capable of distinguishing functionally relevant morphologic differences that might contribute to variability in neural firing patterns and neuronal plasticity. As a first step in this direction we have introduced a variant of the cumulative mass method that can be applied to finding multifractal mass dimensions of aggregates of non-uniformly sized particles. This method was then applied to finding the mass fractal dimension of two functionally distinct types of pyramidal neurons. It was found that differences in complexity between long and short projection neurons which are difficult to appreciate by eye (Fig. 4) can nevertheless be discriminated using the mass fractal dimension.

Our results can be contrasted with a recent study comparing 2-D fractal dimensions of pyramidal neurons in different cortical areas of macaque monkey[17], which found that progressively 'higher order' cortical neurons had progressively higher fractal dimensions. Using the variant of 3-D mass fractal dimension described in this paper, we find that short projecting neurons, which likely subserve more local information binding within a single cortical domain, had higher values of D_m than long projecting neurons, which link more complex and apparently functionally disparate cortical regions. This result reflects the difference between coastline methods of computing fractal dimension, which measure complexity of the perimeter of an object, and mass methods, which measure the rate at which total mass of the object increases with distance from the center of mass. The rate of increase in mass is directly related to the rate of increase in dendritic surface area with distance from the soma. A recent simulation study examined the correlation between different morphometric parameters and efficiency of forward and backpropagation of action potentials in dendritic arbors with different branching topologies[15]. The strongest predictor of propagation failure was the rate of increase in dendritic membrane area with distance from the soma. Functionally, this will result in differences in dendritic integration and synaptic plasticity in neurons with different values of D_m, that are attributable to morphologic variation. As such, D_m represents a valuable independent measure of functionally relevant morphologic differences that are difficult to assess visually.

In future work we plan to investigate the multifractal dimension spectrum as well as detailed branching analysis for these and other neurons.

Acknowledgments

We thank Huiling Duan, Michael Einstein, and Daniil Rolshud for assistance with cell loading and imaging, Douglas Ehlenberger, Kevin Kelliher and Alfredo Rodriguez for technical assistance, and Dr John Morrison for scientific advice. Supported by NIH Grants AG05138, AG06647, MH58911 and DC04632, the Howard Hughes Medical Institute and the Australian Research Council.

References

1. T. Tel, A. Fulop and T. Vicsek, Determination of fractal dimensions for geometrical multifractals, *Physica A*, **159**, 155-166, (1989).
2. T. Vicsek, F. Family and P. Meakin, Multifractal geometry of diffusion-limited aggregates, *Europhys. Letts.*, **12**, 217-222, (1990).
3. C-H Lam, Finite-size effects in diffusion-limited aggregation, *Phys. Rev. E*,

52, 2841-2847, (1995).
4. F. Jestczemski and M. Sernetz, Multifractal approach to inhomogeneous fractals, *Physica A*, **223**, 275-282, (1996).
5. F. Caserta, H.E. Stanley, W.D. Eldred, G. Daccord, R.E. Hausman and J. Nittman, Physical mechanism underlying neurite outgrowth: A quantitative analysis of neuronal shape, *Phys. Rev. Letts.*, **64**, 95-98, (1990).
6. F. Caserta, W.D. Eldred, E. Fernandez, R.E. Hausman, L.R. Stanford, S.V. Bulderev, S. Schwarzer and H.E. Stanley, Determination of fractal dimension of physiologically characterized neurons in two and three dimensions, *J. Neurosci. Methods*, **56**, 133-144, (1995).
7. T.G. Smith Jr., G.D. Lange and W.B. Marks, Fractal methods and results in cellular morphology - dimensions, lacunarity and multifractals, *J. Neurosci. Methods*, **69**, 123-136, (1996).
8. E. Fernandez, J.A. Bolea, G. Ortega and E. Louis, Are neurons multifractals?, *J. Neurosci. Methods*, **89**, 151-157, (1999).
9. E.A. Nimchinsky, P.R. Hof, W.G. Young and J.H. Morrison, Neurochemical, morphologic, and laminar characterization of cortical neurons in the cingulate motor areas of the macaque monkey, *J. Comp. Neurol.*, **374**, 136-160, (1996).
10. P.R. Hof, E.A. Nimchinsky and J.H. Morrison. Neurochemical phenotype of corticocortical connections in the macaque monkey: quantitative analysis of a subset of neurofilament protein-immunoreactive projection neurons in frontal, parietal, temporal, and cingulate cortices, *J. Comp. Neurol.*, **362**, 109-133, (1995).
11. W. Rall, Theoretical significance of dendritic trees for input-output relations, In: *Neural Theory and Modeling*, R.F. Reiss (Ed.), 73-79, Stanford University Press, Stanford, (1964).
12. Z.F. Mainen and T.J. Sejnowski, Influence of dendritic structure on firing patterns in model neocortical neurons, *Nature*, **382**, 363-366, (1996).
13. A. Surkis, C.S. Peskin, D. Tranchina and C.S. Leonard. Recovery of cable properties through active and passive modeling of subthreshold membrane responses from laterodorsal tegmental neurons, *J. Neurophysiol.*, **80**, 2593-2607, (1998).
14. C. Koch, Biophysics of Computation: Information Processing in Single Neurons, Oxford University Press, New York, (1999).
15. P. Vetter, A. Roth and M. Hausser, Propagation of action potentials in dendrites depends on dendritic morphology, *J. Neurophysiol.*, **85**, 926-937, (2001).
16. W.G. Young, E.A. Nimchinsky, P.R. Hof, J.H. Morrison and F.E. Bloom, NeuroZoom Software User Guide and Reference Books, YBM Inc, San Diego, (1997).
17. H.F. Jelinek and G.N. Elston, Pyramidal neurones in macaque visual cortex: interareal phenotypic variation of dendritic branching patterns, *Fractals*, **9**, *In Press*.

SOCIAL, BIOLOGICAL AND PHYSICAL META-MECHANISMS: A TALE OF TAILS

BRUCE J. WEST
Mathematics Division, US Army Research Office, Research Triangle Park, NC, USA
and
Physics Department, Duke University, Durham, NC, USA

The tale concerns the uncertainty of knowledge in the natural, social and life sciences and the tails are associated with the statistical distributions and correlation functions describing these scientific uncertainties. The tails in many phenomena are mentioned, including the long-range correlations in DNA sequences, the long-time memory in human gait and heart beats, the patterns over time in the births of babies to teenagers, as well as in the sexual pairings of homosexual men, and the volatility in financial markets among many other exemplars. I shall argue that these phenomena are so complex that no one is able to understand them completely. However, insights and partial knowledge about such complex mechanistic understanding of the phenomena being studied. These strategies include the development of models, using the fractal stochastic processes, chaotic dynamical systems, and the fractional calculus; all of which are tied together, using the concept of scaling, and therein hangs the tale.

The perspective adopted in this lecture is not the dogmatic presentation often found in text books, in large part because there is no "right answer" to the questions being posed. Rather than answers, there are clues, indications, suggestions and tracks in the snow, as there always are at the frontiers of science. Is is my perspective of this frontier that I will be presenting and which is laid out in detail in *Physiology, Promiscuity and Prophecy at the Millennium: A Tale of Tails* [25].

1 Background

In this lecture we want to lay the foundation for how such concepts as complexity, fractals, diverging sample moments, dynamics and many more are used in the understanding of complex phenomena. Of course, a number of books have been written about any one of these ideas - books for the research expert, books for the informed teacher, books for the struggling graduate student and books for the intelligent lay person. Different authors stress different characteristics of complex phenomena, from the erratic data collected by clinical researchers to the fluctuations generated by deterministic equations used to model such systems. Authors have painted with broad brush strokes, indicating only the panorama that these concepts reveal to us, whereas others have sketched with painstaking detail the structure of such phenomena and have greatly enriched those that could follow the arguments. Here, today, we do not have either the time or the inclination to do either. I view our effort today as a workshop, by which I mean that I intend to provide you with some tools that you may be able to use in the understanding of your own data set (phenomenon).

It seems quite remarkable that it was over thirty years ago that, as a graduate student, I sat in a seminar room at the University of Rochester and listened to Benoit Mandelbrot talk about why the night sky was not uniformly illuminated (Olber's paradox) and how income is distributed in western societies (Pareto's Law). At the time these were quite exotic topics for physics colloquia. It would be more than ten years before he (Mandelbrot) coined the word fractal to take cognizance of the fact that there is a large class of natural and social phenomena that traditional statistical physics is not equipped to describe much less to explain. In the intervening years there has been a blossoming literature on fractal random processes with inverse power-law spectra, characteristic of long-time memory, episodic processes with Lévy stable distribution functions, and the applications of these ideas to phenomena in the physical, social and life sciences [20,18,24,12,3]. Three separate approaches have been used to model such phenomena and it is impossible to list even a representative sample of that work here and so we confine our references to a few of the books and review articles that we have found useful.

One approach to modeling complex phenomena, here represented as fractal time series, is by means of low-dimensional, nonlinear, deterministic dynamical equations having intermittent chaotic solutions [1]. The spectra of such systems spread themselves into broad band, inverse power laws, indicative of fractal random time series. Of course such scaling processes are generated by colored noise as well, which leads us to the second method of modeling, that being stochastic differential equations. In particular, such equations are often generated using random walks with long-time correlations in the random fluctuations, yielding fractional diffusion equations to describe the evolution of the probability density for the random walk variable [2]. The statistics of a system's response to such fluctuations is often found to deviate strongly from that usually expected using the central limit theorem (CLT). For example, the second moment of the random walk variable may diverge. A generalized version of the CLT yields Lévy statistical distributions to describe the system random response to such correlated fluctuations, see for example Montroll and West [17]. This last work, done at the end of the seventies, was the harbinger of what was to be an avalanche of research into the nature of phenomena whose evolution cannot be described by differential equations. Subsequently, it was found that both the inverse power-law spectra and the Lévy statistical distribution are consequences of scaling and fractals [13], as was discussed in the second edition of ref.[17] and in its sequel [16].

The third method for generating fractal time series, and probably the least well known in the physical science literature, is by means of fractional differences in discrete stochastic equations [10,8,5]. This technique has until very recently had only relatively modest acceptance in the field of economics where it was first introduced, see ref. [4] for a historical review and some recent applications. This should come as no surprise, the work of Mandelbrot [15], first published in the economics literature thirty years ago, has only in the past few years begun to strongly influence that community. But as we shall see, fractional differences and fractals serve our purposes in modeling complex phenomena, with long-time memory, very well, independently of their origins.

Each of the above approaches explains the erratic behavior in time series from a particular perspective. Just as a simple random walk is the discrete time analog of Brownian motion [17], a fractional-difference process driven by discrete white noise is the discrete time analog of fractional Brownian motion (*fBm)*, that is, a process with long-time memory and Gaussian statistics [10]. Herein we examine extensions of these latter arguments to non-Gaussian statistics, in particular to α-stable Lévy statistical processes. Before presenting this discussion, however, it is necessary to relate the continuum limits of our fractional differences to fractional derivatives, since it turns out that the evolution of Lévy distributions are described by a fractional partial differential equation as first noted by West and Seshadri [23] and later rediscovered by a number of authors [6,30,7,31].

We shall address the complex phenomena referred to above from the point of view of scaling. Scaling first took the form of allometric relations in biology to describe the growth of various parts of an organism. Julian Huxley wrote a book [11] on this in the thirties and gave a number of examples of such systems in biology and botany. The form of an allometric relation is

$$Y = aX^b \tag{1}$$

where Y and X are two observables in an organism, such as the weight of the antlers (X) and total weight of a deer (Y). In another area, evolutionary biology, a deterministic relation was observed between the central moments of a spatially heterogeneous distribution of the number of species, that is, the variance in the number of species (Y) is proportional to a power of the mean number of species (X), in direct analogy with the allometric relation [22].

We shall review the basis for these allometric relations and examine how such scaling properties also appear in time series. In particular we discuss anomalous diffusion in which the second moment of the time series increases as a power of the time that is not linear. This is shown to be due to the fact that the smallest scale fluctuations are tied to the largest scale fluctuations through a renormalization group relation. We shall see that there are a large number of complex phenomena that are controlled by such scaling relations.

In an attempt to understand how scaling emerges in science and why it seems to be so prevalent in the life and social sciences, we shall examine a number of models having their origins in physics, but which capture the essential features of a great many non-physical phenomena. The first model we describe is that of fractional discrete stochastic equations, which is an extension of the traditional random walk models to include random forces that have a long-time memory and are therefore not Markovian [10,25]. Such models are used to describe certain kinds of fractal stochastic point processes, such as the number of events that occur within a given time interval [26]. The continuum limit of this model is the fractional Langevin equation, which as one would guess, is a fractional stochastic differential equation. We discuss how these models capture the essential features of a number of complex phenomena.

2 Scaling and random walks

Let us begin the discussion of time series for complex phenomena by noting that such series are in general stochastic, which is to say that there are fluctuations in the quantity being measured, that are greater than the resolution interval of the measuring instrument. In Figure 1 we denote three kinds of time series: 1) intervals in human heart beats; 2) the number of births to teenagers in Texas and 3) the stride intervals during walking. Here the heart beat data is the size of the interval between heart beats, from which we can see that the heart beat is not the regular signature taught in school, but is a time series with a great deal of variability. This variability is indicative of health. The teen birth data is the number of births per week to teenagers in Texas for a two year period. Here again we can see a great deal of variability in the data and perhaps some oscillatory regularity in the average number of births. Finally, the stride interval data is for a normal, healthy individual walking in a relaxed manner for 800 steps. In fact, for the sake of comparison we have only indicated the first 800 data points in each of the time series. From these time series it should be evident that variability not regularity is the normal situation. The same observation can be made regarding indices in financial markets [14], the intervals between breaths [24], and so on.

We can see that the fluctuations in these three time series look quite different, with regard to their statistics, the level of correlation, and any possible underlying patterns that may exist in the data. We wish to understand the possible mechanisms that contribute to these processes so we use a sequence of random walk models of increasing complexity. For example one method of generalizing the simple random walk model is by correlating the steps of the walk in space and time, while allowing the walker to step an arbitrary distance during each step. We shall find that this generalization enables us to incorporate long-time memory into the process, which implies a fractional random walk process, or in the continuum leads to a fractional differential stochastic equation of evolution for the dynamics of the process. This latter equation we refer to as the fractional Langevin equation, since it combines the stochastic aspect of the ordinary Langevin equation and a new twist involving fractional derivatives. Of course, like the original Langevin equation, what is written as a differential equation is mathematically an integral equation [25].

2.1 Random walks and anomalous diffusion

The simple random walk model has an equal probability of taking a step to the right or to the left. We denote the process being modeled as a random walk by the position variable X with the subscript j that denotes the discrete time of the step, producing the set of values $\{X_j\}$ with $j = 1, 2, ..., N$. The operator B decreases the time index by one unit, $BX_j = X_{j-1}$, so a simple random walk with no memory is denoted by

$$(1 - B) X_j = \xi_j, \tag{2}$$

where ξ_j is a random force, as in Brownian motion. If ξ_j is delta correlated in time and has a finite second moment, then the sum variable,

$$X(N) = \sum_{j=1}^{N} X_j \tag{3}$$

for N sufficiently large, corresponds to normal diffusion whose mean-square value increases linearly with the number of steps,

$$\left\langle X(N)^2 \right\rangle \propto N. \tag{4}$$

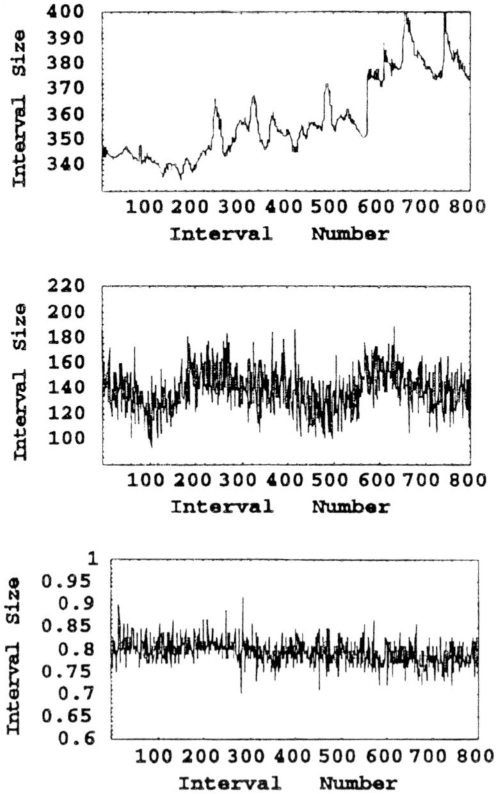

Figure 1. Three different time series are depicted. At the top is the time interval between heart beats in a healthy human neonate. In the middle is the number of births to teenagers in Texas over a two year period. At the bottom is the interval in the stride interval for a normal healthy adult walking in a relaxed manner.

In the theory of simple random walks it is assumed that each step takes the same length of time so the number of steps is proportional to the time. This is the classical, unbiased, diffusion process where the random walk variable has Gaussian statistics, zero mean, and a variance that increases linearly with time.

Consider a time series generated by a random walk model in which successive steps of the walk are not independent. In general the second moment of the displacement of the walk after N steps, or in continuous time after an interval t, is given by

$$\left\langle X(t)^2 \right\rangle \propto t^{2H}. \tag{5}$$

Where, of course, $H = 1/2$ is normal diffusion, so that successive steps are statistically independent of one another in the random walk, and the mean-square displacement increases linearly in time. If $H > 1/2$ the walker, having taken a step in a given direction, is more likely to continue in that direction rather than reversing directions. In this case the mean-square displacement increases faster than linearly and the random walk is persistent. If $H < 1/2$ the walker, having taken a step in a given direction, is more likely to reverse directions than to continue walking in the same direction. In this case the mean-square displacement increases more slowly than linearly and the random walk is anti-persistent.

Such random walks have a correlation function that is an inverse power-law in time, which is to say that the correlation between points in time decreases with increasing time separation

$$C(\tau) \equiv \langle X(t+\tau) X(t) \rangle \propto \tau^{2H-2}. \tag{6}$$

Here we see that the correlation function is constant when the random walk is ballistic with $H = 1$. This behavior is also manifest in the spectrum, the Fourier transform of the correlation function, which is an inverse power law in the frequency

$$S(\omega) \equiv \mathcal{FT}\{C(\tau)\} \propto \frac{1}{\omega^{2H-1}}. \tag{7}$$

But the spectrum becomes a power law if the process is anti-persistent, $H \leq 1/2$. Together these three properties, the algebraic increase in time of the mean-square displacement (5), the inverse power law in time of the correlation function (6) and the inverse power law in frequency of the spectrum (7), are typically observed in anomalous diffusion and are a consequence of long-term memory in the statistical process. These are all manifestations of scaling in the underlying phenomena.

2.2 Fractional random walks and Lévy statistics

One way of incorporating the above memory into the dynamics of the random walk is through the introduction of fractional differences. We again consider the time shift operator B, to write the fractional-difference random walk process as [10,25]

$$(1-B)^{\alpha} X_j = \xi_j \tag{8}$$

where the index α is not an integer, $-1/2 \leq \alpha \leq 1/2$, and the random force driving the system, ξ_j, is a discrete random force. The solution to this equation is formally obtained by expanding the inverse of the fractional-difference operator in a binomial expansion to obtain

$$X_j = \sum_{k=0}^{\infty} \frac{(-1)^k \Gamma(k+\alpha)}{\Gamma(k+1)\Gamma(\alpha)} \xi_{j-k}. \tag{9}$$

The solution to (8) given by (9) at "time" j is tied to fluctuations in the infinitely distant past, that is, to "times " $j-k$ and $0 \leq k \leq \infty$. In fact the correlation function of the solution yields an inverse power law [25]

$$\langle X_j X_{j-k} \rangle \propto k^{2H-2} \tag{10}$$

obtained using the properties of the gamma functions in (9) with $\alpha = H - 1/2$, so that $0 \leq H \leq 1$. The solution to this fractional-difference random walk is therefore determined to be a discrete fractional Brownian motion process [10], which is to say, the statistics of the process are Gaussian and the spectrum is inverse power law.

It is also possible to use the above discrete analysis to generate Lévy, rather than Gaussian, statistics. Let us use (9) to generate a dichotomous random process with long-time memory. We do this using the Heaviside unit step function, $\Theta(x < 0) = 0$ and $\Theta(x \geq 0) = 1$, to define the function

$$\eta_j = 1 - 2\Theta(X_j) \tag{11}$$

which has the desired properties of being +1 when the dynamical variable is negative and -1 when the dynamical variable is positive, and has a long-time memory since the correlation function of X_j is given by (10). The appropriate random walk process is then given by

$$(1 - B) Y_j = \eta_j \tag{12}$$

where we wish to find the statistics of $Y(N)$. The analysis of this random walk process with an inverse power-law, correlated, stochastic force has been done [26] and the probability density for the continuous form of this random walk process is determined by a fractional diffusion equation to be [23]

$$P(y,t) = \frac{1}{2\pi} \int_{-\infty}^{\infty} \exp\left[-bt\,|k|^{\mu} + iky\right] dk \tag{13}$$

which is the symmetric Lévy distribution, with $\mu = 2H$.

The characteristic function, $\phi(k,t)$, defined by the Fourier transform of the probability density, for the symmetric Lévy process is given by

$$\phi(k,t) = e^{-bt|k|^{\mu}}. \tag{14}$$

The equation of evolution for the characteristic function is obtained by taking the time derivative of (14) to obtain

$$\frac{\partial \phi(k,t)}{\partial t} = -b\,|k|^{\mu}\,\phi(k,t). \tag{15}$$

The inverse Fourier transform of (15) yields the fractional diffusion equation first obtained by West and Seshadri [23]

$$\frac{\partial P(y,t)}{\partial t} = \frac{b\Gamma(1+\mu)\sin[\mu\pi/2]}{\pi} \int_{-\infty}^{\infty} \frac{P(y',t)\,dy'}{|y-y'|^{1+\mu}} \tag{16}$$

where the integral is the Reisz fractional derivative with $0 \leq \mu \leq 2$ [21].

2.3 *Fractional Langevin equation and scaling*

In the continuum limit the solution to (8), which is to say the inverse of the discrete fractional operator, is replaced with the fractional integral equation

$$X(t) = X(0) + D_t^{-\alpha}\xi(t) \tag{17}$$

where here the fractional integral is chosen to be of the Riemann-Liouville form [21,25]

$$D_t^{-\alpha}[\xi(t)] = \frac{1}{\Gamma(\alpha)}\int_0^t \frac{\xi(t')\,dt'}{|t-t'|^{1-\alpha}}. \tag{18}$$

We choose $\xi(t)$ to be a delta correlated, Gaussian, random process so that the mean-square displacement can be explicitly evaluated to be

$$\left\langle [X(t) - X(0)]^2 \right\rangle = \frac{\langle \xi^2 \rangle}{(2\alpha - 1)\Gamma(\alpha)} t^{2\alpha-1} \propto t^{2H-2} \tag{19}$$

the same functional dependence on time as (10). Thus, we see the consistency between the discrete fractional random walk result and the fractional stochastic differential equation. Of course we can also write (17) as the fractional Langevin equation

$$D_t^{\alpha}[X(t)] - \frac{t^{1-\alpha}}{\Gamma(1-\alpha)}X(0) = \xi(t), \tag{20}$$

where the dependence on the initial value in the differential equation arises from the fact that the fractional derivative of a constant is not zero. Thus, the fractional derivative in (20) is seen to introduce a scaling into the system response to the random force as manifest in (19).

Another way of understanding scaling in data is by means of a simplified renormalization group argument. Consider an unknown function $Z(t)$ that satisfies a scaling relation of the form

$$Z(bt) = aZ(t). \tag{21}$$

Such scaling relations can be solved in the same way one solves differential equations and that is to guess the form of the solution, substitute it into the equation and see if it works. We assume a trial solution of the form

$$Z(t) = A(t)\,t^{\mu} \tag{22}$$

which when substituted into (21) yields the condition that the function $A(t)$ is periodic in the logarithm of the time with period $\log b$ and the power-law index has the value

$$\mu = \log a / \log b. \tag{23}$$

In the literature the function $Z(t)$ that satisfies (21) is called a homogeneous function.

The homogeneous function $Z(t)$ defines the scaling observed in the moments of the time series with memory, that is to say, for the second moment

$$\left\langle X\left(bt\right)^2\right\rangle = b^{2H}\left\langle X\left(t\right)^2\right\rangle, \tag{24}$$

the correlation function

$$C\left(b\tau\right) = b^{2H-2}C\left(\tau\right), \tag{25}$$

and finally the spectrum

$$S\left(b\omega\right) = b^{1-2H}S\left(\omega\right). \tag{26}$$

The corresponding solutions to each of these scaling equations is precisely the algebraic forms of these quantities observed in equations (6)-(7), respectively; assuming, of course, that the slowly varying term $A\left(t\right)$ is, in fact, constant.

3 Scaling and time series

An apparently different kind of scaling was observed by Taylor in 1961 [22]. He examined the statistical properties of the number of biological species that were heterogeneously distributed in space (a pasture) and found that the mean and variance were not independent quantities. The relation between the central moments has the form

$$VarX = a\overline{X}^b. \tag{27}$$

Note that (27) quite different from normal diffusion where the statistics are Gaussian and the mean $(\overline{X})$ and variance $(VarX)$ are independent of one another. The power-law relation between the mean and variance has come to be called the power curve and is a straight line when plotted on log-log graph paper. The reference curve is that of a Poisson distribution in which case the mean and variance are equal, so that the power curve is a straight line with unit slope. Due to this equality the Poisson case is referred to as random since the spatial distribution in the number of species is homogeneous. If the power-law index is greater than one, $b > 1$, the distribution of species is spatially clumped, if less than one, $b < 1$, the distribution is regular. Thus, the slope of the power curve provides a measure of the degree of spatial heterogeneity of biological species, this is called the "evenness" by the ecological community. The greater the slope the greater the variety of species [25].

Taylor graphed the variance and mean by increasing the resolution of the spatial areas and recalculating these two quantities as a function of the resolution scale size. We follow the same procedure with time series by calculating the variance and mean for N data points, that is, $VarX\,(1)$ and $\overline{X}\,(1)$. We then add adjacent data point together to form a data set with $N/2$ data points, and from this calculate the variance and mean, $VarX\,(2)$ and $\overline{X}\,(2)$. Going back to the original data set we now add three adjacent data point together to form a data set with $N/3$ data points, and from this calculate the variance and mean, $VarX\,(3)$ and $\overline{X}\,(3)$. In this way we continue to aggregate the original data set and after aggregating m neighbors in the original data set, the aggregated variance and mean are $VarX\,(m)$ and $\overline{X}\,(m)$. If the data set is fractal, that is to say the time series scales, then the relative dispersion, the ratio of the standard deviation to the mean, has the inverse power-law form [3,25]. Alternatively we may rewrite the allometric relation between the mean and variance in the form of Taylor's law

$$VarX\,(m) = a\overline{X}\,(m)^b \tag{28}$$

and the fractal dimension can be written

$$D = 2 - b/2. \tag{29}$$

We can interpret such data sets using the correlation coefficient from the simple equation for the correlations between adjacent data points [3]

$$r_1 = 2^{3-2D} - 1 \tag{30}$$

where $r_1 = 0$ for no correlations in the time series, implying that the fractal dimension is given by $D = 1.5$ and the Hurst exponent is $H = 0.5$. In the case of perfect correlations in the time series $r_1 = 1$ so that the fractal dimension is $D = 1.0$ and the Hurst exponent is $H = 1.0$. In the former case we have uncorrelated Brownian motion and the latter a regular curve, or in terms of random walks, the former is normal diffusion and the latter is ballistic motion.

Thus, if we graph the aggregated variance versus the aggregated mean on log-log graph paper, and the underlying time series is fractal, we would obtain a straight line given by

$$\log VarX\,(m) = \log a + b \log \overline{X}\,(m)\,. \tag{31}$$

The slope of the empirical curve, b, would using (29), allow us to read off the fractal dimension.

3.1 Time series data

Let us now examine the three data sets from Figure 1 to see how pervasive the above scaling relations truly are. The three data sets are the number of births to girls between the ages of 10 and 19 in the state of Texas during the years 1980 to 1997 [27]. The second data set consists of the stride intervals of normal healthy individuals undergoing a relaxed walk [28]. The final time series is that of the interbeat interval in a healthy human neonate. We choose these three for a number of reasons. First, as

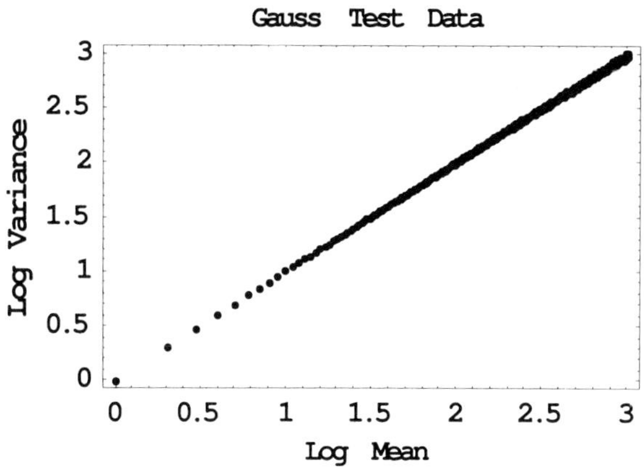

Figure 2. Here is a computer generated data set with Gaussian statistics generated by a random number generating computer code. The aggregation process clearly shows that Taylor's Law is applicable to these data.

one can see in Figure 1, the raw data look quite different. Second, the mechanisms generating the time series have virtually nothing in common, at least not on the surface. Finally, we find that all three time series are fractal.

For orientation, before we examine the experimental data, consider the time series with uncorrelated Gaussian statistics generated by a typical computer program for generating "random" numbers that is commercially available. We generated 10^6 such points using *Mathematica* 4.0 and then applied the aggregation procedure, calculating the variance and mean at each level of aggregation. In Figure 2 we see that the allometric relation persists over more than two orders of magnitude variation. The slope of the computer generated curve is $b = 1$ so that the fractal dimension given by (29) is $D = 1.5$, as it should be for an uncorrelated Gaussian process.

In Figure 1 we showed the heart rate intervals for a normal, healthy, active 36 week old neonate. This time series is not regular, but rather shows apparently erratic fluctuations across multiple time scales. We plot the aggregated variance versus the aggregated mean for these data in Figure 3. The slope of the curve is determined to be $b = 1.83$, using a mean-square minimization code, so that the corresponding fractal dimension is $D \approx 2 - 1.863/2 \approx 1.07$, a value remarkably close to a regular curve. The correlation coefficient in this case is $r_1 \approx 0.82$. We emphasize that this result is typical of neonate time series. We have a dozen or so, such time series, that we have processed and have obtained similar results. It should be noted that this is not so different from the interbeat interval statistics in mature organisms. For mature human adults the fractal dimension is $D = 1.13 \pm .07$ [29] so that the correlation coefficient is in the interval $r_1 = 0.67 \pm 0.17$ indicating a relatively strong correlation between adjacent interval variations. It was also determined that there is less than one chance in a million that the fractal dimension, and the corresponding correlation of the neonate time series, can be explained using an uncorrelated Gaussian process. In fact it was determined that the statistics of

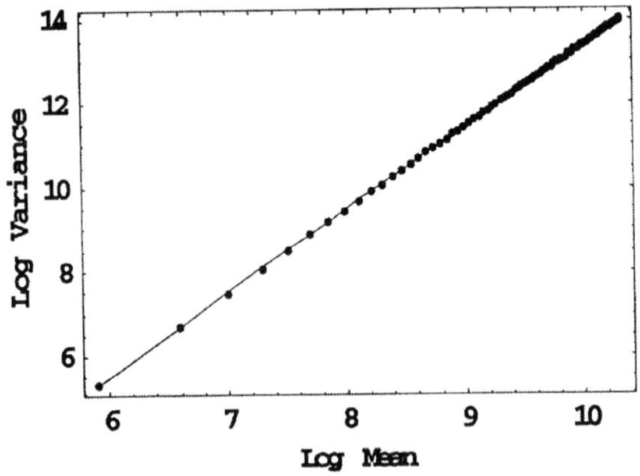

Figure 3. The time series for the interbeat interval of a healthy neonate, the data depicted in Figure 1, is aggregated as described in the text. The plot of the variance versus the mean on log-log graph paper clearly depicts a straight line.

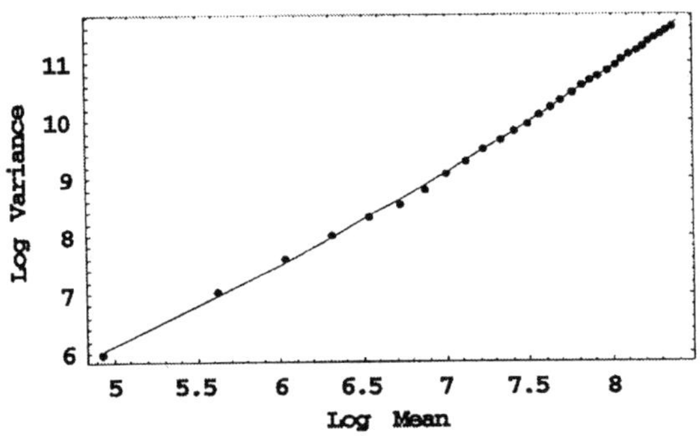

Figure 4. The aggregated variance is plotted versus the aggregated mean for the Texas teen birth data. It is clear that there is curvature to the best fit line joining the aggregated data points.

the variations in the heart beat intervals for adults was determined to be Lévy stable [19].

In Figure 4 we see that the dominant relation between the aggregated variance and the aggregated mean of the number of births to teenagers in Texas during the years of 1980 and 1997 is that of a power law. The best fit curve to the data in this figure has, in addition to the power law, a modulated amplitude. Here we focus on the slope of the power law, which is found to be $b \approx 1.46$. The fractal dimension for this time series is thus $D \approx 1.27$ giving rise to a correlation coefficient $r_1 \approx 0.38$. West, Hamilton and West [27] interpret the modulation in this time series as being the result of scaling in the data, such that the moments satisfy a renormalization group relation. The modulation of the power law is also a consequence of the renormalization group scaling. We point out that these results are not unique to

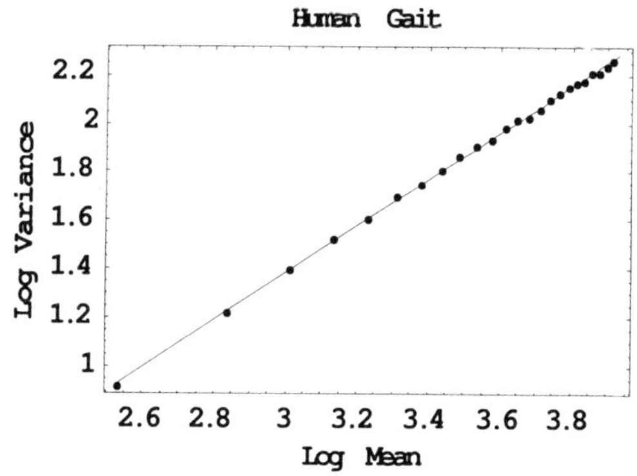

Figure 5. The aggregated variance is plotted versus the aggregated mean for the stride interval data depicted in Figure 1.

the Texas data. Similar analysis has been done on birth data from the states of Oklahoma and Minnesota as well, with virtually the same results. By the "same" we mean a modulated allometric relation in the variance-mean plot, thereby supporting the interpretation as a fractal random time series.

Finally, consider the length of the stride in the walking of a normal, healthy human being. It has been known for over one hundred years that the variability in the length of the stride is on the order of 3% as measured by the standard deviation in the time of a stride interval. The inference drawn from this variability was that there is no information in the fluctuations in human gait. This inference is wrong [28,9]. In fact, if the fluctuations in human gait do scale, then the variance-mean curve should have a slope different from one. In Figure 5 we depict the aggregated variance and aggregated mean from 15 minutes of gait data. The solid curve is the best curve fit to the data on this log-log graph paper. The slope of the curve is given by $b \approx 1.49$ yielding a fractal dimension of $D \approx 1.26$. The relation between the fractal dimension and the Hurst exponent, $H = 2 - D$, yields for the stride interval data $H = 0.74$ [28]. Note that this value of H is completely consistent with the value found using a more sophisticated analysis by Hausdorff *et al.* [9]. Thus, as in the two previous examples, the human gait time series has a long-time memory, that is, the fluctuations in stride interval have an inverse power-law memory, so that a change occuring now influences one that occurs nearly a 100 steps later.

4 Conclusions

The idea of the fractional evolution of a physical process used in the data analysis is that through coarse graining one can determine if the phenomenon under investigation has universality and scaling. By universality we mean that the macroscopic properties of the system are independent of the particular microscopic mechanisms present in the phenomenon. Thus, for the purposes here, the particular values of the parameters determined from the data, are not significant, except in so far as they indicate that the data do scale. Over the long term if we can establish a norm for these parameters, that is, a range of values that can be associated with health and values outside that range can be associated with pathologies, then the values of the scaling parameters for a single individual may be quite important.

We can now draw a number of conclusions from our data sets and the corresponding analyses. First of all, time series from complex phenomena are often erratic and have scaling properties. The scaling is manifest in the second moment that scales algebraically in time, the correlation function that is an inverse power law in time and the spectrum that is an inverse power law in frequency. The inverse power-law nature of these second order measures is the signature of fractal random processes. So we conclude that heart rate variability is a fractal random point process even in neonates, as is the inter-stride intervals in humans, and the complex phenomenon of teen births.

The final conclusion we draw from the scaling properties of the data is that the statistical distribution often found in complex phenomena is Lévy stable, rather than Gaussian or any other, more familiar, distributions we find in simpler physical processes.

The interdependence, organization and concinnity of biophysical processes have traditionally been expressed in biology through the principle of allometry. However, this principle, as usually articulated is static in nature, and it is only recently that an attempt to extend the allometry idea to irregular physiological time series in terms of the properties of feedback control have been made. An allometric control system achieves its purpose through scaling, enabling a complex system such as the regulation to be adaptive and accomplish concinnity of the many interacting subsystems. Allometric control is a generalization of the idea of feedback regulation that was implicit in Cannon's concept of homeostasis. The basic notion is to take part of the system's output and feed it back into the input, thus making the system self-regulating by minimizing the difference between the input and the sampled output. More complex systems such as autoregulation of heart beat variation, human gait and even complex social phenomena such as mating, that involve the elaborate interaction of multiple sensor systems, have more intricate feedback arrangements. In particular, since each sensor responds to its own characteristic set of frequencies, the feedback control must carry signals appropriate to each of the interacting subsystems. The coordination of the individual responses of the separate subsystems is manifest in the scaling of the time series in the output and the separate subsystems select that aspect of the feedback to which they are the most sensitive. In this way an allometric control system not only regulates, but also adapts to changing environmental and biophysical conditions.

Acknowledgement

The views expressed herein are those of the author and do not reflect the views of the Army Research Office.

References

1. E. Ott, *Chaos in Dynamical Systems*, Cambridge University Press, Cambridge (1993); N.B. Abraham, A.M. Albano, A. Passamante, P.E. Rapp and R. Gilmore, *Complexity and Chaos*, World Scientific, Singapore (1993).
2. P. Allegrini, P. Grigolini and B.J. West, "Dynamical approach to Lévy processes", Phys. Rev. E **54**, 4760-67 (1996); B.J. West, P. Grigolini, R. Metzler and T.F. Nonnenmacher, "Fractional diffusion and Lévy stable processes", Phys. Rev. E **55**, 99-106 (1997).
3. J. B. Bassingthwaighte, L.S. Liebovitch, and B.J. West., *Fractal Physiology*, Oxford University Press, Oxford (1994).
4. R. T. Baillie, "Long memory precesses and fractional integration in econometrics", J. Econometrics **73**, 5-59 (1996).
5. J. Beran, *Statistics of Long-Memory Processes, Monographs on Statistics and Applied Probability* **61**, Chapman & Hall, New York (1994).
6. A. Compte, "Stochastic foundations of fractional dynamics", Phys. Rev. E **53**, 4191-93 (1996).
7. H. C. Fogedby, "Aspects of Lévy flights in a quenched random force field", in *Lévy Flights and Related Topics in Physics*, editors M.F. Shlesinger, G.M. Zaslavsky and U. Frisch, Springer, New York (1995).
8. C. W. J. Granger, "Long memory relationships and the aggregation of dynamic models", J. Econometrics **14**, 227 (1980).
9. J. M. Hausdorff, C.-K. Peng, Z. Ladin, J.Y. Wei and A.L. Goldberger, J. Appl. Physiol. **78**, 349 (1995).
10. J.T.M. Hosking, "Fractional Differencing", Biometrika **68**, 165-176 (1981).
11. J.S. Huxley, *Problems of Relative Growth*, The Dial Press, New York (1932).
12. S.B. Lowen and M. C. Teich, "Estimation and Simulation of Fractal Stochastic Point Processes", Fractals **3**, 183-210 (1995); M.O. Vlad, B. Schonfirch and M.C. Mackey, "Self-similar potentials in random media, fractal evolutionary landscapes and Kmura's neutral theory of molecular evolution", Physica **229** A, 343-64 (1996).
13. B.B. Mandelbrot, *Fractals, Form and Chance*, W.H. Freeman, San Francisco (1977).
14. B.B. Mandelbrot, *Fractals and Scaling in Finance: Discontinuity, Concentration, Risk*, Springer-Verlag, New York (1997).
15. B.B. Mandelbrot and J.W. van Ness,"Fractional Brownian motions, fractional noises and applications.", SIAM Rev. **10**, 422 (1968).
16. E.W. Montroll and M.F. Shlesinger, "On the wonderful world of random walks", in *Nonequilibrium Phenomena II: From Stochastics to Hydrodynamics*, edited by E.W. Montroll and J.L. Lebowitz, pp 1-121, North Holland, Amsterdam (1984).
17. E.M. Montroll and B.J. West, "An Enriched Collection of Stochastic Processes", in *Fluctuation Phenomena*, Eds. E.W. Montroll and J. Lebowitz, North-Holland (1979); 2nd Edition, North- Holland Personal Library (1987).
18. M. M. Novak and T. G. Dewey, Editors, *Fractal Frontiers*, World Scientific, Singapore (1997); S. Harefall and M.E. Lee, Editors, *Chaos, Complexity and*

Sociology, SAGE, Thousand Oaks, California (1997); P.M. Iannaccone and M. Khokha, Editors, *Fractal Geometry in Biological Systems*, CRC Press, Boca Raton (1996); T.F. Nonnenmacher, G.A. Losa, E.R. Weibel, Editors, *Fractals in Biology and Medicine*, Birkhauser Verlag, Basel (1994).
19. C.K. Peng, J. Mietus, J.M. Hausdorff, , S. Havlin, H.G. Stanley and A.L. Goldberger, Phys. Rev. Lett. **70**, 1343 (1993).
20. H.G. Stanley, Editor, *Statphys 16*, North-Holland, Amsterdam (1986); H.G. Stanley and N. Ostrowsky, Editors, *Random Fluctuations and Pattern Growth: Experiments and Models*, Kluwer Academic Publishers, Dordrecht, NATO Scientific Affairs Division (1988)
21. S.G. Samko, A.A. Kilbas and O.I. Marichev, *Fractional Integrals and Derivatives*, Gordon and Breach Science Publishers, Switzerland (1993).
22. L.R. Taylor, Nature **189**, 732 (1961).
23. B.J. West and V. Seshadri, "Linear systems with Lévy fluctuations", Physican A **113**, 203-216 (1982).
24. B.J. West and W. Deering, "Fractal Physiology for Physicists: Lévy Statistics", Phys. Rept. **246** (1&2), 1-100 (1994); B.J. West and W. Deering, *The Lure of Modern Science: Fractal Thinking, Studies in Nonlinear Phenomena in Life Science Vol.* **3**, World Scientific, River Edge, New Jersey (1995); B.J. West, *Fractal Physiology and Chaos in Medicine, Studies in Nonlinear Phenomena in Life Science Vol.***1**, World Scientific, River Edge, New Jersey (1990).
25. B.J. West, *Physiology, Promiscuity and Prophecy at the Millennium : A Tale of Tails*, Studies of Nonlinear Phenomena in the Life Sciences Vol. **7**, World Scientific, Singapore (1999).
26. B.J. West and D.R. Bickel, Physics Letters A **256**, 188 (1999).
27. B.J. West, P. Hamilton and D.J. West, Fractals **7**, 113 (1999).
28. B.J. West and L. Griffin, Fractals **6**, 101 (1998).
29. BJ. West, R. Zhang, A.W. Sanders, S. Miniyar, J.H. Zuckerman and B.D. Levine, Physica A **270**, 552 (1999).
30. G.M. Zaslavsky, "Anomalous transport and fractional kinetics", H.K. Moffatt et al., Editors, *Topological Aspects of Dynamics of Fluids and Plasmas*, ppsl 4581-91, Kluwer Academic Publishers, Netherlands (1992).
31. J. Klafter, G. Zumofen, and M.F. Shlesinger, "Lévy description of anomalous diffusion in dynamical systems", in *Lévy Flights and Related Topics in Physics*, editors M.F. Shlesinger, G.M. Zaslavsky and U. Frisch, Springer, New York (1995).

TOWARDS A UNIVERSAL LAW OF TREE MORPHOMETRY BY COMBINING FRACTAL GEOMETRY AND STATISTICAL PHYSICS

J. DUCHESNE, P. RAIMBAULT AND C. FLEURANT

Horticulture National Institut, Landscape Department,
2 rue Le Nôtre, 49045 Angers cedex 01
E-mail: cyril.fleurant@inh.fr

This article aims at establishing a very general law of plant organization. By introducing the notion of hydraulic lengths which are considered as the coordinates of a symbolic space with n-dimensions, a reasoning of statistical physics, derived from Maxwell's method, and combining with the fractal geometry leads to a law of hydraulics lengths distribution which could appear very general because it is the remarkable gamma law form

Key-words : statistical physics, fractal geometry, morphology, tree, scalling

1 Introduction

1.1 The applications of morphometry in geomorphology

Before the conception of the fractal geometry by Mandelbrot (1975)[1], morphometric analysis was at first used by geologists to understand the river systems organization. Horton (1945)[12] links talweg sections by their source point and by their confluence point with an other talweg of similar importance. Horton defines two empirical laws expressed by two ratios:

- the bifurcation ratio, $R_C = \frac{N_{i-1}}{N_i}$, which has a constant value between 3 and 5 for river systems. N_i is the number of i order sections,
- the length ratio, $R_L = \frac{L_i}{L_{i-1}}$, which has a constant value between 1.5 and 3.5 for the rivers. L_i is the average length of i order sections.

Finally La Barbera and Rosso (1982)[14] define fractal dimension for a drainage basin, $D = \frac{\ln R_C}{\ln R_L}$. Weibel and Gomez (1962)[15] used morphometry to model lungs, then numerous studies were carried out on trees.

1.2 Applications of the morphometry and fractal geometry on plants

Fitter (1982)[16] presents a morphometric tree classification inspired by river networks. In his methodological study on root systems of herbaceous species, he shows that one can use Horton's laws by ordering ramifications according to the morphometric order to quantify the root ramification.

Holland (1969)[17] shows that the ramification of several species of Eucalyptus can be described and be explained by Horton's laws and by the effect of the apical control in the young twigs' growth. Leopold (1971)[18] working on different architecture plant (*Abies concolor*, *Pinus taeda*) comes to the same conclusion. He adds that the most likely classification seems to minimize the total length of the branches in the

ramification system. Oohata and Shidei (1971)[19] study with the aid of Horton's method the ramification of four types of ligneous plants among which shrubs with big evergreen leaves (*Cinnamomum camphora*) and conifers with evergreen leaves. He shows that the ramification ratio varies in a range much wider than river systems : from 3.0 to 8.0. This ratio varies according to the plant biologic type. Whitney (1976)[20] shows on 16 ligneous species that the ramification ratio depends mainly on the leaves disposal, on the deciduousity of the leaf and branches and on the needles size, and that it is more characteristic of species and relatively independent of external conditions.
Using the morphometric tree of Strahler (1958)[13] shows that for the birch and the apple tree, the logarithms of the average numbers of terminal branches of every order of ramification, of the average diameter and of the number of buds carried by these branches are aligned compared to the ramification order. The logarithms of the twigs' average length are much more scattered. They deduct that these two species have a fractal ramification and that lengths are more significant of the specific shape of trees. Crawford and Young (1990)[21] show, for oaks (*Quercus robur*) that the branches' distribution lengths follow a simple fractal algorithm. Berger (1991)[22] uses fractals to model the growth of trees (*ficus elastica*), Chen *et al.* (1993)[23] to model the canopy of a poplar population (*Populus sp.*), Macmahon and Kronauer (1976)[24] to model the mechanics of the tree (*Quercus rubra*).

1.3 The invariant structure of plants

Generally, a branching system is constituted bythe subset of branching systems. A branch is the part of a tree included between two successive ramifications. To study the branching organization, we shall use the typology of Strahler (1952)[4] (see Figure 1):

- a bud or a growing shoot is called the first-order branch
- when two branches of order i join, a branch of order $i + 1$ is created,
- when two branches of different orders join, the branch immediately at the junction retains the higher of the two joining branches.

The branching system order is thus the main order found in the plant.

1.4 A universal law of morphology of landscapes

Two attempts have been made to apply a reasoning of statistical physics to hydrography. Lienhardt (1964)[6] is the first to have perceived the interest of the statistical physics and Shreve (1966)[7] has opened an innovative way by making the hypothesis that the law of the stream numbers as a function of the order results from a statistic of a large number of channels branching out at random, as the ideal gas law results from a statistics of a huge amount of molecules colliding at random. Like Mandelbrot (1975)[1] we are convinced that in both geomorphology and biomorphology, a statistical approach can be fruitful. However, one must be sure to respect two conditions that are basic ideas of statistical physics: *i)* the system size must be

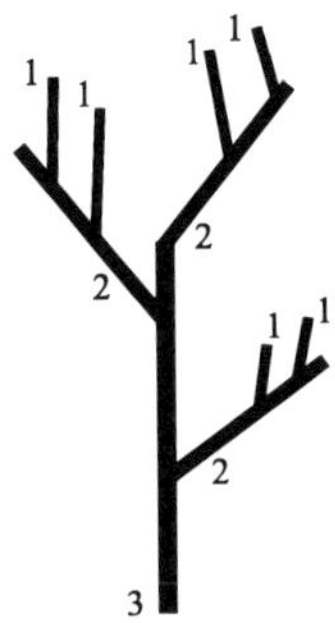

Figure 1. Tree typology and principle of the orders numerotation.

very large compared to the one of the constituent element that will be taken into account, *ii)* the local properties of the system must be homogeneous enough. The validity limits of the law that we are going to establish now is probably very closely linked to the respecting of these two conditions.

2 Demonstration of the law

2.1 Choice of the symbolic space

The difficulty of the extension of such a reasoning in the morphology of trees lies in the choice of the symbolic space as defined by Maxwell (Sears, 1971[11]). The idea of our approach is based on the use of the symbolic space where the velocity components v_x, v_y, v_z, are replaced by *ad hoc* components. Maxwell uses a symbolic space, called velocities space, where each velocity vector ends in a point which is characterized by its coordinates v_x, v_y, v_z. He defines a function of these three coordinates:

$$F(v_x, v_y, v_z) = \frac{d^3N}{N dv_x dv_y dv_z} \tag{1}$$

Where d^3N is the number of molecules whose velocity vector leads to the elementary volume $dv_x dv_y dv_z$, among a total number of N molecules. From the very beginning of the reasoning we decided to take into account the fractal property of our particular system, the branching system, by introducing two differences in comparison with Maxwell's symbolic space:

- we do not use a velocity space which would have no meaning here but a symbolic space, hydraulic length of the plant. We decide to call component of order i, the length of the set of links or part of links having the same order i (we shall note it l_i). So, for any point of the branching system, the hydraulic length is the sum $L = \sum l_i$ of its n-constituents. Where n is the order of the branching system. One can introduce the ratio:

$$r_l = \frac{\overline{l_i}}{\overline{l_{i-1}}} \tag{2}$$

Where numerator and denominator represent respectively the average of all the constituents with order i and the constituents with order $i+1$.

- since we consider that each possible hydraulic length has n-constituents l_i, our symbolic space will no longer have three dimensions, as those of Maxwell, but n-dimensions, n being the order of the branching system. However, to use without any trouble the properties of n-dimensions vectorial space, instead of considering the components l_i, we will consider using their square roots $x_i = \sqrt{l_i}$, $L = X^2 = \sum_{i=1}^{n} x_i^2$. Thus, if we denote by N the total number of hydraulic length, we can define a function $F(l_1, l_2, \ldots, l_n)$ with n-variables l_1, l_2, ..., l_n:

$$F(l_1, l_2, \ldots, l_n) = \frac{d^n N}{N dl_1 dl_2 \ldots dl_n} \tag{3}$$

2.2 Choice of fundamental hypotheses

We adopt the same hypotheses as Maxwell, but by adapting them to our symbolic space and by taking into account consequences of the scalling invariance:

- according to the hypothesis of the independence of the hydraulic length distribution law, the components l_i are independent. One is so led to express function F as a product of n one-variable functions:

$$F(l_1, l_2, \ldots, l_n) = f_1(l_1) f_2(l_2) \ldots f_n(l_n) \tag{4}$$

- as Maxwell did for the velocities distribution we should admit that the distribution law of x_i is isotropic. According to scalling invariance and the relation 2, the ith order component is on average r_l times larger than the $(i-1)^{th}$ component. The hypothesis of isotropy must therefore not be applied to the symbolic space of coordinates $x_i = \sqrt{l_i}$, but z_i defined as reduced hydraulic lengths components $z_i = \frac{l_i}{r_l^{i-1}}$:

$$z_i = \sqrt{\frac{l_i}{r_l^{i-1}}} = \frac{x_i}{r_l^{\frac{i-1}{2}}} \tag{5}$$

So, the corresponding vector magnitude is Z such as:

$$Z^2 = \sum_{i=1}^{n} z_i^2 \tag{6}$$

The isotropy hypothesis entails that the density of points representing the vectors ends in the symbolic space has a spherical symmetry. Thus, we can consider that all the functions $f_i(l_i)$ in the relation 4 are identical provided that F is written as following:

$$F(l_1, l_2, \ldots, l_n) = f(l_1) f(\frac{l_2}{r_l}) \ldots f(\frac{l_n}{r_l^{n-1}}) \tag{7}$$

That is:

$$\phi(z_1, z_2, \ldots, z_n) = \varphi(z_1)\varphi(z_2)\ldots\varphi(z_n) \tag{8}$$

The function ϕ can also be written:

$$\phi(z_1, z_2, \ldots, z_n) = \frac{d^n N}{N dz_1 dz_2 \ldots dz_n} \tag{9}$$

If one moves in the surface of the hypersphere whose equation is:

$$z_1^2 + z_2^2 + \ldots + z_n^2 = C \tag{10}$$

One has:

$$\phi(z_1, z_2, \ldots, z_n) = C \tag{11}$$

C being a constant. These two above hypotheses are sufficent to determine the probability density function (pdf) of the hydraulic lengths.

2.3 Determination of the hydraulic lengths pdf

By taking the derivative of relations 10 and 11 gives:

$$2z_1 dz_1 + 2z_2 dz_2 + \ldots + 2z_n dz_n = 0 \tag{12}$$

and

$$\frac{\partial\phi}{\partial z_1} dz_1 + \frac{\partial\phi}{\partial z_2} dz_2 + \ldots + \frac{\partial\phi}{\partial z_n} dz_n = 0 \tag{13}$$

According to the relation 8:

$$\frac{1}{\phi}\frac{\partial\phi(z_i)}{\partial z_i} = \frac{1}{\varphi(z_i)}\frac{d\varphi(z_i)}{dz_i} \qquad \forall i, 1 \leq i \leq n \tag{14}$$

Which allows us to replace 13 by:

$$\frac{1}{\varphi(z_1)}\frac{d\varphi(z_1)}{dz_1} + \frac{1}{\varphi(z_2)}\frac{d\varphi(z_2)}{dz_2} + \ldots + \frac{1}{\varphi(z_n)}\frac{d\varphi(z_n)}{dz_n} = 0 \tag{15}$$

In this stage of the reasoning, we can use the Lagrange's optimization method (Bruhat, 1968[10], Sears, 1971[11]). It allows us to combine relations 12 and 15 by multiplying 12 by a constant λ and by adding it in the relation 15. This leads to the following equation in which n differentials can be considered as independent:

$$\left[2\lambda z_1 \frac{1}{\varphi(z_1)}\frac{d\varphi(z_1)}{dz_1}\right] dz_1 + \ldots + \left[2\lambda z_n \frac{1}{\varphi(z_n)}\frac{d\varphi(z_n)}{dz_n}\right] dz_n = 0 \tag{16}$$

All the expressions between brackets are simultaneously equal to zero, so that each one can be integrated:

$$\frac{1}{\varphi(z_i)}\frac{d\varphi(z_i)}{dz_i} = -2\lambda z_i \qquad \forall i, 1 \leq i \leq n \tag{17}$$

Whose integral is:

$$\varphi(z_i) = Ae^{-\lambda z_i^2} \tag{18}$$

Where z_i is distributed according to a normal law. The constant A can be calculated because the complete integral of φ between zero and the infinity must be equal to unity:

$$\int_0^\infty \varphi(z_i)dz_i = \int_0^\infty Ae^{-\lambda z_i^2}dz_i = 1 \tag{19}$$

The variables change, by taking into account the relation 5, one sees then appearing the pdf of l_i:

$$f(l_i) = Ae^{-\lambda \frac{l_i}{r_l^{i-1}}} \frac{l_i^{-\frac{1}{2}}}{2\sqrt{r_l^{i-1}}} \tag{20}$$

The graph of this law is a convex shape and quickly decreasing, more exactly, it is about a gamma law of general equation $\left(\frac{1}{\beta}\right)^\alpha \frac{1}{\Gamma(\alpha)} x^{\alpha-1} e^{-\frac{x}{\beta}}$ with parameters $\alpha = \frac{1}{2}$ and $\beta = \frac{r_l^{i-1}}{\lambda}$. Relations 8 and 18 allow us therefore to write:

$$\phi(z_1, z_2, \ldots, z_n) = A^n e^{-\lambda(z_1^2+z_2^2+\ldots+z_n^2)} = A^n e^{-\lambda Z^2} \tag{21}$$

According to the relation 11, one obtains:

$$\frac{d^n N}{N} = A^n e^{-\lambda Z^2} dz_1 dz_2 \ldots dz_n \tag{22}$$

If one focuses on the number dN_Z of vectors which have their arrowhead between the two hyperspheres of Z and $Z + dZ$ radius, one obtains thus:

$$\frac{dN_Z}{N} = A^n e^{-\lambda Z^2} \int \ldots \int dz_1 dz_2 \ldots dz_n \tag{23}$$

The equation above expresses the relative number of hydraulic length included between the hyperspheres of Z and $Z + dZ$ radius. Multiple integral expresses the volume included between these two infinitely close hyperspheres in a n-dimensions space. The element of volume dV is proportional to $Z^{n-1}dZ$:

$$\frac{dN_Z}{N} = A^n e^{-\lambda Z^2} B Z^{n-1} dZ \tag{24}$$

B being a constant. To clarify the variable L, which is the most interesting, one can make a change of variables. Since the relation 2 means that component l_i is on average r_l times larger than l_{i-1}, if we decide to argue about the mean values of z_i^2, relations 5 and 6 give:

$$Z^2 = n\overline{z_i^2} \tag{25}$$

The hypothesis of isotropiy permits in particular cases to write this relation for $i = 1$:

$$Z^2 = n\overline{z_1^2} = n\overline{x_1^2} = nL\left(\frac{r_1 - 1}{r_1^n - 1}\right) \tag{26}$$

Where L is the average hydraulic length corresponding to the vectors which end on the hypersphere of radius Z. Thus, the relation 21 enables to express the relative number of hydraulic lengths $\frac{dN_L}{N}$ whose value is between L and $L + dL$:

$$\frac{dN_L}{N} = Ce^{-\frac{\mu}{L}} L^{\frac{n}{2}-1} dL \tag{27}$$

Where

$$\mu = \lambda n \left(\frac{r_1 - 1}{r_1^n - 1} \right) \tag{28}$$

μ and C being constants. The pdf of hydraulic lengths $\rho(L)$ can now be written:

$$\rho(L) = \frac{dN_L}{NdL} = Ce^{-\mu L} L^{\frac{n}{2}-1} \tag{29}$$

By integrating relation 27, L varying from 0 to ∞, one finds C:

$$C = \frac{\mu^{\frac{n}{2}}}{\Gamma\left(\frac{n}{2}\right)} \tag{30}$$

Where Γ is gamm function. The constant μ can be deducted from the average hydraulic length by complete following:

$$\overline{L} = \int_0^\infty L\rho(L)dL = \frac{n}{2\mu} \tag{31}$$

Which gives:

$$\mu = \frac{n}{2\overline{L}} \tag{32}$$

The pdf of hydraulic lengths $\rho(L)$ can be so written:

$$\rho(L) = \frac{dN_L}{NdL} = \left(\frac{n}{2\overline{L}}\right)^{\frac{n}{2}} \frac{1}{\Gamma\left(\frac{n}{2}\right)} L^{\frac{n}{2}-1} e^{-\frac{nL}{2\overline{L}}} \tag{33}$$

Let us remember that L is the hydraulic length, $\overline{L}$ is the average of all the possible hydraulic lengths on the studied plant, n is the order of the tree, and that Γ is the gamma function. One can easily recognize in this last relation $f(L, \alpha, \beta)$ the gamma law with parameters $\alpha = \frac{n}{2}$ and $\beta = \frac{2\overline{L}}{n}$.

3 Results and discusion

There is a big difference between systems involved in statistical physics and the system we use in our reasoning : Maxwell, for example, does not indicate the shape of the container that contains the N molecules because it does not matter. In our case, the studied plant can have very different shapes, its size is very variable and its order can vary from 1 for a young maiden tree to 8 or 9 for very big trees. Moreover, even though law 33 is very general, it will be all the more respected since the two conditions to apply a reasoning of statistical physics will be respected: a large number of elementary constituents (that is of hydraulic distances) and homogeneity of the population.

Order i	$R_C = \frac{N_{i-1}}{N_i}$	$R_L = \frac{L_i}{L_{i-1}}$
2	12.06	2.16
3	11.5	4.25
Theorical	11.78	3.03

Table 1. average values of R_C and R_L calculated from the experimental data. The last values of the Table are calculated from a logarithmic regression between the order $i-1$ and the large number of experimental values of R_C and R_L. The fractal dimension is thus $D = \frac{\ln(R_C)}{\ln(R_L)} = 2.22$.

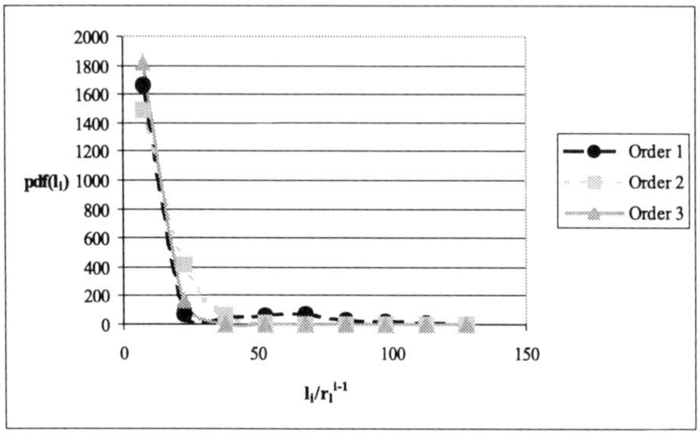

Figure 2. The pdf of the reduced hydraulic components l_i according to equation 20.

So one will either choose a big tree or a population of several trees with the same species having grown in the same environment. The pdf of hydraulic lengths is calculated through the law 33 by *i)* taking as the order n of the specified population value ν of the maximum order observed in the population, *ii)* taking as the average of hydraulic lengths $\overline{L}$ calculated from N measures. We chose 12 apple trees (*Malus pumila (L) Mill.*) four years old and from the same "parents" and with order 3. The hydraulic lengths were measured manually from all the growing shoots.
The average values of the ratios R_C and R_L which are presented in the table 1 are calculated for all the sections. One can notice that the stability of R_C is excellent, but that R_L is a little more variable, as noted by many authors (Horton, 1945[12], Schumm, 1956[5] and Shreve, 1967[8]). The stability of this last parameter, *a priori* considered to define the symbolic space is therefore verified. For each parameter, the accepted value is calculated by logarithmic regression because of the shape of R_C and R_L laws (values on the last Table line). By the way, one can note that the values of R_C and R_L lead to a fractal dimension $D = 2.22$. Figure 2 shows the components of hydraulic lengths as function of $\frac{l_i}{r_l^{i-1}}$. As one can see, considering the limited number of objects in some classes, the distribution law of components 20, as well as the isotropy hypothesis, from which it is deduced, can be considered as well verified. Figure 3 shows the theoretical graph supplied by the equation 33 and the experimental one obtained by measurement. Considering the very general

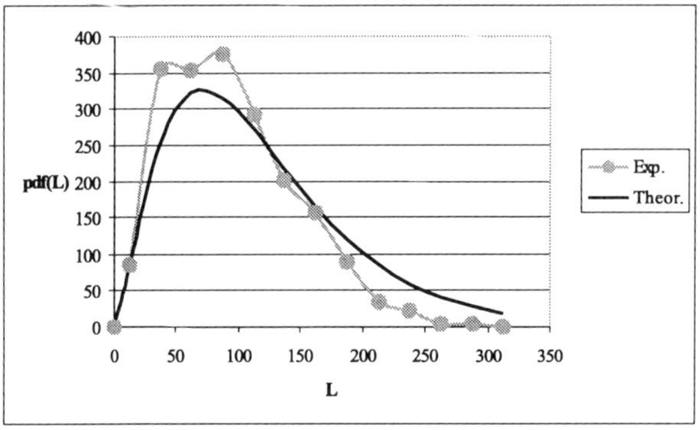

Figure 3. The pdf of the hydraulic lengths (Theor.) of the studied plant compared with the experimental results (Exp.). The pdf parameters are $n = 3$ and $\overline{L} = 47.6$ cm.

hypotheses from which the theoretical graph is deduced, one can be struck by the fact that it coincides correctly with the experimental one.
Of course the theory and the experimental data do not fit as well as in statistical thermodynamics. For example, Maxwell's distribution in a molecular stream can be directly verified by counting the number of molecules that have a given velocity. Miller and Kusch (1956)[25] showed that theoretical prediction was strikingly verified by the experiment. In thermodynamics, considering the huge number of microscopic objects, the relative fluctuations of the molecule numbers are proportional to $\frac{1}{\sqrt{N}}$, where N is the number of molecules in the considered class, so they become imperceptible at the macroscopic scale. In the case of botany, conformity cannot be rigorous because of two reasons which proceed directly from two conditions we had put *a priori* in order to apply a reasoning of statistical physics: *i)* the number of hydraulic lengths, corresponding to apexes, can not exceed a few thousand, or ten thousand, for a given class; thus the statistical fluctuations will always be much more important, compared to the fluctuations one can observe in thermodynamics, even if one can reduce them by widening the classes, *ii)* moreover, the distribution of hydraulic lengths, as well as the distribution of their n components can be more or less influenced by the environment constraints. We based our demonstration on the frame that Maxwell used to study the law of distribution of molecular velocities in a gas. One knows that his theory was reused on other bases by Boltzmann and Gibbs. It would be interesting to apply general formalism of statistical mechanics to the botanical morphology.

4 Conclusion

We have just presented an original reasoning of statistical physics as far as it applies to a macroscopic object made up of elements themselves macroscopic: the plant. The study of the spatial organization of a plant leads to a mathematical description that completes, through the pdf of the hydraulic lengths, a classical morphogenetic description of its architecture. Such a description is essential in botany, whether it

is for the understanding of the functioning of the plant or for the landscape analysis. Moreover, we think that the innovative approach adopted here the introduction of a fractal description into a reasoning of statistical physics could be applied successfully to other physical domains. The application of our results to other branched objects such as river systems or vascular systems can be easily attempted. One can also try to apply them to other objects provided they are fractal as for example a fractured surface in a solid material test.
Besides the interest of a morphological description of the tree, the above theory opens many perspectives of application in the dynamic domain. The mathematical description of the branches' organization indeed allows us to describe the dynamics of the transfers which take place in the plant on one hand and in the modeling of the dynamics of the plant genesis on the other hand. Thus the applications could be immediate in the fields of the complete plant physiology and in the ecophysiology, for example in the mechanism of axes selection. In particular, we shall show in another article Raimbault *et al.* (2001)[3] how function 33 established above can be connected with morphogenesis concepts as the apical control.

References

1. B. Mandelbrot, Les object fractals , Flammarion (1975).
2. L.K. Sherman, English New Record **108**, (1932).
3. P. Raimbault, J. Duchesne and C. Fleurant J. Theor. Biol. , submit. (2001).
4. A.N. Strahler, Bulletin of the Geological Society of America **63**, (1952).
5. S.A. Schumm, Bulletin of the Geological Society of America **87**, (1956).
6. J.H. Lienhardt, Journal of Geophysical Research **69**, (1964).
7. R.L. Schreve, Journal of Geology **74**, (1966).
8. R.L. Schreve, Journal of Geology **75**, (1967).
9. J.E. Nash, AIHS **3-14 sept.**, (1957).
10. G. Bruhat, Masson and Cy , (1968).
11. F.W. Sears, Addison-Wesley Publishing Cy , Reading (1971).
12. R.E. Horton, Bulletin of the Geological Society of America **56**, (1945).
13. A.N. Strahler, Bulletin of the Geological Society of America **69**, (1958).
14. P. La Barbera and R. Rosso, Eos. Trans. AGU **68**, (1982).
15. E.R. Weibel and D.M. Gomez, Science **137**, (1962).
16. A.H. Fitter, Plant Cell and Environment **5**, (1982).
17. P.G. Holland, New Phyto. **68**, (1969).
18. L.B. Leopold, J. Theor. Biol. **31**, (1971).
19. S.I. Oohata and T. Shidei, T. Japanese Journal of Ecology **217**, (1971).
20. G.G. Whitney, Bull. of the Torrey Botanical Club **103**, 2 (1976).
21. J.W. Crawford and I.M. Young, J. Theor. Biol. **145**, (1990).
22. D.S. Berger, J. Theor. Biol. **152**, (1991).
23. S.G. Chen, I. Impens, R. Ceulemans and F. Kockelberg, Agricultural and Forest Meteorology **64**, (1993).
24. T.A. Macmahon and R.E. Kronauer, J. Theor. Biol. **59**, (1976).
25. R.C. Miller and P. Kusch, Journal of Chemical Physics **25**, (1956).

AN ATTEMPT TO CHARACTERIZE HEDGEROW LATTICE BY MEANS OF FRACTAL GEOMETRY

B. ROLAND
Horticultural National Institute, Landscape Department,
2 rue Le Nôtre, 49045 Angers cedex 01
E-mail: bernard.roland@inh.fr

The bocage landscape has evolved very radically during the last fifty years with the agricultural developement. The shapes of the hedgerow lattice have many consequences for example, in hydrological or ecological functions of this landscape. So we attempt to characterize the network structure by means of fractal geometry. The box-counting method permits to get significantly different results to analyse this structure with fractal parameters (dimension, upper and lower cuts off). Combining them with other parameters, we plan to model the functionning of this landscape.

Key-words : bocage landscape, fractal dimension, hedgerow, lattice, box-counting

1 Introduction

The bocage landscape indicates, in the language of the geographers, an agrarian landscape of green enclosures. It is the opposite of the open fields landscape. Enclosures are constituted, in the strict sense, by quickset hedges which surround completely exploited plots of land. By extension, enclosures which are not closed or are composed of low dry stone walls can be qualified as hedged farmland ([10]). The shapes are very different and present a strong irregularity, thus their determination is complex. Historians, geographers, ethnologists, agronomists were interested in it, trying to identify the causes which presided over these forms or to explain their evolution. [4] describes this quest and the difficulties met "To understand genesis, in the long duration, of these complex objects which are not urban landscapes, to distinguish the parameters of their structuralization, to disentangle the mess of the diverse factors which shaped them ..." As such, the hedged farmland presents a particular interest, because structure in network of the meshing hedges accentuates the plot of land plan and can facilitate the analysis of its forms. It materializes the foundations of agricultural structures, the technical and economic organization of a country. Besides, through its variety, it strongly marks the identity of a territory. Finally, the evolution of the forms permits an historic analysis. Particularly in the last fifty years this landscape was profoundly upset by the modernization of agricultural structures and by mechanization. The increase of productivity often came along with the extension of plots of land:

- the hedge becomes an obstacle to the passage of the bigger and bigger machines
- the maintenance of hedges becomes too expensive with the increase of the surfaces of exploitation

- its dimension of several meters on the ground, reduces of so much the surface of production

- authorities and rural development policies favored these practices.

The various specialists who were interested in this landscape show the complexity of these forms and describe it qualitatively without being able to quantify this irregularity. Nevertheless the interest to know the forms which characterize a landscape is evident, notably in the interpretation of the streams which propagate at various scales.

Several mathematical tools are at the disposal of the observers to describe forms, notably geometry, which is based on the characterization of the properties of forms. However, geometry quickly reaches its limits for describing natural forms, as soon as it deals, for example, with lines which have an infinitely variable curve radius. The mathematicians of the beginning of the 20th century spoke about pathological curves. [6] quotes the figures of von Koch, Peano or Sierpinski. They resist any attempt to be described by the mathematical tools known until then. Fractal Geometry which he invents, in the 60s, permits the description of these curves called "not rectifiable". He introduced a characteristic parameter: the fractal dimension. "It is the number which quantifies the degree of irregularity and fragmentation of a geometrical set or a natural object, and which is reduced, in the case of the objects of Euclide's usual geometry, to their usual dimensions" ([6]). Observation shows that broken lines on a plan have a fractal dimension included between 1 and 2. In the introduction of his book "fractal objects", he formulates, in an intuitive way, the relation between fractal dimension and shape of the studied object: "to characterize such figures, one can at first say, very roughly, that a figure the dimension of which is situated between 1 and 2 has to be more 'disentangled' than a common surface, while being more 'massive' than a common line." So one can, by extension, consider that the fractal dimension of an object constituted of lines which present very variable directions by covering the plan in a dense way, approaches 2 and that on the contrary for a plan little occupied by lines of lightly opposed curve, the dimension is close to 1. Between these two limits appears all the range of variety of curves presenting more or less big gaps. [4] talks about "which allows to introduce that of the phenomena of scale, fitting scale and, there , discontinuity." These descriptive elements are characteristic of fractal objects. It seems so possible to measure fractal dimension of a meshing bocage landscape. [2] in the work "Ecologie du paysage", presents several characteristic parameters of forms. Among them, fractal analyses constitutes a tool which describes the occupation of soils. So the fractal dimension of a bocage landscape, represented by an air-view of a meshed network, seems convenient. As [8] proposed for the description of a network of fractures and joints (Yucca Mountain, Nevada), or [1] for a railway network, we are going to analyze the nature of a bocage landscape by means of the fractal tool to try to characterize its shape.

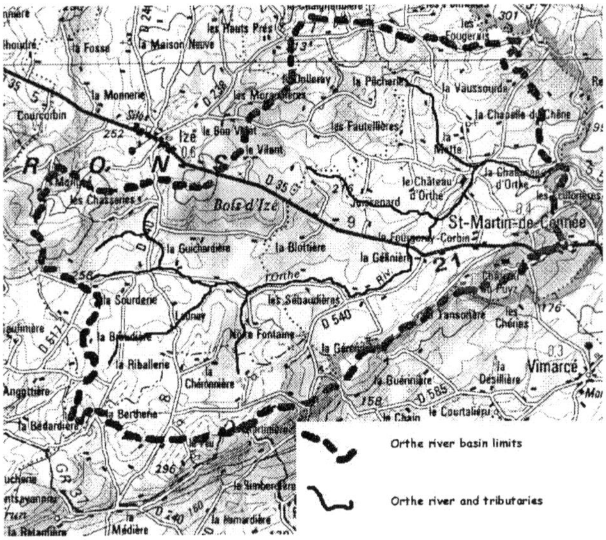

Figure 1. Situation map of the Orthe river basin

2 Material and method

So we looked for a site of study in a bocage landscape still dense enough to present an irregularity of the meshing hedges, with the following characteristics:

- a river basin (with the aim of a following application to hydrology) from 20 to 50 km^2,
- a hilly landscape,
- the remains of a patchwork of hedges which evolved over the last fifty years, through episodes of more or less intense agricultural development,
- the existence of air shots of the sector at different times during this period.

The department of Mayenne (France) presents numerous sites fulfilling these criteria. The one that was selected is situated upstream to the basin of the Orthe river, on the municipalities of Saint-Martin de Connée and Izé (see Figure 1). Furthermore, there is, in this region, a program of rehabilitation and of replantation of hedges which could become a ground of application for the results of this theoretical study. By interpreting of the air photos (missions IGN on 1958 and 1996 - scale : 1/25 000), we situated the plan of the meshing on a background of cadastral plan in 1/5 000, to respect the existing fragmented limits on which hedges are implanted and to take into account the distortion of photographic representations (Figures 2 and 3).

We delimited the outline of the river basin (of a surface of about 30 km^2). To make the analysis of the evolution more elaborate and to characterize the variations in the evolution of the meshing, inside this perimeter, we distinguished the two

Figure 2. Hedge network in 1958

Figure 3. Hedge network in 1996

municipalities of the basin in measures, separated by a "border" zone. Indeed each municipality possesses a particular history in the field of land property: St Martin de Connée : A land reorganization between 1990 and 1994, Izé : privately negotiated exchanges spread out over the duration of the study. Several methods to estimate fractal dimension are proposed in literature. We chose the one that [5]describes because he applies it to urban structure, a fundamentally anthropic environment, as the hedged farmland. [2] also uses it to characterize the forms of landscape or the occupation of soil. It is about the box-counting method, Frankhauser adapts Mandelbrot's formula in the following way :

The relation between the number of full boxes and the unitary size of boxes can be written in the fractal zone:

$$N\epsilon^D = K \tag{1}$$

Where N is the number of full boxes. ϵ the unitary size of the box, D the fractal dimension and K a constant. The relation can also be written like this: (one constant excepted)

$$D = \frac{\ln N}{\ln\left(\frac{1}{\epsilon}\right)} \tag{2}$$

The method consists in applying successively to the space that one studies a more and more precise covering and in counting at each iteration the number of full boxes (i.e. : the number of boxes which contain at least a part of the object to be measured). One draws the graph of points $P_i\left[\ln\left(\frac{1}{\epsilon_i}\right), \ln\left(N_i\right)\right]$. The slope of the regression straight-line, estimated in the cloud of points included in the fractal zone thus corresponds to the fractal dimension of the object, according to the relation 2. We applied the method of counting to the resultant plan, by adapting it to our object. Indeed, to take into account the weak predictable area of the fractal behavior we began counting from an area constituted of 27 boxes of 1 km (ground length). Then, successively, was applied to the zone of study a covering with square stitches corresponding to a complete division of this initial meshing, 1/2 km, 1/3 km, 1/4 km, 1/5 km, 1/6 km, 1/7 km, 1/8 km, 1/9 km, 1/10 km, 1/11 km, 1/12 km, 1/14 km, 1/16 km, 1/24 km, 1/32 km (realized by means of the AUTOCAD^R software). The perimeter of the studied region is so strictly identical at every stage of the counting. This iterative method presents the advantage, with regard to the method proposed by [5] (for whom ϵ is divided by two at every iteration) of increasing the number of points of measure and of making the result more precise in the case of an object presenting a small fractal zone. We counted the boxes by applying the following rule : "a box is full, if it contains a section of hedge ; on the opposite, it is empty if it does not contain any".

3 Results

By putting the results of the counting, on a graph $\ln(N_i)$ according to $\ln\left(\frac{1}{\epsilon_i}\right)$, one obtains Figures 4 and 5.

One obtains graphs of the same type for the set of the studied sectors. The observation of graphs shows that the cloud of points can be separated in 3 sectors(Figure 6):

- before the upper cut, the gradient of the regression straight-line is equal to 2 (all the boxes of the covering are full, their number increases as the square of their size) (represented by a rhomb on graphs)
- in the fractal zone, the gradient is D_f: the fractal dimension (represented by a point on graphs).
- for boxes of a size smaller than the lower cut, the gradient is 1 because the number of full boxes increases proportionally to the size of boxes(represented by a triangle on graphs).

The determination of the fractal dimension thus means identifying the most rigorous zone of adaptation on the graphs presented above. The latter proves to be delicate because the extent of the fractal element is reduced. The determination of

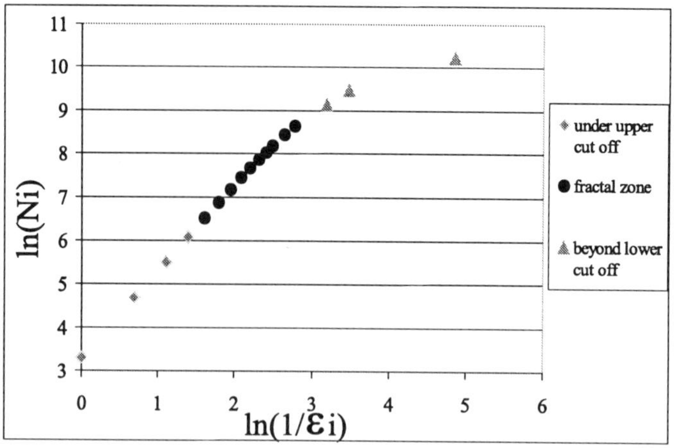

Figure 4. Box-counting method results in 1958 - Orthe river basin.

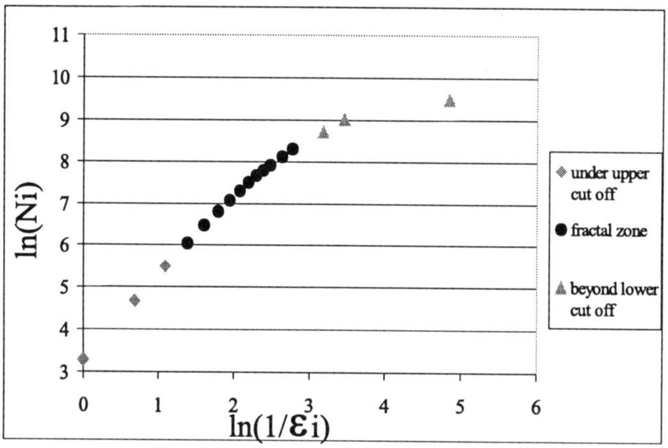

Figure 5. Box-counting method results in 1996 — Orthe river basin.

the limits of the zone characterizing scaling behavior (upper and lower cuts) thus seems essential to identify this area. [7] indicate that measuring, by the box-counting method, of most of natural phenomena, reveals the same type of behavior. He points out that "One of the most unsettled questions when applying the box method to natural objects is what range of box sizes should be considered in generating the regression line in the log-log plot. This range is influenced by the range of self-similarity, the reduction factor and possible lower or upper cut off values of the feature."

It is necessary here to point out the notion of lower or upper cut off and its intuitive interpretation: As [8] indicates, the determination of the lower and upper

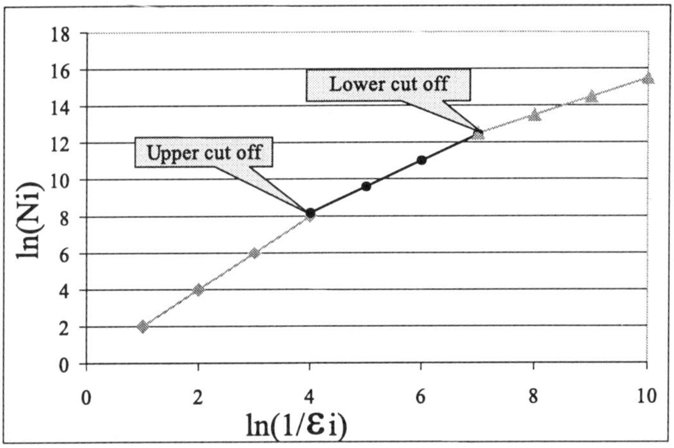

Figure 6. Theorical set of points in a box-counting method

Year	Izé	St Martin	River Basin
1958 (low.)	40 m	40 m	40 m
1958 (up.)	200 m	140 m	200 m
1996 (low.)	70 m	60 m	70 m
1996 (up.)	250 m	200 m	250 m

Table 1. Upper (up.) and lower (low.) cut off.

cuts "that is of the smallest not fractal size and of the size up to which geometry is fractal can establish a useful characterization". In the discussion about the length of the Britanny coastline, [6] defines the notions of external and internal cuts which limit the extent of the scaling behavior of a fractal object. "It is reasonable to suppose that the real coastline is liable to two 'cuts'. The 'external cut' Λ can be measured in dozens or in hundreds of kilometers. For a coastline not joining up, Λ could be the distance between two extremities. For an island, Λ could be the diameter of the smallest circle which contains all the coastline. On the other hand, the 'internal cut' can be measured in centimeters." By analogy to this demonstration, concerning the study of the fragmented hedged farmland, one could suppose that the upper cut (or external) corresponds to a dimension comparable to the largest open spaces in the bocage landscape and that the lower cut (or internal) is of the size of the smallest plots of land surrounded with hedges on the studied fragment. If this hypothesis proves to be exact, the determination of the upper and lower cuts would permit to characterize the structure of a meshing hedgerow lattice in a quantitative way. The observation of the results of the counting permits us to determine the upper cut, because it corresponds to the size of the boxes for which there appear empty boxes in the perimeter of the study. On the cloud of points included between two cuts, one calculates the slope of the regression line and its criterion of dispersal

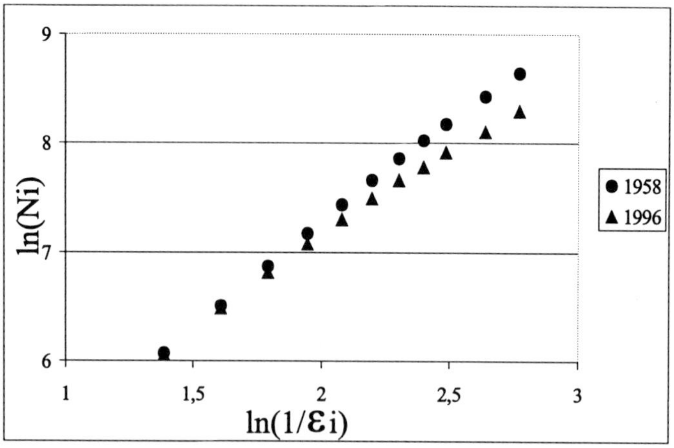

Figure 7. Extract of the cloud of points on the fractal zone

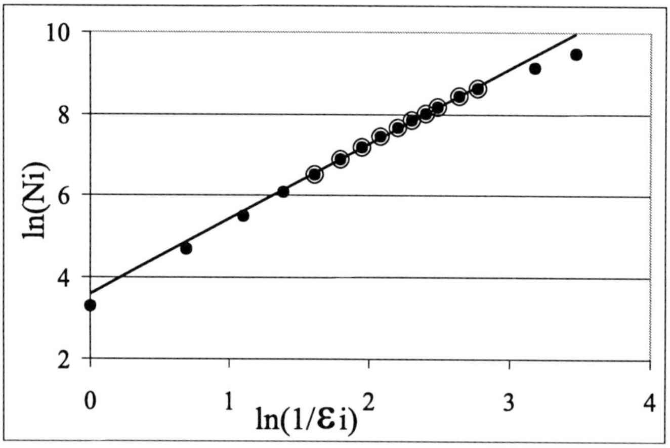

Figure 8. Representation of the slope of regression line of the cloud of points. Resulting from the counting boxes on the Orthe river basin in 1958. $\ln(N_i) = 1.8413 \ln\left(\frac{1}{\epsilon_i}\right) + 3.5875$, $R^2 = 0.9983$.

Year	Izé	St Martin	River Basin
1958	1.82	1.81	1.84
1996	1.69	1.57	1.64

Table 2. Fractal dimensions.

(coefficient R^2) We proceeded in the same way for the municipalities of Izé and St Martin de Connée. The results of the observation are indicated in the Tables 1 and 2.

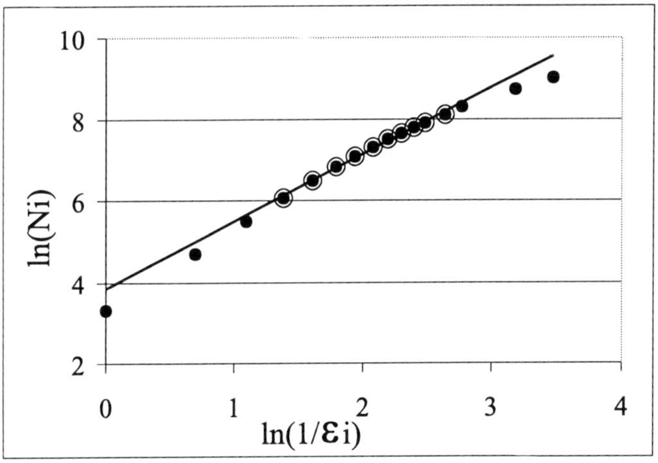

Figure 9. Representation of the slope of regression line of the cloud of points. Resulting from the counting boxes on the Orthe river basin in 1996. $\ln(N_i) = 1.6439 \ln\left(\frac{1}{\epsilon_i}\right) + 3.8424$, $R^2 = 0.9958$.

The coefficient R^2 of the regression line is upper to the value 0.99. These results show a very satisfactory adaptation of points in the fractal zone thus verifying our hypothesis on the determination of the upper and lower cuts of the meshing. The results also permit us to verify the accuracy of the measure and to confirm one of our hypotheses. One indeed obtains a significantly different result, in 1996 , between the municipality of Izé and the municipality of St Martin de Connée, characterizing the history of the fragmented private exchanges of each of the two municipalities. The fractal analysis reveals, on this example, the significant differences which permit to characterize the structure of the meshed network. In 1958, the network is mainly constituted by small plots of land, closed and connected to each other. In 1996, the network is sometimes opened and very disintegrated. Fractal dimension can also be interpreted as an indicator of the lacunarity of the network. [3] proposes tools which enable us to characterize it. So the field of study opened here seems interesting to investigate in order to correlate the value of the fractal dimension with the other characteristic parameters of the structure of the meshing farmland hedgerow.

4 Discussion and conclusion

These first results enable us to show that it is possible to characterize the forms of the hedged farmland by means of fractal geometry. The disintegration of the meshing farmland hedgerow is translated by a decrease of fractal dimension. We also showed that even though the irregularity of the meshing is important, the extent of the scaling behavior is reduced. This leads us to manipulate this tool with precaution. However, we think that the fractal dimension of some areas of hedgerow lattice brings a new element of knowledge. Besides, the fractal analysis permits us to determine a frequency of upper cut off and a frequency of lower cut off, characteristics of a hedgerow lattice area. In our study, in every case, they are characteristically distinct from each other by a dense and well-connected patchwork, before land organization (during the 60' to the 90'), and a disintegrated network which results from increases in the size of agricultural plots of land. The measure of these sizes also consequently brings an element of characterization of the meshed structure of the hedged farmland. This information could be used in the same way as the other parameters, such as the density of hedges, the connectivity of the network in models of functioning of this landscape : geo-physics (hydrology, streaming and erosion), chemical (distribution of micro-pollutants) or biologic (ecological corridors ...). Hydrology, in particular, appears to be a way of possible application. The phenomenon of percolation is modelled in the tiny scale by means of fractal geometry. [9] is interested in the notion of threshold of percolation on a network from the probability of occupation of the stitches of a square area. We intend to apply these works to the macroscopic field of the hedged farmland. This first analysis on a test-territory already opens up interesting perspectives of development and application. The obtained results could also contribute to quantify indicators used in the landscape analysis. The fractal dimension and the upper and lower cuts could be in future used as indicators in the landscape analysis. One will also be able to suggest correlating them with certain visual codes used by landscape architects such as depth, transparency, opening of a landscape. Lacunarity seems to us a good indication to measure these characteristics.

Acknowledgments

To Gérard Clouet (Farmers' association of Mayenne) for the collaboration in the choice of the site of study due to his perfect knowledge of the soil and the hedged farmland. To Jean Duchesne (Professor, director of the Laboratory of the Landscape of the INH - National Institute of Horticulture - Angers, France) and Cyril Fleurant (Lecturer - Laboratory of the Landscape of the INH) for the second reading of this text and its shaping.

References

1. L. Benguigui, Physica A, 191, 1992.
2. F. Burel and J. Baudry, Ecologie du paysage, Paris, 1999.
3. Q. Cheng, Computers and Geosciences, 1999.
4. I. Chiva, Etudes rurales, 1991.
5. P. Frankhauser, La fractalité des structures urbaines, Economica, 1994.
6. B. Mandelbrot, Les Objets Fractals, Flammarion, 1995.
7. H.S.W. Pinnaduwa Kulatikale, R. Fielder and B. Bibhuti, Engineering Geology, 48, 1997.
8. B. Sapoval, Universalités et fractales, Flammarion, 1997.
9. M. Schroeder, Fractals, Chaos, Power laws, Freeman and Company, 1991.
10. D. Soltner, L'arbre et la haie, Sciences et Techniques Agricoles, 1995.

CONTRASTING SELF-SIMILARITY AND RANDOMNESS: SPECIES-AREA RELATIONS IN A CALIFORNIAN SERPENTINE GRASSLAND

J. L. GREEN

Department of Nuclear Engineering, University of California at Berkeley, Berkeley CA 94720, USA[1]

E-mail:jlgreen@nuc.berkeley.edu

Department of Biological Sciences, Macquarie University, NSW 2109, Australia[2]

The relationship between species richness and sampled area - the 'species-area relationship' (SAR), is of central concern in ecology, both as a conservation biology tool and as a theoretical stepping stone to characterize community structure. Using plant census data collected in a Californian serpentine grassland, I compare the form of the SAR under the hypothesis of self-similarity and three different hypotheses of random placement. The serpentine grassland community SAR is well fit by a power-law, suggesting self-similarity in the spatial distribution of species. In contrast, the empirical data does not agree with a theoretical model in which individuals are distributed randomly and independently. In addition, the empirical data does not agree with two different computer simulations where species, and then clusters of species, are distributed randomly. All three random placement hypotheses significantly overestimate species richness at the study site. Thus, a quantifiable distinction is made between the SAR for self-similarly distributed species and the SAR for individuals, species, and clusters of species that are distributed randomly.

1 Introduction

The relationship between species richness and sampled area - the 'species-area relationship' (SAR) - is one of the most widely studied patterns in ecology. According to McGuiness [22], evidence that the number of species within a taxonomic group tends to increase with increasing area was documented as early as 1855 [9]. The SAR has played a major role in ecology, both as a conservation biology tool and as a theoretical stepping stone to characterize community structure. The SAR has been used to estimate species richness at large spatial scales using empirical data from small areas [10, 20], to help in the design and management of nature reserves [12, 21, 30], and to estimate extinction rates due to habitat loss and destruction [24, 25, 32].

Olof Arrhenius was the first to propose a mathematical description of the SAR [1], which he later simplified to a general power-law that can be written in the form:

$$S_i = cA_i^z \qquad (1)$$

where S_i is the number of species, on average, on a census patch of area A_i, and c and z are empirically derived constants [2]. Although other functional forms of the SAR have been proposed [see 7, 18, 22 for reviews], the power-law form is widely accepted in ecology [21, 23, 28, 29, 33]

---

[1] Current address

[2] Address as of January 1, 2002

It has recently been demonstrated that the power-law SAR implies self-similarity in the spatial distribution of species [16, 17]. To see this, consider the following definition of self-similarity for a community of species within a broad taxon (for example birds, mammals or reptiles). Visualize a square biome or habitat patch of area A_0, and let $A_i = A_0/2^i$ be the size of areas obtained from A_0 following nearly 'shape preserving' bisections, such that each A_i area formed at the ith bisection are squares or 2 x 1 rectangles. If a randomly chosen species is known to be in an A_{i-1} patch, and nothing else about that species (such as its abundance) is known, let $\boldsymbol{a}_i$ denote the probability that under bisection it will be found in at least a specific one of the two resulting A_i patches. The distribution of species is self-similar, or scale invariant, if the probability $\boldsymbol{a}_i$, averaged across all of the A_i patches that comprise A_0, is independent of spatial scale i.

If the probability $\boldsymbol{a}_i$ is independent of i, it follows that the mean number of species in A_i is

$$S_i = \boldsymbol{a}^i S_0 \tag{2}$$

and the variance in species richness across the A_i patches is

$$\sigma_i^2 = S_0(\boldsymbol{a}^i - \boldsymbol{a}^{2i}) \tag{3}$$

where S_0 is the total number of species sampled within A_0. Note that Eqs. 2 and 3 implicitly assume that species are self-similarly *and* independently distributed across A_0. From Eq. 2 it follows that $\boldsymbol{a} = S_i/S_{i-1}$, which in turn yields the following expression relating $\boldsymbol{a}$ to the power-law SAR exponent z in Eq. 1:

$$z = -\log_2(\boldsymbol{a}) \tag{4}$$

Hence, the community-level postulate of self-similarity implies the power-law SAR.

In contrast, by assuming individuals are spatially distributed randomly and independently across a biome, Arrhenius [3] derived a SAR that can be condensed into the following functional form:

$$S_i = S_0 - \sum_{j=1}^{S_0} (1 - A_i / A_0)^{n_j} \tag{5}$$

where S_i is the number of species, on average, on a census patch of area A_i residing within a larger habitat patch of area A_0 and n_j is the abundance of the jth species. Later, Coleman [5] derived the following expression for the variance of species richness across the A_i patches:

$$\sigma_i^2 = \sum_{j=1}^{S_0} (1 - A_i / A_0)^{n_j} - \sum_{j=1}^{S_0} (1 - A_i / A_0)^{2n_j} \tag{6}$$

For the remainder of this paper, I refer to Eqs. 5 and 6 as the 'Individuals Random Placement Model'. Because the Individuals Random Placement Model is an explicit function of species abundance, the shape of the species-area curve will depend on the species-abundance distribution on A_0.

The hypothesis of random placement is a widely discussed null model in ecology. The aim of this paper is to compare the form of the SAR under the hypothesis of random placement and the hypothesis of self-similarity. To do this, I examine data collected from a Californian serpentine grassland that maps the spatial location of over 37,000 individually identified plants across spatial scales ranging from ~ 0.01 m^2 to ~ 100 m^2. I compare the observed SAR to a hypothesized SAR assuming: 1) self-similarity, 2) the Individuals Random Placement Model, 3) the random placement of species (versus individuals), and 4) the random placement of clusters of species. The first two hypothesized SARs are generated by applying the theory described above (Eqs. 2, 3, 5 and 6), and the latter two hypothesized SARs are generated via computer simulation, as described below (section 2.3).

2 Methods

2.1 *Study System and Sampling Design*

Field data were collected from serpentine substrates at the Homestake Mine/Donald and Sylvia McLaughlin University of California Natural Reserve (lat 38°51′N, long 122°24′W) in northern Napa and southern Lake Counties, California (U.S.A.). Serpentine soils are characterized by a high magnesium to calcium ratio, and are sometimes rich in heavy metals such as nickel, cobalt and chromium. Serpentine soils exclude most plants found on surrounding nonserpentine soils and harbor a high proportion of endemic species (for example, 10% of the flora endemic to California are restricted to serpentine soils). Wagner and Bortugno [31] and Fox et al. [11] have mapped serpentine areas in the study region; approximately half of the 4200-acre reserve consists of serpentine and the other half consists of sedimentary rocks and soils. D'Appolonia Company [8] provides a detailed description of the soils, vegetation and flora at the reserve. Serpentine flora within and around the reserve have been studied extensively by Koenigs et al. [19], Callizo [4], Harrison [13, 14, 15].

Plant census data were collected from a grassland plant community at a site called Little Blue Ridge. Geologically, Little Blue Ridge is unlike many other serpentine grassland regions at the McLaughlin Reserve because it is on detrital serpentine rock, which occurs in lenses of broken rock, rather than serpentine bedrock. At the site, I laid out one square 64 m^2 plot. The plot was gridded into 256 0.25 m^2 square-shaped areas, and the species in every 0.25 m^2 square was identified and recorded. This type of sampling scheme is often referred to as 'complete nested'. Data were collected in early May through late July 1998. By sampling throughout this time period, it was possible to sample all plant species while they were flowering. Plants were identified with the help of an expert regional botanist (Joseph Callizo, Napa Land Trust and California Native Plant Society).

2.2 *Constructing the Observed Species-Area Curve*

To plot the species-area curve at the site, I took the species presence/absence data from the 256 contiguous 0.25 m^2 quadrats and calculated the mean species richness (S_i) across nonoverlapping square and rectangular (2 x 1) quadrats of increasing area A_i (from A_8 = 0.25 m^2 through A_0 = 64 m^2). Because species richness on rectangular quadrats depends on the orientation of these quadrats within the plot (i.e., vertical versus horizontal alignment), S_i and the variance of species richness (σ_i^2) across the quadrats was calculated by averaging S_i and σ_i^2 for vertically and horizontally aligned nonoverlapping rectangles, respectively. The standard error of the mean species richness at each spatial scale was calculated as $\sigma_i / \sqrt{2^i}$.

This overall method of constructing a species-area curve using data from a complete nested sampling scheme yields a 'complete nested SAR', in contrast to an 'island SAR', where the sampled areas are neither nested nor contiguous.

2.3 *Simulating the Random Placement of Species*

I simulated the random placement of species at Little Blue Ridge two different ways. Imagine looking down onto the A_0 habitat patch at Little Blue Ridge and seeing the species list on each of the 256 A_8 quadrats. In the first method, I randomly shuffled the locations of the 256 A_8 quadrats, while maintaining the original species list on each quadrat. I will refer to this first method of randomization RAND1. For the second method of randomly distributing species, I maintained the species richness on each of the 256 A_8 quadrats at their original values, yet I did not keep species grouped together at the A_8 = 0.25 m^2 spatial scale as they were found naturally at the site. Instead, for every A_8 quadrat, I replaced each species on that quadrat with a randomly chosen species from the entire species pool. I will refer to this second method of randomization RAND2.

I randomly distributed species using methods RAND1 and RAND2 10^4 times. For every iteration, I calculated the mean species richness and the variance at each of the 9 spatial scales (i = 0 through i = 8), and then calculated S_i.and σ_i^2 averaged across all 10^4 simulations.

2.4 *Testing for Self-Similarity versus Randomness*

To test for a self-similar distribution of species at the site, I plotted the species-area curve ($\ln(S_i)$ versus $\ln(A_i)$) across the 9 sampled spatial scales and reported the r^2 of a linear regression through the data. Then, I compared the observed mean species richness at each spatial scale to that predicted from self-similarity (Eq. 3), assuming $\boldsymbol{a} = 1/2^z$ (Eq. 4). I estimated the SAR exponent z as the slope of a weighted least squares linear regression through the $\ln(S_i)$ versus $\ln(A_i)$ data. Due to the nested sampling scheme at the site, there was more data available to estimate mean species richness at smaller spatial scales and hence the estimate of S at smaller spatial scales was more certain. To compensate for this, I chose weights corresponding to the sample size at each data point, ranging from 256 at 0.25 m^2 to 1 at 64 m^2.

To test for a random distribution of individuals and species at the site, I compared the observed mean species richness at each spatial scale to that predicted from the Individuals Random Placement Model (Eq. 5) and the species random placement simulations RAND1 and RAND2.

3 Results

There were a total of 24 species on the 64 m^2 plot at the site ($S_0 = 24$, $A_0 = 64$ m^2). The species-area relationship is well fit by a power-law, indicating that the spatial distribution of species is self-similar at the community-level. A weighted linear regression of ln(S) versus ln(area) yields $r^2 > 0.999$, and the slope of the curve yields a SAR exponent $z = 0.215$, corresponding to $\boldsymbol{a} = 0.862$. The SAR predicted by self-similarity, assuming $\boldsymbol{a} = 0.862$, is almost indistinguishable from the observed SAR (Figure 1).

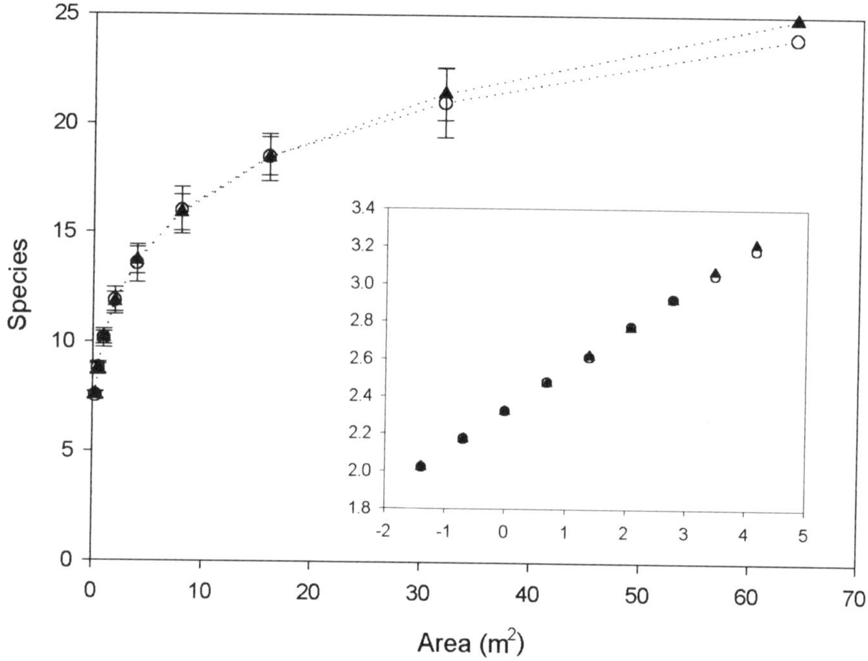

Figure 1. The SAR predicted by self-similarity is almost indistinguishable from the observed SAR at Little Blue Ridge. The open circles illustrate the measured SAR (+ 1 standard error) and the closed triangles illustrate the SAR predicted by self-similarity (± 1 standard error). The inset repeats the graph on ln-ln axes, error bars excluded for visual clarity.

The SAR predicted by the Individuals Random Placement Model significantly overestimates species richness at Little Blue Ridge (Figure 2). For the spatial scales A_8 = 0.25 m^2 through A_2 = 16 m^2, the measured species richness is well outside of the 2 standard error confidence interval given by the random model. At the spatial scale A_0 = 64 m^2, the random model and data agree, as they must.

The resulting SAR curves from randomly distributing the species according to RAND1 and RAND2 also greatly overestimated species richness at Little Blue Ridge (Figure 3). RAND2, which randomly distributed the species in each A_8 0.25 m^2 quadrat, yielded a much higher species richness at the spatial scales A_7 = 0.5 m^2 through A_2 = 32 m^2 than RAND1, which randomly distributed clusters, or groups of species at the A_8 spatial scale. In fact, for $A \geq A_4 = 4$ m^2, the RAND2 simulation resulted in a mean species richness that nearly approached the overall species richness sampled at Little Blue Ridge (24 species).

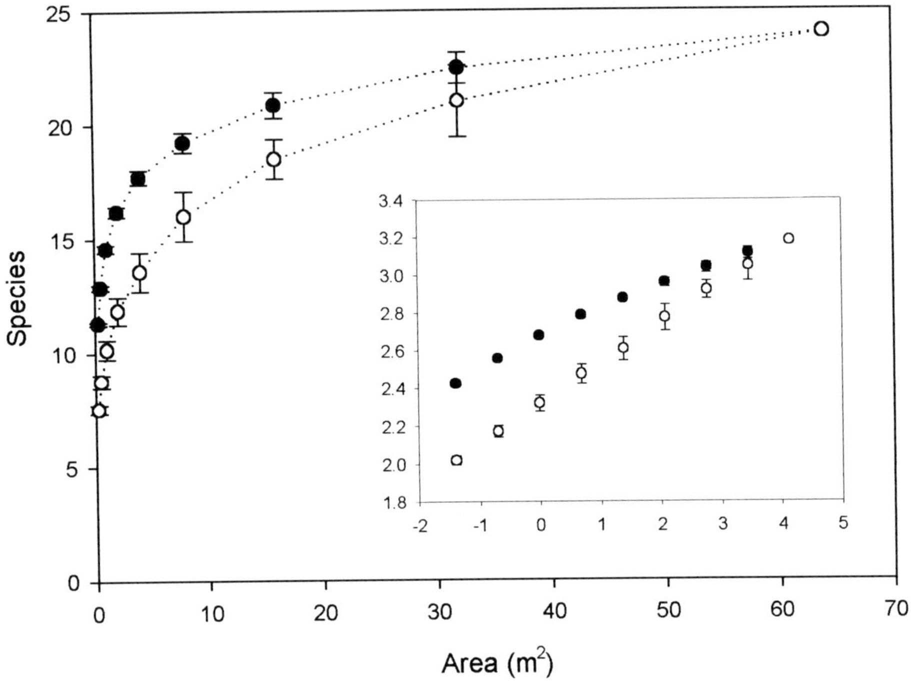

Figure 2 The Individuals Random Placement Model significantly overestimates species diversity at Little Blue Ridge. The open circles represents the measured SAR (± 1 standard error) and the closed circles represents the SAR predicted by the random model (± 1 standard error). The inset repeats the graph on ln-ln axes.

Discussion

The SAR at Little Blue Ridge is well fit by a power law across all sampled spatial scales, indicating that the spatial distribution of species is self-similar. It is therefore not surprising that the Individuals Random Placement Model poorly estimates species richness at the site. According to Coleman [5], the hypothesis of random individuals placement implies that as A_i increases from 0 to A_0, the exponent z of the power-law SAR should decrease from 1 to the fraction of the total species in A_0 consisting of one individual. In other words, the Individuals Random Placement Model produces a concave ln-ln species-area curve, except for the extreme and unrealistic case where each of the S_0 species on A_0 consist of only one individual. Thus, across the spatial scales that a community of species are distributed self-similarly (and hence are characterized by a

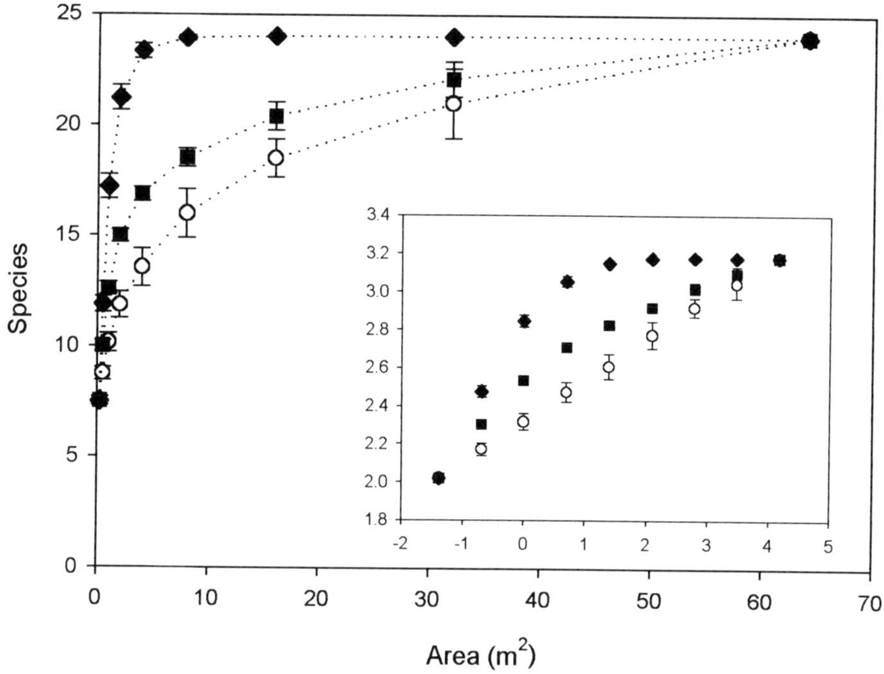

Figure 3. Randomly distributed species (RAND2) and clusters of species (RAND1) significantly overestimate species diversity at Little Blue Ridge. The open circles represent the measured SAR (± 1 standard error), the diamonds represent the SAR predicted by RAND2 (± 1 standard error), and the squares represent the SAR predicted by RAND1 (± 1 standard error). The inset repeats the graph on ln-ln axes.

power-law SAR), theory combined with the serpentine grassland data suggest that the individuals within that community cannot be distributed randomly in space.

Although the Individuals Random Placement Model is discussed extensively throughout the literature, to my knowledge this model has seldom been compared to empirical data. Both Arrhenius [3] and Coleman et al. [6] found empirical evidence supporting the Individuals Random Placement Model, for flora and breeding birds sampled on islands, respectively. However, using tropical forest data collected using a nested sampling scheme, Plotkin et al. [26, 27] found that while the tree species were not distributed self-similarly, the Individuals Random Placement Model also significantly overestimated species richness. An open question is whether or not the type of censusing sampling scheme (a complete nested design versus censusing of isolated islands) influences how well the Individuals Random Placement Model fits empirical data.

The RAND1 simulation is different from randomly placing individuals of each species across A_0. By applying the RAND1 simulation, community structure is maintained at the $A_8 = 0.25$ m^2 spatial scale. Species that naturally co-occur at this spatial scale are allowed to remain clustered. Randomly shuffling groups of species that are clustered together at the A_8 spatial scale yields a concave ln-ln species-area curve which overestimates species richness at the intermediate spatial scales (between 0.5 m^2 and 32 m^2), suggesting that at Little Blue Ridge, interspecific aggregation occurs at spatial scales larger than 0.25 m^2.

The RAND2 simulation does not account for an important aspect of community structure that is explicitly accounted for in the Individuals Random Placement Model and implicitly accounted for (to a lesser degree) in the RAND1 simulation: the relative abundance of species within the community (and hence the species-abundance distribution). Although the RAND1 simulation does not factor in the actual abundance of each species, it accounts for how abundant or rare different species are. Abundant species occur more frequently at the $A_8 = 0.25$ m^2 spatial scale, and rare species occur less frequently. By replacing each species at the A_8 spatial scale with a randomly chosen species from the overall species pool, the RAND2 simulation assumes that each of the S_0 species occurs with the same frequency across A_0. Thus, the RAND2 simulation is essentially employing the Individuals Random Placement Model, using a uniform species abundance distribution. The only additional constraint of the RAND2 simulation is the true species richness on each of the $A_8 = 0.25$ m^2 quadrats. This is why the mean species richness for the RAND2 simulation and the empirical data are identical at the A_8 spatial scale.

These results suggest that self-similarity in the spatial distribution of species is incompatible with a random distribution of individuals, species, or clusters of species within a community.

4 Acknowledgements

I would like to thank John Harte and Annette Ostling for reviewing the manuscript, Joseph Callizo and Nicole Jurjavcic for help identifying plant species at Little Blue Ridge, and the American Association of University Women and the UC Office of the President for financial support.

References

1. Arrhenius, O. 1920. *Oecologische Studien in den Stockholmer Scharen.* Disseration, University of Stockholm.
2. Arrhenius, O. 1920. Distribution of the species over area. *Meddelanden från K. Vetenskapsakademiens Nobelinstitut* 4: 1 – 6.
3. Arrhenius, O. 1921. Species and area. *Journal of Ecology* 9: 95 - 99.
4. Callizo, J. 1992. Serpentine habitats for the rare plants of Lake, Napa and Yolo Counties, California. In Baker, A. J. M., J. Proctor and R. D. Reeves, *The Vegetation of Ultramafic (Serpentine) Soils*. pp. 35-51. Intercept Ltd., Andover, Hants., England.
5. Coleman, B. 1981. Random placement and species-area relations. *Mathematical Biosciences* 54: 191-215.
6. Coleman, B., Mares, M.A., Willig, M.R. & Y. Hsieh. Randomness, area and species richness. *Ecology* 63: 1121 - 1133.

7. Connor, E.F. & E.D. McCoy. 1979. The statistics and biology of the species-area relationship. *American Naturalist* 113: 791 – 833.
8. D'Appolonia Company. 1982. McLaughlin project: proposed gold mine and mineral extraction facility, Homestake Mining Company. Environemental report. D'Appolonia Company, San Francisco, California.
9. DeCandolle, A. 1855. *Géographie botanique raisonnée; ou exposition des faits principaux et des lois concernant la distribution géograhique des plates de l'époque actuelle*. Maisson, Paris.
10. Evans, F.C., Clark, P.J. & R.H. Brand. 1955. Estimation of the number of species present on a given area. *Ecology* 36: 342 - 343.
11. Fox, K.F., J.D. Sims, J.A. Barlow, & E J. Helley. 1973. Preliminary geologic map of eastern Sonoma and western Napa Counties, California. United States Geological Survey, Denver, Colorado.
12. Gilpin, M.E. & J.M. Diamond. 1976. Calculation of immigration and extinction curves from the species-area-distance relation. *Proceedings of the National Academy of Sciences, USA* 73: 4130 - 4134.
13. Harrison, S. 1997. How natural habitat patchiness affects the distribution of diversity in Californian serpentine chaparral. *Ecology* 78: 1898-1906.
14. Harrison, S. 1999. Local and regional diversity in a patchy landscape: native, alien and endemic herbs on serpentine. *Ecology* 80: 70-80.
15. Harrison, S. 1999. Native and alien species diversity at the local and regional scales in a grazed Californian grassland. *Oecologia* 121: 99-106.
16. Harte, J. & Kinzig, A. P. 1997. On the implications of species-area relationships for endemism, spatial turnover, and food web patterns. *Oikos* 80: 417-427.
17. Harte, J., Kinzig, A., and Green, J. 1999a. Self-similarity in the Distribution and Abundance of Species. *Science* 284: 334-336.
18. He, F. & P. Legendre. 1996. On species-area relations. *American Naturalist* 148: 719-737.
19. Koenigs, R. L., W. A. Williams and M. B. Jones, 1982. Factors affecting vegetation on a serpentine soil. *Hilgardia* 50: 1-14.
20. Palmer, M.W. 1990. The estimation of species richness by extrapolation. *Ecology* 71: 1195 - 1198.
21. MacArthur, R.H. and E.O. Wilson. 1967. *The Theory of Island Biogeography*. Princeton Monographs in Population Biology. Princeton University Press, Princeton, N.J.
22. McGuiness, K.A. 1984a. Equations and explanations in the study of species-area curves. *Biological Reviews* 59: 423 – 440.
23. May, R.M. 1975. Patterns of species abundance and diversity. In M.L. Cody and J.M. Diamond, eds., *Ecology and Evolution of Communities*, 81-120. Harvard University Press, Cambridge, Mass.
24. May, R., Lawton, J. & N. Stork. 1995. *Extinction Rates* (eds. J. Lawton and R. May), pp. 1 - 24. Oxford University Press, Oxford.
25. Pimm, S. & P. Raven. 2000. Extinction by numbers. *Nature* 403: 843 - 845.
26. Plotkin et al. 2000a. Predicting species diversity in tropical forests. *Proceeding of the National Academy of Sciences* 97: 10850-10854.

27. Plotkin, J.B., Potts, M.D., Leslie, N., Manokaran, N., LaFrankie, J., Ashton, P.S. 2000b. Species-area curves, spatial aggregation, and habitat specialization in tropical forests. *Journal of Theoretical Biology* 207: 81-99.
28. Preston, F.W. 1962. The canonical distribution of commonness and rarity, Parts I and II. *Ecology* 43: 185-215 and 410-432.
29. Rosenzweig, M.L. 1995. *Species Diversity in Space and Time.* Cambridge University Press, Cambridge.
30. Simberloff, D.S. & L.G. Abele. 1982. Refuge design and island biogeographic theory: effects of fragmentation. *American Naturalist* 120: 41 - 50.
31. Wagner, D.L. and E.J. Bortugno. 1982. Geologica map of Santa Rosa Quadrangle, California, 1:250,000. State of California Resources Agency, Sacramento.
32. Whitmore, T.X. & J.A. Sayer, eds. 1992. *Tropical Deforestation and Species Extinction.* Chapman & Hall, London.
33. Williamson, M. 1988. Relationship of species number to area, distance and other variables. In: *Analytical Biogeography: An Integrated Approach to the Study of Animal and Plant Distributions* (eds. A.A. Myers & P.S. Giller), pp. 91-115. Chapman & Hall, London.

DYNAMICAL REGIMES IN THE METABOLIC CYCLE OF A HIGHER PLANT ARE CHARACTERIZED BY DIFFERENT FRACTAL DIMENSIONS

M.-TH. HÜTT, U. RASCHER, U. LÜTTGE

Institute of Botany, Schnittspahnstr. 3-5,
Darmstadt University of Technology, D-64287 Darmstadt, Germany
E-mail: huett@bio.tu-darmstadt.de, rascher@bio.tu-darmstadt.de, luettge@bio.tu-darmstadt.de

Crassulacean acid metabolism (CAM) serves as a plant model system for the investigation of circadian rhythmicity. Recently, it has been discovered that propagating waves and, as a result, synchronization and desynchronization of adjacent leaf areas, contribute to an observed temporal variation of the net CO_2 uptake of a CAM plant. The underlying biological clock has thus to be considered as a spatiotemporal product of many weakly coupled nonlinear oscillators. Here we study the structure of these spatiotemporal patterns with methods from fractal geometry. The fractal dimension of the spatial pattern is used to characterize the dynamical behavior of the plant. It is seen that the value of the fractal dimension depends significantly on the dynamical regime of the rhythm. In addition, the time variation of the fractal dimension is studied. The implications of these findings for our understanding of circadian rhythmicity are discussed.

1 Introduction

The crassulacean acid metabolism (CAM[1]) plant *Kalanchoë daigremontiana* Hamet et Perrier de la Bâthie shows an endogenous circadian rhythm of net CO_2 exchange (J_{CO_2}) under constant external conditions in continuous light.[2] Previous studies have demonstrated that above a certain threshold temperature changes from rhythmic to arrhythmic behavior of J_{CO_2} occur[3] and that this is reversible when temperature is lowered again.[4] It has been shown that these findings are well reproduced by a system of four coupled nonlinear differential equations with temperature, light intensity and external CO_2 concentration as external parameters.[5,6] Due to the combination of a highly controllable experimental set-up and a consistent theoretical representation we may regard the endogenous circadian CAM rhythm of *K. daigremontiana* as a very suitable model system for the study of the biological clock. Recently, it was established that the spatial organisation of the different leaf areas ("oscillators") is responsible for certain aspects of the functioning of this biological clock.[7] Non-invasive, highly sensitive chlorophyll fluorescence imaging revealed randomly initiated patches of varying photosynthetic efficiency (ϕ_{PSII}) which are propagated within minutes to hours in wave fronts, forming dynamically expanding and contracting clusters and clearly dephased regions of ϕ_{PSII}. This biological clock has to be seen as a spatiotemporal product of many weakly coupled individual oscillators. These oscillators are defined by the metabolic constraints of CAM but operate independently in space and time as a consequence of the dynamics of metabolic pools and limitations of CO_2 diffusion between tightly packed cells.[7] In the present paper we discuss the properties of these spatiotemporal patterns in the light of fractal geometry.

2 Material and methods

2.1 Plants and gas exchange measurements

Plants of *K. daigremontiana* were raised from adventitious plantlets obtained from leaves of the plant collection of the Botanical Garden, Darmstadt University of Technology. They were grown in soil culture in the glasshouse until they had produced six to seven pairs of fully developed leaves, and were about 0.4 - 0.5 m tall. Adaptation of plants and measurements of net CO_2 exchange were performed in a climate-regulated chamber of the phytotron in the Department of Biology in Darmstadt using the minicuvette system of H. Walz (Effeltrich, Germany) as previously described.[3,4] A mature leaf of a plant was enclosed in the gas exchange cuvette while remaining attached to the plant. Gas exchange data were recorded every 5 minutes. The relative humidity of the air inside the cuvette was set at 60 $\pm$ 5 per cent and photosynthetically active irradiance (photon flux density PFD) was set between 100 and 200 μmol m^{-2} s^{-1} depending on the experimental run. Net CO_2 exchange rate was calculated according to a standard procedure.[10]

2.2 Chlorophyll fluorescence imaging

While the leaf remained inside the climate-regulated cuvette, fluorescence of chloropohyll a was measured, using a peltier-cooled digital camera (AP1/14, Apogee Instr., Tucson, Arizona) with computer controlled exposure. The efficiency of PS II (ϕ_{PSII}) was imaged by non-invasive chlorophyll fluorescence measurements at 20 min intervals following the saturating flash method similar to that described in the literature.[8,9] Values were normalized to the maximum obtained during the experiment.

3 Theoretical background

3.1 Fractal dimension of a biological pattern

The remarkable property of fractal geometry is that, seemingly, fractals are capable of representing natural objects in a much more efficient way than classical (Euclidian) geometry. In dynamical (spatiotemporal) patterns fractality can indicate that the system is close to a critical point characterizing a phase transition.[11,12] Apart from the possibility of demonstrating proximity to a phase transition, two more practical aspects come to mind, when discussing the fractal dimension of a biological system:

1. The fractal dimension can be thought of as an ordination number, i.e. as the number of effective nearest neighbors. This is due to the fact that the number N of copies in an iterated function system (IFS) coding for a particular fractal directly enters the fractal dimension via $D_F = \ln N / \ln \varepsilon$, where ε is the inverse scaling factor in the contraction maps of the IFS. This relation is valid only, when the same scaling factor appears throughout the IFS, but similar expressions exist for the more general cases.

2. A substantial change in the fractal dimension, when the external conditions of the system are changed, is evidence for a major qualitative change in the system's pattern of self-organization. The value of the fractal dimension then may serve as a characterization of the state of the system.

On a more general note, the distinction between a classical and a fractal object is important, when one applies statistical methods, as for a fractal object certain statistical properties do not exist (e.g. the mean diverges or goes to zero with increasing number of samples).

3.2 Tests of the box counting algorithm

A structure is called "fractal", when the dimension D_F differs significantly from an integer. One method for determining D_F from experimental data is the box-counting algorithm, which quantifies the dependence of a volume on the scale used to discretize the carrier space of the potential fractal. The use of such algorithms demands high-quality data, as a quantification over several orders of magnitude (in space or time) is necessary. We will now briefly describe the box-counting algorithm and then test our implementation on several examples. Consider an object O (e.g. part of a time series or some spatial structure projected on the state space $\{0, 1\}$), which is embedded in (i.e. fills out partially) a d-dimensional space. This space now is tiled with (d-dimensional) cubes of length r. One then can count, how many such cubes at fixed r contain part of O. This number $N = N(r)$ as a function of r gives access to the fractal dimension. One has

$$N\left(r\right) \propto r^{-D_F} \quad . \tag{1}$$

Fig. 1 shows two spatial structures, together with the results obtained from applying the box-counting algorithm. It is seen that in all three cases (i.e. for the fractal structure, for the boundary of the black area in the classical structure and for the area itself) one obtains to good approximation linear relations between $\log N$ and $\log r$ with only small deviation from the linear fit (Fig. 1 (c)-(e)).

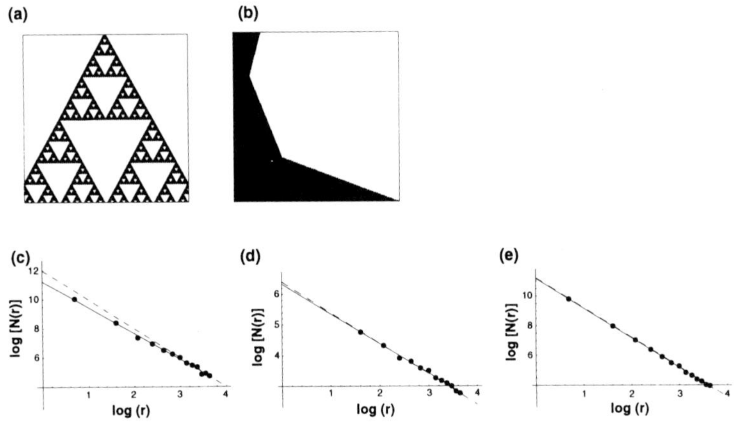

Figure 1 *Two examples of simple spatial structures, which can be used to check the box-counting algorithm. Fig. (a) shows a (cellular automata) approximation of a Sierpinski triangle, while Fig. (b) is a pattern built from elements of classical geometry. The results of the box-counting procedure are shown in Figs. (c)-(e), namely for the Sierpinski triangle (c), for the boundary between black and white image points (d) in the classical structure and for the black area (e). The slopes of the corresponding linear fits are: (a) -1.78, (b) -0.97 and (c) -2.01. In all cases a line with the nearest integer slope is also given for sake of comparison (dashed lines).*

The important difference between the two structures is seen in the extracted fractal dimension, i.e. the negative slope of the linear fit. For the fractal structure (Fig. 1(a)) one has with $D_F = 1.78$ a significant difference to an integer dimension. Fig. 1(a) also shows that the data quality is sufficient to rule out the nearest integer dimension, i.e. $D_F = 2$ (dashed line). The deviation from the theoretical value of the fractal dimension for the Sierpinski triangle, namely $D_F = 1.59$, is a consequence of the approximation with a finite length-scale. For the structure from Fig. 1(b) one has two ways of defining a dimension: one can analyze the boundary (the criterion for contribution to $N(r)$ is that both, black *and* white image points lie in the box of size r) or one considers the (black) area (then all boxes are counted, which contain not solely white points). In both cases the box-counting algorithm yields an integer dimension within statistical accuracy, namely $D_F = 0.97$ for the boundary and $D_F = 2.01$ for the area (Fig. 1 (d) and (e)).

A key feature of the two structures shown in Fig. 1 is that only values 0 and 1 appear, graphically represented as black and white. One method of applying this algorithm to structures with a larger state space Σ is to project Σ onto $\{0,1\}$. In a state space with a distance, e.g. $\Sigma = \{1,...,N\}$ this projection can be achieved by introducing a threshold:

$$\begin{aligned} \Sigma = \quad \{1,...,N\} &\to \{0,1\} \\ k \in \Sigma \quad &\mapsto \begin{cases} 1 & k > S \\ 0 & k \leq S \end{cases} . \end{aligned}$$

In order to minimize dependence on the threshold parameter S one can for example use the average value of the states over the whole spatial structure as threshold.

Note, however, that many different ways of applying Eq. (1) to an image are possible, see e.g. [13]. Fig. 2 shows a snapshot from a two-dimensional diffusion process, which was initialized with a random concentration distribution. The time development has been simulated with a cellular automata implementation of the diffusion equation (cf. e.g. [14] for details). The average concentration over the whole structure has been used as a threshold for projection onto a binary state space. To the eye, the resulting structure, Fig. 2(b), might well be a fractal. However, the analysis, Fig. 2(c), gives a dimension $D_F = 2.02$, which corresponds to the non-fractal character of the pattern.

4 Results

On the basis of these (and other[15]) tests, we can now use our implementation of the box-counting algorithm to analyze the spatial and spatiotemporal patterns observed in CAM. Details on the physiological interpretation of photosynthetic efficiency have been described previously.[7] Large values denote a higher photosynthetic efficiency. The graphical treatment of such a leaf image is shown in Fig. 3. For the analysis a region is chosen, which is not dominated by anatomical features (such as major veins) of the leaf (Fig. 3(b)).

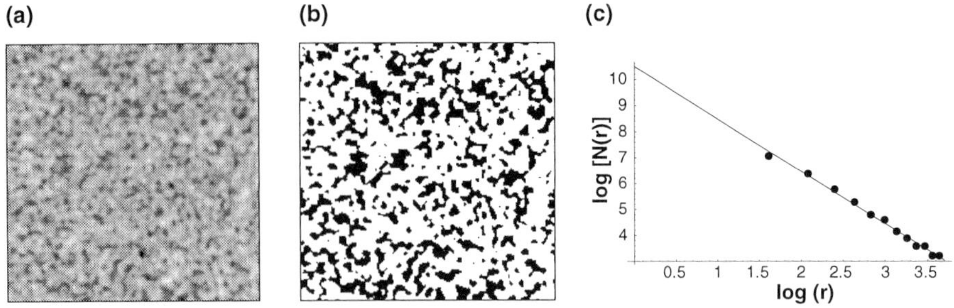

Figure 2 *Analysis of a diffusion pattern with the box-counting algorithm. The structure (a) generated by a discretized diffusion equation is converted into a binary pattern (b) by an average-value threshold (cf. the discussion in the text), for which the dimension can be computed. One obtains a slope of the linear fit of* -2.02 *in the usual representation (c) of pairs* $(\log r, \log N)$.

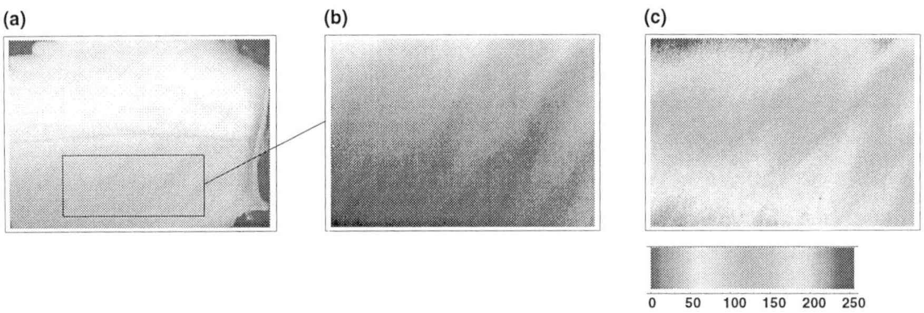

Figure 3 *Graphical treatment of an image (a) displaying the distribution of photosynthetic efficiency* ϕ_{PSII} *over a whole leaf of a CAM plant. A region in the lower half of the leaf is chosen (b) for further analysis and then displayed as a color image (c) by applying a given look-up table.*

In order to enhance visibility of the patterns a color code is then applied to this region (Fig. 3(c)). Typical snapshots for the rhythmic and the arrhythmic case are shown in Fig. 4. It is seen that in the rhythmic case large-scale structures (domains) of approximately equal activity and propagating wave fronts appear, while the arrhythmic behavior is characterized by smaller homogeneous areas and a much smaller range in chlorophyll fluorescence activity.

Our hypothesis is that the borderlines separating one greyscale value from the next could be fractal. For one examplary image of the distribution of photosynthetic efficiency details of the box-counting are shown in Figs. 5 and 6. By introducing a threshold, as described in the previous section, we can now visualize such a borderline (Fig. 5). Application of the box-counting algorithm to this structure is shown in Figs. 5 and 6. An important issue

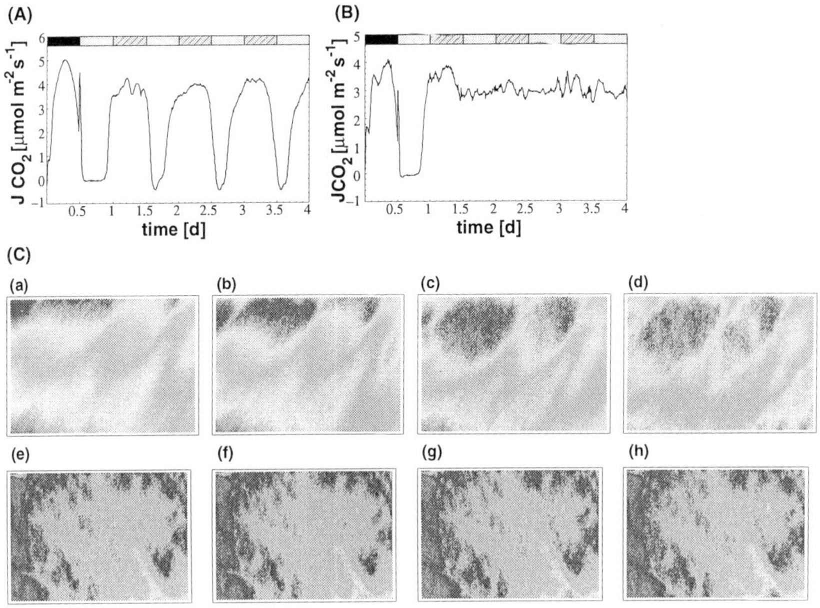

Figure 4 *Examples of the two main dynamical regimes of CAM, i.e. the net CO_2 exchange rates J_{CO_2} for the rhythmic (A) and the arrhythmic (B) case. The black bars in each diagram indicate dark periods, the grey bars continuous illumination where the hatched parts correspond to the subjective dark periods prior ro application of continuous light. In both cases the CAM plant has been exposed to continuous light after a light-dark adaptation cycle. Further details on these expriments and their biological interpretation can be found elsewhere.[3] For these two regimes the corresponding spatial patterns of photosynthetic efficiency can be studied (C). Here we show typical snapshots of the ϕ_{PSII} distribution measured during rhythmic ((a)-(d)) and arrhythmic ((e)-(h)) CAM metabolism in continuous light. The color coding is the same as in Fig. 3. The time difference between neighboring snapshots in each row is 40 minutes.*

is the statistical validity of the pairs (r, N) obtained with this procedure, i.e. the ratio of N and the number of all boxes at a given size. Fig. 5 (c) - (e) shows the binning of image points to boxes of size r for three different values of r. The contribution of these three examples to the fractal dimension is given in Fig. 6, where the usual $(\log r, \log N)$-plot is shown. For the particular structure shown in Fig. 5(b) one obtains a dimension $D_F = 1.46$, which confirms our hypothesis of a fractal borderline between different greyscale values.

The next step is to study the time dependence of this fractal dimension for the two different dynamical regimes of CAM, namely the rhythmic and the arrhythmic behavior. This is shown in Fig. 7. While the function $D_F(t)$ for the rhythmic case displays clearly visible variations, which are related to the circadian oscillation of the plant's metabolism, the corresponding curve

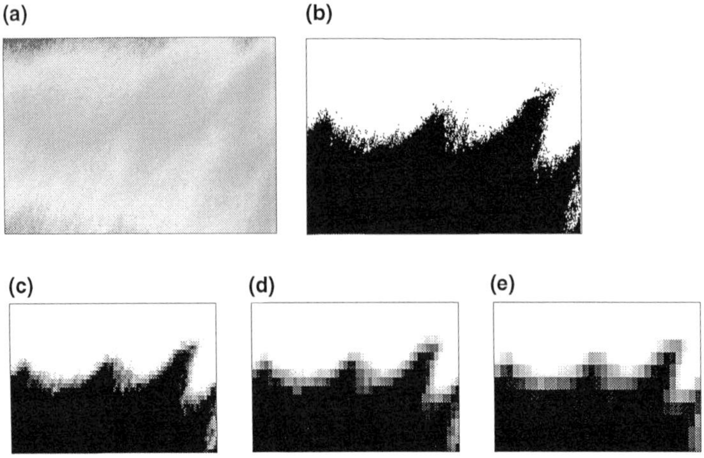

Figure 5 *Visualization of a ϕ_{PSII} borderline. An examplary image (a) of the distribution of photosynthetic efficiency is projected onto a binary state space with an average-value threshold (b). Analysis of the borderline between two greyscale values thus obtained with the box-counting algorithm. For three of the pairs* $(\log r, \log N)$ *the box structure, which is obtained by local averaging ("binning") of the image points during box counting, is shown explicitly ((c) - (e)). Each box, which is neither black nor white is counted as part of the borderline at that particular length scale r and thus contributes to the value of N.*

for the arrhythmic case remains almost constant. Furthermore, the average fractal dimension is significantly higher for the arrhythmic behavior. From this difference one can conclude that the effective spatial organization of the leaf areas (expressed for example in the average number of *relevant* nearest neighbors) changes, when the plant is taken from one dynamical regime to another. The values of D_F in Fig. 7 can further be interpreted by comparing them with typical fractal dimensions found in theoretical model systems. One should keep in mind, however, that we discuss the time course of a spatial pattern with D_F characterizing the spatial structure. Nevertheless, the dependence of the fractal dimension on statistical properties of a random walk can provide some further information. In the case of a correlated random walk, i.e. when a value tends to be close to the preceeding one, a typical value of D_F is 1.1, while an uncorrelated random walk leads to D_F=1.5 and anticorrelation increases D_F further.[13] Applied to Fig. 7 the values of D_F for the arrhythmic case imply a lesser degree of spatial correlation than in the rhythmic case, possibly due to the presence of a significant stochastic component in the system's dynamics leading to a spatial desynchronization. This interpretation is consistent with the understanding of the arrhythmic time series put forward by dynamical systems theory.[17]

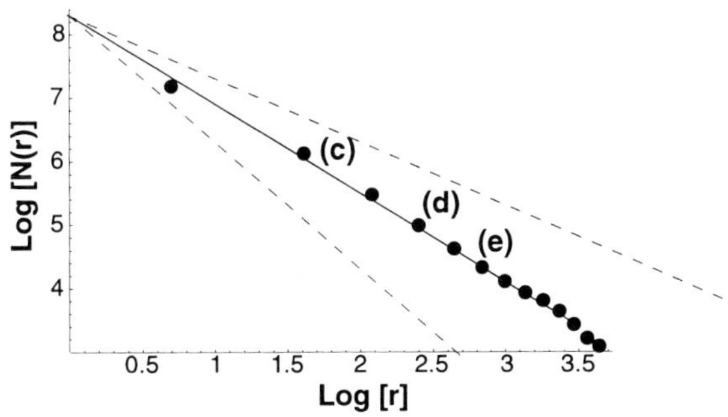

Figure 6 *The pairs* ($\log r, \log N$) *for the exemplary image shown in Figs. 5. The labels (c) - (e) refer to the box distributions from Fig. 5. Linear regression leads to a fractal dimension* $D_F = 1.46$.

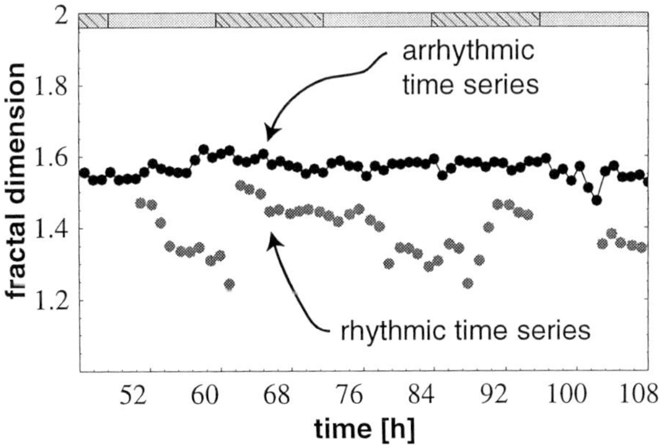

Figure 7 *Time dependence of the fractal dimension* D_F *for the rhythmic and the arrhythmic behavior of the CAM plant* K. daigremontiana. *The experimental data used for the dimension analysis have been published elsewhere.*[7,16].

5 Discussion

As an addition to the recent results on spatiotemporal dynamics in the circadian rhythm of CAM we could establish that the spatial patterns observed are fractal. The value of the corresponding fractal dimension D_F depends on the dynamical regime, with the arrhythmic behavior being characterized by a higher D_F than the rhythmic case.

By varying the threshold we checked that this is an effect arising solely from the dynamics of the plant's metabolism, rather than by the anatomical structure of the leaf. Indeed, when one tunes the threshold S to the grey value of the leaf's vascular system, the dimension drops substantially to a value close to unity (data not shown), identifying the given anatomical structures as non-fractal elements in the fractal metabolic dynamics of the leaf tissue.

For the purely temporal part of CAM dynamics a mathematical model exists, which yields numerical predictions well in agreement with the observed experimental CO_2-exchange patterns.[5,6] A next step would be to couple these CAM oscillators of the model as a spatial array and study their synchronization behavior for different physiological realizations of the coupling. Comparing theory and experiment on the level of fractal dimensions for the different dynamical regimes might be an important step to reveal the physiological nature of the coupling.

References

1. Winter K. and Smith J.A.C., eds, (1996) Crassulacean acid metabolism. Biochemistry, ecophysiology and evolution. Ecological Studies Vol. 114 eds. Springer Verlag , Heidelberg
2. Lüttge, U. and Ball, E. (1978) Free running oscillations of transpiration and CO2 exchange in CAM plants without a concomitant rhythm of malate levels. Z. Pflanzenphysiol. 90, 69-77.
3. Lüttge, U. and Beck, F. (1992) Endogenous rhythms and chaos in crassulacean acid metabolism. Planta 188, 28-38.
4. Grams, T.E.E., Beck, F. and Lüttge, U. (1996) Generation of rhythmic and arrhythmic behavior of crassulacean acid metabolism in Kalancho daigremontiana under continuous light by varying the irradiance or temperature: Measurements in vivo and model simulations. Planta 198, 110-117.
5. Blasius, B., Beck, F. and Lüttge, U. (1997) A model for photosynthetic oscillations in crassulacean acid metabolism (CAM). J. Theor. Biol. 184, 345-351.

6. Blasius, B., Neff, R., Beck, F. and Lüttge, U. (1999) Oscillatory model of crassulacean acid metabolism with a dynamic hysteresis switch. Proc. R. Soc. Lond. B 266, 93-101.
7. Rascher, U., Hütt, M.-T., Siebke, K., Osmond, B., Beck, F. and Lüttge, U. (2001), Spatio-temporal variation of metabolism in a plant circadian rhythm: the biological clock as an assembly of coupled individual oscillators, Proc. Natl. Acad. Sci. (USA) 98, 11801-11805.
8. Siebke, K. and Weis, E. (1995) Assimilation images of leaves of Glechoma hederacea: Analysis of non-synchronous stomata related oscillations. Planta 196, 155-165.
9. Genty, B., Briantais, J.-M. and Baker, N.R. (1989) The relationship between the quantum yield of photosynthetic electron transport and quenching of chlorophyll fluorescence. Biochim. Biophys. Acta 990, 87-92.
10. Farquhar G.D. and Sharkey T.D. (1982) Stomatal conductance and photosynthesis. Annu. Rev. Plant Physiol. 33, 317-345.
11. Bak, P., Tang, C. and Wiesenfeld, K. (1988), Self-organized criticality, Phys. Rev. **A 38**, 364.
12. Solé, R.V., Manrubia, S.C., Luque, B., Delgado, J. and Bascompte, J. (1996), Phase Transitions and Complex Systems, Complexity **2**, 13.
13. Bassingthwaighte, J.B., Liebovitch, L.S. and West, B.J. (1994), Fractal physiology, Oxford Univ. Press.
14. Hütt, M.-Th. and Neff, R. (2001), Quantification of spatiotemporal phenomena by means of cellular automata techniques, Physica **A 289**, 498.
15. Hütt, M.-Th. (2001), Datenanalyse in der Biologie, Springer Verlag, Berlin.
16. Rascher, U. (2001), PhD thesis, TU Darmstadt.
17. Beck, F., Blasius, B., Lüttge, U., Neff, R. and Rascher, U. (2001), Stochastic noise interferes coherently with biological clocks and produces specific time structures, Proc. Roy. Soc. Lond. B, 268, 1307-1313.

APPLICATION OF THE JOINT MULTIFRACTAL THEORY TO STUDY RELATIONSHIPS BETWEEN CROP GRAIN YIELDS, SOIL ELECTRICAL CONDUCTIVITY AND TOPOGRAPHY

A. N. KRAVCHENKO

Department of Crop and Soil Sciences, Michigan State University, East Lansing, MI 48824-1325, USA

E-mail: kravche1@msu.edu

Estimation and quantification of yield spatial variability and evaluation of spatial aspects of yield affecting factors are important issues in precision agriculture. In this study, joint multifractal theory was applied to analyze variability of crop grain yields and relationships between the yields, terrain elevation, and soil electrical conductivity (EC). Corn and soybean yield data from 1996 to 1999 were collected from a 20 ha agricultural field in Illinois, USA, along with elevation and soil EC measurements. Joint multifractal theory allowed successful delineation of the ranges of elevation and EC values that were of particular influence on crop yields. It was found to be an efficient tool for analysis of the yield spatial variability and is recommended for studying the relationships between scaling properties of two and more variables.

1 Introduction

Efficient agricultural management on a site-specific basis requires a thorough quantitative knowledge of (i) spatial and temporal variability of crop yields within an agricultural field and (ii) factors and interactions influencing crop yields. The main factors influencing crop yields are field topography and soil properties. Topography is regarded as one of the most important yield affecting factors and with development of digital elevation models obtaining topographical information became even easier than before. Soil electrical conductivity (EC) also can be easily measured in the field using recently developed fast and non-destructive methods[1]. Hence, dense topographical and soil EC data can be used for explaining and predicting field variability of crop yields.

Complexity of crop yield variability often could not be fully characterized by traditional statistical approaches. A multifractal analysis[2] can provide detailed additional information about data spatial variability[3] and has been utilized for analysis of various spatially distributed natural phenomena[4]. An extension of multifractal theory for analyzing more than one variable called joint multifractal theory was developed by Meneveau et al.[5] Joint multifractal theory can be used for conducting simultaneous analyses of several multifractal measures existing on the same geometric support and for studying relationships between the measures. The objective of this research was to apply joint multifractal theory to analyze relationships between crop yield, topography and soil electrical conductivity.

2 Theory

2.1 *Multifractal Spectrum*

Let us consider a distribution of a certain variable on a geometric support of the studied field. Let us normalize the distribution of the variable by introducing a new variable, $\mu_i(\epsilon)$, that describes the portion of the total mass of the studied variable contained in each map cell i of the size ϵ:

$$\mu_i(\varepsilon) = \mu_i / \mu_{sum} \tag{1}$$

where μ_i is the data value from the map cell, μ_{sum} is a sum of all μ_i values from the studied field, and the cell size ϵ is calculated as a ratio of the actual cell size to the total size of the field. The variable $\mu_i(\epsilon)$ changes with the cell size and the manner of the change is defined by the data spatial variability.

For multifractal measures, $\mu_i(\epsilon)$ scales with the cell size as

$$\mu_i(\varepsilon) \propto \varepsilon^{\alpha} \tag{2}$$

where α is called a Hölder exponent or a singularity strength. The number of cells of size ϵ with α values falling within α to $\alpha + d\alpha$ interval, $N_\alpha(\epsilon)$, scales with the cell size as

$$N_\alpha(\varepsilon) \propto \varepsilon^{-f(\alpha)} \tag{3}$$

where the exponent $f(\alpha)$ characterizes abundance of the cells with a certain α. Parameters α and $f(\alpha)$ characterize spatial variability of the variable by describing its local scaling properties (α) and numbers of locations where certain scaling properties are observed ($f(\alpha)$). A plot of $f(\alpha)$ versus α is called a multifractal spectrum. In this study, the maximum value of $f(\alpha)$ was equal to 2 (box-counting dimension of a plane).

In general terms, the meaning of the α and $f(\alpha)$ values for characterizing data distributions can be inferred from Eqs. (2) and (3) as follows: larger α values correspond to the locations where $\mu_i(\epsilon)$ values are small at small initial ϵ values but increase rapidly with increasing cell size. Smaller α values correspond to the locations where $\mu_i(\epsilon)$ values are high at small initial ϵ values and their increase with increasing cell size is relatively slow. If cells with very high (very low) α values are spread relatively homogeneously through the studied field at the initial cell size, then, at large cell sizes the number of cells with this α value will decrease rapidly due to averaging of the extreme data values resulting in high $f(\alpha)$ values. If the cells with extreme data values are concentrated in certain locations, then, at large cell sizes, the number of the cells with extreme data values will still be relatively high, hence, lower $f(\alpha)$ values. The larger are the deviations of the $f(\alpha)$ at high or low q from the maximum $f(\alpha)$ value of 2, the more pronounced is the multifractality of the data, i.e. the fractal dimensions of the data with high/low values are substantially different from the dimension of the whole data set, while if $f(\alpha)$ at high or low q are similar to the maximum $f(\alpha)$ value, then, the multifractal nature of the data is less pronounced.

In this study, we analyze $\mu_i(\epsilon)$ distributions at five actual cell sizes ranging from 6.6 m to 105.6 m. For the cells of the smallest size (actual cell size of 6.6 m), μ_i, are obtained directly from the map. For the following cell sizes, variable values for each cell are calculated as the sum of μ_i values of the cells included in the cell of that size[6].

A method of moments was used to compute multifractal spectra of the data in this study[7]. The method of moments estimates α and $f(\alpha)$ as

$$\alpha(q) = -d\tau(q)/dq \tag{4}$$

$$f(\alpha(q)) = q\alpha(q) + \tau(q) \tag{5}$$

where q is a real number ranging from $-\infty$ to ∞, and $\tau(q)$ is the mass exponent[7]. Parameters $\alpha(q)$ and $f(\alpha(q))$ at low q are mainly influenced by cells with low variable values, and at high q they are defined primarily by the properties and abundance of cells with high variable values. In this study we considered q values ranging from -3 to 3 in 0.2 increments.

2.2 *Joint Multifractal Spectrum*

Joint multifractal spectra were obtained based on the procedure proposed by Meneveau et al.[5] Probability mass functions for two measures coexisting on the same geometrical support (e.g., an agricultural field) are defined as

$$\mu_i^1(\varepsilon) = \mu_i^1 / \mu_{sum}^1 \tag{6}$$

$$\mu_i^2(\varepsilon) = \mu_i^2 / \mu_{sum}^2 \tag{7}$$

where superscripts 1 and 2 refer to the first and the second studied variables, respectively. Since both studied variables are hypothesized to be mutifractal measures, their probability mass functions scale with cell size as

$$\mu^1{}_i(\varepsilon) \propto \varepsilon^{\alpha^1} \tag{8}$$

$$\mu^2{}_i(\varepsilon) \propto \varepsilon^{\alpha^2} \tag{9}$$

where α^1 and α^2 are corresponding singularity strengths of the two measures. The number of cells with a singularity strength for the first variable within a range from α^1 to $\alpha^1 + d\alpha^1$ and with a singularity strength for the second variable ranging from α^2 to $\alpha^2 + d\alpha^2$, $N_\varepsilon(\alpha^1, \alpha^2)$, scales with the cell size as

$$N_\varepsilon(\alpha^1, \alpha^2) \propto \varepsilon^{-f(\alpha^1, \alpha^2)} \tag{10}$$

where the parameter $f(\alpha^1, \alpha^2)$ is a characteristic describing the abundance of cells with common α^1 and α^2 values.

Extension of the method of moments to more than one variable calculates a joint partition function from the probability mass functions of the two variables weighted by the real numbers q^1 and q^2:

$$\chi_{q^1q^2}(\varepsilon) = \sum_{i=1}^{n} [\mu_i^1(\varepsilon)]^{q^1} [\mu_i^2(\varepsilon)]^{q^2} \tag{11}$$

At q^1 (or q^2) equal to 0, the joint partition function in Eq. (11) becomes equal to the partition function of a single variable, and the joint spectrum reduces to a single multifractal spectrum. As for a single multifractal spectrum, at low q^1 (or q^2) joint partition function values are defined mostly by low values of the first (or the second) variable, while at high q^1 (or q^2) the partition function depends mostly on high data values. The maximum $f(\alpha^1,\alpha^2)$ value of the joint multifractal spectrum is equal to the box-counting dimension of the geometrical support and is reached when both q^1 and q^2 are equal to zero. The mass exponent of order q^1 / q^2, $\tau(q^1,q^2)$, is obtained by analyzing the scaling properties of the joint partition function:

$$\chi_{q^1q^2}(\varepsilon) \propto \varepsilon^{\tau(q^1,q^2)} \tag{12}$$

The $\tau(q^1,q^2)$ values were obtained as slopes of log-transformed joint probability mass functions plotted versus the log-transformed cell sizes. Double Legendre transformations of the $\tau(q^1,q^2)$ curve result in expressions for calculating singularity strengths for the two variables and the joint $f(\alpha^1,\alpha^2)$ value as

$$\alpha^1(q^1,q^2) = -d\tau(q^1,q^2)/dq^1 \tag{13}$$

$$\alpha^2(q^1,q^2) = -d\tau(q^1,q^2)/dq^2 \tag{14}$$

$$f(\alpha^1,\alpha^2) = q^1\alpha^1 + q^2\alpha^2 - \tau(q^1,q^2) \tag{16}$$

3 Materials and Methods

The studied area was a square 20 ha central portion of a 259 ha agricultural field located in Central Illinois, USA. Corn and soybean grain yield data were collected with yield monitors (Ag Leader Technology, Ames, IA) during the years 1996-1999. The yield measurements were taken every second by grain sensors mounted on a combine with each site measurement covering an area of about 2x5 m. Simultaneously, the site's coordinates were determined by a GPS unit. Inverse distance interpolation method provided by ArcView Spatial Analyst (Environmental Systems Research Institute, Redlands, CA) was used to convert grain yield point data into cell-based maps. Interpolated yield maps for the four studied years are shown in Fig. 1. Survey grid GPS (Leica 500 RTK) was used to take about 1,500 elevation measurements on a semi-regular grid with a mean distance between measurements of approximately 10 m.

Soil electrical conductivity data were collected using a Veris 3100 sensor cart (Division of Geoprobe® Systems, Salina, Kansas). Geo-referenced electrical conductivity measurements were taken every 3 to 5 m with the distance between cart passes about 10 m, resulting in about 6900 EC measurement points for the studied area. The EC measurements corresponded to soil depth of approximately 0-90 cm.

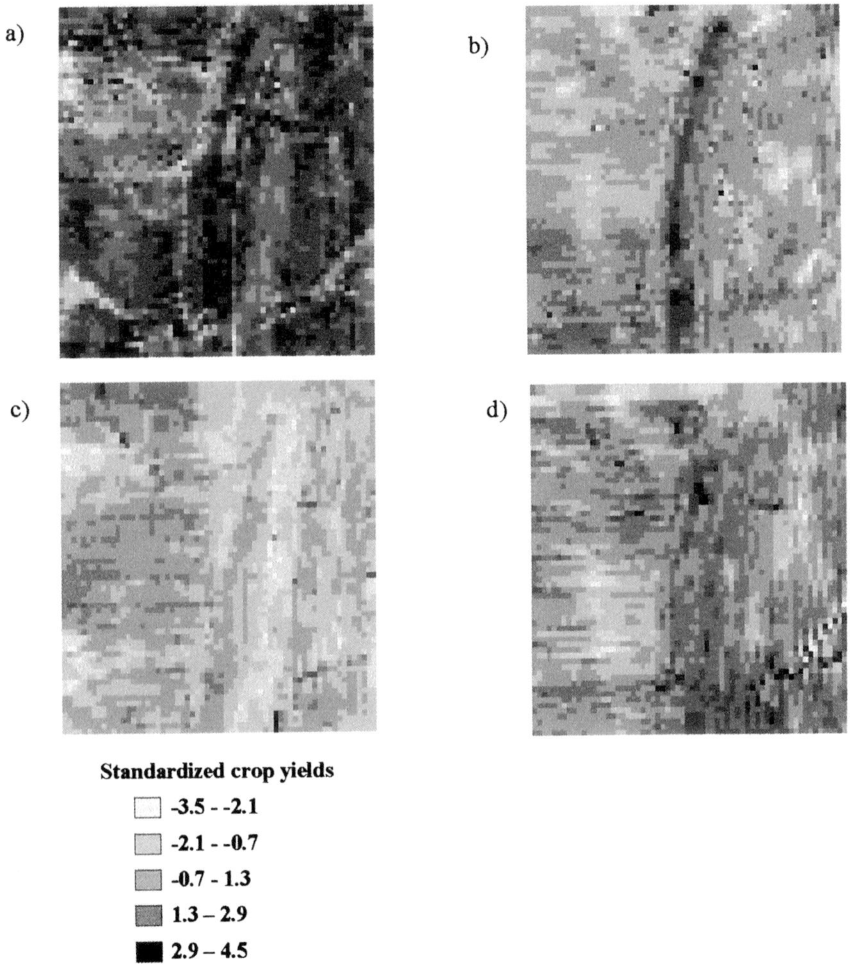

Figure 1. Interpolated maps of the standardized crop yields for a) soybean of 1996, b) corn of 1997, c) soybean of 1998, and d) corn of 1999.

4 Results and Discussion

Analysis of joint multifractal spectra of crop yields at different soil EC values and different elevations depicts the factors affecting yield variability and the relationship between yield variability and scale. The multifractal spectra of yields at q_{EC} of -3 describe scaling properties of the yields located at sites with low soil EC values, and yield spectra at q_{EC} of 3 describe yields located at sites with high soil EC (Fig. 2). The effect of EC on yield multifractal spectra differs for corn and soybean yields. Multifractal spectra of corn yields at high and low EC were relatively similar in terms of shape and range of α and $f(\alpha)$ values and the similarities were consistent in both 1997 and 1999 (Figs. 2b and 2d). Hence, soil EC did not affect the distribution and scaling properties of corn yield values in the studied years. Scaling properties of the soybean yields, however, were substantially different at locations with low and high soil EC. At locations with high EC (q_{EC}=3) the multifractal spectra of the soybean yields had much longer right sides with lower $f(\alpha)$ values and higher α values, than at locations with low EC (q_{EC}=-3).

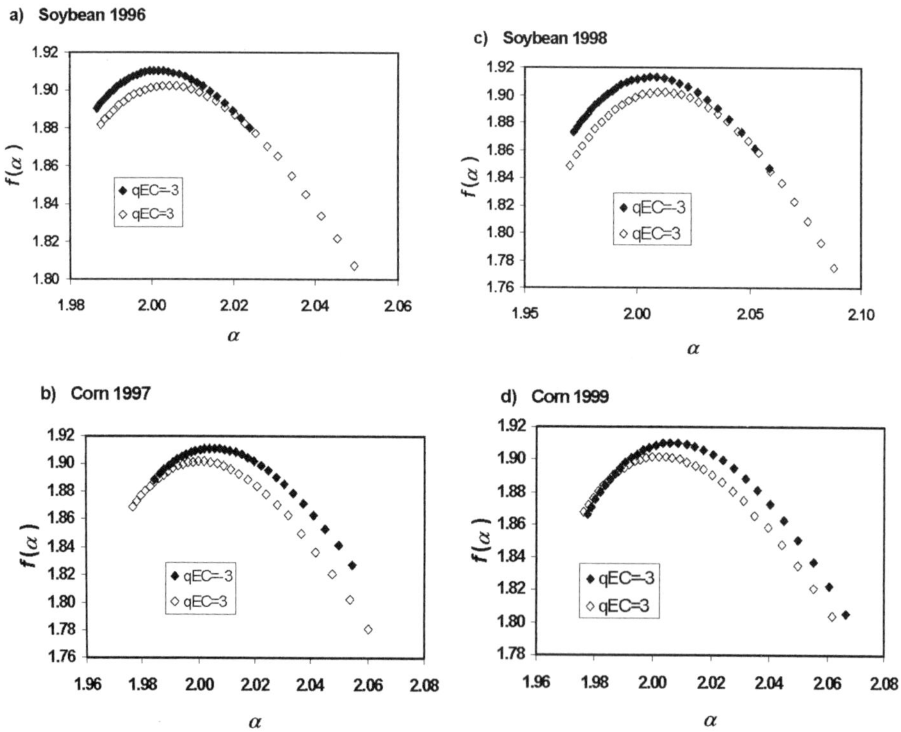

Figure 2. Multifractal spectra of crop grain yields at low (q_{EC}=-3) and high (q_{EC}=3) soil electrical conductivity.

Similar relationships between soybean multifractal spectra at high and low q_{EC} values were observed in both 1996 and 1998 (Figs 2a and 2c). These observations indicate that a portion of the low soybean yield values was located at sites with high soil EC and that sites with higher EC were of somewhat lesser benefit for soybean growth.

Multifractal spectra of yields at different elevations revealed similar trends in yield/elevation relationships for both corn and soybean in all four studied years (the graphs are not shown). Multifractal spectra of the yields at low elevations (q_{Elev}=-3) were asymmetrical with left side dominated by low $f(\alpha)$ values. Low $f(\alpha)$ values of yields at low elevations indicate that the large clusters of high yields were present at low elevation locations.

When the yield multifractal spectra were analyzed at varying elevation and soil EC, the relationships between yield and soil EC data distributions differed at low (q_{Elev}=-3) and high (q_{Elev}=3) elevations (Fig. 3). At low elevations, the multifractal spectra of yields were of a similar shape for both low (q_{EC}=-3) and high (q_{EC}=3) EC, and the yield $f(\alpha)$ values for q_{EC}=-3 were higher than those for q_{EC}=0 and q_{EC}=3. Similar results were obtained for yields from all four studied years (graphs are not shown). For high elevation (q_{Elev}=3), different trends in shapes of multifractal spectra were observed for soybean and corn yields. For corn yields in 1997 and 1999, yield multifractal spectra had similar shape and similar $\alpha/f(\alpha)$ values at both low and high EC (Figs. 3b and 3d). For soybean yields in 1996 and 1998 at high EC (q_{EC} =3) the yield spectra had asymmetrically long right part, and relatively high α values, which is an indication that clustered low yields were present at sites with high elevation and high EC (Figs. 3a and 3c).

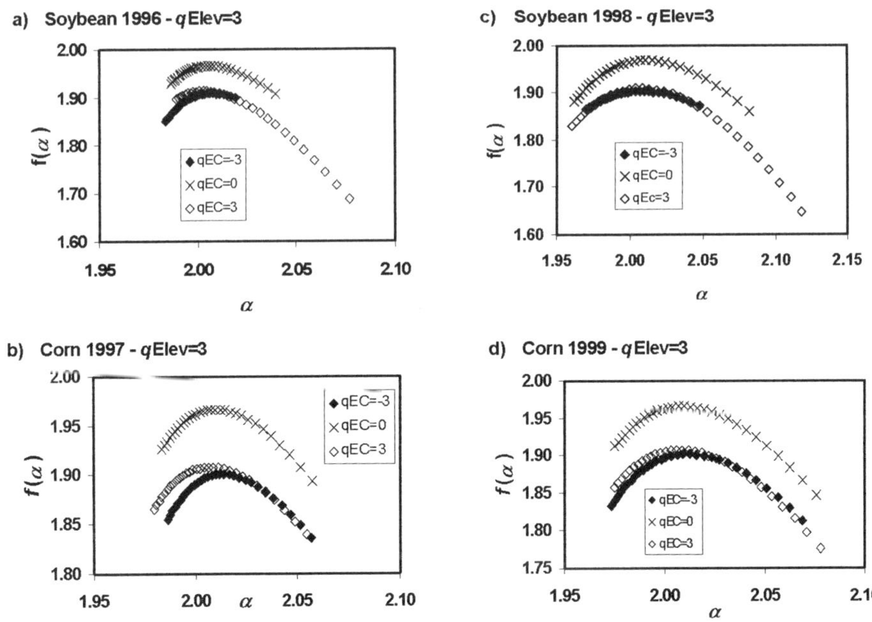

Figure 3. Multifractal spectra of crop yields at high elevation (q_{Elev}=3) and various values of soil electrical conductivity.

5 Conclusions

In this study we used joint multifractal analysis to examine variability of corn and soybean yields and relationships between the yields and the terrain elevation and soil EC. The analysis results indicated that (i) high yields prevailed at low elevations for both corn and soybean yields and (ii) that at high elevations lower soybean yields were associated with high EC values. The first observation is an expected result of combined effect of higher water availability, higher organic matter content, and higher nutrient contents in soil at lower located sites. The second observation can be explained by the negative effect of soil erosion observed at higher elevations on soybean yield. Based on the findings, further research can be conducted to study the plant growth conditions at higher located areas.

The results of the study indicate that joint multifractal approach is a very useful tool for studying spatial aspects of the relationships between two or more variables. The level of details in a quantitative characterization of the relationships between the variables at different ranges of variable values that can be provided by joint multifractal approach surpasses that of traditional methods of quantitative data analysis. ur results are representative of different real ecological situations.

References

1. N.R. Kitchen, K.A. Sudduth, and S.T. Drummond, *J. Prod. Agric.* **12**, 607 (1999).
2. B.B. Mandelbrot, *J. Fluid Mech.* **62**, 331 (1974).
3. O.A. Folorunso, C.E. Puente, D.E. Rolston, and J.E. Pinzon, *Soil Sci. Soc. Am. J.* **58**, 284 (1994).
4. H. Grout, A.M. Tarquis, and M.R. Wiesner, *Environ. Sci. Technol.* **32**, 1176 (1998).
5. C. Meneveau, K.R. Sreenivasan, P. Kailasnath, and M.S. Fan. *Physical Review* A **41**, 894 (1990).
6. A.N. Kravchenko, and D. G. Bullock, *Agronomy J.* **91**, 1033 (1999).
7. C.J.G. Evertsz, and B.B. Mandelbrot in *Chaos and Fractals*, eds. H.-O. Peitgen, H. Jurgens, and D. Saupe (Springer-Verlag, New York, NY, 1992).

A HOMOGRAPHIC-WEIBULL MODEL FOR RARENESS

THIERRY HUILLET
LPTM, CNRS-Université de Cergy-Pontoise, ESA 8089
5, mail Gay-Lussac, Neuville sur Oise,
95031 Cergy-Pontoise Cedex
E-mail: huillet@ptm.u-cergy.fr

Heavy-tailed random variables constitute a source of various investigations at the moment, with views in applications as diverse as Finance, Hydrology, Seismology, Geology and Telecommunications. Here, the problem is to evaluate the probability of an outstanding (extreme) event to occur as the damage involved could be huge. In this note, we suggest that the problem (borrowed from Geology or Environment Sciences) of deciding whether a natural resource (such as a particular ore or pollutant) is abundant or sparse in Nature is intimately related to these tails' questions; we give a natural connection between the rareness/abundance question and the one arising from the distinction between heavy or light tails of its "value". On this basis, we suggest that a "good" statistical model for "rare ore" could be an homographic-*Weibull* distribution for its grade. In sharp contrast to this model, ore whose "value" is *Pareto* or *log-normal* distributed should represent data where ore and body of ore are strongly mixed. One way to address the estimation problem from data under the Homographic-*Weibull* model is briefly discussed, exploiting a logratio transformation which appears natural. This is a preliminary step before the enrichment process of rare ore can be posed.

1 Introduction

It has recently been claimed (see, e.g., [23], [22], [1]) that a wide range of natural or social phenomena, as modelled by a positive random variable with *pdf* (probability distribution function) F, can be accurately fitted by using the so-called "heavy-tailed" (*HT*) power-law distributions of the form $F(v) = 1 - (v/v_0)^{-a}$, $a > 0$. A simple distinctive feature of this class of distributions is that the log-log plot of the complementary *pdf* (*cpdf*) $\overline{F}(v) := 1 - F(v)$ is a straight line with negative slope $-a$. *HT* distributions are specific in the sense that the largest event in a sample dominates the others so that the practitioner's problem of predicting values far beyond the maximum of available data may be an uneasy task. To take an example, in a recent investigation [5], this model has been fitted, in the context of Geology, to data as diverse as earthquake magnitude data, diamond values and crater size distribution on various planets. It turns out that empirical *pdf*s sometimes exhibit at most a limited quasi-linear regime followed by significant curvature, which is the hint that, in some sense, the power-law class should be "*unstable*" and that lighter tails should result. In [19], the authors argued that such departures from the power-law description should not necessarily be explained by the finite size of the data, but could result from a deeper departure from the power-law hypothesis. Using rank-ordering statistics to back up their claim, they suggested that occurrences of numerous phenomena, ranging from earthquake death tolls and energies [18], [24], to radio light emissions in galaxies, apparently fit the so-called "stretched exponential" (*Weibull*) subexponential distribution $F(v) = 1 - \exp[-(v/v_0)^a]$, $a < 1$. In the context of earthquake magnitude data, the so-called "soft" magnitude cut-off model [17], [20], based on a *Gamma*-like distribution has also shown worth considering. Recently, [13], [14], similar problems were addressed using extreme value (EV) theory; here fitting a *Fréchet/Weibull* or log-*Weibull* distribution on data was the central point in order to decide on the light/heavy or even very heavy (*i.e.* with tail index 0) earthquake magnitudes. As these notions form the body of this article, the needed material in the sequel on distributions tails is compiled in Section 2. For reasons to appear later, special attention is paid in this respect to the *Fréchet/Weibull*, *Pareto* and *log-normal* models.

Here, we would like to investigate a problem which intuitively (and indeed) is connected to the previous one and which arises from the Environmental Sciences. The problem is the following: *Can one give a statistical content to the notion of a rare natural resource?*

Heavy-tailedness is naturally connected to the rareness problem in the following "frequentist" way: the statistical interpretation of the sentence "events occur *rarely* in *time*" may be interpreted as: "the random inter-arrival times between consecutive events are *HT*, possibly with infinite mean residual time". Such (renewal) processes endeavour special statistical properties [25], [16], and are certainly ubiquitous in Nature [26]. However, the rareness problem of a natural resource is of a different nature, as "time" is obviously absent in the following assertions: "Gold is a rare mineral, the concentration of a particular pollutant in air is negligible or infinitesimal, only traces of a particular chemical are found in certain areas, the water or moisture content is very low... " Such affirmation evoking sparseness and scatteredness of matter within a raw material are part of every day life readings and concern. Rare natural resources, whatever this means, have received particular attention as their "value" or price are expected to be high, either directly or in terms of indirect costs. Assertions evoking rareness (say of a particular ore) are most of the time based on the collection of data presenting in the form of a sample sequence of (low?) grade records of the ore of interest within its bulk. Grade records are $(0, 1)$-valued variables so that at first sight there seems to be no question of tails in the rareness problem...however, such datasets inevitably display variability which requires an appropriate statistical methodology for subsequent geological inferences. Can one then give a precise statistical meaning to these rareness advocations, in other words what is a "good" statistical model for rareness of a natural resource and what is a sample of *low* grade records? Is there any connection of this problem with the one advocated above of heavy/light-tailedness of a distribution? Formulating a satisfactory answer to these questions is one of the objective of this article. In Section 3, we indeed supply such possible connection between the rareness/abundance investigation of some resource and questions arising from the distinction between heavy or light tails of its "value", which is modelled as a ratio of grades. On this basis, we suggest that a "good" statistical model for "rare ore" could be (among others) an homographic-*Weibull* distribution for its grade. Exploiting this connection, it happens that the *Beta* and homographic *log-normal* models for grade appear "ambiguous" concerning the abundance/rareness problem. Rather, they should serve to represent data where ore and body of ore are strongly *mixed*. Strong mixing should normally lead to separation difficulties. In Section 4, we briefly investigate the fluctuations of the sample *maxima* and *minima* in the three cases envisaged and show that they can serve as a discriminating criterium between these models. The homographic-*Weibull* model is a two-parameters distribution for grade of rare ore, say R. Similarly, the grade of what is not ore (say "anti-ore"), which is $\overline{R} := 1 - R$, is homographic-*Fréchet* distributed. In Section 5, we first address the problem of estimating these parameters from data. It turns out that solving both the parameter identification problem and the compatibility tests is made a lot easier by using a logratio transformation $X = \log\left[\overline{R}/R\right]$, which therefore emerges as the "natural" choice of coordinates for this class of problems: the distribution of X is now a *Gumbel* distribution, with range the whole real line and remarkable statistical features. The impact of this transformation on large deviation (LD) from the mean reinforces the "logratioing" techniques in compositional data analysis. Finally, one possible way to model the enrichment process of rare ore is posed on statistical grounds. The problem consists in determining the "optimal" cut-off value for grade, say r_c, in the following sieve process: only rocks whose grade of ore exceed the threshold r_c are selected, discarding the others.

2 Random models for the "value" of a natural resource: heavy or light tails?

2.1 *The Weibull-Fréchet model*

Consider the class of random variables ($r.v.$) (to be thus identified later as the "value" of some natural resource), defined through $\overline{V} = (sS)^{1/a}$ where $a > 0$, $s > 0$ and S is an exponentially distributed $r.v.$ with mean 1. The $r.v.$ $\overline{V}$ can be seen as the output of some deterministic "machine", with parameters (s, a), triggered by some stochastic source of disorder S. While s is simply a scaling factor, the "structure" parameter a defines, roughly speaking, the way in which the disorder generated by the source S is spread over the positive real axis. For positive v, the density function (df) and complementary distribution function ($cpdf$) of $\overline{V}$ are obtained easily to be the *Weibull distribution:*

$$f_{\overline{V}}(v) = \frac{a}{s} v^{a-1} \exp\left(-\frac{1}{s} v^a\right), \; \overline{F}_{\overline{V}}(v) := 1 - F_{\overline{V}}(v) = \exp\left(-\frac{1}{s} v^a\right) \tag{1}$$

The *Weibull* $r.v.$ $\overline{V}$ is special case of the so-called Von Mises $r.v.$s [8], whose $cpdf$ can be written in the form $\overline{F}_{\overline{V}}(v) = \overline{F}_{\overline{V}}(v_0) \exp\left[-\int_{v_0}^{v} h_{\overline{V}}(v)\, dv\right]$ and where the *hazard energy density* $h_{\overline{V}}$ defined by this formula verifies $\lim_{v\uparrow+\infty} v h_{\overline{V}}(v) = +\infty$. The $cpdf$ of a *Von Mises* $r.v.$ decreases towards 0 faster than any power-law, so that these distributions are light-tailed (or rapidly varying) with moments of any arbitrary positive order. If in addition the function $h_{\overline{V}}$ verifies $\lim_{v\uparrow+\infty} h_{\overline{V}}(v) = 0$, the $r.v.$ $\overline{V}$ is said to be sub-exponential or with moderately heavy tails. Otherwise, it is super-exponential, as it has tails lighter than exponential ones. In the *Weibull* example, we get $h_{\overline{V}}(v) = \frac{a}{s} v^{a-1}$. Thus, when $0 < a < 1$ the *Weibull* $r.v.$ $\overline{V}$ is sub-exponential, whereas for $a \geq 1$ it is super-exponential. Next consider the *inverse* $r.v.$ $V := 1/\overline{V}$. It has df and pdf given by

$$f_V(v) = \frac{a}{s} v^{-(a+1)} \exp\left(-\frac{1}{s} v^{-a}\right), \; F_V(v) = \exp\left(-\frac{1}{s} v^{-a}\right) \tag{2}$$

The *r.v.* V is known as *Fréchet*'s [8]. Let us now recall [9] that a distribution is said to be *HT* (or *regularly varying*) it there exists some finite strictly positive constant a (the tail exponent) such that its *cpdf* satisfies $\overline{F}(v) \underset{v\uparrow+\infty}{\sim} v^{-a}L(v)$, where L is some slowly varying function for which $\lim_{v\uparrow+\infty}\frac{L(tv)}{L(v)}=1$, $t>0$. These have no moments of order larger than a. Clearly, for the *Fréchet* *r.v.* $V=1/\overline{V}$, $\overline{F}_V(v)=1-\exp\left(-\frac{1}{s}v^{-a}\right) \underset{v\uparrow+\infty}{\sim} \frac{1}{s}v^{-a}$ and the *Fréchet* distribution F_V is *HT*, with tail exponent a. Note that when $a\in(0,1)$, V does not even have a mean value. With $\mathbf{E}$ standing for mathematical expectation, it may be shown [13] that the moment generating function is $\mathbf{E}\left(\overline{V}^{\beta}\right):=\int_0^{\infty}v^{\beta}f_{\overline{V}}(v)\,dv$ is $\mathbf{E}\left(\overline{V}^{\beta}\right)=s^{\beta/a}\Gamma(1+\beta/a)$ where Γ is *Euler*'s function. Hence, β-moments for $\overline{V}$ exist as soon as $\beta>-a$; in particular the mean value of $\overline{V}$ is $m_{\overline{V}}=s^{1/a}\Gamma(1+1/a)$. Similarly, we get $\mathbf{E}\left(V^{\beta}\right)=s^{-\beta/a}\Gamma(1-\beta/a)$and β-moments for V are finite as soon as $\beta<a$. Note that the median value, say $\overline{m}_{\overline{V}}$, defined as the solution of $F_{\overline{V}}\left(\overline{m}_{\overline{V}}\right)=1/2$, is $\overline{m}_{\overline{V}}=(s\log(2))^{1/a}$. Its median value is $\overline{m}_V=1/\ \overline{m}_{\overline{V}}$. Finally, the distribution of $\overline{V}$ has a non-zero mode, say $m^*_{\overline{V}}$, at the only condition that $a>1$, and $m^*_{\overline{V}}=\left(s\frac{a-1}{a}\right)^{1/a}$. On the contrary, the mode of V is always non-zero and is $m^*_V=\left(s\frac{a+1}{a}\right)^{-1/a}$.

2.2 *The Pareto and log-normal models*

Pareto. Let us now skip to different models. Let $a>0$ and $b>0$. Define the (generalized) *Pareto* *r.v.* as a positive *r.v.* with density $f_{\overline{V}}(v)=\frac{\Gamma(a+b)}{\Gamma(a)\Gamma(b)}v^{a-1}(1+v)^{-(a+b)}$. Note that $\overline{F}_{\overline{V}}(v) \underset{v\uparrow+\infty}{\sim} v^{-b}L(v)$ so that $\overline{V}$ is *HT* with tail index $b>0$. Next consider the inverse *r.v.* $V=1/\overline{V}$. The density of V is obtained from the one of $\overline{V}$ simply by exchanging the parameters a and b, and $\overline{F}_V(v) \underset{v\uparrow+\infty}{\sim} v^{-a}L(v)$ so that V is again *Pareto* distributed, hence *HT*, but with tail index $a>0$. Note that $\overline{V}\overset{d}{=}V$ if and only if $a=b$ and that if $a+b=1$, both a and b are less than one, so that $\mathbf{E}\left(\overline{V}\right)=\mathbf{E}(V)=+\infty$.

Lognormal. Finally, for a log-normal model [2], $\overline{V}=\exp N$ with N a normal *r.v.* with (say) mean value m and variance σ^2. Thus $f_{\overline{V}}(v)=\frac{1}{\sigma\sqrt{2\pi}}\exp-\left[\frac{1}{2\sigma^2}(\log(v)-m)^2+\log(v)\right]$ and $\overline{F}_{\overline{V}}(v) \underset{v\uparrow+\infty}{\sim} \exp\left[-\frac{1}{2\sigma^2}(\log(v))^2\right]$. The log-normal model is in the *Von Mises* subexponential class with moderately heavy tails. Note that $V:=1/\overline{V}$ is also a log-normal *r.v.* and that some symmetry is found when $m=0$, in the sense that $\overline{V}\overset{d}{=}V$. In sharp contrast to *Weibull*, both *Pareto* and log-normal *r.v.*s are thus "inverse-stable".

3 Random models for the grade of ore: towards a definition of rareness

Consider a collection of rocks (say Pitchblende) eventually with different (large enough) sizes and possibly enclosing some ore (say Uranium) of interest. For each rock, the grade of ore is the ratio $r=w/(w+\overline{w})$ of ore's tonnage w to the total body of ore's tonnage in this particular rock. One may assume the sample data to be a realization of *iid r.v.*s, say R. Grade of ore *r.v.*s have compact support $(0,1)$, as a ratio $R = W/\left(W+\overline{W}\right)$ of ore's tonnage W to the total tonnage. These *r.v.*s are interesting as they are scale-invariant (which is not true of tonnage), assuming some homogeneity of ore's distribution within the raw material. However, there seems to be no question concerning their distribution's tails, rather the behavior of their density in the vicinity of *both* 0 and 1 seems to be the central point in the apprehension of rareness. The key transformation to pass from positive "value" *r.v.*s $\overline{V}$ to grade *r.v.*s, say $R \in (0,1)$, is the homographic (increasing) transform $\overline{V} = R/(1-R)$. Conversely, R may be expressed in terms of $\overline{V}$ through the inverse homography $R = \overline{V}/\left(1+\overline{V}\right) = 1 - 1/\left(1+\overline{V}\right)$. Thus $\overline{V} > 0$ simply is the ratio of grade R and (say) "anti-grade" $\overline{R} := 1 - R$ of the rock under study. We note that $V = \overline{R}/R = 1/R - 1$ is a decreasing function of the grade of ore *r.v.* R and that, in terms of tonnage, $V = \overline{W}/W$ is itself scale invariant. Thus V may be interpreted as the price (or value) of ore, giving a precise quantitative meaning to the common-sense sentence: "*what is rare is valuable*". Conversely, in this interpretation, $\overline{V} = 1/\overline{R} - 1$ may be interpreted as the value of what is not ore, *i.e.*, say, of "*anti-ore*". Note that the *total* value $T > 0$ of the rock in this interpretation is the sum of the value V of ore and the one $\overline{V}$ of anti-ore $T = V + \overline{V} = V + 1/V$. The distribution of this *r.v.* can easily be computed in each specific case. We shall next study the grade *r.v.* R when the ratio $R/\overline{R}$ is assumed either *Weibull, Pareto* or *log-normal.*

*3.1 The homographic-*Weibull *model for rareness*

Suppose one is interested by a grade *r.v.* R which is such that $R/\overline{R} = \overline{V}$, with $\overline{V}$ *Weibull.* We note that $\overline{R}/R = V$ is alternatively a *Fréchet r.v.*. For $r \in (0,1)$, the *df* and *cpdf* of R are now easily obtained, yielding the *homographic-Weibull* (*HW*) distribution:

$$f_R(r) = \frac{a}{s}\left(\frac{r}{1-r}\right)^{a-1}(1-r)^{-2}\overline{F}_R(r)\,,\ \overline{F}_R(r) = \exp\left[-\frac{1}{s}\left(\frac{r}{1-r}\right)^{a}\right] \qquad (3)$$

At $r=1$, this density satisfies $f_R(r) \underset{r\uparrow 1^-}{\sim} (1-r)^{-(a+1)}\exp\left[-\frac{1}{s}(1-r)^{-a}\right]$. Although the algebraic prefactor is a divergent one, the vanishing "exp-algebraic" postfactor tends to annihilate this effect; globally, the density vanishes at $r = 1^-$ and is even very "flat" there: it has vanishing derivatives of any order. Thus a characteristic feature of such distributions is that samples with grade close to one are highly improbable. Next, at $r=0$, the density satisfies $f_R(r) \underset{r\uparrow 0^+}{\sim} r^{a-1}$ and thus has algebraic behavior there: there is significant probability mass in the vicinity of $r=0$. It diverges if $a<1$ (supercritical grade) and vanishes if $a>1$ (subcritical grade). The critical value is $a=1$ for which $f_R(0) = \frac{a}{s}$. For supercritical grade models, samples with grade close to 0 are the most probable which is an additional sign of rareness; however, it should be emphasized here that the smaller the structure parameter a is, the more probability mass is assigned at both ends (0 and 1) of the support. Significant (*i.e.* algebraic) probability mass at lowest grade $r = 0^{+}$ and insignificant (*i.e.* vanishing "exp-algebraic" fast) probability mass at highest grade $r = 1^{-}$ may serve as a quantitative definition of *rareness* of the natural resource under study. In this sense, an HW model for the grade of ore is a "good" model for its rareness. Note that the (*Fréchet* distributed) value V of a "rare" natural resource in this sense is high as a result of its "heavy-tailedness". More generally, we may suggest that any light-tailed at infinity *r.v.* $\overline{V} > 0$ such that $V := 1/\overline{V}$ is HT induces through the homographic transform described above a suitable model for rareness as well. Returning now to our specific model, natural macroscopic *r.v.*s which could serve as quantifiers of the grade of ore *r.v.* scattering are its median value $\overline{m}_R = \left[1 + (s \log(2))^{-1/a}\right]^{-1}$, defined by $F_R(\overline{m}_R) = 1/2$ and "most probable fluctuation" d_R defined by $\mathbf{P}(|R - \overline{m}_R| > d_R) = F_R(\overline{m}_R + d_R) - F_R(\overline{m}_R - d_R) = 1/2$. Note finally that if ore is rare in the previous sense, "anti-ore", whose grade is $\overline{R}$, is abundant. Its median value $\overline{m}_{\overline{R}}$ is $\overline{m}_{\overline{R}} = 1 - \overline{m}_R$. Its value $\overline{V}$ is low (*i.e.* light-tailed). As a result, an homographic-*Fréchet* model for the grade *r.v.* may as well serve as a "good" statistical model for abundance. There is a natural connection between the rareness-abundance question of a natural resource and the ones arising from the distinction between heavy or light tails of its value.

3.2 The ambiguity of the Beta *model for the grade of ore*

Suppose one is now interested by a grade *r.v.* R which is such that $R/\overline{R} = \overline{V}$, with $\overline{V}$ *Pareto* with parameter (a, b) in the above sense. We note that if this is so, $\overline{R}/R = V$ is alternatively a *Pareto r.v.* with parameter (b, a). For $r \in (0,1)$, the density function (*df*) of R is now obtained easily, yielding the *Beta* distribution: $f_R(r) = \frac{\Gamma(a+b)}{\Gamma(a)\Gamma(b)} r^{a-1} (1-r)^{b-1}$. Note that if $a + b = 1$, the distribution of R is the one of a generalized *Arcsine* law which appears here in there in fluctuation theory (if in addition $a = b = 1/2$, the distribution of R is strictly the one of the *Arcsine* law). At $r = 1$, (respectively $r = 0$), this density exhibits an algebraic decay if $b > 1$ (respectively $a > 1$), whereas it exhibits an algebraic divergence if $b < 1$ (respectively $a < 1$).

Assuming a *Beta* model for the grade of ore *r.v.* amounts to assign significant (*i.e.* algebraic) probability mass at *both* lowest and highest grades $r = 0^{+}$ and $r = 1^{-}$. Significant probability mass at $r = 1^{-}$ is a sign of abundance of ore whereas significant probability mass at $r = 1^{-}$ is a sign of its rareness in the previous sense. Thus, there is some ambiguity there. In other equivalent terms, the (*Pareto* distributed) value of ore V is high as *Pareto* distributions are HT, but so is the value $\overline{V}$ of anti-ore: in this model, both ore and anti-ore are abundant. Of course, the condition $a < b$ (respectively $a > b$) is still a hint of rareness (respectively abundance) as probability mass at 0 (respectively 1) is larger than the one at $r = 1^{-}$ (respectively $r = 0^{+}$).

3.3 The homographic-Log-normal model: strong mixing

When R is such that $R/\overline{R} = \overline{V}$, with $\overline{V}$ *log-normal*, the density function (*df*) of R is obtained to be: $f_R(r) =$

$\frac{1}{\sigma\sqrt{2\pi}}(1-r)^{-2}\exp\left[-\frac{1}{2\sigma^2}\left(\log\left(\frac{r}{1-r}\right)-m\right)^2+\log\left(\frac{r}{1-r}\right)\right]$, with $r \in (0,1)$. At $r=1$, the ore density satisfies $f_R(r) \underset{r\uparrow 1^-}{\sim} (1-r)^{-2}\exp\left[-\frac{1}{2\sigma^2}\left[\log(1-r)\right]^2\right]$. Although the algebraic prefactor is a divergent one, the vanishing "*exp – log*^{2}" postfactor tends to annihilate this effect; globally, the density vanishes at $r=1^-$ and is very "flat" there too (decay is faster than algebraic), although not as flat as in the first *Weibull* case. Next, at $r=0$, the density satisfies $f_R(r) \underset{r\uparrow 0^+}{\sim} \exp\left[-\frac{1}{2\sigma^2}\left[\log(r)\right]^2\right]$ and vanishes there also, in a similar "*exp – log*^{2}" way. The *Log-normal* model is thus also ambiguous concerning the abundance/rareness problem and for two reasons: the behavior of the density of grade is similar at both ends 0 and 1 of the support making it hard to decide between rareness and abundance; in addition, this behavior is neither algebraic nor "exp-algebraic" but indeed in between. It should therefore serve as a "good" model in situations when ore and anti-ore are intimately mixed in the sense that in this model, both ore and anti-ore are neither rare nor abundant.

4 Statistics of extremes

In this Section, we suggest that the sample maxima and minima for grade records should serve as a discriminating test between the three cases envisaged. Let $(\overline{V}_1, ..., \overline{V}_n)$ be an n-sample of positive *iid* *r.v.*s. When dealing with extreme events [7, 8, 12, 11], it may first be useful to understand the way the maximum $\overline{V}_{n:n} := \max(\overline{V}_1, ..., \overline{V}_n)$ behaves as the sample size grows. First observe the obvious fact that $\overline{V}_{n:n} \overset{a.s.}{\to} +\infty$, as $n \uparrow +\infty$. This observation does not enclose too much information and one would like a deeper insight on how the order of magnitude of the maximum evolves, as $n \uparrow \infty$. This is the purpose of what follows.

Define the increasing quantile sequence $(v_n, n \geq 1)$ by $n\overline{F}_{\overline{V}}(v_n) = 1$. With $h_{\overline{V}}$ the energy function defined in Subsection 2.1. and $\sigma_{\overline{V}} := 1/h_{\overline{V}}$, the *Fisher-Tippett* theorem [8] for *Von Mises' r.v.* yields the following convergence in distribution for the maximum $\overline{V}_{n:n}$: $\frac{(\overline{V}_{n:n}-v_n)}{\sigma_{\overline{V}}(v_n)} \overset{d}{\to} G$ as $n \uparrow \infty$ where G is a *Gumbel r.v.*, *i.e.* with *cpdf:* $P(G \leq t) = e^{-e^{-t}}$. These constructions therefore allow for an approximation of the width of the confidence interval of the maximum around its "typical" value v_n, as the sample size n grows.

In sharp contrast to this situation, the *Fisher-Tippett* theorem for *HT r.v.*s with tail-exponent α yields the following convergence in distribution for the maximum $\overline{V}_{n:n}$: $n^{-1/\alpha}\overline{V}_{n:n} \overset{d}{\to} F_\alpha$ as $n \uparrow \infty$, where F_α is a *Fréchet r.v.*, *i.e.* with *pdf:* $P(F_\alpha \leq t) = \exp(-t^{-\alpha})$.

Concerning the minimum of n *iid* observed *r.v.*s, say $\overline{V}_{1:n} := \min(\overline{V}_1, ..., \overline{V}_n)$, observing that, with $V = 1/\overline{V}$, we get $\overline{V}_{1:n} = 1/V_{n:n}$ so that it suffices to reason in a similar way as for the maximum, but working with $V = 1/\overline{V}$.

Let us now translate these results in terms of grade *r.v.*s. As the homographic transform is increasing, we easily get: if R is a grade *r.v.* such that $R/\overline{R} = \overline{V}$ is

Von Mises, then, with $R_{n:n} := \max(R_1, ..., R_n)$

$$\frac{(R_{n:n} - r_n)}{\sigma_R(r_n)} \xrightarrow{d} G \text{ as } n \uparrow \infty \tag{4}$$

Here, with v_n defined above, the "typical" value for $R_{n:n}$ is $r_n = 1 - 1/(1 + v_n)$. With $h_{\overline{V}}$ defined in Subsection 2.1. and $\sigma_{\overline{V}} := 1/h_{\overline{V}}$, the scaling factor is $\sigma_R(r_n) = r_n\left(\frac{r_n}{1-r_n}\right)^{-2}\sigma_{\overline{V}}\left(\frac{r_n}{1-r_n}\right)$. The typical value r_n of $R_{n:n}$ tends to 1: the approach to 1 is "slow" because there is insignificant probability mass at highest grade $r = 1^-$. If R is a grade *r.v.* such that $R/\overline{R} = \overline{V}$ is *HT* with exponent $\alpha > 0$, then, $n^{1/\alpha}(1 - R_{n:n}) \xrightarrow{d} W_\alpha$ as $n \uparrow \infty$ where W_α is a *Weibull r.v.*, *i.e.* with *cpdf:* $\mathbf{P}(W_\alpha > t) = \exp(-t^\alpha)$. Here, the "typical" value of $R_{n:n}$ is 1 directly, but with very large fluctuations: the approach to 1 is much faster because there is now significant (*i.e.* algebraic) probability mass at highest grade $r = 1^-$. Let us briefly particularize these facts in the three examples studied in this article.

- $\overline{V}$ **Weibull.** In this *Von Mises* case, one may check that $v_n \underset{n\uparrow+\infty}{\sim} [s\log(n)]^{1/a}$ and $\sigma_{\overline{V}}(v_n) \underset{n\uparrow+\infty}{\sim} [s\log(n)]^{1/a-1}$. As a result, the maximum $\overline{V}_{n:n}$ typically grows like $\log(n)$ raised to the power $1/a$. Remark also that the quantity $\sigma_{\overline{V}}(v)$ can be interpreted as the *absolute* standard fluctuation for the maximum around $v = v_n$: from the previous expression of $\sigma_{\overline{V}}(v_n)$, this *absolute* fluctuation tends to $+\infty$ (0) as n grows depending on $a < 1$ ($a > 1$) and thus showing two distinct regimes when crossing the critical value. However, in any case, the *relative* fluctuation for the maximum: $\sigma_{\overline{V}}(v_n)/v_n$ tends to 0 like $1/\log(n)$ as $n \uparrow \infty$.

With such v_n and $\sigma_{\overline{V}}(v_n)$, r_n and $\sigma_R(r_n)$ are available and the limit law for the maximal grade *r.v.* $R_{n:n}$ is given by (4). This *r.v.* approaches 1 slowly. However in this model, $V = 1/\overline{V}$ is *Fréchet*, hence *HT* with exponent a. As a result, $n^{1/a}R_{1:n} \xrightarrow{d} W_a$ as $n \uparrow \infty$ and, on the contrary, the minimum grade $R_{1:n}$ approaches 0 fast.

- $\overline{V}$ **Pareto.** Both $\overline{V}$ and V are *HT* (*Pareto*) with respective tail exponent b and a. We thus have $n^{1/b}(1 - R_{n:n}) \xrightarrow{d} W_b$ and $n^{1/a}R_{1:n} \xrightarrow{d} W_a$ as $n \uparrow \infty$ so that $R_{n:n}$ (respectively $R_{1:n}$) approaches 1 (respectively 0) fast.

- $\overline{V}$ **Log-normal.** In this *Von Mises* case, one may check that $v_n \underset{n\uparrow+\infty}{\sim} \exp\left[2\sigma^2\log(n)\right]^{1/2}$ and $\sigma_{\overline{V}}(v_n) \underset{n\uparrow+\infty}{\sim} \exp\left[2\sigma^2\log(n)\right]^{1/2} \cdot [\log(n)]^{-1/2}$.

In this case, the relative fluctuation for $\overline{V}_{n:n}$ tends to 0 like $[\log(n)]^{-1/2}$ as $n \uparrow \infty$. The limit law for the maximal grade *r.v.* $R_{n:n}$ is given by (4). However, in this case, $V = 1/\overline{V}$ is again log-normal, hence *Von Mises* in such a way that the limit law for the minimal grade $R_{1:n}$ is again given by (4). Both $R_{n:n}$ and $R_{1:n}$ approach 1 and 0 slowly.

As a partial conclusion, a characteristic feature of what we think fair to call a random model for rareness of ore (*e.g.* the *HW* model) is that the maximum value of the grade sample approaches the value 1 much slower than the minimum approaches 0 in a sense which is now precise.

We now briefly indicate how these limit results can be used in practice, testing the adequacy of some guessed model for grade from the behavior at the end points.

Suppose the model for grade is a *HW* model with known parameters (s, a), hence with known *pdf* F_R. Fix a small real number, say $\alpha = .05$. We would like to compute the number $\epsilon_n(\alpha)$ defined by $\mathbf{P}\{|R_{n:n} - r_n| > \epsilon_n(\alpha)\} = \alpha$. The number $\epsilon_n(\alpha)$ is therefore the radius of the ball centered at r_n which is likely (at level α) to enclose the n-th largest value $R_{n:n}$, were F_R to be a good model. Under our assumption, from (4), $\epsilon_n(\alpha)$ can be effectively computed by $\mathbf{P}\{|G| > \epsilon_n(\alpha) h_R(r_n)\} \sim \alpha$. This construction therefore yields an approximation of the width of the confidence interval of the maximum around r_n and the adequacy test works as follows: a) evaluate the empirical maximum $r_{n:n}$ from the sample and compute r_n; b) select some α, and determine the level value $\epsilon_n(\alpha)$ for which the above inequality holds; c) if $|r_{n:n} - r_n| > \epsilon_n(\alpha)$, reject the hypothesis that the tail of the sample has been generated with the theoretical *pdf* F_R, otherwise accept it; the probability to take the wrong decision is α. In case of rejection, try another model F_R and apply the adapted limit theorem. This approach supposes that the parameters (s, a) are known which may not be the case, so that an estimator must prior be designed.

5 Data analysis and the enrichment process of rare ore

In the sequel, we shall assume that the content of ore model is given by a homographic-*Weibull HW* model, assuming rareness. The problem of estimating the parameter set (s, a) from data will now be investigated: completely identifying the model is essential for further treatments. Among these, the ore enrichment process (under the identified model) appears unavoidable. This problem will finally be briefly addressed.

5.1 Estimating the parameter set

In many physical situations, the *r.v.* R cannot be directly observed. Rather, the observed *r.v.* could for instance be $R_{obs} := U.R \in (0,1)$ where U is uniformly distributed. This seems to be a realistic way to take into account that the ore of interest may be scattered at random within its mould. However, we shall not enter into this model and rather assume that R is directly observed. If such is the case, we are left with an n-sample $(r_1, ..., r_n)$ of "observed" grade records with *cpdf* (3). The proposed strategy which allows to identify the parameter set (s, a) from the sample will now be briefly discussed. To achieve this estimation program, and for reasons to appear later, we shall rather work with the logratio *r.v.* $X = \log(\overline{R}/R)$ *i.e.* with the logarithm of ore's value. So the working hypothesis is rather now: with $x_m := \log(\overline{r}_m/r_m)$, $m = 1, ..., n$, the vector $(x_1, ..., x_n)$ is an n-sample of observations with common *cpdf* the one of X. Before entering into the estimation problem itself, let us then compute the distribution of X and underline some of its remarkable properties. From (2) the *pdf* for the *r.v.* X is found to be $F_X(x) = \exp\left(-\frac{1}{s}e^{-ax}\right)$. We identify this *pdf* as the one of a *Gumbel* distribution [11]: the "exp-*Fréchet*" *r.v.* X is a *Gumbel r.v.* on the extended domain $(-\infty, +\infty)$. For any choice of (s, a), the *r.v.* X is *Von Mises*'; in addition, it is super-exponential, which means that the tails of its *pdf* decrease towards 0 at exponential rate or faster at both extremities $\pm\infty$ of the support. Hence, the distribution is "thin", although very asymmetric (exponential at $x = +\infty$ and doubly exponential at $x = -\infty$). The *Laplace* transform $Z_X(\beta)$ of X is given by $Z_X(\beta) := \mathbf{E}\left(e^{\beta X}\right) = \mathbf{E}\left(V^{\beta}\right) = s^{-\beta/a}\Gamma(1-\beta/a)$. This function is thus defined on the range $\beta < a$, therefore containing the origin $\beta = 0$, as required. As a result, in sharp contrast to the *Fréchet* distributed value of ore V itself, the *r.v.* X always has convergent moments of arbitrary integral order, which can be obtained as the *Taylor* coefficients of $Z_X(\beta)$ at $\beta = 0$. For instance, denoting as γ the *Euler*'s constant, the mean value of X is found to be

$$m_X = \mathbf{E}(X) = -\frac{1}{a}(\log(s) - \gamma) \simeq -\frac{1}{a}(\log s - 0.5772) \tag{5}$$

In addition, the median value of X is

$$\overline{m}_X = -\frac{1}{a}(\log(s) + \log\log(2)) \simeq -\frac{1}{a}(\log(s) - 0.3665) \tag{6}$$

It should be emphasized that the mean and median have a simple expression in terms of the pair (s, a). In addition, (see e.g. [10] for an exploitation of this fact) the distribution X is always unimodal, and even strongly unimodal, which means that the information (log-density) function $I_X(x) := -\log f_X(x)$ is strictly convex; its mode is $m_X^* = -\frac{1}{a}\log(s)$. A remarkable feature of the mean-median-mode trio in the *Gumbel* model is thus that $m_X > \overline{m}_X > m_X^*$. Since the empirical mean and median provide almost surely $(a.s.)$ convergent estimators for m_X and $\overline{m}_X$, they can be used to estimate the pair (s, a). This estimation problem will now be discussed.

Let us now consider the problem of deciding whether, and for what choice of the parameter pair (s, a), the distribution function F_X is a good statistical model for a particular data set $(x_1, .., x_n)$; or equivalently, whether F_R is a good model for the grade data set $(r_1, .., r_n)$. The solution advocated here will be a two-step procedure. The first step is to identify the value of the parameter pair (s, a) under the hypothesis that $(x_1, .., x_n)$ is a realization of an *iid* sequence $(X_1, .., X_n)$ with *pdf* F_X. The second step is to decide whether the identified distribution fits the data globally.

Step 1. One possible approach would be to compute the maximum likelihood estimator for (s, a). This estimator is defined as the value of the parameter pair $(\widehat{s}_n, \widehat{a}_n)$ which maximizes the likelihood function $L_n := \prod_{m=1}^{n} f_X(x_m)$; the corresponding distribution, whenever it exists, is precisely the one for which the realization $(x_1, x_2, \ldots, x_n)$ is the more likely to occur [6]. However, it turns out that we have no guarantee that the likelihood function L_n attains its maximum at some point inside its domain, which is unfortunately open, so that the very maximum likelihood approach may not make sense; furthermore, we cannot even guarantee that the likelihood function L_n is convex, so that it could be difficult to compute in a reliable way the maxima of L_n, should they exist.

As V is a *Fréchet r.v.*, thus in the "Generalized Extreme Value Distribution" class, other methods such as the method of Probability-Weighted Moments or quantile estimation techniques could be used as well (see [8] p. 321-323). Also, *Pickands* or *Hill's* strongly consistent estimators could be used to estimate the shape parameter (and also here tail exponent) a, using "upper" order statistics' techniques from the value sample $(v_1, .., v_n)$ (see [8] p. 325-345 for a survey and [5] for some additional features on what "upper" means).

An alternative simple approach in our particular case is to substitute the (strongly consistent) empirical mean and median of the sample $(x_1, x_2, \ldots, x_n)$ in (5)-(6), and to solve these equations in (s, a), fully exploiting the logratio transform.

Denote as $\overline{x}_n$ the cumulative sum of the sample, *i.e.* $\overline{x}_n = \sum_{m=1}^{n} x_m$ and as $(x_{1:n}, .., x_{n:n})$ the ordered version of $(x_1, .., x_n)$, which means $x_{1:n} \leq .. \leq x_{m:n} \leq .. \leq x_{n:n}$. The empirical mean and median are then respectively $\frac{1}{n}\overline{x}_n$ and $x_{[n/2]:n}$. Proceeding in this way, we suggest the following estimator pair for the scale and shape parameters:

$$\widehat{a}_n = \frac{\gamma + \log\log(2)}{\frac{1}{n}\overline{x}_n - x_{[n/2]:n}} \text{ and } \widehat{s}_n = \exp\left[\frac{\gamma x_{[n/2]:n} + \log\log(2)\,\frac{1}{n}\overline{x}_n}{\frac{1}{n}\overline{x}_n - x_{[n/2]:n}}\right] \tag{7}$$

Step 2. Model data fitness: the adequacy of the identified distribution corresponding to $(\widehat{s}_n, \widehat{a}_n)$, say $\widehat{F}_X = F_X$, with the empirical distribution can be tested from three different point of views, depending on what practical questions the model is supposed to answer: globally (*Kolmogorov-Smirnov* test); in the central body of the data (order statistics [7], [24], [13]); for its tails or endpoints (see Section 4). The statistics associated with these tests are based on the ordered sample $(x_{1:n}, .., x_{n:n})$ and on the quantile distribution function (*qdf*) of X: $F_X^{-}(p) := \inf(x : F_X(x) > p)$. This *qdf* is easily computed. It is $F_X^{-}(p) = -\frac{1}{a}\log[s\log(1/p)]$. The *Kolmogorov-Smirnov* (*KS*) test [4] enables to decide whether an *iid* sample $(x_1, x_2, \ldots, x_n)$ has been generated with some guessed (theoretical) probability distribution function, say $\widehat{F}_X = F_X$. Denote respectively as F_n and F_n^{-} the empirical *pdf* and quantile distribution functions of the sample, *i.e.* $F_n(x) := \frac{1}{n}\sum_{m=1}^{n} \mathbf{1}(x_{m:n} \leq x)$ and $F_n^{-}(p) := \inf(x : F_n(x) > p)$. The *KS* test is based on the quantity $\sup_x |F_n(x) - F_X(x)|$ which measures some distance between the empirical and theoretical *pdfs*. Using the transformation $x = F_X^{-}(p)$, this is also $\sup_{p\in[0,1]} \left|F_n^U(p) - p\right|$ where F_n^U is the empirical *pdf* of an *iid* uniform sequence on the interval $(0,1)$, so that $F_n^U(p) := F_n\left(F_X^{-}(p)\right) = \frac{1}{n}\sum_{m=1}^{n} \mathbf{1}(u_{m:n} \leq p)$ with $u_{m:n} := F_X(x_{m:n})$. Using this notation, it is known [4] that $\sqrt{n}\sup_{p\in[0,1]} \left|F_n^U(p) - p\right| \underset{n\uparrow\infty}{\overset{d}{\to}} M$ where the *r.v.* M is the absolute supremum of a *Brown*ian bridge with *pdf* $F_M(x) = 1 - 2\sum_{k\geq 1}(-1)^{k-1}\exp\left(-2k^2x^2\right)$. Hence, searching for the level value $\gamma_n(\alpha)$ such that $\mathbf{P}\left\{\sup_{p\in[0,1]} \left|F_n^U(p) - p\right| > \gamma_n(\alpha)\right\} = \alpha$ for small α (say $\alpha = 0.05$) yields $\gamma_n(\alpha) \simeq \frac{1}{\sqrt{n}}\left[\frac{\log(2/\alpha)}{2}\right]^{1/2}$. Thus, the *KS* test procedure may be summarized as follows: a) convert the original sample $(x_1, .., x_n)$ into $(u_1, .., u_n) := (F_X(x_1), .., F_X(x_n))$; b) compute $\max_{m=1,..,n} |m/n - u_{m:n}| = \sup_{p\in[0,1]} \left|F_n^U(p) - p\right|$ c) if this number exceeds $\gamma_n(\alpha)$, reject the hypothesis that the sample has been generated with the theoretical *pdf* $F_X(x)$, otherwise accept it. α is the probability to decide that the sample is not a realization of the distribution F_X when it really is.

This procedure may be applied, in our case study, for the estimated *pdf*, say $\widehat{F}_X(x)$, replacing (s, a) by the estimated parameter set $(\widehat{s}_n, \widehat{a}_n)$. However, the sample which was used to estimate (s,a) and the sample that should serve to test the model data fitness should be two distinct hopefully *iid* samples.

5.2 *Large deviation from the mean: the urge for logratios in compositional data analysis*

In the sequel, we shall assume that the homographic- *Weibull* model fits the available data for some known (or identified) parameter set (s,a). We shall here investigate some additional properties of the HW model for rare ore, underlining the additional importance of logratioing, independently of the estimation problem. First observe that, from the law of large numbers, the empirical mean $\frac{1}{n}\sum_{m=1}^{n} R_m$ converges almost surely $(a.s.)$ to the theoretical mean $m_R = \mathbf{E}(R)$, known to exist as R is $(0,1)$-valued. Large deviation (LD) theory is concerned with the evaluation of the (small) probability $\mathbf{P}\left(\frac{1}{n}\sum_{m=1}^{n} R_m \to r\right)$ as $r \in (0,1)$. More precisely, it is concerned with the rate at which this probability tends to 0, as the sample size n tends to infinity. It may be shown in this case that this probability tends to 0 exponentially fast at a rate which is under control. Note that, with $\overline{R}_m := 1 - R_m$, $m = 1...n$, the anti-grade sequence, this probability is also $\mathbf{P}\left(\frac{1}{n}\sum_{m=1}^{n} \overline{R}_m \to 1-r\right)$ and that there is no information here on the way $\left(\frac{1}{n}\sum_{m=1}^{n} R_m, \frac{1}{n}\sum_{m=1}^{n} \overline{R}_m\right)$ *jointly* behave as $n \uparrow +\infty$. We shall see how this drawback can be by-passed. Back to the *r.v.* X itself, from the law of large numbers, the empirical mean $\frac{1}{n}\overline{X}_n$ also converges almost surely $(a.s.)$ to the theoretical mean whose expression is $m_X = \mathbf{E}(X) = -\frac{1}{a}(\log(s) - \gamma)$. Here again, LD theory is concerned with the evaluation of the (small) probability $\mathbf{P}\left(\frac{1}{n}\overline{X}_n \to \alpha\right)$ as $\alpha \in (-\infty, +\infty)$. Let us now compute the rate at which this probability tends to 0, as a function of the sample size n.

Recall the *Laplace* transform of X, say $Z_X(\beta)$, is $Z_X(\beta) := \mathbf{E}\left(e^{\beta X}\right) = \mathbf{E}\left(V^{\beta}\right) = s^{-\beta/a}\Gamma(1-\beta/a)$ with real β defined in a open neighborhood of $\beta = 0$. As a result, the exact LD result is

$$n^{-1}\log \mathbf{P}\left(\frac{1}{n}\overline{X}_n \to \alpha\right) \underset{n\uparrow\infty}{\to} -s(\alpha) \leq 0 \tag{8}$$

In this expression, $s(\alpha)$ is the convex *Legendre-Cramér-Chernoff* transform of the convex "free energy" $\log Z_X(\beta)$: $s(\alpha) = \sup_{\beta<a}(\alpha\beta - F_X(\beta))$ satisfying $s(m_X) = 0$ and $s'(\alpha) = \beta$. Consequently, if $\alpha \neq m_X$, the probability that the empirical mean deviates from the theoretical mean tends to 0 exponentially fast as n grows, at rate $s(\alpha) > 0$. Note now that the *arithmetic* mean for the sequence $(X_1, ..., X_n)$, which is $\frac{1}{n}\sum_{m=1}^{n} X_m$, corresponds to the logratio of the *geometric* means of the anti-grade and grade of ore records $(R_1, ..., R_n)$. Indeed $\frac{1}{n}\overline{X}_n = \log\left[\prod_{m=1}^{n}\overline{R}_m^{1/n} / \prod_{m=1}^{n} R_m^{1/n}\right]$. Thus, under our hypothesis, the law of large numbers reads $\prod_{m=1}^{n}\overline{R}_m^{1/n} / \prod_{m=1}^{n} R_m^{1/n} \overset{a.s.}{\to} \exp m_X$.

With $v > 0$, (8) now reads $n^{-1}\log \mathbf{P}\left(\prod_{m=1}^{n}\overline{R}_m^{1/n} / \prod_{m=1}^{n} R_m^{1/n} \to v\right) \underset{n\uparrow\infty}{\to} -s(\log(v)) \leq 0$. In sharp contrast to the *LD* rate function of the arithmetic mean value of $(R_1, .., R_n)$, the function $s(.)$ now provides useful *joint* informations on the way ore and anti-ore are mixed (through geometric averaging). This observation is strongly reminiscent of multifractal analysis of singular measures with LD theory at the heart of it [21], [15] (although here, one rather considers a *Hölder* ratio of logarithms): we shall therefore call this function the *mixing spectrum* of ore and anti-ore. This reinforces the current trend supporting the constructive role of logratio analysis in geological compositional problems (see [3] and the references therein for both defenders and detractors of the approach), as opposed to raw compositional data; however, the exact role of this spectrum in the quantization of *mixing* is not fully nor deeply understood at the moment.

Remark: To emphasize the differences between the *HW* model and (say) the *Beta* model, let us examine the *mixing spectrum* when R is *Beta*. In this case, it may be checked that the logratio X admits the density $f_X(x) = e^{bx}/(1+e^x)^{a+b}$, with exponential equivalent at *both* $x = \pm\infty$. The *Laplace* transform $Z_X(\beta)$ in this case is: $Z_X(\beta) = \mathbf{E}\left(e^{\beta X}\right) = \mathbf{E}\left(V^\beta\right) = \mathbf{E}\left(\frac{(1-R)^\beta}{R^\beta}\right)$, which may be checked to be $Z_X(\beta) = \frac{\Gamma(a-\beta)\Gamma(b+\beta)}{\Gamma(a)\Gamma(b)}$, $\beta \in (-b, a)$ so that the derivative of the *mixing spectrum* is bounded from above *and* from below.

5.3 *Modelling the enrichment process*

In the sequel, we shall again assume that the homographic-*Weibull* model fits the available data for some known (identified) parameter set (s, a). We now say a few words on the enrichment process which chiefly consists in selecting those rocks whose ore content is large enough while neglecting the others. This is a very standard problem in Separation Engineering (think of Uranium enrichment). It is especially meaningful in the rare ore context. Fix a "cut-off" value for grade at $r = r_c$. Define the truncated grade after "cut-off" at r_c as $R_c := R\cdot\mathbf{1}(R > r_c)$. The enrichment process may simply be summarized as follows: the rocks whose grade of ore are lower than r_c are discarded, whereas those rocks with content exceeding r_c are considered and preciously stored. The grade of the enriched rocks is then obviously given by the truncated grade after "cut-off" *r.v.*. From (3), the *cpdf* of R_c is, with $0 > r > r_c > 1$, $\mathbf{P}(R_c > r) := \overline{F}_{R_c}(r) = \frac{\mathbf{P}(R>r)}{\mathbf{P}(R>r_c)} = \frac{\overline{F}_R(r)}{\overline{F}_R(r_c)}$. From the expression (3) of $\overline{F}_R(r)$, the median value of R_c, say $\overline{m}_{R_c}$, defined by $\overline{F}_{R_c}(\overline{m}_{R_c}) = 1/2$, is $\overline{m}_{R_c} = \left[1 + \left(s\log(2) + \left(\frac{1}{r_c} - 1\right)^{-a}\right)^{-1/a}\right]^{-1}$. Of course, $\overline{m}_{R_c} > \overline{m}_R$, as the median value for grade after "cut-off" exceeds the median value without "cut-off". Define then the enrichment degree as $\tau(r_c) = \frac{\overline{m}_{R_c}}{\overline{m}_R} > 1$. The "cut-off" value r_c which maximizes the enrichment degree of course is $r_c^* = 1$, in which case $\tau(r_c^*) = 1/\overline{m}_R$: actually, in this case, $R_c = 1$, *i.e.* R_c is degenerate. This optimization program is of course incomplete as it does not take into account an additional feature which is essential in the enrichment process, namely the cost of the action of discarding rocks with grades below the cut-off. This event is nicely represented by the *Bernoulli r.v.* $\mathbf{1}(R \le r_c)$. The average "discarding" cost therefore is $F_R(r_c) := \mathbf{P}(R \le r_c) = \mathbf{E}[\mathbf{1}(R \le r_c)]$ and the proposed trade-off strategy is the following: Find the "cut-off" value r_c^* which minimizes both $1/\tau(r_c)$ (as the median grade after cut-off should be maximum) and the average discarding cost. In absence of any information on the relative weights of each of these individual cost functions, the searched "optimal" cut-off value could be the one which solves the min-max trade-off problem, *i.e.* $r_c^* = \arg\min_{r_c} C(r_c)$ with $C(r_c) := \max(1/\tau(r_c), F_R(r_c))$. Here $C(r_c)$ is the total cost of the enrichment process. Here, r_c^* is the abscissa at which the decreasing function $1/\tau(r_c)$ intersects the increasing function $F_R(r_c)$.

References

1. A. Aharony and J. Feder (Eds.), Fractals in physics, *Physica D* **38**, 1 (1989).
2. J. Aitchison and J.A.C. Brown, *The log-normal distribution,* (Cambridge University Press 1957).
3. J. Aitchison, *Math. Geol.*, **31**, no 5, 563 (1999).
4. P. Billingsley, *Convergence of probability measures, (*J. Wiley and Sons 1968).
5. J. Caers, J. Beirlant and M.A. Maes, *Math. Geol.* **31**, no 4, 391- and 411- (1999).
6. H. Cramér, *Mathematical Methods of Statistics,* (Princeton University Press 1946).
7. H.A. David, *Order statistics:* 2nd edition, (Wiley, New York 1981).
8. P. Embrechts, C. Klüppelberg and T. Mikosh, *Modelling extremal events, (*Springer-Verlag **33,** 1997).
9. W. Feller, *An introduction to probability theory and its applications,* **2**, (Wiley, New York 1971).
10. U. Frisch and D. Sornette, *J. Phys. I* **7**, 1155 (1997).
11. E.J. Gumbel, *Statistics of Extremes*, (Columbia University Press 1958).
12. J. Galambos, *The Asymptotic theory of Extreme Order Statistics,* (J. Wiley and Sons 1978).
13. T. Huillet and H.F. Raynaud, *J. Phys. A: Math. Gen.* **32,** 1099 (1999).
14. T. Huillet and H.F. Raynaud, *The European Physical Journal B* **12**, 457 (1999).
15. T. Huillet and B. Jeannet, *J. Phys. A: Math. Gen.* **31,** 2567 (1998).
16. T. Huillet and H.F. Raynaud, *Chaos, Solitons and Fractals* **12**, 823 (2001).
17. Y.Y. Kagan, *Physica D*, 160 (1994).
18. L. Knopoff and D. Sornette, *J. Phys I* **5**, 1681 (1995).
19. J. Laherrère and D. Sornette, *The European Physical Journal B* **2**, 525 (1998).
20. I. Main, *Review of Geophysics* **34**, 433 (1996).
21. C. Evertsz and B Mandelbrot, in *Chaos and fractals*, ed. H. Peitgen, H. Jürgens, D. Saupe (1992).
22. B.B. Mandelbrot, *The Fractal geometry of nature*, (W.H. Freeman, New York 1983).
23. V. Pareto, *Cours d'Economie politique* (1896); reprinted as *Oeuvres complètes*, (Droz, Geneva 1965).
24. D. Sornette, L. Knopoff, Y.Y. Kagan and C. Vanneste, *J. of Geophys. Res.* **101** (B6), 13883 (1996).
25. D. Sornette and L. Knopoff, *Bull. Seism. Soc. Am.*, 789 (1997).
26. L. Telesca, V. Cuomo, M. Lanfredi, V. Lapenna and M. Macchiato, *Fractals* **7**, no. 3, 221 (1999).

POWER-LAWS AND SCALING IN FINANCE: EMPIRICAL EVIDENCE AND SIMPLE MODELS

JEAN-PHILIPPE BOUCHAUD
Service de Physique de l'État Condensé, Centre d'études de Saclay, Orme des Merisiers, 91191 Gif-sur-Yvette CEDEX, *FRANCE*
E-mail: bouchau@spec.saclay.cea.fr

CFM-Science & Finance, 109-111 rue Victor Hugo, 92532 Levallois CEDEX, *FRANCE; http://www.science-finance.fr*

We discuss several models that may explain the origin of power-law *distributions* and power-law *correlations* in financial time series. From an empirical point of view, the exponents describing the tails of the price increments distribution and the decay of the volatility correlations are rather robust and suggest universality. However, many of the models that appear naturally (for example, to account for the distribution of wealth) contain some multiplicative noise, which generically leads to *non universal exponents*. Recent progress in the empirical study of the volatility suggests that the volatility results from some sort of multiplicative cascade. A convincing 'microscopic' (i.e. trader based) model that explains this observation is however not yet available. We discuss a rather generic mechanism for long-ranged volatility correlations based on the idea that agents constantly switch between active and inactive strategies depending on their relative performance.

1 Introduction

Physicists like using power-laws to fit data. The reason for this is that complex, collective phenomena do indeed give rise to power-laws which are furthermore often *universal*, that is to a large degree independent of the microscopic details of the phenomenon. These power-laws emerge from collective action and transcend individual specificities. As such, they are unforgeable signatures of a collective mechanism. Examples in the physics literature are numerous. A well known example is that of phase transitions, where a system evolves from a disordered state to an ordered state: many observables behave as universal power laws in the vicinity of the transition point [1]. This is related to an important property of power-laws, namely scale invariance:[a] the characteristic length scale of a physical system at its critical point is infinite, leading to self-similar, scale-free fluctuations. Another example is fluid turbulence, where the statistics of the velocity field has scale invariant properties, to a large extent independent of the nature of the fluid, of the power injected, etc. [2].

Power-laws are also often observed in economic and financial data [3,4,5,6,7]. Compared to physics, however, much less effort has been devoted to understand these power-laws in terms of 'microscopic' (i.e. agent based) models and to relate the value of the exponents to generic mechanisms. The aim of this contribution is to give a short review of diverse power-laws observed in economics/finance, and to

[a]Power-law distributions are scale invariant in the sense that the relative probability to observe an event of a given size and an event ten times larger is independent of the reference scale.

discuss several simple models (most of them inspired from physics) which naturally lead to power-laws and could serve as a starting point for further developments. It should be stressed that none of the models presented here are intended to be fully realistic and complete, but are of pedagogical interest: they nicely illustrate how and when power-laws can arise.

2 Empirical power-laws: a short review

2.1 Distributional power-laws

The oldest and most famous power-law in economics is the Pareto distribution of wealth [3]. The distribution of individual wealths $P(W)$ is often described, in its asymptotic tail, by a power law [b]:

$$P(W) \simeq \frac{W_0^\mu}{W^{1+\mu}} \qquad W \gg W_0, \tag{1}$$

where μ characterizes the decay of the distribution for large W's: the smaller the value of μ, the slower the decay, and the larger the contrast between the richest and the poorest. More precisely, in a Pareto population of size N, the ratio of the largest wealth to the typical (e.g. median) wealth grows as $N^{1/\mu}$. In the case $\mu < 1$, the average wealth diverges: this corresponds to an economy where a finite fraction of the total wealth is in the hands of a few individuals, even when $N \to \infty$. On the contrary, when $\mu > 1$, the richest individual only holds a zero fraction of the total wealth (again in the limit $N \to \infty$). Empirically, the exponent μ is in the range $1-2$. This Pareto tail also describes the distribution of income, the size of companies, of pension funds,etc. [9,10,11].

In financial markets, the distribution of the price increments $\delta x_t = x(t') - x(t)$ over different time scales $t' - t$ is important both for risk control purposes and for derivative pricing models [12]. The availability of long series of high frequency data has motivated many empirical studies in the past few years. By pooling together the statistics of a thousand U.S. stocks, it is possible to study quite accurately the far tail of the distribution of intra-day price increments δx, which can be fitted as a power law [13]:

$$P(\delta x) \simeq \frac{\delta x_0^\mu}{|\delta x|^{1+\mu}}, \tag{2}$$

where μ is found to be close to 3. Similar values have also been reported for Japanese stocks [13], German stocks [14], currencies [15,16], bond markets [17], and perhaps also the distribution of the (time dependent) daily volatility σ, defined as an average over high frequency returns [18] (although other works report a log-normal distribution [23,26]).[c] This suggests that the value of μ could be universal. Note however that the value of μ depends somewhat on the stock and on the period of time studied,

[b] The bulk of the distribution, containing most of the population, however seems to behave as a pure exponential: see [8]

[c] Note that the tails of a log-normal distribution can be fitted (over a restricted interval) by a power-law. Therefore it is not always easy to distinguish between true power-laws and effective power-laws.

and that the error bar is quite large. For example, the value of μ for the S&P500 during the years 1991-1995 is found to be close to 5. Furthermore, as the time lag used for the definition of δx increases, the effective exponent describing the tail of the distribution increases as the distribution becomes more and more Gaussian like [12,13].

2.2 Temporal power-laws

Actually, the price increment at time t can usefully be thought of as the product of a *sign part* ϵ_t and an *amplitude part* (or volatility) σ_t:

$$\delta x_t = \epsilon_t \times \sigma_t. \tag{3}$$

The random variable ϵ_t has short ranged temporal correlations, extending over a few minutes or so on liquid markets [12,6]. The volatility, on the contrary, has very long ranged correlations, which can be fitted as a power- law with a small exponent ν [19,20,15,21,22,23]:

$$C_1(\tau) = \langle \sigma_t \sigma_{t+\tau} \rangle - \langle \sigma_t \rangle^2 \simeq \frac{c_1}{\tau^\nu}, \tag{4}$$

with ν of the order of 0.1. This behaviour is, again, seen on many different types of markets, and quantifies the intermittent activity of these markets: volatility tends to cluster in bursts which persist over very different time scales, from minutes to months. A similar power-law behaviour of the temporal correlation of the volume of transactions (number of trades) is also observed [24,18]. This is not surprising, since volatility and volume are strongly correlated. From an empirical point of view, the intermittent nature of the activity in financial markets is similar to the energy dissipation in a turbulent fluid [2,25].

In fact, the distribution of $\log \sigma$ is not far from being Gaussian [23,26]. Therefore, it is natural to study the temporal correlation of $\log \sigma$ [27,28]:

$$C_0(\tau) = \langle \log \sigma_t \log \sigma_{t+\tau} \rangle - \langle \log \sigma_t \rangle^2. \tag{5}$$

This correlation is also found to decay very slowly with τ. This decay can be fitted by a logarithm [27,28]: $C_0(\tau) = \lambda^2 \log(T/\tau)$, with a rather small value of $\lambda^2 \simeq 0.05$. This, together with the assumption that $\log \sigma$ is exactly Gaussian, leads to a *strict multifractal* model for the price changes [28], in the sense that different moments of the price increments scale with different powers of time [25,29,30]. Within this model, one finds [28]:

$$\langle |x(t+\tau) - x(t)|^q \rangle \propto \tau^{\zeta_q} \qquad \zeta_q = \frac{q}{2}\left[1 - \lambda^2(q-2)\right], \tag{6}$$

for $\tau \ll T$ and $q\lambda^2 < 1$ (for higher values of q, the corresponding moment is divergent). It is easy to show that in this model, the exponent ν defined by Eq. (4) is equal to λ^2. On the other hand, one can also fit $C_0(\tau)$ by a power-law with a small exponent, in which case the model would only be approximately multifractal, in the sense that the quantity $\langle |x(t+\tau) - x(t)|^q \rangle$ is the *sum* of different powers of τ, which can also be fitted by an effective, q-dependent exponent $\zeta_q < q/2$ [31].

3 Simple models of wealth distribution

3.1 A model with trading and speculative investment

As a simple dynamical model of economy, one can consider a stochastic equation for the wealth $W_i(t)$ of the ith agent at time t, that takes into account the exchange of wealth between individuals through trading, and is consistent with the basic symmetry of the problem under a change of monetary units. Since the unit of money is arbitrary, one indeed expects that the equation governing the evolution of wealth should be invariant when all W_i's are multiplied by a common (arbitrary) factor. A rather general class of equation which has this property is the following [32,33]:

$$\frac{dW_i}{dt} = \eta_i(t)W_i + \sum_{j(\neq i)} J_{ij}W_j - \sum_{j(\neq i)} J_{ji}W_i \ , \tag{7}$$

where $\eta_i(t)$ is a Gaussian random variable of mean m and variance $2\sigma^2$, which describes the spontaneous growth or decrease of wealth due to investment in stock markets, housing, etc. The term involving the (asymmetric) matrix J_{ij} describes the amount of money that agent j spends buying the production of agent i. We assume that this production is consumable, and therefore must not be counted as part of the wealth of i once it is bought. The equation (7) is obviously invariant under the scale transformation $W_i \to \lambda W_i$.

The simplest trading network one can think of is when all agents exchange with all others at the same rate, i.e $J_{ij} \equiv J/N$ for all $i \neq j$. Here, N is the total number of agents, and the scaling J/N is needed to make the limit $N \to \infty$ well defined. In this case, the equation for $W_i(t)$ becomes:

$$\frac{dW_i}{dt} = \eta_i(t)W_i + J(\overline{W} - W_i), \tag{8}$$

where $\overline{W} = N^{-1}\sum_i W_i$ is the average overall wealth. This is a 'mean-field' model since all agents feel the very same influence of their environment. It is useful to rewrite eq. (8) in terms of the normalised wealths $w_i \equiv W_i/\overline{W}$. This leads to:

$$\frac{dw_i}{dt} = (\eta_i(t) - m - \sigma^2)w_i + J(1 - w_i), \tag{9}$$

to which one can associate the following Fokker-Planck equation for the evolution of the density of wealth $P(w,t)$:

$$\frac{\partial P}{\partial t} = \frac{\partial[J(w-1) + \sigma^2 w]P}{\partial w} + \sigma^2 \frac{\partial}{\partial w}\left[w \frac{\partial wP}{\partial w}\right]. \tag{10}$$

The equilibrium, long time solution of this equation is easily shown to be:

$$P_{eq}(w) = \frac{(\mu-1)^\mu}{\Gamma[\mu]} \frac{e^{-\frac{\mu-1}{w}}}{w^{1+\mu}} \qquad \mu \equiv 1 + \frac{J}{\sigma^2}. \tag{11}$$

Therefore, one finds in this model that the distribution of wealth indeed exhibits a Pareto power-law tail for large w's. Interestingly, however, the value of the exponent μ is *not universal*, and depends on the parameter of the model (J and σ^2, but not on the average growth rate m). In agreement with intuition, the exponent μ grows (corresponding to a narrower distribution), when exchange between agents is more active (i.e. when J increases), and also when the success in individual investment strategies is more narrowly distributed (i.e. when σ^2 decreases). In this model, the exponent μ is always found to be larger than one. This means that the wealth is not too unevenly distributed within the population.

Let us now describe more realistic trading network, where the number of economic neighbours to a given individual is finite. We will assume that the matrix J_{ij} is still symmetrical, and is either equal to J (if i and j trade), or equal to 0. A reasonable assumption is that the graph describing the connectivity of the population is completely random, i.e. that two points are neighbours with probability c/N and disconnected with probability $1 - c/N$. In such a graph, the average number of neighbours is equal to c. We have performed some numerical simulations of Eq. (7) and have found [32] that the wealth distribution still has a power-law tail, with an exponent μ which only depends on the ratio J/σ^2. However, one finds that the exponent μ can now be smaller than one for sufficiently small values of J/σ^2 [32]. In this model, one therefore expects 'wealth condensation' when the exchange rate is too small, in the sense that a finite fraction of the total wealth is held by only a few individuals.

Although not very realistic, one could also think that the individuals are located on the nodes of a d-dimensional hyper-cubic lattice, trading with their neighbours up to a finite distance. In this case, one knows that for $d > 2$ there exists again a phase transition between a 'social' economy where $\mu > 1$ and a rich dominated phase $\mu < 1$. On the other hand, for $d \leq 2$, and for large populations, one is always in the extreme case where $\mu \to 0$ at large times. In the case $d = 1$, i.e. operators organized along a chain-like structure (as a simple model of intermediaries), one can actually compute exactly the distribution of wealth [34]. One finds for example that the ratio of the maximum wealth to the typical (e.g. median) wealth behaves as $\exp \sqrt{N}$, where N is the size of the population, instead of $N^{1/\mu}$ in the case of a Pareto distribution with $\mu > 0$.

The conclusion of the above results is that the distribution of wealth tends to be very broadly distributed when exchanges are limited, either in amplitude (i.e. J too small compared to σ^2) or topologically (as in the above chain structure). Favoring exchanges (in particular with distant neighbours) seems to be an efficient way to reduce inequalities.

3.2 Two related models

Let us now interpret W_i as the size of a company. The growth of this company can take place either from internal growth, depending on its success or failure. This leads to a term $\eta_i(t)W_i$ much as above. Another possibility is merging with another company. If the merging process between two companies is completely random and takes place at a rate Γ per unit time, then the model is exactly the same as the one considered by Derrida and Spohn [35] in the context of 'directed polymers in random media', and bears strong similarities with the model discussed in the previous section. One again finds that the distribution of W's is a power-law with a non universal exponent, which depends on the values of Γ and σ^2, and can be smaller than one.[d]

One can also consider a model where companies grow at a random rate η, but may also suddenly die at a rate Γ' per unit time, and be replaced by a new (small) company. There again, one finds a stationary Pareto distribution, with a non universal exponent which depends continuously on m, σ and Γ'.

4 Simple models for herding and mimicry

We now turn to simple models for thick tails in the distribution of price increments in financial markets. An intuitive explanation is herding: if a large number of agents acting on markets coordinate their action, the price change is likely to be huge due to a large unbalance between buy and sell orders [36]. However, this coordination can result from two rather different mechanisms.

- One is the feedback of past price changes onto themselves, which we will discuss in the following section. Since all agents are influenced by the very *same* price changes, this can induce non trivial collective behaviour: for example, an accidental price drop can trigger large sell orders, which lead to further downward moves.

- The second is direct influence between the traders, through exchange of information that leads to 'clusters' of agents sharing the same decision to buy, sell, or be inactive at any given instant of time.

4.1 Herding and percolation

A simple model of how herding affects the price fluctuations was proposed in [37]. It assumes that the price increment δx depends linearly on the instantaneous offset between supply and demand [37,38]. More precisely, if each operator in the market i wants to buy or sell a certain fixed quantity of the financial asset, one has [37]:[e]

$$\delta x = \frac{1}{\lambda} \sum_i \varphi_i, \tag{12}$$

where φ_i can take the values -1, 0 or $+1$, depending on whether the operator i is selling, inactive, or buying, and λ is a measure of the market depth. Note that the linearity of this relation, even for small arguments, has been questioned by Zhang [40]. Recent empirical analysis however seems to confirm that the relation is indeed linear for small arguments, but bends down and even saturates for larger arguments [41].

Suppose now that the operators interact among themselves in an heterogeneous manner: with a small probability c/N (where N is the total number of operators on

[d] In the language of disordered systems, this corresponds to the 'glassy' phase of the directed polymer, where the partition function is dominated by a few paths only.

[e] This can alternatively be written for the relative price increment $\delta x/x$, which is more adapted to describe long time scales. On short time scales, however, an additive model is preferable. See the discussion in [12,39].

the market), two operators i and j are 'connected', and with probability $1 - c/N$, they ignore each other. The factor $1/N$ means that on average, the number of operators connected to any particular one is equal to c (the resulting graph is precisely the same as the random trading graph of Section 3.1). Suppose finally that if two operators are connected, they come to agree on the strategy they should follow, i.e. $\varphi_i = \varphi_j$.

It is easy to understand that the population of operator clusters into groups sharing the same opinion. These clusters are defined such that there exists a connection between any two operators belonging to this cluster, although the connection can be indirect and follow a certain 'path' between operators. These clusters do not have all the same size, i.e. do not contain the same number of operators. If the size of cluster $\mathcal{C}$ is called $S(\mathcal{C})$, one can write:

$$\delta x = \frac{1}{\lambda} \sum_{\mathcal{C}} S(\mathcal{C}) \varphi(\mathcal{C}), \tag{13}$$

where $\varphi(\mathcal{C})$ is the common opinion of all operators belonging to $\mathcal{C}$. The statistics of the price increments δx therefore reduces to the statistics of the size of clusters, a classical problem in percolation theory [42]. One finds that as long as $c < 1$ (less than one 'neighbour' on average with whom one can exchange information), then all $S(\mathcal{C})$'s are small compared to the total number of traders N. More precisely, the distribution of cluster sizes takes the following form in the limit where $1-c = \epsilon \ll 1$:

$$P(S) \propto_{S \gg 1} \frac{1}{S^{5/2}} \exp -\epsilon^2 S \qquad S \ll N. \tag{14}$$

When $c = 1$ (percolation threshold), the distribution becomes a pure power-law with an exponent $\mu = 3/2$, and the Central Limit Theorem tells us that in this case, the distribution of the price increments δx is precisely a pure symmetrical Lévy distribution of index $\mu = 3/2$ [12] (assuming that $\varphi = \pm 1$ play identical roles, that is if there is no global bias pushing the price up or down). If $c < 1$, on the other hand, one finds that the Lévy distribution is truncated exponentially, leading to a larger effective tail exponent μ [37]. If $c > 1$, a finite fraction of the N traders have the same opinion: this leads to a crash. This simple model has been extended in several directions by Stauffer and collaborators [43]. Very recently, a somewhat related model was studied in [44] where each agent probes the opinion of a pool of m randomly selected agents. The agent then chooses either to conform to the majority opinion or to be contrarian if the majority is too strong. This interesting model leads to various types of behaviour, including a chaotic phase.

4.2 Avalanches of opinion changes

The above simple percolation model is interesting but has one major drawback: one has to assume that the parameter c is smaller than one, but relatively close to one such that Eq. (14) is valid, and non trivial statistics follows. One should thus explain why the value of c spontaneously stabilises in the neighbourhood of the critical value $c = 1$. Furthermore, this model is purely static, and does not specify how the above clusters evolve with time. As such, it cannot address the problem of volatility clustering. Several extensions of this simple model have been proposed

[43,45], in particular to increase the value of μ from $\mu = 3/2$ to $\mu \sim 3$ and to account for volatility clustering.

One particularly interesting model is inspired by the recent work of Dahmen and Sethna [46], that describes the behaviour of random magnets in a time dependent magnetic field. Transposed to the present problem (as first suggested in [12]), this model describes the collective behaviour of a set of traders exchanging information, but having all different *a priori* opinions. One trader can however change his mind and take the opinion of his neighbours if the coupling is strong, or if the strength of his *a priori* opinion is weak. More precisely, the opinion $\varphi_i(t)$ of agent i at time t is determined as:

$$\varphi_i(t) = \operatorname{sign}\left(h_i(t) + \sum_{j=1}^{N} J_{ij}\varphi_j(t)\right), \tag{15}$$

where J_{ij} is a connectivity matrix describing how strongly agent j affects agent i, and $h_i(t)$ describes the *a priori* opinion of agent i: $h_i > 0$ means a propensity to buy, $h_i < 0$ a propensity to sell. We assume that h_i is a random variable with a time dependent mean $\overline{h}(t)$ and root mean square Δ. The quantity $\overline{h}(t)$ represents for example confidence in long term economy growth ($\overline{h}(t) > 0$), or fear of recession ($\overline{h}(t) < 0$, leading to a non zero average pessimism or optimism. Suppose that one starts at $t = 0$ from a 'euphoric' state, where $\overline{h} \gg \Delta, J$, such that $\varphi_i = 1$ for all i's.[f] Now, confidence is decreased progressively. How will sell orders appear ?

Interestingly, one finds that for small enough influence (or strong heterogeneities of agents' anticipations), i.e. for $J \ll \Delta$, the average opinion $O(t) = \sum_i \varphi_i(t)/N$ evolves continuously from $O(t = 0)) = +1$ to $O(t \to \infty) = -1$. The situation changes when imitation is stronger since a discontinuity then appears in $O(t)$ around a 'crash' time t_c, when a *finite* fraction of the population simultaneously change opinion. The gap $O(t_c^-) - O(t_c^+)$ opens continuously as J crosses a critical value $J_c(\Delta)$ [46]. For J close to J_c, one finds that the sell orders again organise as avalanches of various sizes, distributed as a power-law with an exponential cut-off. In the 'mean-field' case where $J_{ij} \equiv J/N$ for all ij, one finds $\mu = 5/4$. Note that in this case, the value of the exponent μ *is universal*, and does not depend, for example, on the shape of the distribution of the h_i's, but only on some global properties of the connectivity matrix J_{ij}.

A slowly oscillating $\overline{h}(t)$ can therefore lead to a succession of bull and bear markets, with a strongly non Gaussian, intermittent behaviour, since most of the activity is concentrated around the crash times t_c. Some modifications of this model are however needed to account for the empirical value $\mu \sim 3$ observed on the distribution of price increments (see the discussion in [43]).

Note that the same model can be applied to other interesting situations, for example to describe how applause end in a concert hall [47] (here, $\varphi = \pm 1$ describes, respectively, a clapping and a quiet person, and $O(t)$ is the total clapping activity). Clapping can end abruptly when imitation is strong, or smoothly when many fans are present in the audience. A static version of the same model has been proposed to describe rational group decision making [48].

---

[f] Here J denotes the order of magnitude of $\sum_j J_{ij}$

5 Models of feedback effects on price fluctuations

5.1 *Risk-aversion induced crashes*

The above average 'stimulus' $\overline{h}(t)$ may also strongly depend on the past behaviour of the price itself. For example, past positive trends are, for many investors, incentives to buy, and vice-versa. Actually, for a given trend amplitude, price drops tend to feedback more strongly on investors' behaviour than price rises. Risk-aversion creates an asymmetry between positive and negative price changes [49]. This is reflected by option markets, where the price of out-of-the-money puts (i.e. insurance against crashes) is anomalously high.

Similarly, past periods of high volatility increases the risk of investing in stocks, and usual portfolio theories then suggest that sell orders should follow. A simple mathematical transcription of these effects is to write Eq. (12) in a linearized, continuous time form:[g]

$$\frac{dx}{dt} \equiv u = \frac{1}{\lambda}\overline{h}(t), \tag{16}$$

and write a dynamical equation for $\overline{h}(t)$ which encodes the above feedback effects [49,38]:

$$\frac{d\overline{h}}{dt} = au - bu^2 - cu + \eta(t), \tag{17}$$

where a describes trends following effects, b risk aversion effects (breaking the $u \to -u$ symmetry), c is a mean reverting term which arises from market clearing mechanisms (the very fact that the price moves clears a certain number of orders), and η is a noise term representing random external news [49]. Eliminating $\overline{h}$ from the above equations leads to:

$$\frac{du}{dt} = -\gamma u - \beta u^2 + \frac{1}{\lambda}\eta(t) \equiv -\frac{\partial V}{\partial u} + \frac{1}{\lambda}\eta(t) \tag{18}$$

where $\gamma = (c-a)/\lambda$ and $\beta = b/\lambda$. Equation (18) represents the evolution of the position u of a viscous fictitious particle in a 'potential' $V(u) = \gamma u^2/2 + \beta u^3/3$. If trend following effects are not too strong, γ is positive and $V(u)$ has a local minimum for $u = 0$, and a local maximum for $u^* = -\gamma/\beta$, beyond which the potential plummets to $-\infty$.[h] A 'potential barrier' $V^* = \gamma u^{*2}/6$ separates the (meta-)stable region around $u = 0$ from the unstable region. The nature of the motion of u in such a potential is the following: starting at $u = 0$, the particle has a random harmonic-like motion in the vicinity of $u = 0$ until an 'activated' event (i.e. driven by the noise term) brings the particle near u^*. Once this barrier is crossed, the fictitious particle reaches $-\infty$ in finite time. In financial terms, the regime where u oscillates around $u = 0$ and where β can be neglected, is the 'normal' fluctuation regime. This normal regime can however be interrupted by 'crashes', where the time derivative of the price becomes very large and negative,

[g] In the following, the herding effects described by J_{ij} are neglected, or more precisely, only their *average effect* encoded by $\overline{h}$ is taken into account.

[h] If γ is negative, the minimum appears for a positive value of the return u^*. This corresponds to a speculative bubble. See [49].

due to the risk aversion term b which destabilizes the price by amplifying the sell orders. The interesting point is that these two regimes can be clearly separated since the average time t^* needed for such crashes to occur is exponentially large in V^* [50], and can thus appear only very rarely. A very long time scale is therefore naturally generated in this model.

Note that in this line of thought, a crash occurs because of an improbable succession of unfavorable events, and not due to a single large event in particular. Furthermore, there are no 'precursors' in this model: before u has reached u^*, it is impossible to decide whether it will do so or whether it will quietly come back in the 'normal' region $u \simeq 0$. Solving the Fokker-Planck equation associated to the Langevin equation (18) leads to a stationary state with a power law tail for the distribution of u (i.e. of the instantaneous price increment) decaying as u^{-2} for $u \to -\infty$. More generally, if the risk aversion term took the form $-bu^{1+\mu}$, the negative tail would decay as $u^{-1-\mu}$.

5.2 *Dynamical volatility models*

The simplest model that describes volatility feedback effects is to write an ARCH like equation [51], which relates today's activity to a measure of yesterday's activity, for example:

$$\sigma_k = \sigma_{k-1} + K(\sigma_0 - \sigma_{k-1}) + g|\delta x_{k-1}|, \tag{19}$$

where σ_0 is an average volatility level, K a mean-reverting term, and g describes how much yesterday's observed price change affects the behaviour of traders today. Since $|\delta x_{k-1}|$ is a noisy version of σ_{k-1}, the above equation is, in the continuous time limit, a Langevin equation with *multiplicative* noise:

$$\frac{d\sigma}{dt} = K(\sigma_0 - \sigma) + g\sigma\eta(t), \tag{20}$$

which is, up to notation changes, exactly the same equation as (9) above. Therefore, the stationary distribution of the volatility in this model is again given by Eq. (11), with the tail exponent now given by $\mu - 1 \propto K/g^2$: over-reactions to past informations (i.e. large values of g) decreases the tail exponent μ. Therefore, one again finds a non universal exponent in this model, which is bequeathed to the distribution of price increments if one assumes that the 'sign' contribution to δx_k (see Eq. (3)) has thin tails.

Note that the temporal correlation function of the volatility σ can be exactly calculated within this model [52], and is found to be exponentially decaying, at variance with the slow power-law (or logarithmic) decay observed empirically. Furthermore, the distribution (11) does not concur with the nearly log-normal distribution of the volatility that seems to hold empirically [23,26].

At this point, the slow decay of the volatility can be ascribed to two rather different mechanisms. One is the existence of traders with many different time horizons, as suggested in [53,22]. If traders are affected not only by yesterday's price change amplitude $|\delta x_{k-1}|$, but also by price changes on coarser time scales $|x_k - x_{k-p}|$, then the correlation function is expected to be a sum of exponentials with decay rates given by p^{-1}. Interestingly, if the different p's are uniformly distributed on a log scale, the resulting sum of exponentials is to a good approximation decaying as a logarithm. More precisely:

$$C(\tau) = \frac{1}{\log(p_{\max}/p_{\min})} \int_{p_{\min}}^{p_{\max}} d(\log p) \exp(-\tau/p) \simeq \frac{\log(p_{\max}/\tau)}{\log(p_{\max}/p_{\min})}, \tag{21}$$

whenever $p_{\min} \ll \tau \ll p_{\max}$. Now, the human time scales are indeed in a natural pseudo-geometric progression: hour, day, week, month, trimester, year [22].

Yet, some recent numerical simulations of traders allowed to switch between different strategies (active/inactive, or chartist/fundamentalist) suggest strongly intermittent behaviour [54,55,45,56,57], and a slow decay of the volatility correlation function without the explicit existence of logarithmically distributed time scales. Is there a simple, universal mechanism that could explain these ubiquitous long range volatility correlations ?

A possibility, discussed in [58], is that the volume of activity exhibits long range correlations because agents switch between different strategies depending on their relative performance. Imagine for example that each agent has two strategies, one active strategy (say trading every day), and one inactive, or less active strategy. The 'score' of the inactive strategy (i.e. its cumulative profit) is constant in time, or more precisely equal to the long term average growth rate. The score of active strategy, on the other hand, fluctuates up and down, due to the fluctuations of the market prices themselves. Since to a good approximation the market prices are not predictable, this means that the score of any active strategy will behave like a random walk, with an average equal to that of the inactive strategy (assuming that transaction costs are small). Therefore, on some occasions the score of the active strategy will happen to be higher than that of the inactive strategy and the agent will be active, before the score of the active strategy crosses that of the inactive strategy. The time during which an agent is active is thus a random variable with the same statistics as the return time to the origin of a random walk (the difference of the scores of the two strategies). Interestingly, the return times of a random walk are well known to be very broadly distributed: the *average* return time is actually infinite. Hence, if one computes the correlation of activity in such a model, one finds long range correlations due to long periods of times where many agents are active (or inactive). We have considered a specific model (the 'Grand Canonical Minority Game' [59]) where this scenario can be studied more quantitatively, and have found that indeed long range correlations in the volume of activity are observed. This model even allows one to reproduce quantitatively the volume of activity correlations observed on the New-York stock exchange market: see Fig. 1. This mechanism is very generic and probably also explains why this effect arises in more realistic market models [54,57].

As shown in [58], this mechanism can thus explain the long-range volatility correlations observed on all financial markets. However, this interpretation is quite different from the 'cascade' picture proposed in [25,29,28], where the volatility results from some sort of multiplicative random process (which actually naturally leads to the log-normal volatility distribution actually observed empirically). More precise statistical tests allowing one to discriminate between these two scenarii would be welcome.

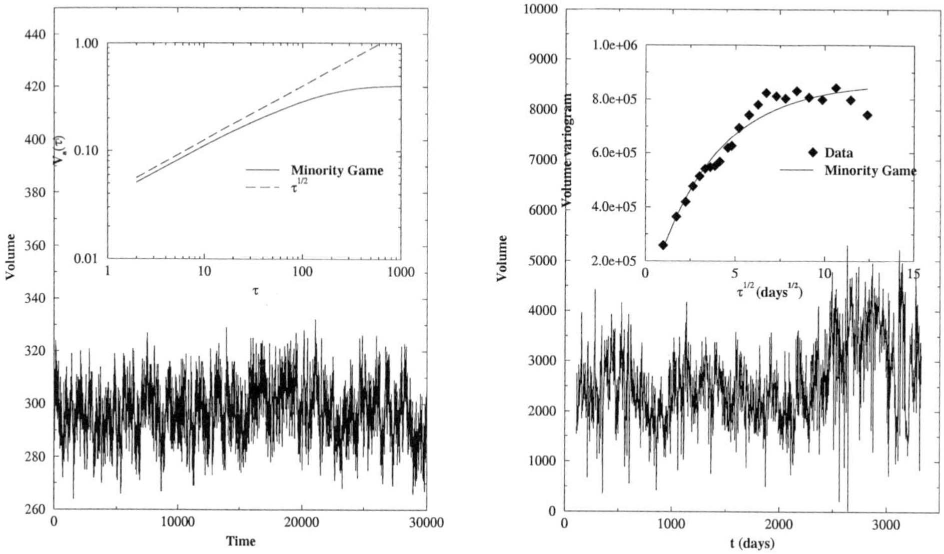

Figure 1. 1-a: Volume of activity $v(t)$ (number of active agents) as a function of time for the Grand Canonical Minority Game (GCMG). Inset: The corresponding activity variogram $\langle (v(t+\tau) - v(t))^2 \rangle$, as a function of the lag τ, in a log-log plot to emphasize the predicted $\sqrt{\tau}$ singularity at small τ's. 1-b: Total daily volume of activity (number of trades) on the S&P 500 futures contracts in the years 1985-1998. Inset: Corresponding variogram (diamonds) as a function of the square-root of the lag. Note the clear linear behaviour for small $\sqrt{\tau}$. The full line is the GCMG fit, with both axis rescaled and a constant added to account for the presence of 'white noise' trading.

6 Concluding remarks

Many of the ideas presented above will perhaps turn out to be wrong, but will hopefully motivate some further work to understand the origin of power-law *distributions* and power-law *correlations* in financial time series. From an empirical point of view, the exponents describing the tails of the price increments distribution and the decay of the volatility correlations are rather robust and suggest some kind of universality, probably related to the fact that all speculative markets obey common rules where simple human psychology (greed and fear) coupled to basic mechanisms of price formation ultimately lead to the emergence of scaling and power-laws. Still, many points remain obscure: we have seen above that models that appear naturally in the context of economics and finance contain multiplicative noise, which is a simple mechanism for power-law distributions (as emphasized in, e.g. [33,43]). However, these models generically lead to *non universal exponents* (as discussed above in the context of the Pareto tails). So, are the exponents observed on empirical data *universal* and related to some kind of criticality of the underlying process, or non universal, and arising from multiplicative like effects ? Why is the volatility distribution so close to a log-normal ? Is this an accidental coincidence or does this reveal some multiplicative cascade mechanism, the interpretation of which in terms of agent based models being far from obvious.

Acknowledgments

I want thank in particular R. Cont, I. Giardina, A. Matacz, M. Mézard, and M. Potters for many discussions. This text is a modified and updated version of an article that appeared in Quantitative Finance, **1**, 105 (2001)

References

1. see e.g. N. Goldenfeld, *Lectures on phase transitions and critical phenomena*, Frontiers in Physics, 1992, for an excellent introduction.
2. U. Frisch, *Turbulence: The Legacy of A. Kolmogorov*, Cambridge University Press (1997).
3. V. Pareto, Cours d'économie politique. Reprinted as a volume of *Oeuvres Complètes* (Droz, Geneva, 1896-1965).
4. B.B. Mandelbrot, Int. Eco. Rev. **1** (1960) 79, B. B. Mandelbrot, *Journal of Business* **36**, 394 (1963); ibid. **40**, 394 (1967).
5. B.B. Mandelbrot, *Fractals and Scaling in Finance*, Springer (1997).
6. R. Mantegna & H. E. Stanley, *An introduction to Econophysics*, Cambridge University Press, 1999.
7. J. D. Farmer, *Physicists attempt to scale the ivory towers of finance*, in Computing in Science and Engineering, November 1999.
8. A. Dragulescu, V. M. Yakovenko, *Exponential and power-law probability distributions of wealth and income in the United Kingdom and the United States*, e-print cond-mat/0103544, to appear in Physica A, proceedings of NATO workshop Applications of Physics in Economic Modeling, Prague, February 2001.
9. A. B. Atkinson, A. J. Harrison, *Distribution of total wealth in Britain* (Cambridge University Press, 1978), Y. Ijri, H. A. Simon, *S kew distribution of sizes of Business Firms* (North-Holland, Amsterdam),
10. M. H. R. Stanley, S. Buldyrev, S. Havlin, R. Mantegna, M. Salinger, H. E. Stanley, Eco. Lett. **49** (1995) 453.
11. H. Aoyama, Y. Nagahara, M. Okasaki, W. Souma, H. Takayasu, M. Takayasu, *Pareto's law for income of individuals and debt of bankrupt companies*, e-print cond-mat/0006038.
12. J.-P. Bouchaud and M. Potters, *Théorie des Risques Fin anciers*, Aléa-Saclay, 1997; *Theory of Financial Risks*, Cambridge University Press, 2000.
13. V. Plerou, P. Gopikrishnan, L.A. Amaral, M. Meyer, H.E. Stanley, Phys. Rev. **E60** 6519 (1999).
14. T. Lux, Applied Financial Economics, **6**, 463, (1996).
15. M. M. Dacorogna, U. A. Müller, R. J. Nagler, R. B. Olsen and O. V. Pictet, *J. Inter. Money and Finance* **12**, 413 (1993); D. M. Guillaume, M. M. Dacorogna, R. D. Davé, U. A. Müller, R. B. Olsen and O. V. Pictet, *Finance and Stochastics* **1** 95 (1997).
16. F. Longin, Journal of Business, **69** 383 (1996)
17. J. Nuyts, I. Platten, *Phenomenology of the term structure of interest rates with*

Padé approximants, e-print cond-mat/9901096.
18. V. Plerou, P. Gopikrishnan, L.A. Amaral, X. Gabaix, H.E. Stanley, e-print cond-mat/9912051.
19. A. Lo, Econometrica, **59**, 1279 (1991).
20. Z. Ding, C. W. J. Granger and R. F. Engle, *J. Empirical Finance* **1**, 83 (1993).
21. R. Cont, M. Potters, J.-P. Bouchaud, *Scaling in stock market data: stable laws and beyond*, in *Scale invariance and beyond*, B. Dubrulle, F. Graner, D. Sornette (Edts.), EDP Sciences (1997).
22. M. Potters, R. Cont, J.-P. Bouchaud, *Europhys. Lett.* **41**, 239 (1998).
23. Y. Liu, P. Cizeau, M. Meyer, C.-K. Peng, H. E. Stanley, *Physica A***245** 437 (1997), P. Cizeau, Y. Liu, M. Meyer, C.-K. Peng, H. E. Stanley, *Physica A***245** 441 (1997).
24. G. Bonnano, F. Lillo, R. Mantegna, *Dynamics of the number of trades in financial securities*, e-print cond-mat/9912006
25. The first paper to suggest this analogy is S. Ghashghaie, W. Breymann, J. Peinke, P. Talkner, Y. Dodge, *Nature* **381** 767 (1996). See however [6,27,28], and A. Arneodo, J.P. Bouchaud, R. Cont, J.F. Muzy, M. Potters, D. Sornette, cond-mat/9607120, unpublished.
26. The distribution of individual S&P stock volatilities is also found to be nearly log-normal: Science & Finance unpublished report, (2000).
27. A. Arnéodo, J.-F. Muzy, D. Sornette, Eur. Phys. J. **B 2**, 277 (1998)
28. see in particular J.-F. Muzy, J. Delour, E. Bacry, Eur. Phys. J. B 17, 537-548 (2000), and to appear in Physica A.
29. A. Fisher, L. Calvet, B.B. Mandelbrot, 'Multifractality of DEM/$ rates', Cowles Foundation Discussion Paper 1165; B.B. Mandelbrot, *Scientific American*, Feb. (1999).
30. F. Schmitt, D. Schertzer, S. Lovejoy, Applied Stochastic Models and Data Analysis, **15** 29 (1999); M.-E. Brachet, E. Taflin, J.M. Tchéou, *Scaling transformation and probability distributions for financial time series*, e-print cond-mat/9905169
31. J.P. Bouchaud, M. Potters, M. Meyer, Eur. Phys. J. **B 13**, 595 (1999).
32. J.P. Bouchaud, M. Mézard, Physica A, **282**, 536 (2000).
33. Similar models were discussed in: S. Solomon, in *Computational Finance 97*, eds. A-P. N. Refenes, A.N. Burgess, J.E. Moody (Kluwer Academic Publishers 1998); cond-mat/9803367; O. Malcai O. Biham and S. Solomon, Phys. Rev. E, 60, 1299 (1999); M. Marsili, S. Maslov, Y.C. Zhang, Physica A 253, 403 (1998); S. Solomon, P. Richmond, e-print/cond-mat 0102423.
34. D.S. Fisher, C. Henley, D.A. Huse, Phys. Rev. Lett. **55** 2924 (1985).
35. B. Derrida, H. Spohn, J. Stat. Phys. **51** (1988) 817
36. For an introduction and references, see the very interesting book of R. Schiller, *Irrational Exuberance*, Princeton University Press (2000).
37. R. Cont, J.P. Bouchaud, Macroeconomics Dynamics, **4**, 170 (2000). See the numerous references of this paper for other works on herding in economics and finance.
38. J.D. Farmer, *Market Force, Ecology and Evolution*, e-print adap-org/9812005.
39. J.P. Bouchaud, A. Matacz, M. Potters, *The leverage effect in financial markets:*

retarded volatility and market panic, e-print cond-mat/0101120, to appear in Phys. Rev. Lett. (2001)
40. Y.C. Zhang, Physica **A269** 30 (1999).
41. V. Plerou, P. Gopikrishnan, X. Gabaix, H. E. Stanley *Quantifying Stock Price Response to Demand Fluctuations*, e-print cond-mat/0106657.
42. D. Stauffer, A. Aharony, *Introduction to percolation*, Taylor-Francis, London, 1994.
43. D. Sornette, D. Stauffer, H. Takayasu, *Market fluctuations, multiplicative and percolation models, size effects and predictions*, e-print cond-mat/9909439, and references therein.
44. A. Corcos, J.-P. Eckmann, A. Malaspinas, Y. Malevergne, D. Sornette, *Imitation and contrarian behavior: hyperbolic bubbles, crashes and chaos*, e-print cond-mat/0109410
45. G. Iori, *A microsimulation of traders activity in the stock market: the role of heterogeneity, agents' interactions and trade frictions*, e-print adap-org/9905005.
46. K. Dahmen, J. P. Sethna, Phys. Rev. **B 53**, 14872 (1996), J. P. Sethna, K. Dahmen, C. Myers, Nature, **410**, 242 (2001).
47. R. da Silvera, J.P. Bouchaud, in preparation.
48. S. Galam, Physica **A238**, 66 (1997).
49. J.-P. Bouchaud, R. Cont, European Journal of Physics **B 6**, 543 (1998).
50. S. Chandraseckar, Rev. Mod. Phys. **15** 1 (1943), reprinted in 'Selected papers on noise and stochastic processes', Dover (1954).
51. see, e.g. C. Gourieroux, A. Montfort, *Statistics and Econometric Models*, Cambridge University Press, 1996.
52. R. Graham, A. Schenzle, Phys. Rev. **A 25** (1982) 1731.
53. U.A. Muller, M.M. Dacorogna, R. Davé, R. B. Olsen, O.V. P ictet, Journal of Empirical Finance, **4**, 213 (1997).
54. T. Lux, M. Marchesi, *Nature* **397**, 498 (1999).
55. P. Bak, M. Paczuski, and M. Shubik, Physica A **246**, 430 (1997)
56. C. Hommes, Quantitative Finance, **1**, 149 (2001)
57. I. Giardina, J.P. Bouchaud, M. Mézard, e-print/cond-mat/0105076, Physica A, in press, and I. Giardina, J.P. Bouchaud, in preparation.
58. I. Giardina, J.P. Bouchaud, M. Mézard, Quantitative Finance, **1**, 212 (2001)
59. For papers on the Minority Game, see e.g. D. Challet, A. Chessa, M. Marsili, Y.C. Zhang, Quantitative Finance, **1**, 168 (2001) and refs. therein.

FACING NON-STATIONARY CONDITIONS WITH A NEW INDICATOR OF ENTROPY INCREASE: THE CASSANDRA ALGORITHM

P. ALLEGRINI

Istituto di Linguistica Computazionale del Consiglio Nazionale delle Ricerche, Area della Ricerca di Pisa–S. Cataldo, Via Moruzzi 1, 56124, Pisa, Italy
Email: allegrip@ilc.pi.cnr.it

P. GRIGOLINI

Center for Nonlinear Science, University of North Texas, P.O. Box 5368, Denton, TX 76203;
Istituto di Biofisica del Consiglio Nazionale delle Ricerche, Area della Ricerca di Pisa–S. Cataldo, Via Moruzzi 1, 56124, Pisa, Italy
Email: grigo@unt.edu

P. HAMILTON

Center for Nonlinear Science, Texas Woman's University, P.O. Box 425498, Denton, Texas 76204
Email: phamilton@twu.edu

L. PALATELLA, G. RAFFAELLI, M. VIRGILIO

Dipartimento di Fisica dell'Università di Pisa and INFM Piazza Torricelli 2, 56127 Pisa, Italy
Email: palatell@df.unipi.it, jms@katamail.com, michele555@katamail.com

We address the problem of detecting non-stationary effects in time series (in particular fractal time series) by means of the Diffusion Entropy Method (DEM). This means that the experimental sequence under study, of size N, is explored with a window of size $L << N$. The DEM makes a wise use of the statistical information available and, consequently, in spite of the modest size of the window used, does succeed in revealing local statistical properties, and it shows how they change upon moving the windows along the experimental sequence. The method is expected to work also to predict catastrophic events before their occurrence.

1 Introduction

The main aim of this paper is to illustrate a promising strategy to study non-stationary processes. We prove that the method is efficient by means of the joint study of real and artificial sequences, and we reach a conclusion that makes it plausible to imagine the method at work to successfully predict the time of occurrence of catastrophic events.

The method here illustrated is a suitable extension of the Diffusion Entropy Method (DEM). The DEM is discussed in details in other publications[1,2,3]. Here we limit ourselves to give a concise illustration of this technique so as to allow the reader to understand the spirit of the method of this paper, at least through a first reading, without consulting these earlier publications. The first step of this technique is the same as that of the pioneering work of Refs. [4,5]. This means that the experimental sequence is converted into a kind of Brownian-like trajectory. The second step aims at deriving many distinct diffusion trajectories with the technique of moving windows of size l. The reader should not confuse the mobile vindow of size l with the mobile window of value L that will be used later on in this paper to detect non-stationary properties. For this reason we shall refer to the mobile windows of size L as large windows, even if the size of L is relatively small, whereas the mobile windows of size l will be called small windows. The large mobile window has to be interpreted as a sequence with statistical properties to reveal, and will be analyzed by means of small windows of size l, with $l < L$. The success of the method depends on the fact that the DE makes a wise use of the statistical information available. In fact, the small windows overlap among themselves and are obtained by locating their left border on the first site of the sequence, on the second second site, and so on. The adoption of overlapping windows is dictated by the wish to establish a connection with the Kolomogorov-Sinai (KS)[6,7] method. This yields also, as a further beneficial effect, many more trajectories than the widely used method of Detrended Fluctuation Analysis (DFA)[4,5].

In conclusion, we create a conveniently large number of trajectories by gradually moving the small window from the first position, with the left border of the small window coinciding with the first size of the sequence, to the last position, with the right border of the small window coinciding with the last site of the sequence. Atfter this stage, we utilize the resulting trajectories, all of them with the initial position located at $x = 0$, to produce a probability distribution at "time" l. We evaluate the Shannon entropy of this distribution, $S_d(l)$, and with easy mathematical arguments we prove that, if the diffusion process undergoes a scaling of intensity δ, then

$$S_d(l) = A + \delta ln(l). \tag{1}$$

Thus the parameter δ of the scaling condition, if this condition applies, can be measured without recourse to any form of detrending.

This is the original motivation for the DEM[1,2,3]. In this paper we want to prove that the DEM does much more than detecting scaling. In a stationary condition the DEM not only detects with accuracy the final scaling but it also affords a way to monitor the regime of transition to the final *thermodynamic* condition. If the sequence under study is affected by biases and non-stationary perturbations, the attainment of the final regime of steady scaling can be cancelled, and replaced by an *out of equilibrium* regime changing in time under the influence of time dependent biases. We want to prove that the DEM can be suitably extended to face this challenging non-stationary condition.

The outline of the paper is as follows. In Section 2 we shortly review the main tenets of the DEM so as to make this paper, as earlier mentioned, as self-contained as possible. In Section 3 we illustrate the extension of the DEM and we discuss the fundamental problem of assessing which is the shortest portion, of size L, of the sequence under study, which is still large enough as to make the DEM work. In Section 3 we express our hope that this new method might serve prediction purposes.

2 Diffusion Entropy

Let us consider a sequence of M numbers ξ_i , with $i = 1, \ldots, M$. The purpose of the DEM algorithm is to establish the possible existence of a scaling, either normal or anomalous, in the most efficient way as possible, without altering the data with any form of detrending. Let us select first of all an integer number l, fitting the condition $1 \leq l \leq M$. As earlier mentioned, we shall refer ourselves to l as "time". For any given time l we can find $M - l + 1$ sub-sequences defined by

$$\xi_i^{(s)} \equiv \xi_{i+s}, \qquad s = 0, \ldots, M - l. \tag{2}$$

For any of these sub-sequences we build up a diffusion trajectory, labelled with the index s, defined by the position

$$x^{(s)}(l) = \sum_{i=1}^{l} \xi_i^{(s)} = \sum_{i=1}^{l} \xi_{i+s}. \tag{3}$$

Let us imagine this position as referring to a Brownian particle that at regular intervals of time has been jumping forward or backward according to the prescription of the corresponding sub-sequence of Eq.(2). This means that the particle before reaching the position that it holds at time l has been making l jumps. The jump made at the i-th step has the intensity $|\xi_i^{(s)}|$ and is forward or backward according to whether the number $\xi_i^{(s)}$ is positive or negative.

We are now ready to evaluate the entropy of this diffusion process. To do that we have to partition the x-axis into cells of size $\epsilon(l)$. When this partition is made we have to label the cells. We count how many particles are found in the same cell at a given time l. We denote this number by $N_i(l)$. Then we use this number to determine the probability that a particle can be found in the i-th cell at time l, $p_i(l)$, by means of

$$p_i(l) \equiv \frac{N_i(l)}{(M - l + 1)}. \tag{4}$$

At this stage the entropy of the diffusion process at the time l is determined and reads

$$S_d(l) = -\sum_i p_i(l) ln[p_i(l)]. \tag{5}$$

The easiest way to proceed with the choice of the cell size, $\epsilon(l)$, is to assume it independent of l and determined by a suitable fraction of the square root of the variance of the fluctuation $\xi(i)$.

Before proceeding with the illustration of how the DEM method works, it is worth making a comment on the way we use to define the trajectories. The method we are adopting is based on the idea of a moving window of size l that makes the $s-th$ trajectory closely correlated to the next, the $(s+1)-th$ trajectory. The two trajectories have $l-1$ values in common. A motivation for our choice is given by our wish to establish a connection with the Kolmogorov Sinai (KS) entropy[6,7]. The KS entropy of a symbolic sequence is evaluated by moving a window of size l along the sequence. Any window position corresponds to a given combination of symbols, and, from the frequency of each combination, it is possible to derive the Shannon entropy $S(l)$. The KS entropy is given by the asymptotic limit $lim_{l\to\infty}S(l)/l$. We believe that the same sequence, analyzed with the DEM method, at the large values of l where $S(l)/l$ approaches the KS value, must yield a well defined scaling δ. To realize this correspondence we carry out the determination of the Diffusion Entropy using the same criterion of overlapping windows as that behind the KS entropy.

Details on how to deal with the transition from the short-time regime, sensitive to the discrete nature of the process under study, to the long-time limit where both space an time can be perceived as continuous, are given in Ref. [1,3]. Here we make the simplifying assumption of considering so large times as to make the continuous assumption valid. In this case the trajectories, built up with the above illustrated procedure, correspond to the following equation of motion:

$$\frac{dx}{dt} = \xi(t), \tag{6}$$

where $\xi(t)$ denotes the value that the time series under study gets at the $l-th$ site. This means that the function $\xi(l)$ is depicted as a function of t, thought of as a continuous time $t = l$. In this case the Shannon entropy reads

$$S_d(t) = -\int_{-\infty}^{\infty} dx\, p(x,t) ln[p(x,t)]. \tag{7}$$

We can derive with a simple treatment an analytical solution for Diffusion Entropy when the process is characterized by *scaling*, namely when

$$p(x,t) = \frac{1}{t^{\delta}} F\left(\frac{x}{t^{\delta}}\right). \tag{8}$$

Let us plug Eq.(8) into Eq.(7). After a simple algebra, we get:

$$S_d(\tau) = A + \delta(\tau)\tau, \tag{9}$$

where

$$A \equiv -\int_{-\infty}^{\infty} dy\, F(y)\, ln[F(y)] \tag{10}$$

and

$$\tau \equiv \ln(t). \tag{11}$$

It is evident that this kind of technique to detect scaling does not imply any form of detrending, and this is one of the reasons why some attention should be devoted to it. It is also worth mentioning, as we prove now, that it yields the correct scaling values even for the so-called *Lévy walks*, where the time dependence of the second moment with respect to time has an exponent which is different from the scaling exponent of the Lévy process.

We therefore check the efficiency of this technique by the studying the artificial sequence of Refs. [8,9]. This sequence is built up in such a way as to realize long sequences of either +'s or −'s. The probability of finding a sequence of only +'s or only −'s of length t is given by

$$\psi(t) = (\mu - 1)\frac{T^{\mu-1}}{(t+T)^{\mu}}. \tag{12}$$

Here we focus our attention on the condition $\mu < 3$ and we raise the reader's attention on the interval $2 \leq \mu \leq 3$. In fact, this kind of sequence is the same as that adopted in earlier work [9] for a dynamic derivation of Lévy diffusion, which shows up when the condition $2 < \mu < 3$ applies. It corresponds to a particle travelling with constant velocity throughout the whole time interval corresponding to either only +'s or only −'s, and changing direction with no rest, at the end of any string with the same symbols.

We will refer to this model as *Symmetric Velocity Model* (SVM). We know from the theory of Ref.[9] that the scaling of the resulting diffusion process when $2 < \mu < 3$ is

$$\delta = \frac{1}{\mu - 1}. \tag{13}$$

Note, however, that this diffusion process has a finite propagation front, with ballistic peaks showing up at both $x = t$ and $x = -t$. The intensity of these peaks is proportional to the correlation function

$$\Phi_{\xi}(t) = \left(\frac{T}{T+t}\right)^{\mu-2}. \tag{14}$$

As a consequence of this fact, the whole distribution does not have a single rescaling. In fact, the distribution enclosed between the two peaks rescales with δ of Eq.(13) while the peaks are associated to $\delta = 1$. Furthermore, it is well known[9] that the scaling of the second moment is given by

$$\delta_H = \frac{4-\mu}{2}. \tag{15}$$

Thus, it is expected that the scaling detected by the DE method might not coincide with the prediction of Eq.(13) for the whole period of time corresponding to the presence of peaks of significant intensity. We think that the Lévy scaling of Eq.(13) will show up at long times, when the peak intensity is significantly reduced. This conjecture seems to be supported by the numerical results illustrated in Fig.1. We see in fact that the scaling predicted by Eq.(13) is reached after an extended transient, of the order of about 20,000 in the scale of Fig.1. This time interval is about 2000 larger than the value assigned to the parameter T, of Eq.(12), which is, in fact, in the case of Fig.1, T= 10.

In conclusion, this section proves that the DE method applied to the SVM yields, for the scaling parameter δ, the correct value of Eq.(13), rather than the value that would be obtained measuring the variance of the diffusion process, Eq.(15). However, the time necessary to make this correct value emerge is very large. Furthermore, as shown in Fig. 2, the adoption of SVM would make the scaling parameter δ insensitive to μ in the whole interval $1 \leq \mu \leq 2$. This means that the adoption of the DE method would not allow us to distinguish a process with μ very close to

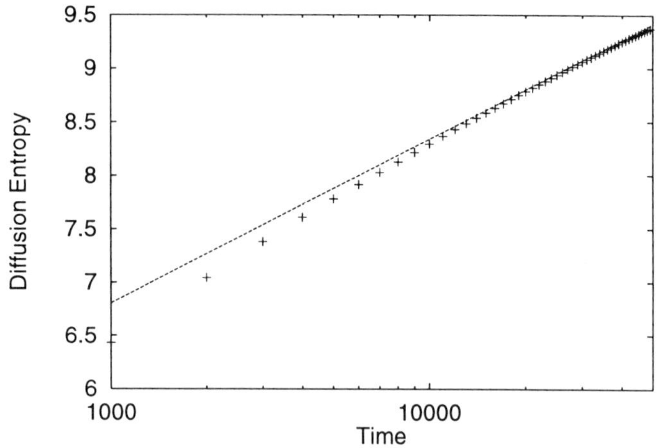

Figure 1. The diffusion entropy as a function of time. The numerical method is applied to the artificial sequence described above, with $\mu = 2.5$, studied according to the SVM prescription. According to the theoretical arguments of the text the scaling parameter δ is the slope of the straight line fitting the numerical results at large times, which yields in this case $\delta = 2/3 = 1/(\mu - 1)$ $(T = 10)$.

1 from one with μ very close to 2. This problem can be solved using different rules for the diffusion process [3]: the random walker can, for instance, walk always in the same direction, and at the "time" when there is a passage from a laminar region of $+$'s to one of $-$'s and vice-versa. If this latter rule is adopted, then it is easy to prove [3] that the resulting $p(x,t)$ is an asymmetric Lévy distribution, with a scaling $\delta = \mu - 1$ for $1 \leq \mu \leq 2$.

Throughout this paper, however, apply the DEM with the rule corresponding to the symmetric velocity model, with δ depending on μ as in Fig. 2. In the regime of ordinary statistical mechanics ($\mu \gg 3$) the ordinary scaling is quickly attained, while the condition of anomalous statistical mechanics $\mu < 3$ is characterized by a long transient regime, which is carefully recorded by the DEM. In this paper we want to use the DEM to monitor the time dependence of the "rules" generating the sequences under study.

3 The new method at work with nonstationary sequences

To illustrate the ideas that led us to propose the method of analysis with two moving window, let us begin with discussing the artificial sequence given by

$$\xi_b(t) = \kappa\xi(t) + Acos(\omega t). \tag{16}$$

The second term on the right hand side of this equation is a deterministic contribution that might mimic, for instance, the season periodicity of Ref. [1]. The first term on the right hand side is a fluctuation with no correlation that can be

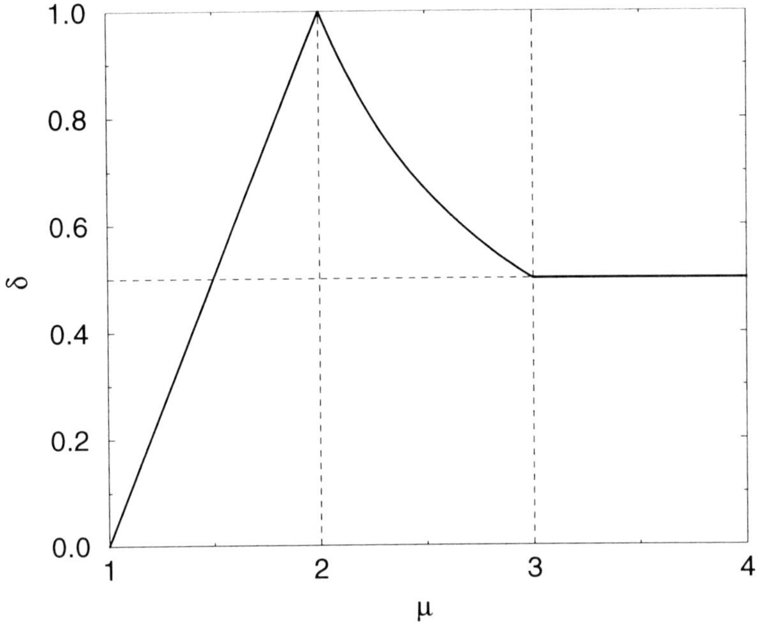

Figure 2. δ as a function of μ according to the prescriptions of Ref. [3].

correlated or not to the harmonic bias.

Fig. 3 refers to the case when the random fluctuation has no correlation with the harmonic bias. It is convenient to illustrate what happens when $\kappa = 0$. This is the case where the signal is totally deterministic. It would be nice if the entropy in this case did not increase upon increasing l. However, we must notice that the method of mobile windows implies that many trajectories are selected, the difference among them being, in the determinist case where $\xi_b(t) = Acos(\omega t)$, a difference on initial conditions. Entropy initially increases. This is due to the fact that the statistical average on the initial conditions is perceived as a source of uncertainty. However, after a time of the order of the period of the deterministic process a regression to the condition of vanishing entropy occurs, and it keeps repeatedly occurring for the multiple times. Another very remarkable fact is that the maximum entropy value is constant, thereby signalling correctly that we are in the presence of a periodic signal, where the initial entropy increase, due to the uncertainty on the initial conditions, is balanced by the recurrences. Let us now consider the effect of a non vanishing κ. We see that the presence of an even very weak random component makes an abrupt transition to occur from the condition where the diffusion entropy is bounded from above, to a new condition where the recurrences are limited from below by an entropy increase proportional to $0.5 \ln l$. In the asymptotic time regime the DEM yields, as required, the proper scaling $\delta = 0.5$. However, we notice that it might be of some interest for a method of statistical analysis to give information on the extended regime of transition to the final *thermodynamic* condition. We notice

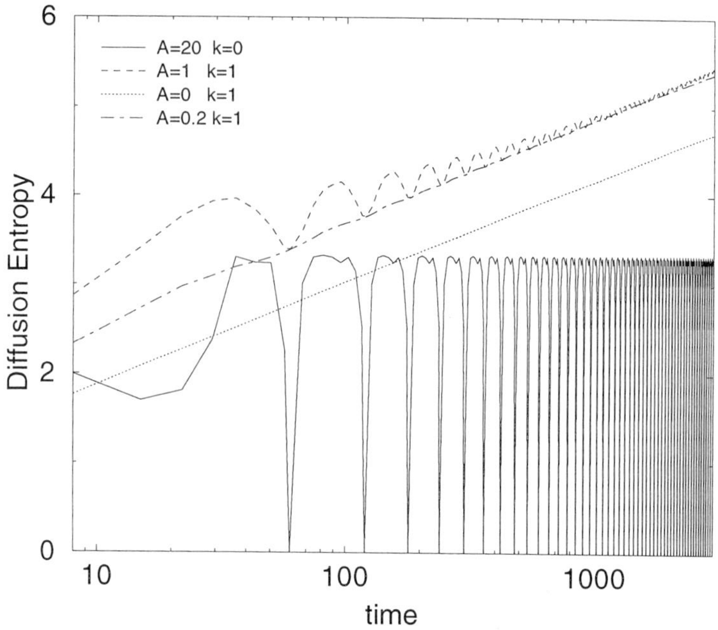

Figure 3. The diffusion entropy $S_d(l)$ as a function of time l for different sequences of the type of Eq. (12).

that if the DEM is interpreted as a method of scaling detection, it might also give the impression that a scaling faster than the ballistic δ is possible. This would be misleading. However, this aspect of the DEM, if conveniently used, can become an efficient method to monitor the non-stationary nature of the sequence under study.

In the special case where the fluctuation $\xi(t)$ is correlated or anticorrelated to the bias, the numerical results illustrated in Fig. 4 show that the time evolution of the diffusion entropy is qualitatively similar to that of Fig. 3. The correlation between the first and the second term on the right hand side of Eq. (16) is established by assuming

$$\xi(t) = \xi_0(t) cos(\omega t), \tag{17}$$

where $\xi_0(t)$ is the genuine independent fluctuation, without memory, whose intensity is modulated to establish a correlation with the second term. It is of some interest to mention what happens when $A = 0, \kappa = 1$, and consequently $\xi_b(t)$ coincides with $\xi(t)$ of Eq. (17). In this case we get the straight (solid) line of Fig. 4. This means that the adoption of the assumption that the process is stationary yields a result that is independent of the modulation.

We use this interesting case to illustrate the extension of the DEM, which is the main purpose of this paper. As earlier mentioned, this extension is based on the use of two mobiles windows, one of length L and the traditional ones of length $l \ll L$. This means that a large window of size L, with $L \ll T = 2\pi/\omega$, is located in a given position i of the sequence under study, with $i \leq N - L$, and the portion

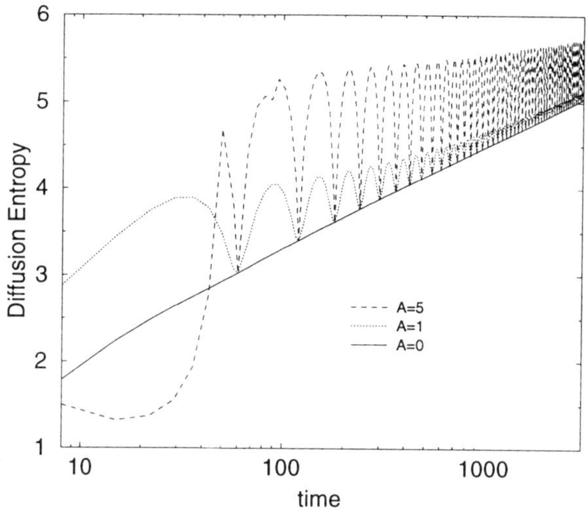

Figure 4. The diffusion entropy $S_d(l)$ as a function of time l for different sequences of the type of Eq. (12) with the prescription of Eq. (13) for the random component.

of the sequence contained within the window is thought of as being the sequence under study. We record the resulting δ (obtained with a linear regression method) and then we plot it as a function of the position i. We show in Fig. 5 that this way of proceeding has the nice effect of making the periodic modulation emerge.

Let us now improve the method to face non-stationary condition even further. As we have seen, the presence of time dependent condition tends to postpone or to cancel the attainment of a scaling condition. Therefore, let us renounce to using Eq. (9) and let us proceed as follows. For any large mobile window of size L let us call l_{max} the maximum size of the small windows. Let us call n the position of the left border of the large window, and and let us evaluate the following property

$$I(n) \equiv \sum_{l=2}^{l_{max}} \frac{S_d(l) - [S_d(1) + 0.5 \ln l]}{l}. \tag{18}$$

The quantity $I(n)$ detects the deviation from the condition of increase that the diffusion entropy would have in the random case. Since in the regime of transition the entropy increase can be much slower than in the corresponding random case, the quantity $I(n)$ can also bear negative values. This indicator affords a satisfactory way to detect local properties. As an example, Fig. 6 shows a case based on the DNA model of Ref. [8], called Copying Mistake Map (CMM). This is a sequence of symbols 0 and 1 obtained from the joint action of two independent sequences, one equivalent to tossing a coin and the other equivalent to establishing randomly a sequence of patches whose length is distributed as an inverse power law with index μ fitting the condition $2 < \mu < 3$. The probability of using the former sequence is $1 - \epsilon$ and that of using the latter is ϵ. Note that at long times an α-Lévy stable

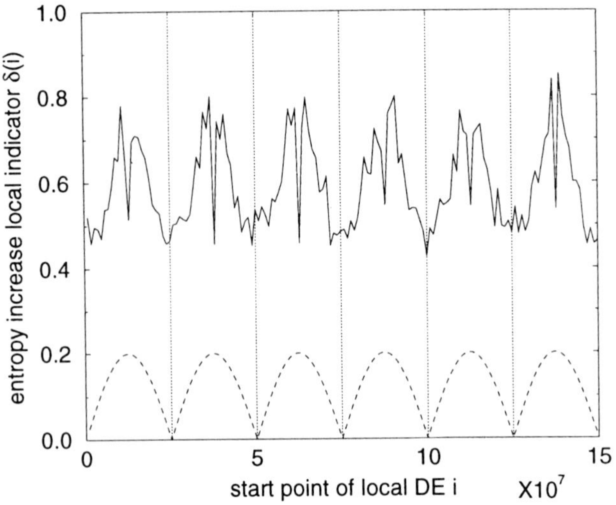

Figure 5. The method of the two mobile windows applied to a sequence given by Eq. (12) with $A = 0$ and $\xi(t)$ given by Eq. (13). The dashed line represents the sinus' amplitude (not in scale) corresponding to the position i of the left border of the large moving window.

distribution with $\alpha = \mu - 1$ is expected. We choose a time dependent value of ϵ

$$\epsilon = \epsilon_0[1 - cos(\omega t)]. \tag{19}$$

In Fig. 6 we show how this periodicity is perceived by using the two-windows generalization, proposed in this paper, of the DEM.

As a final example to show the efficiency of the new method of analysis, let us address the problem of the search of hidden periodicities in DNA sequences. We remind that DNA sequences actually carry long-range correlations and have been modelled as examples of CMM [10]. Fig. 7 shows a distinct periodic behavior for the human T-cell receptor alpha/delta locus. A period of about 990 base pairs is very neat in the first part of the sequence (promoter region), while several periodicities of the order of 1000 base pairs are distributed along the whole sequence.

These periodicities, probably due to DNA-proteins interactions in active eukaryotic genes, are expected by biologists, but the current technology is not yet adequate to deal with this issue, neither from the experimental nor from the computational point of view: such a behavior cannot be analyzed by means of crystallographic or structural NMR methods, nor would the current (or of the near future) computing facilities allow molecular dynamics studies of systems of the order of 10^6 atoms or more.

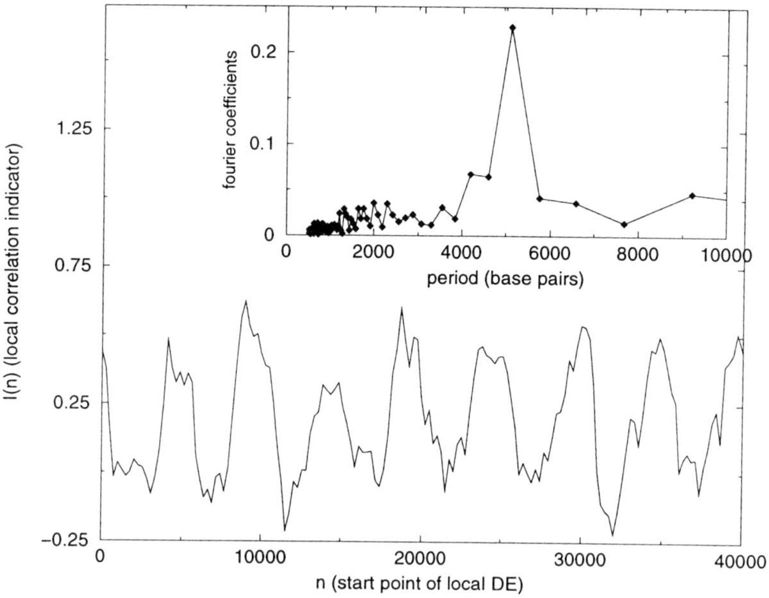

Figure 6. The method of the two moving windows with $l_{max} = 30$ applied to the analysis of an artificial CMM sequence with periodic parameter ϵ. The period of the variation of ϵ is 5000 bps and the analysis is carried out with moving windows of size 2000 bps. Inset: Fourier spectral analysis of $I(n)$.

4 Conclusions

The research work illustrated in this paper shows that the DEM is a very efficient way to detect the departure from ordinary Brownian motion with the shortest sequence as possible. On the basis of these results we are confident that it will be possible to predict the occurrence of catastrophic events, heart-quakes, heart attacks, stock-market crashes, and so on. We think that if all these misfortune events are anticipated by a correlation change, lasting for a fairly extended time period, then the DEM, within the double window procedure here illustrated, will signal in time their later occurrence. We refer to this method of analysis as Complex Analysis of Sequences via Scaling AND Randomness Assessment (CASSANDRA), and we hope to prove by means of future research work that its *prophetic* power is worth of consideration. We wish that the CASSANDRA algorithm will have more fortune and will receive more credit than the daughter of Priam and Hecuba.

References

References

1. N. Scafetta, P. Hamilton, P. Grigolini, *Fractals*, **9**, 193 (2001).
2. N. Scafetta, V. Latora, P. Grigolini, cond-mat/0105041.

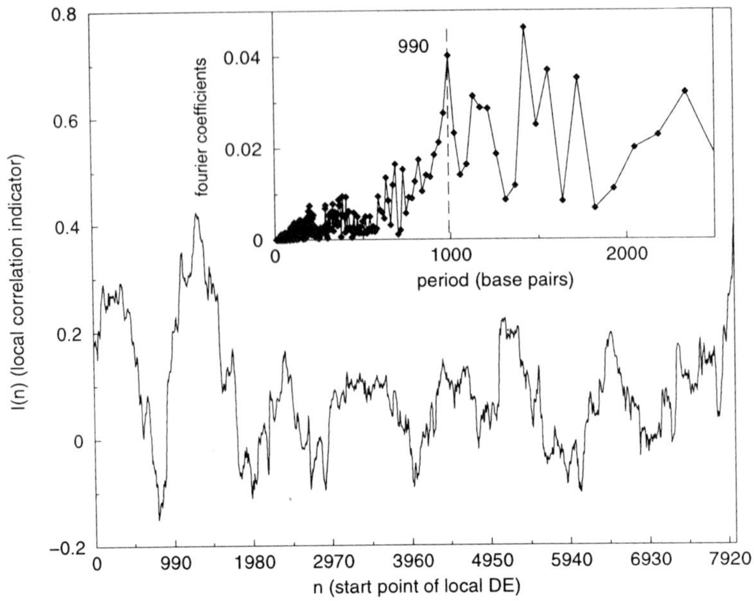

Figure 7. The method of two mobile windows applied to the analysis of the human DNA sequence. The method of two mobile windows ($l_{max} = 20$ $L = 512$) detects a periodicity of 990 bps. Inset: Fourier spectral analysis of $I(n)$.

3. P. Grigolini, L. Palatella, G. Raffaelli, cond-mat/0104166, in press on *Fractals.*
4. C.-K. Peng, S.V. Buldyrev, S. Havlin, M. Simons, H.E. Stanley, and A.L. Goldberger, *Phys. Rev.* E, **49**, 1685 (1994).
5. C.-K. Peng, S. Havlin, H.E. Stanley, A.L. Goldberger, *Chaos* **5**, 82 (1995).
6. C. Beck, F. Schlögl, *Thermodynamics of Chaotic Systems*, Cambdridge University Press, Cambridge (1993)
7. J.R. Dorfman, *An Introduction to Chaos in Nonequilibrium Statistical Mechanics*, Cambridge University Press, Cambridge (1999).
8. M. Buiatti, P. Grigolini, L. Palatella, *Physica* A **268**, 214 (1999).
9. P. Allegrini, P. Grigolini, B.J. West, *Phys. Rev.* E **54**, 4760 (1996).
10. P. Allegrini, M. Barbi, P. Grigolini and B.J. West, *Phys. Rev.* E **52**, 5281 (1995).

RANDOM WALK MODELS FOR TIME-FRACTIONAL DIFFUSION

F. MAINARDI
Dipartimento di Fisica, Università di Bologna, and INFN, Sezione di Bologna, Via Irnerio 46, I-40126 Bologna, Italy
E-mail: mainardi@bo.infn.it

R. GORENFLO
Erstes Mathematisches Institut, Freie Universität Berlin, Arnimallee 3, D-14195 Berlin, Germany
E-mail: gorenflo@math.fu-berlin.de

D. MORETTI
ALINET S.p.A., Via Lame 15, I-40122 Bologna, Italy
E-mail: dmoretti11@supereva.it

P. PARADISI
DIENCA, Dipartmento di Ingegneria Energetica, Nucleare e del Controllo Ambientale, Università di Bologna, Viale Risorgimento 2, I-40136 Bologna, Italy
E-mail: paolo.paradisi@mail.ing.unibo.it

A dynamics approach to anomalous diffusion may be based on generalized diffusion equations (of fractional order) and related random walk models. In particular, the time-fractional diffusion equation is obtained from the standard diffusion equation by replacing the the first-order time derivative with a fractional derivative of order $\beta \in (0,1)$. The fundamental solution (for the Cauchy problem) of this fractional evolution equation is interpreted as a probability density of a self-similar non-Markovian stochastic process which exhibits a variance consistent with a phenomenon of slow anomalous diffusion. By adopting an appropriate finite-difference scheme of solution, we generate discrete models of random walk suitable for simulating random variables whose spatial probability density evolves in time according to this fractional diffusion equation.

1 Introduction

Generalized master equations of fractional order for random walk models are usually introduced and investigated to describe phenomena of *anomalous diffusion* and transport dynamics in complex and/or disordered systems including fractal media. In this respect an interesting review by Metzler and Klafter [16] has recently appeared to which (and references therein) we refer the interested reader. All the proposed models of random walk turn out to be beyond the classical Brownian motion, which is based on the standard diffusion equation, see *e.g.* Klafter *et al* [11].

In this paper we intend to complement the approach started by Gorenflo and Mainardi [4,5,6], see also [2,20], in proposing finite-difference schemes of solution for *space-fractional diffusion* equations [a] that serve as discrete models for *Lévy flights.* These special random walks are related to *self-similar Markovian* stochastic processes which exhibits infinite space variance. Here we consider a *time-fractional*

[a] Related pre-prints can be downloaded from our WW pages at: http://www.fracalmo.org

diffusion equation whose fundamental solution is interpreted as a probability density of a *self-similar non-Markovian* stochastic process which exhibits a space variance consistent with a phenomenon of *slow anomalous diffusion.* By adopting an appropriate finite-difference scheme of solution, we generate discrete models of random walk suitable for simulating random variables whose spatial probability density evolves in time according to this fractional diffusion equation. By properly scaled transition to vanishing space and time steps, these discrete models converge to the corresponding continuous processes, as it will be shown in another paper. Here our finite-difference scheme is adopted in some case-studies for producing *sample paths* and the corresponding *space-increment* of individual particles performing the random walks and for producing *histograms* of the approximate realization of the corresponding probability densities by simulating many individual paths with the same number of time steps and making statistics of the final positions of the particles.

2 The Time-Fractional Diffusion Equation

By *time-fractional diffusion equation* we mean the evolution equation

$$\frac{\partial^\beta}{\partial t^\beta} u(x,t) = \frac{\partial^2}{\partial x^2} u(x,t)\,, \quad 0<\beta<1\,, \quad x \in \mathbf{R}\,, \quad t \in \mathbf{R}_0^+\,, \tag{2.1}$$

where the time-fractional derivative is intended in the Caputo sense, namely,

$$\frac{\partial^\beta}{\partial t^\beta} u(x,t) := \frac{1}{\Gamma(1-\beta)} \int_0^t \left[\frac{\partial}{\partial \tau} u(x,\tau)\right] \frac{d\tau}{(t-\tau)^\beta}\,, \quad 0<\beta<1\,. \tag{2.2}$$

When $\beta = 1$ we recover in the limit the well-known diffusion equation, namely:

$$\frac{\partial}{\partial t} u(x,t) = \frac{\partial^2}{\partial x^2} u(x,t)\,, \quad x \in \mathbf{R}\,, \quad t \in \mathbf{R}_0^+\,. \tag{2.3}$$

The Cauchy problem for the above evolution equations requires the knowledge of the initial condition $u(x,0^+) = f(x)$, where $f(x)$ denotes a given real function defined on $\mathbf{R}$ that we assume to be Fourier transformable in ordinary or generalized sense.

Eqs. (2.1) and (2.3) can be seen as particular cases of the *space-time fractional diffusion* equation recently treated by Mainardi *et al*[14],

$${}_xD_\theta^\alpha\, u(x,t) \;=\; {}_tD_*^\beta\, u(x,t)\,, \quad x \in \mathbf{R}\,, \quad t \in \mathbf{R}^+\,, \tag{2.4}$$

where the α, θ, β are real parameters restricted as follows

$$0<\alpha\le 2\,, \quad |\theta| \le \min\{\alpha, 2-\alpha\}\,, \quad 0<\beta\le 2\,. \tag{2.5}$$

In (2.4) ${}_xD_\theta^\alpha$ is the *Riesz-Feller* space-fractional derivative of order α and skewness θ, and ${}_tD_*^\beta$ is the Caputo time-fractional derivative of order β. These fractional derivatives are integro-differential operators whose definition is briefly recalled hereafter.

For a sufficiently well-behaved function $f(x)$ whose Fourier transform is

$$\widehat{f}(\kappa) = \mathcal{F}\{f(x);\kappa\} = \int_{-\infty}^{+\infty} \mathrm{e}^{+i\kappa x} f(x)\,dx\,, \quad \kappa \in \mathbf{R}\,,$$

we define the *Riesz-Feller* space-fractional derivative of order α and skewness θ through

$$\mathcal{F}\{ {}_xD_\theta^\alpha f(x);\kappa\} = -|\kappa|^\alpha\, \mathrm{e}^{i(\mathrm{sign}\,\kappa)\theta\pi/2}\, \widehat{f}(\kappa)\,, \tag{2.6}$$

where α and θ are restricted as in (2.5). Thus the *Riesz-Feller* derivative is the pseudo-differential operator whose symbol is the logarithm of the characteristic function of a general *Lévy strictly stable* probability density with *index of stability* α and asymmetry parameter θ (improperly called *skewness*) according to Feller's parameterization [1], as revisited by Gorenflo and Mainardi [4,5,6].

For a sufficiently well-behaved function $f(t)$ whose Laplace transform is

$$\widetilde{f}(s) = \mathcal{L}\{f(t);s\} = \int_0^\infty \mathrm{e}^{-st} f(t)\,dt\,, \quad s \in \mathbf{C}\,,$$

we define the *Caputo* fractional derivative of order β through

$$\mathcal{L}\left\{ {}_tD_*^\beta f(t);s\right\} = s^\beta\, \widetilde{f}(s) - \sum_{k=0}^{m-1} s^{\beta-1-k} f^{(k)}(0^+)\,, \quad m-1 < \beta \le m\,, \tag{2.7}$$

where $m \in \mathbf{N}$. This leads in the time domain to, see *e.g.* [3,21]

$${}_tD_*^\beta f(t) := \begin{cases} \dfrac{1}{\Gamma(m-\beta)} \displaystyle\int_0^t \dfrac{f^{(m)}(\tau)\,d\tau}{(t-\tau)^{\beta+1-m}}\,, & m-1<\beta<m\,,\\[2ex] \dfrac{d^m}{dt^m} f(t)\,, & \beta = m\,. \end{cases} \tag{2.8}$$

The reader should observe that the *Caputo* fractional derivative differs from the usual *Riemann-Liouville* fractional derivative, that is here denoted as ${}_tD^\beta f(t)$, when $m-1<\beta<m$ and in the limit $\beta \to (m-1)^+$ [b].

The fundamental solution of the general space-time fractional diffusion equation (2.4)[c] has been obtained in [14] in computational form by using the Mellin-Barnes

[b]We have, see *e.g.* [23],

$${}_tD^\beta f(t) := \frac{d^m}{dt^m}\left[\frac{1}{\Gamma(m-\beta)}\int_0^t \frac{f(\tau)\,d\tau}{(t-\tau)^{\beta+1-m}}\right]\,, \quad m-1<\beta<m\,,$$

Gorenflo and Mainardi [3] have shown the following relationships between the two fractional derivatives (when both of them exist) for $m-1<\beta<m$,

$${}_tD_*^\beta f(t) = {}_tD^\beta\left[f(t) - \sum_{k=0}^{m-1} f^{(k)}(0^+)\,\frac{t^k}{k!}\right] = {}_tD^\beta f(t) - \sum_{k=0}^{m-1} f^{(k)}(0^+)\,\frac{t^{k-\beta}}{\Gamma(k-\beta+1)}\,.$$

[c]The fundamental solution or Green function of Eq. (2.4) is intended to be the solution of the Cauchy problem corresponding to the initial condition is $u(x,0^+) = \delta(x)$ when $0<\beta\le 1$ or $u(x,0^+) = \delta(x)\,,\; u_t(x,0^+) = 0$ when $1<\beta\le 2$.

integral representation. There this solution, denoted by $G^{\theta}_{\alpha,\beta}(x,t)$, has been shown to be a *probability density function* in the ranges $\{0 < \alpha \le 2\} \cap \{0 < \beta \le 1\}$ and $\{1 < \beta \le \alpha \le 2\}$, satisfying the *similarity law* for $x > 0$

$$G^{\theta}_{\alpha,\beta}(x,t) = t^{-\gamma}\, K^{\theta}_{\alpha,\beta}(x/t^{\gamma})\,, \quad \gamma = \beta/\alpha\,. \tag{2.9}$$

Here $K^{\theta}_{\alpha,\beta}(-x) = K^{-\theta}_{\alpha,\beta}(x)$ may be considered as the reduced Green function.

We note that from the view-point of the random walk models related to Eq. (2.4) only the cases with $\beta = 1$ (space-fractional diffusion) have been treated so far by our research group, see *e.g.* [2,4,5,6,20]. A Markovian description of Lévy flights has been obtained both for the symmetric ($\theta = 0$) and the asymmetric ($\theta \neq 0$) cases. Here we limit our analysis to the cases of *time-fractional diffusion* ($\alpha = 2$, $\theta = 0$) with the restriction $\beta \le 1$. Several authors see *e.g.* [9,10,15,16,22,24], have shown that in this case the time-fractional diffusion equation can be derived from the Continuous-Time Random Walk Model (known simply as CTRW) that was formerly introduced and analyzed by Montroll and his associates, see *e.g.* [17,18,19]. It is thus very interesting to provide the discrete random-walk models related to this equation, whose fundamental solution is here recalled for the reader's convenience. Following the works by Mainardi [12,13] (see also [14]), the fundamental solution of our time fractional diffusion equation (2.1) turns out to be for $0 < \beta \le 1$:

$$G^{0}_{2,\beta}(x,t) = G^{0}_{2,\beta}(|x|,t) = \frac{1}{2\,t^{\beta/2}}\, M_{\beta/2}(r)\,, \quad r = |x|/t^{\beta/2}\,, \tag{2.10}$$

where r acts as a *similarity variable*, and $M_{\beta/2}(r)$ is a transcendental function of Wright type[d]. Being non-negative and normalized, such fundamental solution can be interpreted as a peculiar (space-symmetric) *pdf* whose main properties are herewith outlined. When $\beta = 1$ (standard diffusion) formula (2.10) reduces to the Gaussian *pdf* with variance $\sigma = 2t$, namely

$$G^{0}_{2,1}(x,t) = \frac{1}{2\sqrt{\pi}}\, t^{-1/2}\, \mathrm{e}^{-x^2/(4t)}\,. \tag{2.11}$$

From the exponential decay of the function $M_{\nu}(r)$ as $r \to +\infty$ we can prove that for $0 < \beta \le 1$ all the moments of the *pdf* turn out to be

$$\int_{-\infty}^{+\infty} x^{2n}\, G^{0}_{2,\beta}(x,t)\, dx = \frac{\Gamma(2n+1)}{\Gamma(\beta n+1)}\, t^{\beta n}\,, \quad n \in \mathbf{N}_0\,. \tag{2.12}$$

We recognize that the variance associated to the *pdf* is proportional to t^{β}, which for $0 < \beta < 1$ implies a phenomenon of *slow anomalous diffusion*.

---

[d] It turns out that, for $0 < \nu < 1$, $M_{\nu}(z)$ ($z \in \mathbf{C}$) is an entire function of order $\rho = 1/(1-\nu)$, defined as

$$M_{\nu}(z) := \sum_{n=0}^{\infty} \frac{(-z)^n}{n!\,\Gamma[-\nu n + (1-\nu)]} = \frac{1}{2\pi i}\int_{Ha} \mathrm{e}^{\,\sigma - z\sigma^{\nu}}\, \frac{d\sigma}{\sigma^{1-\nu}}\,,$$

where Ha denotes the Hankel path.

3 The Discrete Non-Markovian Random Walk

We now sketch a discrete *redistribution scheme* and a related *random-walk model* for our *time-fractional diffusion* equation (2.1) including, in the limit for $\beta = 1$, the particular case of the standard diffusion (2.3), which leads to a discrete model of the classical Brownian motion. For this purpose we discretize space and time by grid points and time instants as follows

$$x_j = j\,h\,, \quad h > 0\,, \quad j = 0, \pm 1, \pm 2, \ldots\,; \quad t_n = n\,\tau\,, \quad \tau > 0\,, \quad n = 0, 1, 2, \ldots$$

where the steps h and τ are assumed to be small enough. The dependent variable u is then discretized by introducing

$$y_j(t_n) \approx \int_{x_j - h/2}^{x_j + h/2} u(x, t_n)\, dx \approx h\, u(x_j, t_n)\,. \tag{3.1}$$

With the quantities $y_j(t_n)$ so intended, we replace the time-fractional diffusion equation (2.1) (after multiplication by the spatial mesh-width h) by the finite-difference equation

$$_\tau D_*^\beta\, y_j(t_{n+1}) = \frac{y_{j+1}(t_n) - 2y_j(t_n) + y_{j-1}(t_n)}{h^2}\,, \quad 0 < \beta \leq 1\,, \tag{3.2}$$

where the difference operator $_\tau D_*^\beta$ is intended to converge to $_t D_*^\beta$ as $\tau \to 0$. As usual, we have adopted a symmetric second-order difference quotient in space at level $t = t_n$ for approximating the second-order space derivative. For $_\tau D_*^\beta$ we require a scheme which must reduce as $\beta = 1$ to a forward difference quotient in time at level $t = t_n$, which is usually adopted for approximating the first-order time derivative, namely

$$_\tau D_*^1\, y_j(t_{n+1}) = \frac{y_j(t_{n+1}) - y_j(t_n)}{\tau}\,. \tag{3.3}$$

Then, for approximating the time fractional derivative (in Caputo's sense), we adopt a backward Grünwald-Letnikov scheme [e] (starting at level $t = t_{n+1}$), which reads:

$$_\tau D_*^\beta\, y_j(t_{n+1}) = \sum_{k=0}^{n+1} (-1)^k \binom{\beta}{k} \frac{y_j(t_{n+1-k}) - y_j(0)}{\tau^\beta}\,, \quad 0 < \beta \leq 1\,. \tag{3.4}$$

Combining (3.2) and (3.4) and introducing the scaling parameter

$$\mu := \frac{\tau^\beta}{h^2}\,, \quad 0 < \beta \leq 1\,, \tag{3.5}$$

we obtain

$$\begin{aligned} y_j(t_{n+1}) = {} & y_j(t_0) \sum_{k=0}^{n} (-1)^k \binom{\beta}{k} + \sum_{k=1}^{n} (-1)^{k+1} \binom{\beta}{k} y_j(t_{n+1-k}) \\ & + \mu \left[y_{j+1}(t_n) - 2y_j(t_n) + y_{j-1}(t_n) \right]\,, \end{aligned} \tag{3.6}$$

[e] Grünwald (1867) and Letnikov (1868) independently developed an approach to fractional differentiation for which the definition of the (Riemann-Liouville) fractional derivative is the limit of a fractional difference quotient, see *e.g.* [21,23].

where we have used the "empty sum" convention. Thus, (3.6) provides the universal *transition law* from t_n to t_{n+1} valid for all $n \geq 0$.

For convenience let us introduce the coefficients c_k and b_m

$$\begin{cases} c_k = (-1)^{k+1} \binom{\beta}{k} = \left|\binom{\beta}{k}\right|, & k \geq 1, \\ b_m = \sum_{k=0}^{m} (-1)^k \binom{\beta}{k}, & m \geq 0. \end{cases} \tag{3.7}$$

For $\beta = 1$ (standard diffusion) we note that all these coefficients are vanishing except $b_0 = c_1 = 1$. For $0 < \beta < 1$ the coefficients are all positive and possess the following properties

$$\begin{cases} \sum_{k=1}^{\infty} c_k = 1, \\ 1 > \beta = c_1 > c_2 > c_3 > \ldots \to 0, \end{cases} \tag{3.8}$$

$$\begin{cases} b_0 = 1 = \sum_{k=1}^{\infty} c_k, \quad b_m = 1 - \sum_{k=1}^{m} c_k = \sum_{k=m+1}^{\infty} c_k, \\ 1 = b_0 > b_1 > b_2 > b_3 > \ldots \to 0. \end{cases} \tag{3.9}$$

We thus observe that the c_k and the b_m form sequences of positive numbers, not greater than 1, decreasing strictly monotonically to zero.

Thanks to the introduction of the above coefficients the universal transition law (3.6) can be written in the following noteworthy form

$$y_j(t_{n+1}) = b_n\, y_j(t_0) + \sum_{k=1}^{n} c_k\, y_j(t_{n+1-k}) + \mu\left[y_{j+1}(t_n) - 2y_j(t_n) + y_{j-1}(t_n)\right], \tag{3.10}$$

with the empty sum convention convention if $n = 0$. Observe that $c_1 = \beta$.

The scheme (3.10) *preserves non-negativity*, if all coefficients are not negative, hence if

$$0 < \mu = \frac{\tau^\beta}{h^2} \leq \frac{\beta}{2}. \tag{3.11}$$

It is also *conservative*, as we can prove by induction: assuming $\sum_{j=-\infty}^{+\infty} |y_j(t_0)| < \infty$, we find $\sum_{j=-\infty}^{+\infty} y_j(t_n) = \sum_{j=-\infty}^{+\infty} y_j(t_0)$, $n \in \mathbf{N}$.

Nonnegativity preservation and conservativity mean that our scheme can be interpreted as a *redistribution scheme* of clumps $y_j(t_n)$. The interpretation of our *redistribution* scheme is as follows: the clump $y_j(t_{n+1})$ arises as a weighted-memory average of the (previous) $n+1$ values $y_j(t_m)$, with $m = n, n-1, \ldots, 1, 0$, with positive weights

$$\beta = c_1,\ c_2, \ldots, c_n,\ b_n = 1 - \sum_{k=1}^{n} c_k \tag{3.12}$$

(note $c_1 > c_2 > \ldots > c_n$) whose sum being one, then by subtraction of $2\mu\, y_j(t_n)$ which is given in equal parts to the neighboring point x_{j-1} and x_{j+1}.

For probabilistic or *random walk* interpretation we consider the $y_j(t_n)$ as probabilities of sojourn at point x_j in instant t_n. Then we require the normalization condition $\sum_{j=-\infty}^{+\infty} y_j(t_0) = 1$.

For $n = 0$, Eq. (3.10) means (by appropriate re-interpretation of the spatial index j): A particle sitting at x_j in instant t_0 jumps, when t proceeds from t_0 to t_1, with probability μ to the neighbor point x_{j+1}, with probability μ to the neighbor point x_{j-1}, and with probability $1-2\mu$ it remains at x_j. For $n \geq 1$ we write (3.10), using $\beta = c_1$, as follows:

$$y_j(t_{n+1}) = \left(1 - \sum_{k=1}^{n} c_k\right) y_j(t_0) + c_n\, y_j(t_1) + c_{n-1}\, y_j(t_2) + \ldots + c_2\, y_j(t_{n-1}) \tag{3.13}$$

$$+(c_1 - 2\mu)\, y_j(t_n) + \mu\, [y_{j+1}(t_n) + y_{j-1}(t_n)]\,.$$

Obviously, all coefficients (probabilities) are non negative, and their sum is 1. But what does it mean? Having a particle, sitting in x_j at instant t_n, where will we find it with which probability at instant t_{n+1} ? From (3.13) we conclude, by reinterpretation of the spatial index j, *considering the whole history* of the particle, *i.e.* the particle path $\{x(t_0)\,,\, x(t_1)\,,\, x(t_2)\,,\, \ldots\,,\, x(t_n)\}$, that if at instant t_n it is in point x_j, there is the contribution $c_1 - 2\mu$ to be again at x_j at instant t_{n+1}, the contribution μ to go to x_{j-1}, the contribution μ to go to x_{j+1}. But the sum of these contributions is $c_1 = \beta \leq 1$. So, excluding the case $\beta = 1$ in which we recover the standard diffusion (*Markovian* process), for $\beta < 1$ we have to consider the previous time levels (*non-Markovian* process). Then, from level t_{n-1} we get the contribution c_2 for the probability of staying in x_j also at time t_{n+1}, from level t_{n-2} we get the contribution c_3 for the probability of staying in x_j at time $t_{n+1}, \ldots$, from level t_1 we get the contribution c_n for the probability of staying in x_j at time t_{n+1}, and finally, from level $t_0 = 0$ we get the contribution b_n for the probability of staying in x_j at time t_{n+1}. Thus, the whole history up to t_n decides probabilistically where the particle will be at instant t_{n+1}.

Comment: Besides the diffusive part (μ, $c_1 - 2\mu$, μ) which lets the particle jump at most to neighboring points, we have for $\beta < 1$ the memory part which gives a tendency to return to former positions even if they are far away. Due to (3.8-9), of course, the probability to return to a far away point gets smaller and smaller the larger the time lapse is from the instant when the particle was there.

Herewith we present some results on the simulation of the sample paths and histograms corresponding to some typical values of the index β, namely $\beta = 1, 0.75, 0.50$, in Figs 1, 2, 3, respectively. Our simulations are based on 10 thousand realizations. The sample paths and the corresponding increments are plotted against the time steps up to 500 while the histograms refer to densities at $t = 1$ for $|x| \leq 5$. The relevant parameters μ, h and τ used in the Figures are reported in Table I.

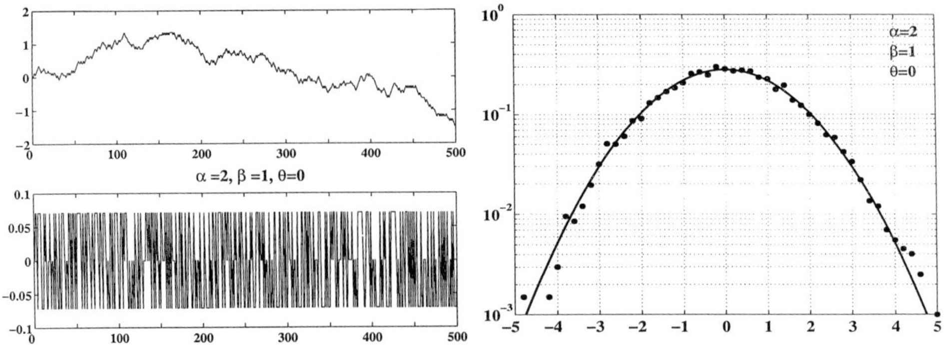

Fig. 1

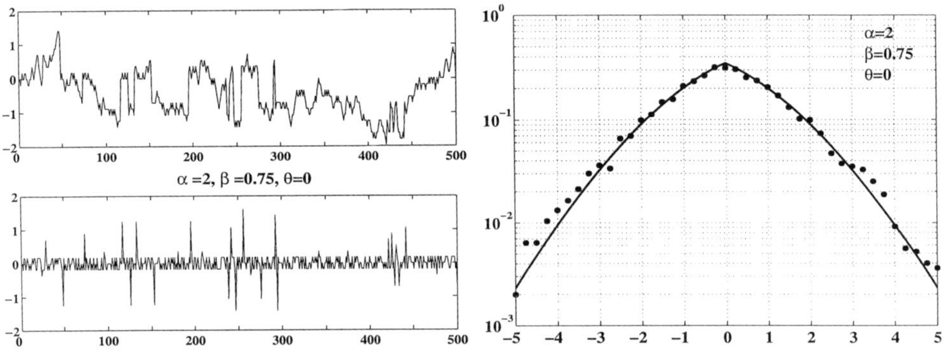

Fig. 2

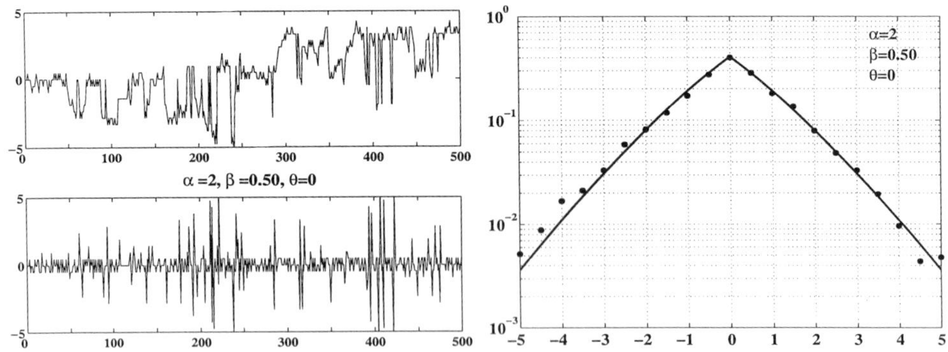

Fig. 3

β	$\mu < \beta/2$	h_S	τ_S	h_H	τ_H
1	$0.4 < 1/2$	0.07	$2.0\,10^{-3}$	0.20	$2.5\,10^{-2}$
0.75	$0.3 < 3/8$	0.17	$2.0\,10^{-3}$	0.25	$5.0\,10^{-3}$
0.50	$0.2 < 1/4$	0.47	$2.0\,10^{-3}$	0.50	$2.5\,10^{-3}$

Table I

β = fractional order, μ = scaling parameter,
h_S = space-step, τ_S = time-step for sample paths,
h_H = space-step, τ_H = time-step for histograms.

4 CONCLUSIONS

Anomalous diffusion processes have in recent years gained revived interest among physicists, and methods of fractional calculus have shown their usefulness for purposes of modelling. In the time-fractional case one is naturally led to a generalization of the classical diffusion equation with respect to the first-order time operator. One arrives at non-Markovian processes in which space-probability distributions evolve in time consistently with the phenomenon of *slow anomalous diffusion* (with variance $\sigma^2 \propto t^\beta$, $0 < \beta < 1$).

In this paper we have provided a discrete random walk approach to this phenomenon. Let us stress the fact that our a random walk model is obtained by discretizing the time-fractional diffusion equation (2.1) in the most straightforward way. The difference scheme so obtained, with the scaling restriction (3.11), imitates on a discrete time-space grid the most essential properties of the continuous process, namely conservativity and preservation of non-negativity. Analogously, natural discretizations have likewise been successful in our previous papers on rather general space-fractional diffusion equations, see *e.g.* [2,4,5,6,20] we will demonstrate their use in general diffusion processes, fractional in space as well as in time, see the forthcoming papers[7,8].

As indicated in the Introduction we have produced *sample paths* with the corresponding *space-increments* and *histograms* of the random walks, see Figures 1-3. In the case $\beta < 1$ the paths exhibit the memory effect visible in a kind of stickiness combined with occasional jumps to points previously occupied, in distinct contrast to the rather tame behaviour in case $\beta = 1$ (simulation of Brownian motion).

Acknowledgments

We are grateful to the Research Commissions of the Free University of Berlin (Convolution Project) and of the University of Bologna (MIUR funds) for supporting joint efforts of our research groups in Berlin and Bologna. Our paper is one of the fruits of this collaboration.

References

1. W. Feller, *An Introduction to Probability Theory and its Applications*, Vol. 2 (Wiley, New York, 1971).
2. R. Gorenflo, G. De Fabritiis and F. Mainardi, Discrete random walk models for symmetric Lévy-Feller diffusion processes, *Physica A* **269** (1999) 79–89.
3. R. Gorenflo and F. Mainardi, Fractional calculus: integral and differential equations of fractional order, in: A. Carpinteri and F. Mainardi (Editors), *Fractals and Fractional Calculus in Continuum Mechanics* (Springer Verlag, Wien, 1997), pp. 223–276. [Reprinted in NEWS 010101 `http://www.fracalmo.org`]
4. R. Gorenflo and F. Mainardi, Random walk models for space-fractional diffusion processes, *Fractional Calculus and Applied Analysis* **1** (1998) 167–191.
5. R. Gorenflo and F. Mainardi, Approximation of Lévy-Feller diffusion by random walk, *Journal for Analysis and its Applications (ZAA)* **18** (1999) 231-146.
6. R. Gorenflo and F. Mainardi, Random walk models approximating symmetric space-fractional diffusion processes, in: J. Elschner, I. Gohberg and B. Silbermann (Editors), *Problems in Mathematical Physics* (Birkhäuser Verlag, Basel, 2001), pp. 120-145.
7. R. Gorenflo, F. Mainardi, D. Moretti, G. Pagnini and P. Paradisi, Fractional diffusion: probability distributions and random walk models, *Physica A*, submitted.
8. R. Gorenflo, F. Mainardi, D. Moretti and P. Paradisi, Discrete random walk models for space-time-fractional diffusion, *Chemical Physics*, submitted.
9. R. Hilfer and L. Anton, Fractional master equations and fractal time random walks, *Phys. Rev. E* **51** (1995) R848-R851.
10. R. Hilfer, Fractional time evolution, in: R. Hilfer (Editor), *Applications of Fractional Calculus in Physics* (World Scientific, Singapore, 2000), pp. 87-130.
11. J. Klafter, M. F. Shlesinger and G. Zumofen, Beyond Brownian motion, *Physics Today* **49** No 2 (1996) 33–39.
12. F. Mainardi, Fractional relaxation-oscillation and fractional diffusion-wave phenomena, *Chaos, Solitons & Fractals* **7** (1996) 1461–1477.
13. F. Mainardi, Fractional calculus: some basic problems in continuum and statistical mechanics, in: A. Carpinteri and F. Mainardi (Editors), *Fractals and Fractional Calculus in Continuum Mechanics* (Springer Verlag, Wien and New-York, 1997), pp. 291–248.
14. F. Mainardi, Yu. Luchko and G. Pagnini, The fundamental solution of the space-time fractional diffusion equation, *Fractional Calculus and Applied Analysis* **4** (2001) 153-192. [Reprinted in NEWS 010401 `http://www.fracalmo.org`]
15. F. Mainardi, M. Raberto, R. Gorenflo and E. Scalas, Fractional calculus and continuous-time finance II: the waiting-time distribution, *Physica A* **287** (2000) 468–481.
16. R. Metzler and J. Klafter, The random walk's guide to anomalous diffusion: a fractional dynamics approach, *Phys. Reports* **339** (2000) 1-77.

17. E.W. Montroll and M.F. Shlesinger, On the wonderful world of random walks, in: J. Leibowitz and E.W. Montroll (Editors), *Nonequilibrium Phenomena II: from Stochastics to Hydrodynamics* (North-Holland, Amsterdam, 1984), pp. 1-121.
18. E.W. Montroll and G.H. Weiss, Random walks on lattices, II, *J. Math. Phys.* **6** (1965) 167–181.
19. E.W. Montroll and B.J. West, On an enriched collection of stochastic processes, in: E.W. Montroll and J. Leibowitz (Editors), *Fluctuation Phenomena* (North-Holland, Amsterdam, 1979), pp. 61-175.
20. P. Paradisi, R. Cesari, F. Mainardi and F. Tampieri, The fractional Fick's law for non-local transport processes, *Physica A* **293** (2001) 130-142.
21. I. Podlubny, *Fractional Differential Equations* (Academic Press, San Diego, 1999).
22. A. Saichev and G. Zaslavsky, Fractional kinetic equations: solutions and applications, *Chaos* **7** (1997) 753-764.
23. S.G. Samko, A.A. Kilbas and O.I. Marichev, *Fractional Integrals and Derivatives: Theory and Applications.* New York and London, Gordon and Breach Science Publishers (1993). [Translation from the Russian edition, Minsk, Nauka i Tekhnika (1987)]
24. V.V. Uchaikin and V.M. Zolotarev, *Chance and Stability. Stable Distributions and their Applications* (VSP, Utrecht, 1999).

DYNAMICS OF SOLAR MAGNETIC FIELD FROM SYNOPTIC CHARTS

N. G. MAKARENKO AND L. M. KARIMOVA
Institute of Mathematics, 125 Pushkin Street, 480100 Almaty, Kazakhstan
E-mail: makarenko@math.kz

M. M. NOVAK
School of Mathematics, Kingston University, Surrey KT1 2EE, England
E-mail: novak@kingston.ac.uk

The paper focuses on diagnosis of extended nonlinear dynamical systems arising in the global solar magnetic field evolution. The methods of mathematical morphology, namely, Minkowski functionals and dimension are applied to analyzing topology of magnetic field cross-sections (synoptic charts). Time series of Euler characteristics obtained from solar magnetic field charts is used to investigate the Solar activity by the embedding methods.

1 Introduction

Nowadays, there are well-developed tools for the reconstruction of the dynamics from scalar time series [1,2]. The heuristic idea of the reconstruction was introduced by Packard *et al.* [3] Takens [4] formalized this heuristics and studied the delay-coordinate map, which generically was a time series embedding in R^m when $m \geq 2d+1$, and where d was the dimension of the attractor of the system. This technique works well under some conditions [5] and real system has a local (not spatially extended) representation.

However, most natural systems *are* spatially extended. Typically such systems show disorder in both space and time and are said to exhibit spatio-temporal chaos. [6] At an instant, the state of the system, sometimes called *snapshot*, can be described as a vector-function $f(\mathbf{x}), \mathbf{x} \in \mathbf{R^n}$. Then an evolution of the system is a sequence of the snapshots. Direct generalization of Takens algorithm to temporal sequences of such spatial patterns (snapshots) leads to computational difficulties, as a phase space becomes a space of matrix instead of vector space. [7]

That is why, it is important to be able to extract dynamical information from such data, which usually have a form of maps. [8] In many cases, nonlinearity and heterogeneity of natural processes produce a field with stochastic properties. The most general description of these fields is a statistical, but one can obtain analytical expressions of field extremum probabilities in rare cases of stationary fields. [9] However, for largest values of $f(\mathbf{x})$ this probability can be accurately approximated by the average Euler characteristic of the excursion set of a random field. [10] Excursion set technique is used by Morphological Image Analysis (MIA) which allows characterizing the geometry and topology of any pattern. [11,12] For a given image the first step in MIA approach is to compute the Minkowski functionals. [11,13] The next step is to study their behaviour as a function of some control parameters, such as time, density, resolution etc. An appealing feature of the MIA is that the image functionals have a clear geometrical and topological interpretation. Thus, this technique allows us to replace time sequence of images by time series of geometrical and topological characteristics to reveal the underlying dynamics.

In this paper we use such approach for exploration of solar magnetic field dynamics. It is known that solar magnetic field has complex spatial structure and exhibits nontrivial temporal behaviour. This field is produced by diffuse components and by magnetic structures of different sizes and different life times and, in rough approximation, can be regarded as a structured random field. In particular, it has been found [14] that magnetic flux might have intermittent (multifractal) structure on a wide range of scales, however, robust estimations of fractal dimensions have not been obtained.

At large scales magnetic structures can be regarded as background magnetic field. [15] This coarse grain field becomes apparent in distribution of unipolar areas, which evolve during approximately 11 years. [16] Neutral magnetic lines on the so-called H_α synoptic magnetic charts separate these areas. Such charts give unique information about dynamics of large-scale solar field, averaged over one 27-days-long rotation of the Sun. The sequence of maps consists of synoptic charts dated from 1915 to 2000. [17] So far, these data were used only for heuristic analysis. [16] Detailed information about solar magnetic field is found in Wilcox Solar Observatory Synoptic Charts, which are available for the period from 1976 year [18]. They are assembled from individual magnetograms observed over the course of solar rotation.

Since the charts represent real structures, at least homeomorphically topological properties are normally preserved. So, the physical information about the field can be derived from topology of isolines and can be described by morphological functionals, [19] whose time dependence facilitates our understanding of the solar activity. Using topological dynamics methods, "attractors" from time series are reconstructed by embedding technique and a nonlinear interrelationship between these attractors and a causal (driver/response) relationship is estimated.

The structure of this paper is as follows. In Sec. 2 we describe briefly the morphological measures, the Minkowski dimension and present the results of applying these tools to solar synoptic charts. Sec. 3 describes the results of reconstruction of the dynamics from spatio-temporal time series and outlines the relationship between the global magnetic field and the local field of sunspots. The summary is found in the concluding section.

2 Minkowski Functionals and Dimension of Solar Synoptic Charts

2.1 Minkowski Functionals

We are interested in the morphological measures of charts, i.e. the measures which are invariant under translations and rotations and which are additive. The branch of mathematics known as integral geometry provides natural tools for this characterization, which are known as the Minkowski functionals. [20,21]

Consider a convex set K in $\mathbf{R}^d$. The parallel set of distance ϵ to K is the set

$$K_\epsilon = \bigcup_{x \in K} B(x, \epsilon), \tag{1}$$

where $B(x,\epsilon)$ is the closed ball of radius ϵ centered at the point x. A relation called the Steiner formula can be taken as the definition of the Minkowski functionals W_i:

$$V(K_\epsilon) = \sum_{i=0}^{d} \binom{d}{i} W_i(K)\epsilon^i, \tag{2}$$

where V denotes the (d-dimensional) volume. For low-dimensional spaces, the Minkowski functionals can be expressed simply in terms of the geometrical and topological quantities, in particular, for $d = 2$ area $A(K)$ and boundary length $L(K)$ and the Euler-Poincare characteristic or connectivity number χ are:

$$W_0(K) = A(K), \qquad W_1(K) = \frac{1}{2}L(K), \qquad W_2(K) = \pi\chi \tag{3}$$

Next, let $\mathbf{x}, \mathbf{y} \in R^2$. Magnetic field of synoptic chart is simply a real value function $f(\mathbf{x})$, given on a grid G:

$$G = \{\mathbf{x} \in R^2 | (x_1 - y_1, x_2 - y_2) = (m, n)\} \tag{4}$$

where $m, n = ..., -1, 0, 1, 2, ...$ It is convenient to represent the value $f(\mathbf{x})$ as a point along the gray scale. Each lattice cell centered at a lattice point is called a pixel and marked by pixel center $\mathbf{x}$.

To analyze a chart in terms of Minkowski functionals, we consider the excursion sets S_u : $S_u = \{\mathbf{x} : f(\mathbf{x}) > u\}$, where u is a threshold. Thus, the excursion set is defined as the set of those pixels from the region of the lattice S where $f(\mathbf{x})$ exceeds u. This set can be represented as union of a finite number of so-called basic sets. [21,13]

Defining the indicator function $N_u(\mathbf{x}) = 1$ if $f(\mathbf{x}) > u$ and $N_u(\mathbf{x}) = 0$ otherwise, the excursion set is symbolically represented as a union [11,22]

$$S_u = \bigcup_{\mathbf{x} \in S} \mathbf{x} N_u(\mathbf{x}), \tag{5}$$

where implicitly only terms with $N_u(\mathbf{x}) = 1$ are present in the union.

Because of additivity of Minkowski functionals [20], the decomposition formula follows for any W_i :

$$\begin{aligned} W_i(S_u) = {} & \textstyle\sum_{\mathbf{x}\in S} N_u(\mathbf{x}) W_i(\mathbf{x}) - \sum_{\mathbf{x}_1 \neq \mathbf{x}_2} N_u(\mathbf{x}_1) N_u(\mathbf{x}_2) W_i(\mathbf{x}_1 \cap \mathbf{x}_2) \\ & + \textstyle\sum_{\mathbf{x}_1 \neq \mathbf{x}_2 \neq \mathbf{x}_3} N_u(\mathbf{x}_1) N_u(\mathbf{x}_2) N_u(\mathbf{x}_3) W_i(\mathbf{x}_1 \cap \mathbf{x}_2 \cap \mathbf{x}_3) - \ldots, \end{aligned} \tag{6}$$

where the sums are taken over all different pairs, triples etc. of lattice elements s. The intersections $\mathbf{x}_1 \cap \mathbf{x}_2 \cap ...$ of pixels are understood in the simple geometric sense, as intersections of polygons.

Direct application of Eq. (6) can be used to calculate the Minkowski functionals for a given map at a given level u by simply evaluating N_u $(\mathbf{x})$ for each pixel $\mathbf{x}$. [11,12,22]

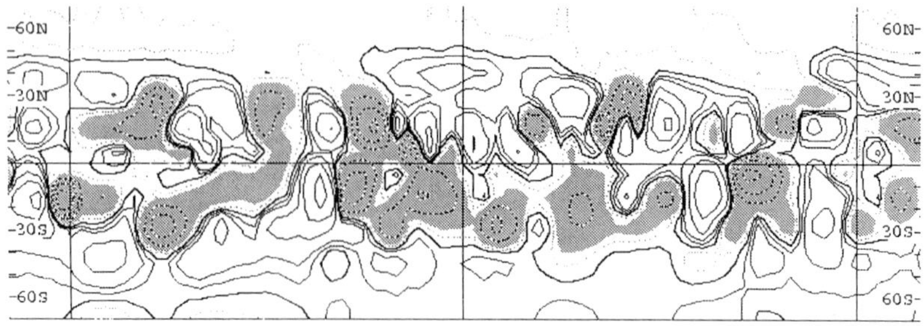

Figure 1. Example of Wilcox solar charts. Carrington Rotation 1700.

2.2 Boulingand-Minkowski Dimension

The Minkowski dimension [23] refers to the idea of parallel set K_ϵ and allows the evaluation of the fractal feature of the set K. Let $vol(K_\epsilon)$ be a volume of ϵ-parallel body to K. The Minkowski dimension of the $K \in R^d$ is

$$d_M(K) = \limsup_{\epsilon \longrightarrow 0} \{d - \frac{\log vol(K_\epsilon)}{\log \epsilon}\}. \tag{7}$$

Numeric estimate of d_M can be made by means of the Minkowski addition (dilation) and subtraction (erosion) MIA operators [12].

With this approach, a binary version of the analyzed image is obtained and circles with increasing radii are superposed onto each of the foreground pixels. The problem is that not all radii are possible in the orthogonal lattice underlying digital images. Recently Costa and *et al* [24] have suggested the method named Sorted Exact Distance Representation (SEDR) which stores all valid distances to guarantee correct dilatation on the orthogonal lattice and permits to estimate Minkowski dimension accurately.

2.3 Characteristics of Solar Synoptic Charts

We use two types of Solar synoptic charts. The first type contains H_α synoptic charts based both on the synoptic charts of the Meudon Observatory and daily H_α and $CaII$ spectroheliograms of the Kodaikanal Observatory. They cover 15-22 Solar Cycles. [17] It is known that the large-scale H_α chromospheric structure reveals a distribution of the neutral line of the large-scale magnetic field in detail and with a good precision. The solar magnetic structures, called filaments and filament channels, look as dark and light areas and form a continuous close contours, which outline the regions of one predominant polarity of the large-scale magnetic field. Within such approximately unipolar region, fine structure elements of the magnetic field of the opposite polarity are always present.

The second type is a daily magnetogram observation of the large-scale photospheric magnetic field made at the John M. Wilcox Solar Observatory at Stanford since 1976. [18] These measurements provide a homogeneous record of the solar field variation through Solar Cycles 21, 22 and 23. Magnetograms with three arc-minute

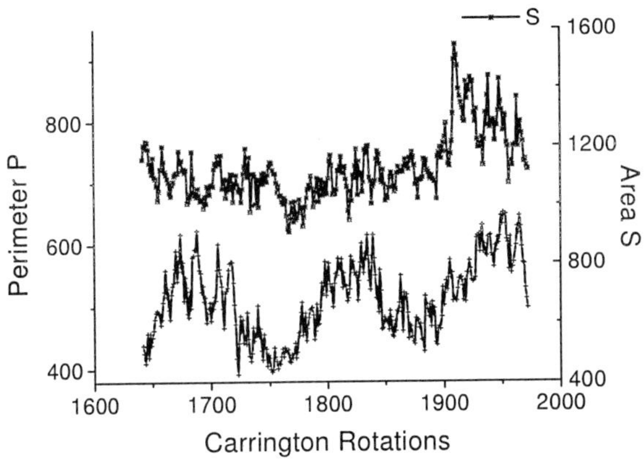

Figure 2. Two Minkowski functionals, namely, perimeter P (lower curve) and area S for Wilcox solar charts as a function of Solar rotation number.

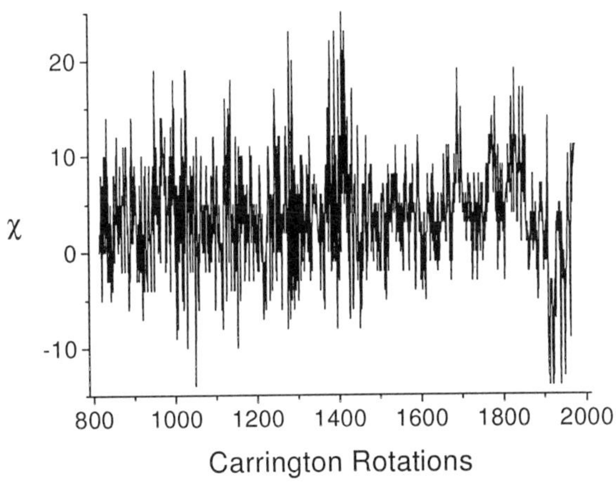

Figure 3. The Euler characteristic χ of the time series of H_α and Wilcox solar charts covering 802-1972 Carrington Rotations.

resolution are taken at Stanford every day. Then the daily magnetograms are interpolated onto a finer grid based on the Carrington coordinate. In the final grid there are 29 equally spaced points between latitudes 70° North and 70° South at 5° intervals. The resolution in longitude is also 5° (Fig. 1).

Due to weather and equipment failure, Stanford magnetograms are not available every day. This results in 11% of the charts being unsuitable for calculation of morphological characteristics. Consequently, to recover missing values in characteristics

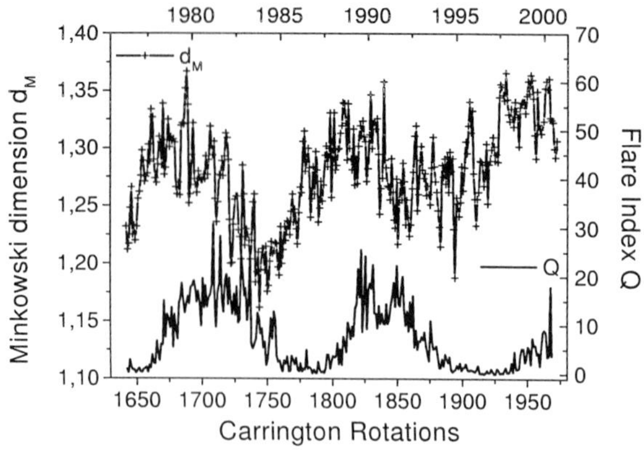

Figure 4. The variation of Minkowski dimension d_M and flare index Q depending on Solar rotation number.

time series, we use the neural network method of data recovery. [25] The idea of the method is as follows. Firstly, we estimate the embedding space dimension R^m from time series. For this purpose we temporarily substitute interpolated values for the missing ones. Then we construct a table of data where each row is m-dimensional delay vector. A sequence of the rows models the attractor trajectories in R^m. Let some of the rows have $k < m$ gaps. The gaps in the vector are considered as k-dimensional linear manifold L_x parallel to k coordinate axes. The problem of gaps recovery consists in proper approximation of L_x by means of some manifold of a given small dimension, for example, by a curve. Neural network is used to find this manifold. [26,27] The results, demonstrated below, have been obtained by this technique.

The Minkowski functionals and dimension were calculated for Stanford synoptic charts. Additionally, the W_2 functional (the Euler characteristic) has been estimated for H_α charts. The results were obtained for cross-section level of synoptic charts corresponding to the sign reversal line and shown in Figures 2, 3 and 4. The morphological characteristics and fractal dimension have time variations which correspond to Solar Cycles. The average value of the Euler characteristic is approximately 4 , that points to non-trivial topology of large scale magnetic field.

We have discovered that global magnetic field exhibits fractal scaling varying from $d_M = 1.17$ to $d_M = 1.37$ (Fig. 4). The behaviour of fractal dimension is similar to perimeter changes, that is not surprising, as sometimes d_M is called the "length of the boundary set". [23]

Moreover, there are relations between the perimeter P and dimension d_M and the solar flares. A solar flare is an enormous explosion in the solar atmosphere which is defined as a sudden and intense variation in brightness. It is believed to result from the sudden release of energy stored in magnetic fields. Power flares are key factors governing space weather, but there does not exist a complete theory of

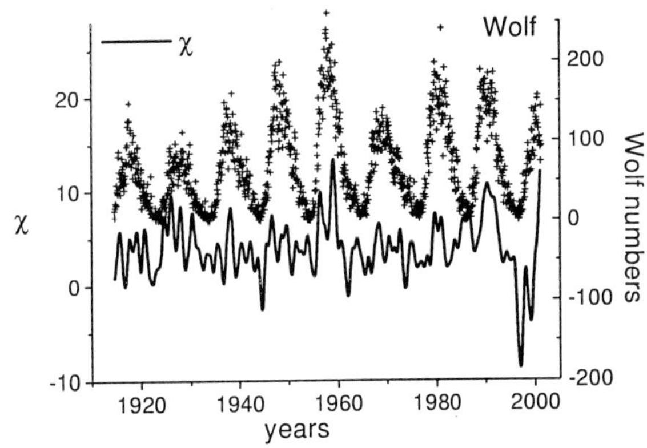

Figure 5. The variation of smoothed Euler characteristic χ and 27 days Wolf number.

flares yet. However, there exists an assumption that flares are governed by a unified mechanism, which affects the global rebuilding of sign reversal line separating magnetic field areas of different polarity. [28]. Kleczek [29] was the first to introduce the flare index Q which gives roughly the total energy emitted by the flares over a day. The daily flare indexes for the 21, 22, and 23 Solar Cycles are compiled by National Geophysical Data Center. The monthly indexes are shown in Fig. 4 together with the fractal dimension.

3 Magnetic Field Dynamics from The Euler Characteristic

We used the Euler characteristic time series in the reconstruction of the "Solar magnetic attractor". Because the synoptic chart is not an instantaneous picture of magnetic patterns, the topology of adjacent charts is unmatched along the boundary. This might produce sharp variations in the behaviour of the Euler characteristic. To reduce this effect we have applied smoothing by means of a statistical curve. [30] One can see the time series of the smoothed Euler characteristic and the solar activity index (the well-known sunspot or Wolf number) in Fig. 5.

The filtered characteristics were used for the reconstruction of the "magnetic attractor" using the Takens algorithm. The estimation of correlation dimension was carried out by means of the Gaussian kernel algorithm developed by Diks. [31] The algorithm is adapted for experimental data corrupted by noise. The idea of the method consists in a replacement of the Heaviside kernel in the Grassberger and Procaccia correlation integral

$$C_m(\epsilon) = \int d\mathbf{x}\rho_m(\mathbf{x}) \int d\mathbf{y}\rho_m(\mathbf{y})\Theta(\epsilon - ||\mathbf{x} - \mathbf{y}||) \tag{8}$$

onto the Gaussian kernel with bandwidth h:

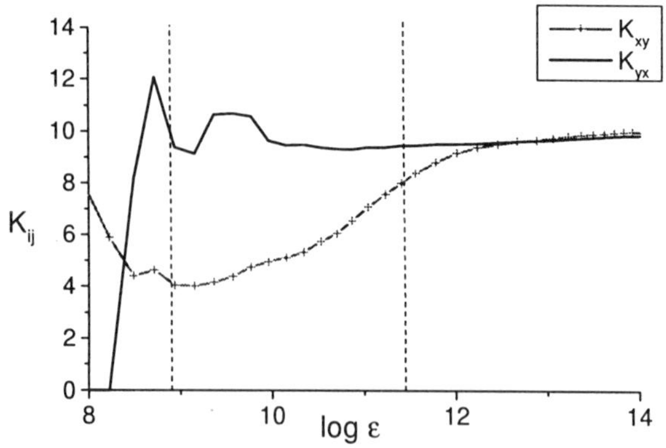

Figure 6. The correlation ratio of interrelation between smoothed Euler characteristics (X system) and Wolf numbers (Y system).

$$T_m(h) = \int d\mathbf{x}\rho_m(\mathbf{x}) \int d\mathbf{y}\rho_m(\mathbf{y}) exp[-||\mathbf{x}-\mathbf{y}||^2/4h^2]. \tag{9}$$

For the fixed m and for the noise-free case there is the scaling [31]

$$T_m(h) \sim exp[-mK\tau](h/\sqrt{m})^{d_c} \tag{10}$$

where τ is lag and K is correlation entropy. In presence of the Gaussian noise the distribution function $\widetilde{\rho_m}(\mathbf{y})$ can be expressed in terms of a convolution between the noise-free function $\rho_m(\mathbf{x})$ and a normalized Gaussian distribution function with standard deviation σ. In this case d_c, K and noise level $\widehat{\sigma}$ can be estimated.

For parameters of embedding $m = 5, 6, 7$ and $\tau = 5$ we have calculated correlation dimension $d_c = 1.95 \pm 0.02$, noise level $\widehat{\sigma} = 0.5\%$ and correlation entropy $K = 0.18$ bit per rotation. The d_c estimate is lower than the correlation dimension of the Wolf number series $d_c = 2.5$ obtained earlier. [32]

3.1 Driver-Response Relationships between Global Field and Sunspots

Global solar cycle is considered as a manifestation of three types of magnetic activity: polar magnetic activity, large-scale weak magnetic and sunspot strong fields. According to modern theories of solar cycle, the large-scale magnetic field of the Sun is formed by the redistribution and decay of old strong magnetic sunspot flux. Preliminary results about the crucial role of the large-scale magnetic field in the solar cycle have been obtained recently [19] on the basis of H_α synoptic charts. Consequently, it is necessary to verify the interdependence between the large-scale magnetic fields traced by synoptic charts and the strong sunspot fields i.e. the Wolf numbers series. Such interdependence is also called *synchronization* of directionally-coupled systems, and different terms have been introduced, namely *identical* for identical individual dynamics and *generalized* for cases of non-identical dynamics. [33] Assuming that the systems under consideration can have non-identical individual dynamics we work in the frame of *generalized* synchronization.

To test this interdependence, the "attractors" from the Euler characteristics and sunspot time series are reconstructed by embedding technique. Causal (driver-response) relationship might be estimated then by the cross-correlation sums method [34,35]

Let $\{x_i\}, \{y_i\}, i = 1, 2, ...N$ be two different time series produced by two dynamical systems X and Y having chaotic attractors of low dimensions d_x and d_y, respectively. Then one can realize embedding into $R^m, m \geq (d_x + d_y)$ and to define cross correlation sum by

$$C_{\mathbf{xy}}(\epsilon) = N^{-2} \sum_{\mathbf{x}} \sum_{\mathbf{y}} \Theta(\epsilon - ||\mathbf{x} - \mathbf{y}||), \tag{11}$$

where $\mathbf{x}$ and $\mathbf{y}$ are delay vectors of X and Y systems. To estimate a power of interrelations, the cross correlation ratio is used [36]

$$K_{\mathbf{xy}}(\epsilon) = \sqrt{\frac{\sum_{i\neq j} ||\mathbf{y}(i) - \mathbf{y}(j)||^2 \Theta(\epsilon - ||\mathbf{x}(i) - \mathbf{x}(j)||)}{\sum_{i\neq j} \Theta(\epsilon - ||\mathbf{x}(i) - \mathbf{x}(j)||)}}. \tag{12}$$

If the X and Y systems are related then one can expect that $||\mathbf{x}(i) - \mathbf{x}(j)|| < \epsilon \Longrightarrow ||\mathbf{y}(i) - \mathbf{y}(j)|| \approx \epsilon$. If this is not so, than $K_{\mathbf{xy}}$ does not depend on ϵ. Moreover, the expression K in (12) is not symmetric for $\mathbf{x}$ and $\mathbf{y}$ permutation and this makes it possible to estimate which system is the driver.

For the reconstruction of attractors from the Euler characteristics time series (X system) and the series of Wolf numbers averaged over 27 days (Y system) we used the embedding dimension $m > d_c(X) + d_c(Y) = 7$ and chose $\tau = 10$, according to the mutual information function. Both time series were randomized to avoid trend effects connected with 11-years cycle. In each time series there remained not more than 40-50 records with initial correlations. Moreover, to improve statistical features of the time series, the "Gaussianization" algorithm was applied. [37] As a result, the processed time series has the unimodal distribution close to the Gaussian. The curves $K_{\mathbf{xy}}(\epsilon)$ and $K_{\mathbf{yx}}(\epsilon)$ are shown in Fig. 6. One can see that the curve $K_{\mathbf{xy}}(\epsilon)$ has a sharp dependence on the scale and this indicates that the system X, i.e. the global magnetic field, drives the strong sunspot field.

4 Conclusion

The spatio-temporal dynamics of the Solar activity is encoded in the Solar synoptic charts. The Minkowski functionals and the dimension of Solar magnetic charts were calculated first, to transform the information contained in the charts to the scalar time series of various characteristics. The time series of perimeter, area, Euler characteristic and fractal dimension were evaluated for the level corresponding to the sign reversal line. Then the Euler characteristic time series, being most representative, was used to analyze the the dynamics of solar activity.

The results have pointed out the fractality of the magnetic structures on the magnetic neutral line. The Minkowski dimensions demonstrate that the time variation is related to the phase of Solar cycle. Both fractal dimension d_M and perimeter P correlate with Solar flares. It is interesting that the Euler characteristic estimated over all available synoptic charts has mean value $< \chi > \sim 4$ that is in agreement with non-trivial topology of the magnetic field.

The correlation dimension of the attractor reconstructed from the Euler characteristic time series is lower than dimension of the Wolf numbers series, nevertheless the estimation of causal relationship between two types of the field has shown the dominant role of the global magnetic field.

References

1. T. Sauer, J. A. Yorke and M. Casdagli, *J. Stat. Phys.* **65**, 579 (1991)
2. R. Gilmore, *Rev. Mod. Phys.* **70**, 1455 (1998)
3. N. H. Packard, J. P. Crutchfield, J. D. Farmer and R. S. Shaw, *Phys. Rev. Lett.* **45**, 712 (1980)
4. F. Takens, in *Lecture Notes in Math.* **898**, 366 (1981)
5. V. S. Afraimovich and A. M. Reimen in *Nonlinear waves, Dynamics and Evolution* , Nauka, Moscow, 1989
6. M. C. Cross and P. C. Hohenberg, *Science* **263**, 1569 (1994)
7. L. Fabrikant, M. I. Rabinovich, L. Sh. Tsimring, *Uspekhi Fiz. Nauk*, **162**, 1 (1992)
8. F. C. Adams, *Asrophys. J.* **387** 572 (1992)
9. R. J. Adler *The Geometry of Random Fields*, (Wiley, New York, 1981)
10. K. J. Worsley, *Chance* **9**, 27 (1996)
11. K. Michielsen and H. De Raedt, *Comp. Phys. Commun* **132**, 94 (2000)
12. K. Michielsen and H. De Raedt, *Phys. Rep.* **347**, 461 (2001)
13. N. Makarenko, L. Karimova, A. Terekhov and M. M. Novak, *Physica A* **289**, 278 (2001)
14. J. K. Lawrence, A. C. Cadavid and A. A. Ruzmaikin, *Asrophys. J.* **465**, 425 (1996)
15. V. I. Makarov and K. R. Sivaraman, *Solar Phys.* **123**, 367 (1989)
16. Z. Mouradian and I. Soru-Escaut, *Astron. and Astroph.* **251**, 649 (1991)
17. V. I. Makarov, K. R. Sivaramanma *H_α-Synoptic charts, Solar Cycle 19 (1955-1964)* Acad. Sci. of the USSR, Moscow, 1984
18. J. T. Hoeksema available in http://quake.stanford.edu/‘wso
19. N. G. Makarenko and V. I. Makarov in *JENAM-2000, May 29-June 3* Moscow, 2001, (in Russian)
20. D. Stoyan, W. S. Kendall and J. Mecke *Stochastic Geometry and its Application*, (Wiley, New York, 1995)
21. H. Hadwiger *Vorlesungen uber Inhalt, Oberflache und Isoperimetric*, (Springer, Berlin, 1957)
22. S. Winitzki, *New Astron.* **3**, 75 (1998)
23. J. Serra *Image Analysis and Mathematical Morphology*, Vol. 1 Academic Press, London, 1982.

24. L. da F. Costa and L. F. Estrozi, *Electronics Letters* **35**, 1829 (1999)
25. V. A. Dergachev, Makarenko N. G. , Kuandykov E. B. and Rossiev A. A. *Izv. Acad. Nauk, ser. Fiz.* **65**, 391 (2001) (in Russian)
26. A. A. Gorban, A. A. Rossiev and D. C. Wunsch II in *Neuroinformatics-2000, ch.1*, MIFI,Moscow, 2000, (in Russian)
27. E. B. Danilkina, Y. B. Kuandykov, N. G. Makarenko in *Neuroinformatics-2001, ch.1* MIFI,Moscow, 2001, (in Russian)
28. V. V. Kasinsky, *Astron. and Astrophys. Transact.* **17**, 341 (1999)
29. J. Kleczek, *Publ. Inst. Centr. Astron., Prague* **22**, (1952)
30. E. T. Whittaker and G. Robinson *The Calculus of Observations. A Treatise on Numerical Mathematics*, Blackie and Son,Limited, 1928
31. C. Diks, *Physica E* **53**, R4263 (1996)
32. V. M. Ostryakov and I. G. Usoskin, *Solar Phys.* **127**, 405 (1990)
33. R. Brown and L. Kocarev, *Chaos* **10**, 344 (200)
34. P. Schneider and P. Grassberger, *Nonlinearity* **10**, 749 (1997)
35. P. Grassberger, J. Arnhold, K. Lehnerts and C. E. Elger, *Physica D* **134**, 419 (1999)
36. G. Lasiene and K. Pyragas, *Physica D* **120**, 369 (1998)
37. D. Weinberg, *Mon. Not. R. astr. Soc.* **254**, 315 (1992)

OBSERVING EXTREME VARIABILITY IN NONLINEAR SYSTEMS

KONSTANTINE P. GEORGAKAKOS[1,2]

[1]*Hydrologic Research Center, 12780 High Bluff Drive, Suite 250, San Diego, CA 92130*
E-mail: kgeorgakakos@hrc-lab.org

[2]*Scripps Institution of Oceanography, UCSD, La Jolla, CA 92093-0224*

ANASTASIOS A. TSONIS

University of Wisconsin, Department of Mathematical Sciences, Atmospheric Sciences Group, P.O. Box 413, Milwaukee, WI 53201
E-mail: aatsonis@uwm.edu

Reconstructing the dynamics of a chaotic system from observations requires the complete knowledge of its state space. In most cases this is either impossible or at best very difficult. Here, using the Lorenz system, we investigate the possibility of deriving useful insights about the system's variability from only a part of the complete state vector. We show that while some of the details of the variability might be lost, other details, especially extreme events, are successfully recovered. This research may have implications for using incomplete state spaces to identify and predict extreme events in physical systems.

1 Introduction

In inference problems concerning spatially extensive physical systems it is often the case that available remotely-sensed spatially-extensive data do not directly measure system state variables. In such cases, a complete characterization of these systems in state space is not available and estimation of system response must be done through surrogate state measures. Along these lines remotely-sensed observations are combined with in situ or remotely-sensed observations of system response to form observation-response relationships (typically based on regression analyses). These relationships are then used in areas and spatial scales where response observations are non existent to estimate system response. For example, in the atmosphere, infrared (IR) and visible (VIS) observations have been used to estimate response, such as surface precipitation, in this manner (e.g., Scofield and Oliver, 1977, Tsonis and Isaac, 1985, Arkin and Meisner, 1987, Adler and Negri, 1988).

In all such cases the observables depend on a portion of the state vector of the system flow while the response does not necessarily depend on the same portion of the state vector. The question is whether the observational problem as described leads to useful data and response estimates. The underlying state vector is not known but it may allow regions in state space where the system flow evolves about strange attractors. The present study focuses on fundamental issues of the observation problem by considering the idealized model for convection presented first by Saltzman (1962a) and Lorenz (1963). The question posed is: *For a nonlinear system with chaotic dynamics, are indirect noisy observations of a part of the system state vector adequate to identify desired characteristics of the system response such as extreme response variability?*

In the next section we present the mathematical formulation of the problem and in section 3 we discuss the results of numerical experiments. Concluding remarks are in section 4.

2 Mathematical Formulation

The physical situation considered is a layer of fluid of infinite horizontal extent, which is subject to a temperature-difference forcing of $\Delta T (> 0)$ between the lower and upper surface. Lorenz (1963) approximates the set of partial differential equations describing the coupled transport of the momentum and heat of the fluid with the set of three non linear ordinary differential equations:

$$\frac{dX}{d\tau} = -\sigma X + \sigma Y \tag{1}$$

$$\frac{dY}{d\tau} = -XY + rX - Y \tag{2}$$

$$\frac{dZ}{d\tau} = XY - bZ \tag{3}$$

where X, Y, and Z are the state variables of the approximate system. The approximation was obtained by considering a two-dimensional description of the original system and a truncated double Fourier expansion of the velocity components and temperature. Temperature is expressed as a deviation from the steady state solution obtained for constant ΔT. The steady state temperature varies linearly with depth in the layer of fluid. In the idealized system of Lorenz (1963), X is proportional to the intensity of the convective motion, Y is proportional to the temperature difference between the ascending and descending currents, and Z is proportional to the distortion of the vertical temperature profile from linearity. The dimensionless time τ is given by

$$\tau = \pi^2 H^{-2}(1 + a^2)kt \tag{4}$$

and the parameters σ, r and b in (1) - (3) are given by:

$$\sigma = \frac{\nu}{k} \tag{5}$$

$$r = \frac{R_a}{R_c} \tag{6}$$

and

$$b = \frac{4}{(1 + a^2)} \tag{7}$$

with σ being the Prandtl number, R_a the Rayleigh number, ν the kinematic viscocity of the fluid, k the thermal conductivity of the fluid, R_c a critical value of R_a, H the depth of the fluid layer, and a the coefficient of thermal expansion of the fluid. It is known that for $\sigma = 10$, $b = 8/3$ and r in the range 24.74 to 31.10, the Lorenz system possesses chaotic dynamics and a strange attractor for large integration times (Berge, et al. 1984, 301-312).

The observation problem may be stated as follows: Given observations, possibly noisy, of one or two of the states of the Lorenz system, estimate system response which may be a function of one or more states (some unobservable). In the simplest case, we postulate observations that are linear functions of certain system states:

$$O_1 = \epsilon X + v_1 \tag{8}$$

$$O_2 = \delta Z + v_2 \tag{9}$$

where ϵ and δ are coefficients, and v_1 and v_2 are independent random processes with uniform distribution functions in the intervals $[-V_1, +V_1]$ and $[-V_2, +V_2]$, respectively. We also postulate a positive response function, which is a linear function of the remaining system state:

$$P = c_2 Y + w^{\cdot}, c_2 > 0; Y > Y_c \tag{10}$$

with w possessing a uniform distribution function in the interval $[-P_2, +P_2]$. The noise terms v_1, v_1, and w represent the effects of non-modeled components in the observation process and the system response function. The only assumption we will make for the response noise w is that P_2 is inversely proportional to Y for large Y. That is, the contribution of non-modeled effects diminishes for high response and, for such a regime, the Lorenz system is largely driving the response function. It is then postulated that w possesses a uniform distribution function in the interval $[-\frac{D}{Y}, +\frac{D}{Y}]$, with $Y \geq Y_c$, and D a scale parameter.

Other combinations of observation and response functions are possible with results analogous to those obtained from the set (8)-(10). Note also that only one of (8) or (9) may be used as an observation equation to estimate the response P as defined in (10).

3 Numerical Experiments

Numerical experiments were performed following the procedure outlined next. The Lorenz system (1)-(3) is simulated numerically for a large number N_E of iterations. For given V_1, V_2, D, and Y_c, observations O_1, O_2 are simulated using (8) and (9) and response values P greater than zero are simulated using (10) for a number of time steps ($N_0 < N_E$). It is supposed that there is no knowledge of the underlying nonlinear system in real cases, and we wish to estimate the response from the observations. To simulate a realistic scenario, a multiple linear regression relationship is then established between P and (O_1, O_2), and the regression correlation coefficient, $\mathcal{R}$, is recorded (the square of this coefficient is the portion of variance in the response explained by the observations). The relationship used for our analysis is:

$$P = \alpha_1 O_1 + \alpha_2 O_2 + \alpha_0 + e \tag{11}$$

where α_0, α_1, and α_2 are regression parameters and e is the regression error. The analysis is done for various threshold values P_T for which $P > P_T$ in order to probe the reliability of estimating extreme values of P. The sensitivity analysis examines the behavior of $\mathcal{R}^2$ when varying the quantities V_1, V_2, Y_c, N_0, and P_T.

The constant coefficients used in the simulations are: $\sigma = 10, b = 8/3, r = 28, \epsilon = 13, \delta = 5, c_2 = 3.5$. Without loss of generality, the values of the constants ϵ, δ, and c_2 are estimated so that the values of O_1 and O_2 represent counts in the range [0-255] and the values of P represent the range [0-100].

The most significant finding is depicted in Figure 1.

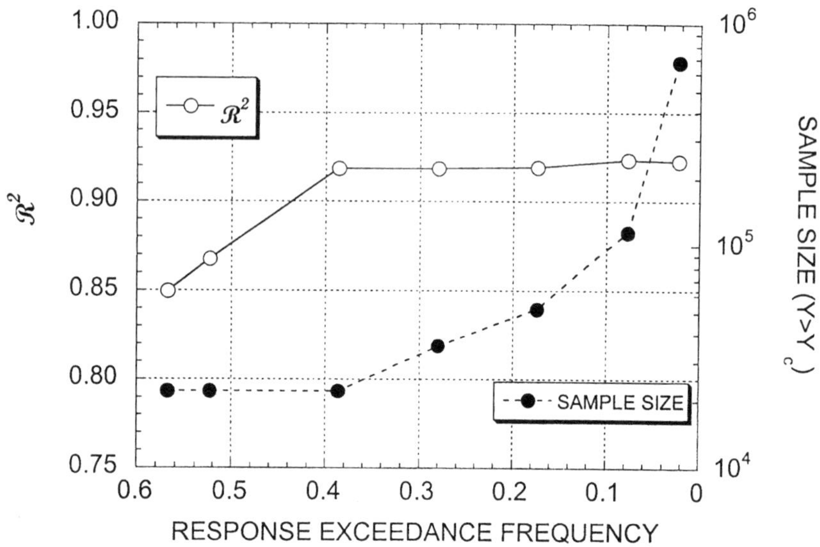

Figure 1. Regression correlation coefficient and sample size for $Y > Y_c$ as functions of response exceedance frequency.

The Figure shows $\mathcal{R}^2$ as a function of the response exceedance frequency for the case of $V_1 = V_2 = P_2 = 0, N_0 = 8000$ and $Y_c = 5$. In the same plot (read along the right ordinate) the sample size of time steps for which $Y > Y_c$ is also shown as a function of the exceedance frequency. It is evident that as the exceedance frequency decreases the observables O_1 and O_2 explain a larger portion of the response variability (from 85% to 92% of response variance). Also, as the exceedance frequency decreases, a progressively larger sample size is needed to assure a total of 8000 points for the multiple linear regression analysis (N_0). Table 1 shows the values of the regression parameters and relevant statistics. A decrease of the intercept α_0 estimates is noted as P_T increases toward more extreme response values, while $\alpha_1, \alpha_2 < 0$.

Dependence of the result on Y_c may be discerned from Figure 2, which shows $\mathcal{R}^2$ as a function of Y_c.

Table 1. Multiple Linear Regression Parameters and Error Statistics
$\mathcal{R}^2$: Multiple linear regression correlation coefficient
SE : Standard error of regression

EXCEEDANCE FREQUENCY (P_T)	α_0 ±(95% Bounds)	α_1 ±(95% Bounds)	α_2 ±(95% Bounds)	$\mathcal{R}^2$	SE
0.568(0.1)	29.30 ± 0.38	0.48 ± 0.005	-0.39 ± 0.005	0.85	6.07
0.523(20)	28.71 ± 0.35	0.51 ± 0.005	-0.42 ± 0.005	0.87	5.42
0.387(30)	25.67 ± 0.27	0.58 ± 0.004	-0.48 ± 0.004	0.92	3.45
0.280(40)	24.66 ± 0.27	0.59 ± 0.004	-0.47 ± 0.004	0.92	2.76
0.174(50)	23.80 ± 0.30	0.59 ± 0.004	-0.46 ± 0.004	0.92	2.17
0.077(60)	22.22 ± 0.31	0.60 ± 0.004	-0.46 ± 0.004	0.92	1.56
0.022(70)	19.52 ± 0.37	0.62 ± 0.004	-0.46 ± 0.003	0.92	1.16

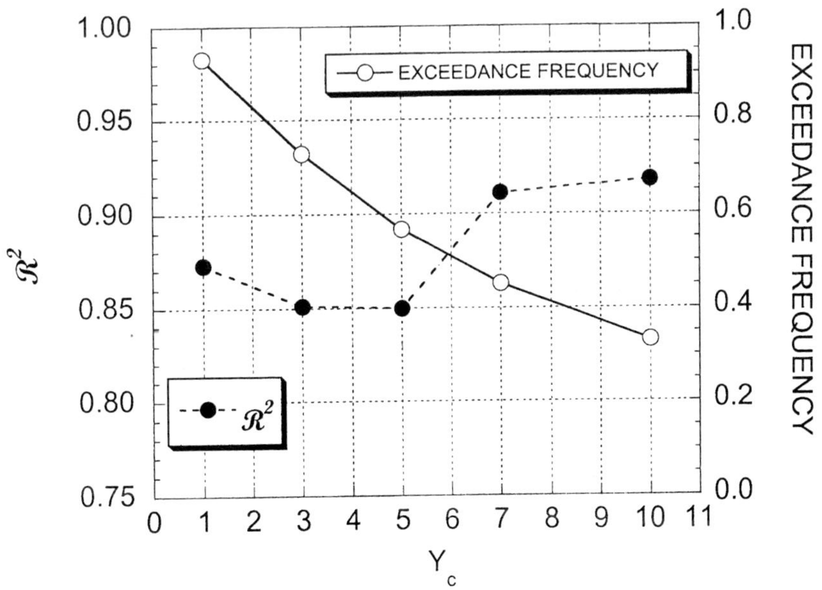

Figure 2. Regression correlation coefficient and response exceedance frequency as functions of parameter Y_c.

The exceedance frequency resulting from a certain value of Y_c is also shown. For the results shown in Figure 2, the rest of the parameters were set to the values used to produce the results of Figure 1, with $P_T = 0.1$. The increase of the explained portion of response variance with increasing Y_c is evident (from about 87% to about 92%). This result corroborates that of Figure 1 in that in both cases for a reduction of response exceedance frequency there is an increase of $\mathcal{R}^2$.

The character of the results (better reproduction of the system response by the observables for extreme cases than otherwise), was preserved when other response functions and observables were used. For example, when the response P was defined as a linear function of the system state Z, with $Z > Z_c$ in analogy to (10), and the observable O_2 was defined as a linear function of Y, in analogy to (9), the analysis produced the results shown in Figure 3 (analogous to Figure 1).

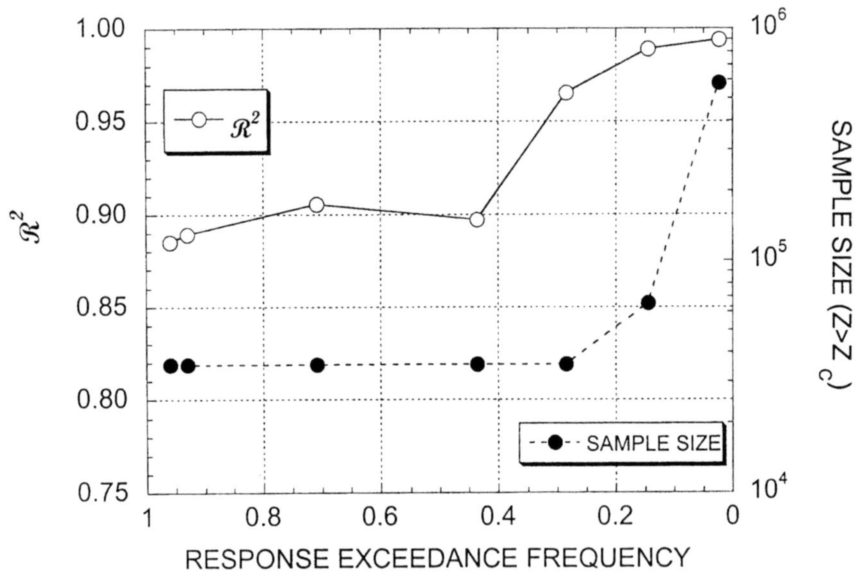

Figure 3. Regression correlation coefficient and sample size for $Z > Z_c$ as functions of response exceedance frequency.

The presence of observation and response noise alters the character of the results. Figures 4a-4c show the dependence of $\mathcal{R}^2$ on the observation noise parameters $V_1 = V_2$ for three values of the response noise scale parameter D and for certain values of the response threshold P_T. The results correspond to the set of observables and response defined in (8) - (10). There are three features that are important in the three panels of Figure 4: (a) there is a crossover of the curves corresponding to increasing values of P_T, for values $V_1 = V_2$ near 5 and for $D \leq 30$ (Figures 4a and 4b); (b) large response noise dominates the estimation of the response by observables, with very different values of P_T producing similar results (Figure 4c); and (c) increased observation noise yields poor response reproduction by the observables (negative slope of curves in all Figures). To put the response noise strength in perspective, it is noted that the response noise w has a uniform density function in an interval with width that is a function of the system state $Y(\geq Y_c)$. For $Y = Y_c = 5$, for the rightmost panel the interval is $[-40, +40]$ and for the middle panel it is $[-6, +6]$. For $Y = 20$, for the rightmost panel the interval is $[-10, +10]$ and for the middle panel it is $[-1.5, +1.5]$. The maximum value of Y in these experiments is about 23.75.

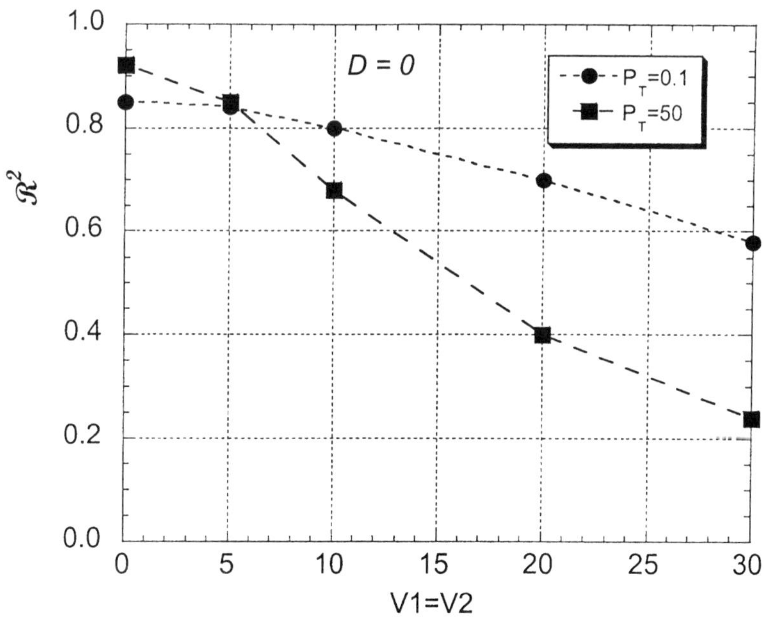

Figure 4a. Sensitivity of regression correlation coefficient with respect to observation noise strength $V_1 = V_2$ for two different exceedance thresholds P_T. Case: D=0.

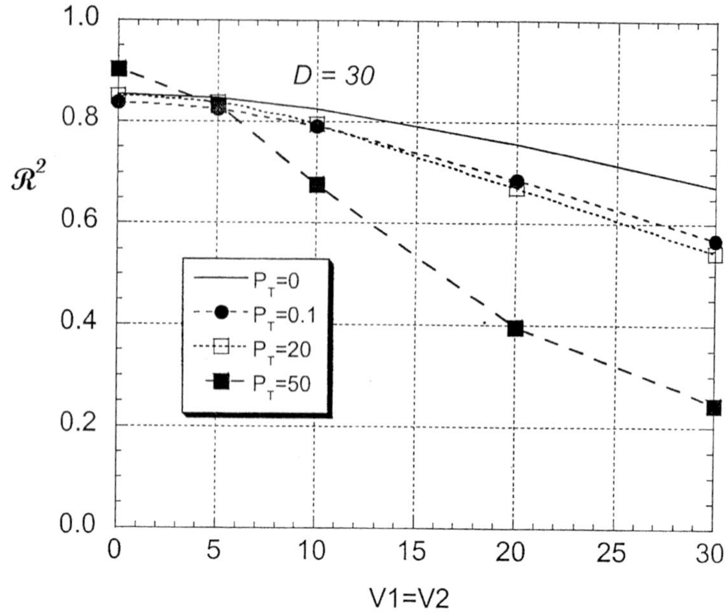

Figure 4b. Sensitivity of regression correlation coefficient with respect to observation noise strength $V_1 = V_2$ for various exceedance thresholds P_T. Case: D=30.

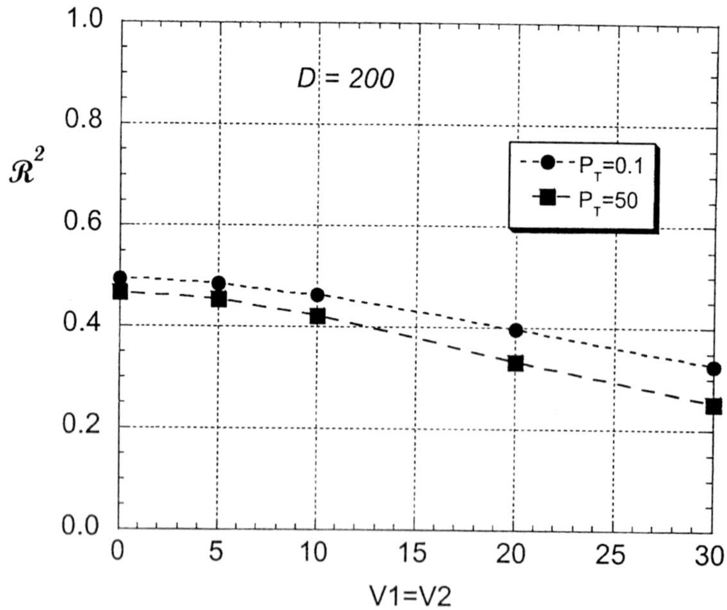

Figure 4c. Sensitivity of regression correlation coefficient with respect to observation noise strength $V_1 = V_2$ for two different exceedance thresholds P_T. Case: 200.

To illustrate the precision of these experiments we show in Figure 5 the computation of the response exceedance frequency for $P_T = 0.1$ and $P_T = 50$, as a function of the logarithm of the number of simulation steps of the Lorenz system (N_E). In the same Figure, read along the right ordinate axis, we also show the resultant N_0. A change in exceedance frequency less than 0.01 results when N_E is greater than a few tens of thousands time-steps.

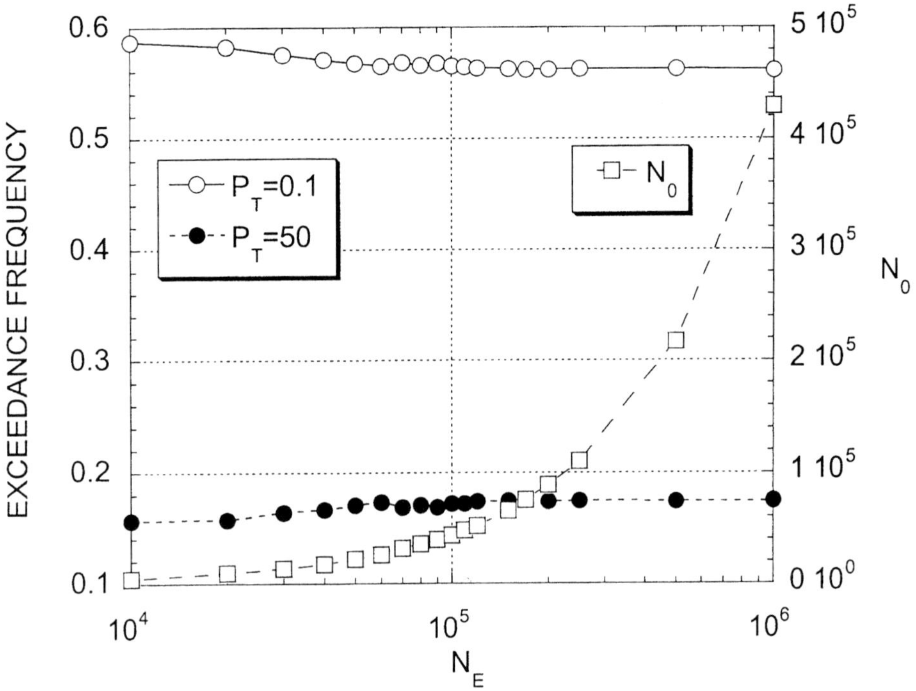

Figure 5. Response exceedance frequency and N_0 as functions of the simulation steps N_E of the Lorenz system equations for two different exceedance thresholds P_T.

For the dependence of the results of multiple linear regression ($\mathcal{R}^2$) on N_0 we include Figure 6 as an example.

The value of $\mathcal{R}^2$ is shown as a function of the number N_0 of time steps used for the estimation of the regression parameters. Results for two extreme values of P_T and for the case of response noise with $D = 30$ are shown. The estimates of $\mathcal{R}^2$ differ by less than 0.01 for N_0 greater than 6000, with greater variability observed for low values of N_0 for the case of $P_T = 0.1$. It is concluded that the significant findings of this study are not artifacts of imprecise computations or small sample sizes.

In cases with single observables (either one of O_1 or O_2) the reproduction of the response may be shown to be poor throughout the range of response magnitude, and especially for the extremes. Figure 7 shows results for the case of single observables.

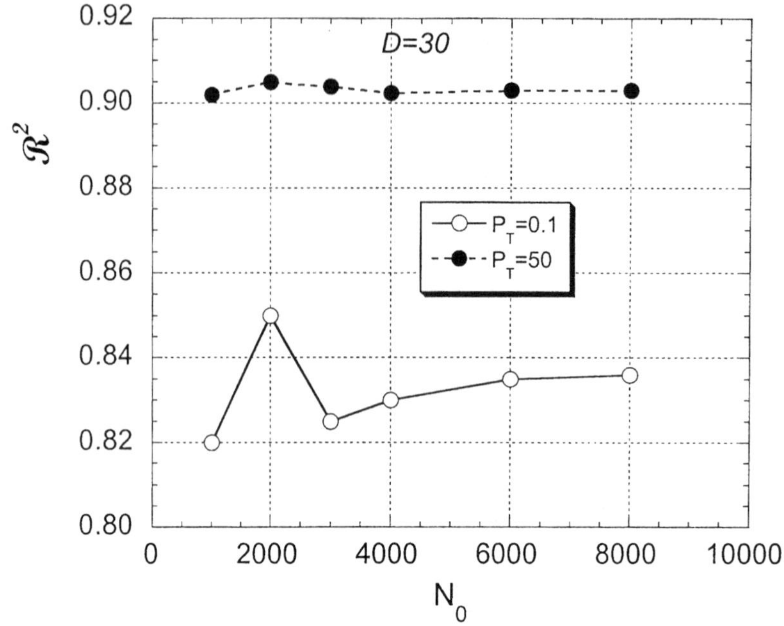

Figure 6. Regression correlation coefficient as a function of sample size N_0 for two different response exceedance thresholds P_T.

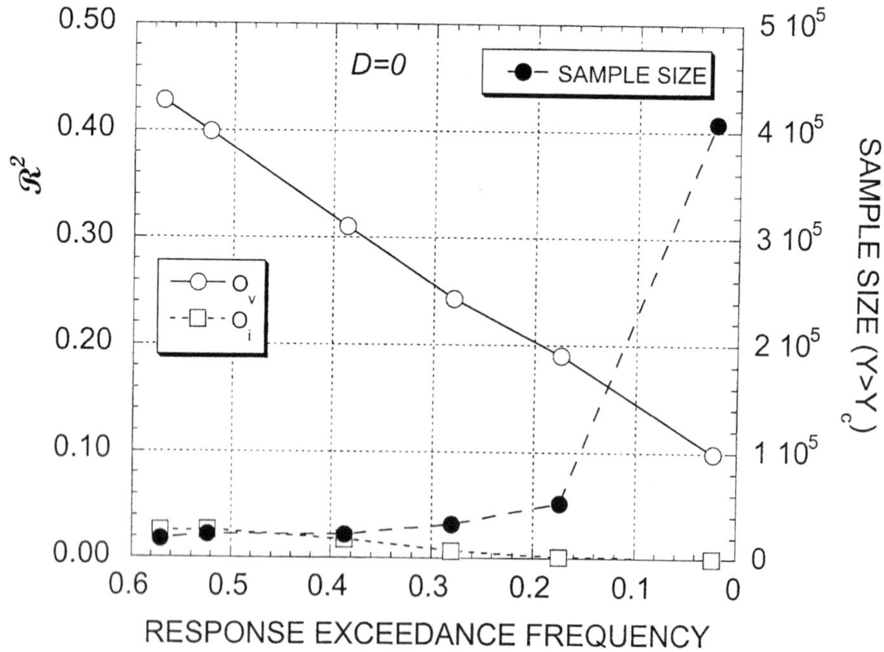

Figure 7. Regression correlation coefficient and sample size for $Y > Y_c$ as functions of response exceedance frequency for single observations 0_v or 0_i.

The parameter values used are: $V_1 = V_2 = 0, D = 0, Y_c = 5$, and $N_0 = 8000$. As P_T increases, the exceedance frequency of the response decreases. The abscissa axis of Figure 7 shows the exceedance frequency of the response, while the left ordinate axis shows $\mathcal{R}^2$ and the right ordinate axis shows sample size. There are two curves corresponding to O_1 and O_2 considered individually as observables. It is notable that the results in this case are much worse than those obtained when both O_1 and O_2 were used as observables (at best only about 40% of the response variance is explained in the present case). Also, O_1 is a much more suitable observable than O_2 for cases when a single observable is used to reproduce the system response P. Additional results were obtained (but not shown) for a number of values of the parameters. It was found that the character of the results in Figure 7 is preserved for other values of the parameters. The introduction of moderate noise ($D = 30$) does not influence the results of this experiment greatly but it does reduce $\mathcal{R}^2$ somewhat, especially for the higher exceedance frequencies.

4 Concluding Discussion

In this numerical study an idealized model of thermally driven convection in a layer of fluid was used as the mathematical model of a low dimensional system which possesses chaotic dynamics for a range of parameter values. A low order observable vector and a system response were postulated as linear functions of portions of the state vector. The ability of the observables to reproduce system response variability was studied under a variety of scenarios of observation and response noise with specific focus on the reproduction of response extremes. The main conclusions drawn for the particular system studied are:

(a) The presence of a strange attractor, causing a contraction in state space flow, allows for reasonably good reproduction of response of extreme variability from observable vectors of lower dimensionality than the underlying system and which are not functions of the state dominating the system response. In our study a two-dimensional observation vector estimated well the response of the system which was a function of the remaining state.

(b) The presence of high noise in observations or response due to non-modeled effects substantially deteriorates the ability of observables to reproduce system response variability. This is attributed to the substantial change (caused by the presence of high noise) in the morphology of the flow on the strange attractor when mapped onto the response-observables space.

The system studied is but a very simple representation of nonlinear aspects of convection in the real atmosphere (see discussion in Lorenz, 1963, and Saltzman, 1962b). It is similar to systems studied to assess predictability in the atmosphere (e.g., Trevisan and Pancotti, 1998, and Benzi et al. 1999). However, should there be chaotic dynamics in the development of convective regimes or in other physical systems, it is conjectured that the general conclusions (a) and (b) would hold true as they only depend on the state space contraction properties of the flow. Recent studies lend support to the suggestion that low dimensional attractors are likely in atmospheric subsystems (e.g., Lorenz, 1991, Barker and Van Zyl, 1993, Tsonis, 1996, Tsonis et al. 1993).

Acknowledgments

The work of the first author was sponsored by the NASA Grant NAG8-1418 and by the Experimental Climate Prediction Center of the Scripps Institution of Oceanography. The work of the second author was partly supported by NSF grant ATM-9727329.

References

1. R.F. Adler and A.J. Negri, A Satellite Infrared Technique to Estimate Tropical Convective and Stratiform Rainfall. *Journal of Applied Meteorology*, **27**, 30-51 (1988).
2. P.A. Arkin and B. Meisner, The Relationship Between Large-Scale Convective Rainfall and Cold Cloud Over the Western Hemisphere During 1982-1984. *Monthly Weather Review*, **115**, 51-74 (1987).
3. H.W. Barker and B. Van Zyl, Correlation Dimensions of Local, Short-Term Climatic Attractors from Observations and a Global Climate Model. *Journal of Climate*, **6**, 858-861 (1993).
4. R. Benzi, M. Marrocu, A. Mazzino, and E. Trovatore, Characterization of the Long-Time and Short-Time Predictability of Low-Order Models of the Atmosphere. *Journal of the Atmospheric Sciences*, **56**, 3495-3507 (1999).
5. P. Berge, Y. Pomeau and C. Vidal, *Order Within Chaos, Towards a Deterministic Approach to Turbulence.* John Wiley & Soms, New York, 329 pp (1984).
6. E.N. Lorenz, Deterministic Nonperiodic Flow. *Journal of the Atmospheric Sciences*, **20**, 130-141 (1963).
7. E.N. Lorenz, Dimension of Weather and Climate Attractors. *Nature*, **353**, 241-244 (1991).
8. B. Saltzman, Finite Amplitude Free Convection as an Initial Value Problem - I. *Journal of the Atmospheric Sciences*, **19**, 329-341 (1962a).
9. B. Saltzman, ed., *Selected Papers on the Theory Convection with Special Application to the Earth's Planetary Atmosphere.* Dover Publications, Inc. New York, 461 pp (1962b).
10. R.A. Scofield and V.J. Oliver, A Scheme for Estimating Convective Rainfall from Satellite Imagery. *NOAA Technical Memorandum NESS 86*, Washington, D.C., 47 pp (1977).
11. A. Trevisan and F. Pancotti, Periodic Orbits, Lyapunov Vectors in the Lorenz System. *Journal of the Atmospheric Sciences*, **55**, 390-398 (1998).
12. A.A. Tsonis, J.B. Elsner and K.P. Georgakakos, Estimating the Dimension of Weather and Climate Attractors: Important Issues about the procedure and interpretation. *Journal of the Atmospheric Sciences*, **50**, 2549-2555 (1993).
13. A.A. Tsonis and G.A. Isaac, On a New Approach for an Instantaneous Rain

Area Delineation in the Midlatitudes Using GOES Data. *Journal of Climate and Applied Meteorology*, **24**, 1208-1218 (1985).

14. A.A. Tsonis, Dynamical Systems as Models for Physical Processes. *Complexity*, **1(5)**, 23-33 (1996).

GAMMA/HADRON SEPARATION USING THE MULTIFRACTAL SPECTRUM FROM 1/F FLUCTUATIONS IN SIMULATED EXTENSIVE AIR SHOWERS

E. FALEIRO

Departamento de Física Aplicada, E.U.I.T. Industrial,
Universidad Politécnica de Madrid, Ronda de Valencia 3, 20012-Madrid, Spain
E-mail: efaleiro@fais.upm.es

J.M.G. GÓMEZ AND A. RELAÑO

Departamento de Física Atómica, Molecular y Nuclear,
Universidad Complutense de Madrid, 28040-Madrid, Spain
E-mail: gomezk@nucl.fis.ucm.es

High-energy interactions of γ rays and protons and also helium, oxygen and iron nuclei with the earth atmosphere have been simulated by means of the CORSIKA Monte Carlo code, and the secondary-particle density distributions in the resulting extensive air showers, at ground level, have been studied. It is shown that the fluctuations of the particle density distributions have features typical of a $1/f$ noise. The multifractal spectrum of the samples is obtained and is found to have different features for different primary cosmic rays. This property is applied to the separation of electromagnetic from hadronic extensive air showers. A cutting parameter related to the multifractal spectrum is calculated and the efficiency of the cutting procedure for gamma/hadron separatrion is evaluated.

1 Introduction

Extensive air showers (EAS) result from the interaction of high energy particles and nuclei arriving from space, collectively called cosmic rays, with the high altitude Earth atmosphere. Successive interactions of the primary particles and their secondaries give rise to a multiplicative cascade of elementary particles. If the energy of the incident cosmic ray is high enough, a considerable number of particles may reach the Earth surface, where they are detected. At this point the secondaries are mostly e^+, e^-, γ and μ. A comprehensive description of EAS physics can be found in references [1,2].

The analysis of the the EAS parameters supplies information on the nature, energy and direction of the original cosmic ray and its interactions in the atmosphere. It is thus known that at energies greater than 1 TeV, the great majority of the cosmic rays are nuclei of atomic elements, mostly protons, but also helium, oxygen and iron among others, with a small, less than 1% proportion of electrons and photons. As is well known, charged particles and nuclei travelling through space constantly change their direction due to the space magnetic fields, but in the case of γ rays, their direction points to the source where they were produced and can thus be used in high energy astronomy.

In real experiments it may be quite difficult to distinguish an electromagnetic shower (generated by a primary γ ray) from a hadronic shower (generated by a primary proton or any other atomic nucleus). This separation constitutes a basic tool in high energy astronomy when sources of high energy γ rays are sought.

Therefore, several γ/hadron separation methods [3] are used to help to identify the meager proportion of γ inicitated EAS among all the showers. Recently, some rather simple and efficient γ/hadron separation methods based on features of the secondary particle distribution at the detection level have been proposed [4,5].

In this paper we present a γ/hadron discrimination method based on the recently reported multifractal structure of the particle density distribution in EAS [6,7,8]. In Section 2 the basic formulas of our multifractal analysis are introduced. In section 3 the cosmic ray simulation is described, and the power spectrum of the particle density distribution is analysed. In Section 4 the scaling properties of the syatistical moments are studied, and the multifractal spectum is calculated for EAS generated by different primary cosmic rays. It is found that the multifractal spectrum is different for electromagnetic showers and hadronic showers, and a cutting procedure is proposed for identification of the γ ray showers. Finally, the conclusions are summarized in section 5.

2 Basic multifractal analysis formulas

The multifractal formalism [9,10] has been established to account for the statistical scaling properties of singular measures arising in various physical situations. Known examples include the distribution of spectral line series [11], the distribution of growth probabilities on the interface of a diffusion-limited aggregate [12] and the dissipation field in fully developed turbulent flows [13]. This formalism lies upon the determination of the $f(\alpha)$ singularity spectrum [9] which associates the Haussdorff dimension $f(\alpha)$ to the subset of the support of the measure μ where the singularity strength is α,

$$f(\alpha) = dim_H[x|\mu(B_x(\delta)) \sim \delta^\alpha, \quad for\ \delta \to 0], \tag{1}$$

where dim_H denotes the Hausdorff dimension and $B_x(\delta)$ is an δ-box centered at x. The thermodynamical analogy provides a natural connection between the $f(\alpha)$ spectrum and an observable spectrum $\tau(q)$ defined from the power law behavior of a partition function (in the limit $\delta \to 0$)

$$Z_q(\delta) = \sum_i \mu(B_i(\delta))^q \sim \delta^{\tau(q)}, \tag{2}$$

where the sum is taken over a partition of the support of the singular measure μ into boxes of size δ. This $\tau(q)$ spectrum is directly related to the generalized fractal dimensions [14]

$$D_q = \frac{\tau(q)}{q-1}, \tag{3}$$

and it is estimated in the limit $\delta \to 0$ by

$$\tau(q) = min_\alpha[q\alpha - f(\alpha)]. \tag{4}$$

The spectrum $\tau(q)$ and $f(\alpha)$ are thus related by a Legendre transform. Actually, the variables q and $\tau(q)$ play the same role as the inverse of temperature and the free energy in thermodynamics [9,10], while the inverse Legendre transform

$$f(\alpha) = min_q[q\alpha - \tau(q)\tau(q)]. \tag{5}$$

indicates that instead of the energy and the entropy, we have α and $f(\alpha)$ as the thermodynamical variables conjugate to q and $\tau(q)$. For practical purposes, the spectrum $\tau(q)$ defined in Eq. (2) is calculated by the multifractal moments

$$G_q = \sum_{j=1}^{M} p_j^q \quad \sim \delta^{\tau(q)}, \tag{6}$$

where p_j, the measure, is the relative number of particles arriving to the j-bin, the δ-box, and the summation is carried over the non-empty bins only. In this paper we adopted an alternative method of estimating $\tau(q)$ based on the calculation of the moment scaling function [6,7] $K(q)$, obtained from the average

$$< \epsilon_\lambda^q > \sim \lambda^{K(q)}, \tag{7}$$

where ϵ_λ represents the normalized number of particles in a generic bin when the support is divided into λ equal bins. The introduced λ and δ parameters are inverse related and their product is the support length. The average is calculated over all the bins belonging to a specific event and over all the available events of the sample at scale ratio λ. It is easy to prove that

$$\tau(q) = q - 1 - K(q). \tag{8}$$

The moment scaling function $K(q)$ is widely used in stochastic multifractality where multiplicative cascade models are proposed as random multifractals [6,7]. Some of this models provide analytical expressions of $K(q)$ for $q \geq 0$ in terms of a finite set of parameters [15]. Thus, the spectrum $\tau(q)$ could be also parametrised at least for $q \geq 0$.

3 Cosmic ray simulation and power spectrum

To simulate EAS we have used the CORSIKA code [16]. We generated high energy γ rays, protons and also helium, oxygen and iron nuclei with energies ranging from 10 to 70 TeV. The number of events generated at each energy ranges from 100 to 500. As a simplification all the showers have been generated vertically. We shall follow the unidimensional event representation proposed in references [5,6,7].

With the aim of eliminating the global fluctuations, the generated events are classified by the number of secondary particles N_s that reach a ring centered at the core of the event, between 50 and 100 m radii. This ring is divided into 256 equal sectors in the azimuthal angle. In order to minimize global fluctuations we have considered intervals ΔN_s and we assume that each interval contains events without global fluctuations whose intrinsic fluctuations have the same nature. For simplicity we have considered only three of such intervals whose amplitude is equivalent to 2500 secondary particles. Table 1 shows the simulated event distribution according to the classification of the intervals. Following also references [6,7], we consider each event as the concrete realization of a wide sense stationary stochastic process (WSS). This process is characterized in the frequency domain by means of the mean power spectrum density (PSD) [17] denoted$P(k)$. The PSD is usually estimated from $P(k) \approx \; < |g(k)|^2 >$, where angular brackets represent the ensemble averaging

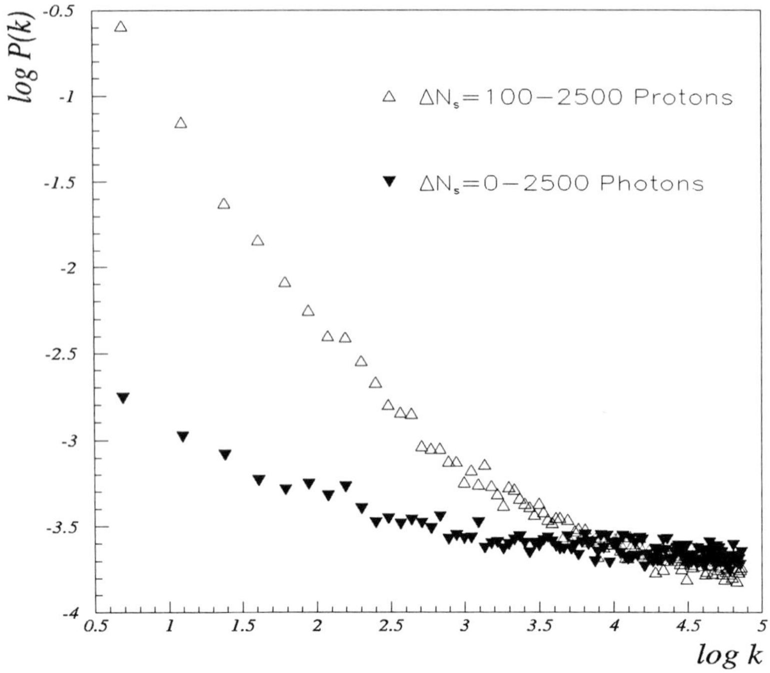

Figure 1. *Power spectrum density P(k) for two of the considered samples. The spatial frequency k is in fundamental frequency units.*

of the quantity inside them, and $g(k)$ stands for the amplitude corresponding to frequency k in the Fourier transform of the original data.

In Fig. 1 the natural logarithm of $P(k)$ for our events is represented versus $\log k$. The plots show two important facts. In the first place, instead of finding a constant $P(k)$, as would be the case for a signal whose fluctuations were uncorrelated, a PSD showing a complex structure is found. As a second observation, the shape of $P(k)$ suggest a simple model of functional form

$$P(k) \sim \frac{A}{k^B} + C \ , \tag{9}$$

where A, B and C are free parameters. The constant term C represents a white noise, the need for it is clearly seen in the constant behavior at high k, while the first term stems from the nearly linear behavior in $1/k$ seen at low k. Table 1 also contains the best fit parameters A, B and C obtained from our model power spectrum. The fit errors affect the last digit of the parameter values given in the table. We see that the model power spectrum (9) fits the data quite well. Thus it can be concluded that the fluctuations of the particle density are composed of a white noise plus a scaling noise $1/f$ which gives rise to its complex structure. This

Table 1. Main features of the Monte Carlo simulation of several types of primary samples ranging from 10 to 70 TeV. Here N_p is the number of primary particles whose number of secondary particles arriving to the ring is within the bin ΔN_s; A, B and C are the parameters of the power spectrum density proposed model, and H is the first-order structure function exponent.

Primary	N_p	ΔN_s	A	B	C	H
γ	514	0-2500	0.05	1.00	0.0235	0.129
	359	2500-5000	0.12	1.00	0.0532	0.128
	201	5000-7500	0.21	1.00	0.0929	0.130
proton	701	100-2500	0.62	1.14	0.0209	0.133
	572	2500-5000	1.61	1.12	0.0569	0.130
	450	5000-7500	2.51	1.13	0.0970	0.130
Helium	176	0-2500	0.96	1.21	0.0292	0.177
	443	2500-5000	2.00	1.08	0.0557	0.151
	355	5000-7500	2.72	1.11	0.0957	0.156
Oxygen	806	0-2500	0.87	1.04	0.0255	0.138
	432	2500-5000	2.15	1.09	0.0518	0.152
	194	5000-7500	3.6	1.13	0.0965	0.163
Iron	1829	0-2500	0.80	1.00	0.0173	0.132
	294	2500-5000	2.56	1.06	0.0526	0.146
	38	5000-7500	5.20	1.13	0.0780	0.170

kind of noise is usually called a *flicker noise* in the literature [18]. The value found for the parameter B, nearly one, shows that the scaling component of the signal can *a priori* be assumed to be *conservative* [15]. However, in this paper we have analysed the first order structure function of this signal component as a test of the conservative assumption. Therefore we have filtered the white noise component of the signal using a real non random filter with transfer function

$$h(k) = \sqrt{1 - \frac{C}{\frac{A}{k^B} + C}} \quad . \tag{10}$$

The filter (10) is a variant of the Wiener filter [17], which is known to be adequate for elimination of the white noise in the signal provided that it is uncorrelated to the $1/f$ component. Although the values for the parameter B of the proposed model are nearly one, and therefore stand for conservativity, the study of structure functions of the scaling component give rise to scaling power laws of the type $< |\epsilon(t+\tau) - \epsilon(t)|^q > \sim \tau^{\zeta(q)}$. We note that a scaling law is fulfilled by the first order structure function with exponent $\zeta(1) = H$, which varies from conservativity, $H = 0$, to absolute non conservativity, $H = 1$, and it is a measure of the degree of non conservativity of the noise [15]. The calculated values for H for all the samples are also displayed in Table 1. It is found that in all cases H is near to zero and therefore a conservative approach may be adopted as a reasonable approximation. Finer results can be found if we consider that the small values of H arise of an H-order fractional integration of the conservative associated signal. To obtain this conservative signal we should have to H-order fractionally derivate the non

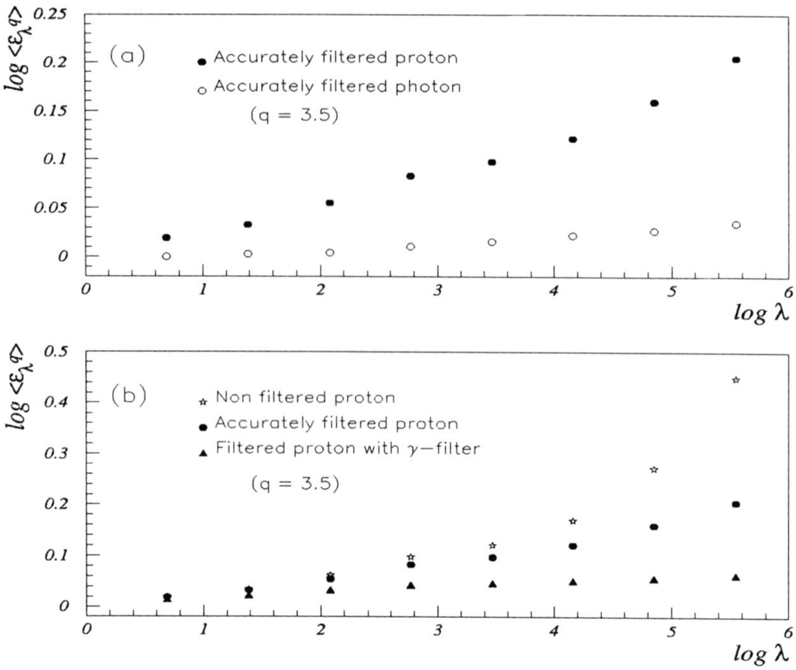

Figure 2. *a) Scaling of the statistical moment q=3.5 for gamma-induced and proton-induced events belonging to the database* $\Delta N_s = 2500 - 5000$. *b) Scaling of the same statistical moment for the proton-induced events when different filtering procedures are applied.*

conservative one. This is equivalent to filtering the signal with a non random filter whose transfer function is [15] $h(k) = k^H$. Yet the issue is full of complexities and therefore, in this paper, the conservative approach is adopted as an aproximation justified by the small values of H.

4 Multifractal spectrum calculations

Considering the resulting $1/f$ noise as conservative, we study the scaling properties of its statistical moments $< \epsilon_\lambda^q >$, where ϵ_λ, as pointed in Section 2, is the content of a generic sector at scale ratio λ and q is a real number here ranging from -1.5 to 3.5. The signal at other scale ratios is obtained by a simple coarse-graining procedure [7]. The scaling law for statistical moments obeys Eq. (7), where the moment scaling function $K(q)$ characterizes the scaling behavior of the data.

Fig. 2a shows the scaling behavior of statistical moments for two different filtered databases, one for proton and the other for γ ray primaries. Accurate filtering means that the parameters of our power spectrum model are obtained from a fit to the power spectrum of each databasis, and therefore the white noise

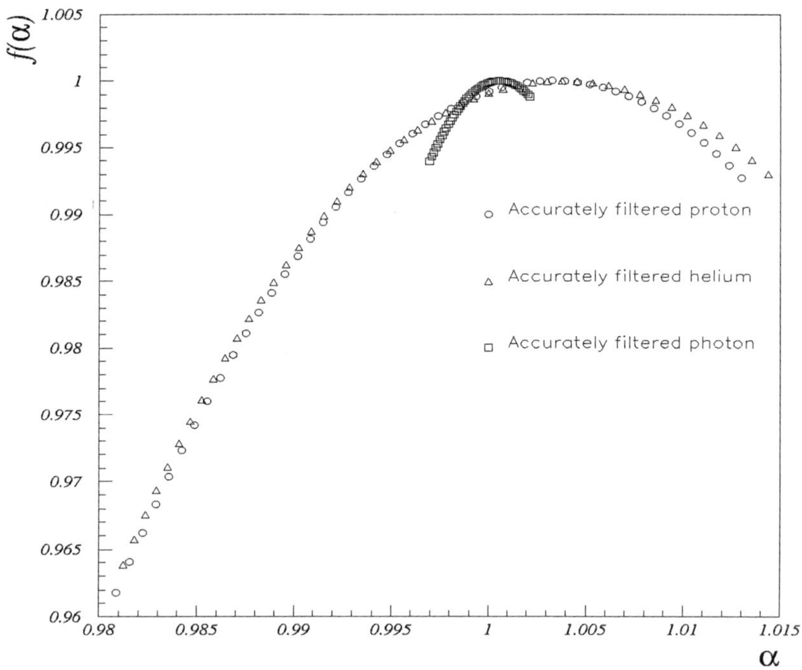

Figure 3. *Multifractal spectrum $f(\alpha)$ for accurately filtered events belonging to several databases with $\Delta N_s = 2500 - 5000$. It is seen that the length Λ of the $f(\alpha)$ curve for primary photons is much shorter than for primary hadrons. Thus Λ is a good candidate to be a cutting parameter.*

is removed as well as possible. It is seen that the scaling is reasonably good.

Fig. 2b shows the scaling behavior of statistical moments for proton induced showers in three different situations. When the power spectrum is not filtered for the white noise, we see that a scaling law at all scales is not fulfilled by the statistical moments. The scaling becomes good after accurate filtering of the white noise component is applied. Finally, we tried filtering the proton shower with the *gamma* ray filter, and still obtained a good power scaling law of statistical moments. This is an interesting result, because in real EAS experiments one does not know *a priori* the nature of primary cosmic ray.

From the moment scaling functions $K(q)$ obtained, the function $\tau(q)$ and the multifractal spectrum $f(\alpha)$ have been calculated for all databases. In Fig. 3 the multifractal spectrrum for events belonging to accurately filtered samples of several different primary cosmic rays are displayed. It is observed that the $f(\alpha)$ function is different for photons and hadrons. Clearly, the length Λ of the $f(\alpha)$ curve is much shorter for photons than for the other primary cosmic rays. This peculiarity is used here to propose a simple γ/hadron separation method. It is clear that using the length of the $f(\alpha)$ curve Λ as a cutting parameter for a sample of EAS with

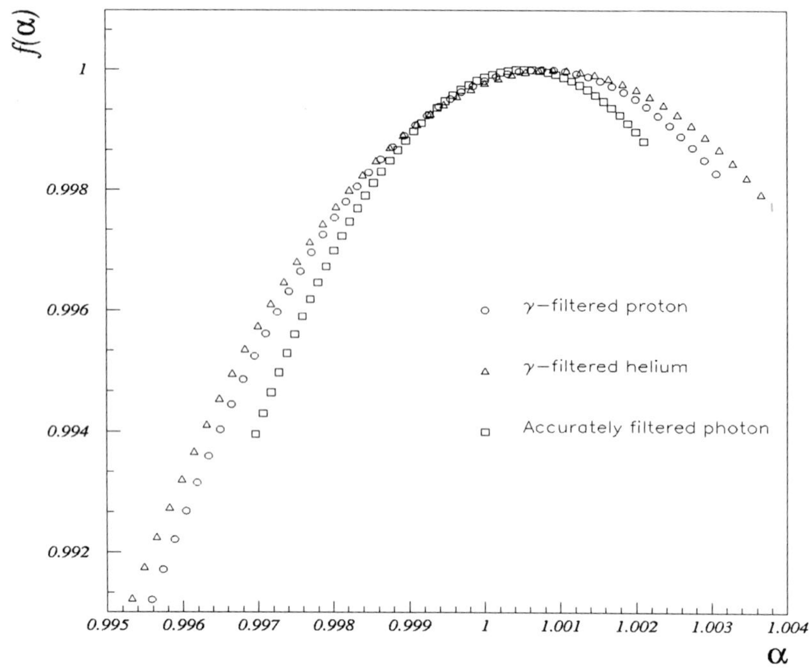

Figure 4. *Multifractal spectrum $f(\alpha)$ using the same gamma-filter for events belonging to several databases with $\Delta N_s = 2500 - 5000$. The length Λ of the $f(\alpha)$ curve still appears as a good candidate to cutting parameter.*

unknown primaries, it is possible to select samples significantly enriched in photons. However, some crucial aspects must be previously solved.

Since the nature of the primary cosmic ray is not known *a priori*, we can no longer use independent parameters A, B and C defined in Eq. (10) to filter hadronic and γ EAS. Instead, we have to transform the whole sample with the same filter. We have tried filters obtained from the whole sample, using an average of the A, B and C coefficients over the events belonging to the same class ΔN_s. We also tried filters obtained independently from the γ and proton sub-samples. The best results were obtained when the γ filter was applied. Fig. 4 shows the multifractal spectrum $f(\alpha)$ for γ-filtered events belonging to databases of different primary particles. Comparing with Fig. 3, we see that the difference in the value of the parameter Λ for helium and γ rays is now smaller. However the remaining difference still makes possible an effective separation.

In γ/hadron separation methods, the gain obtained by a method is usually measured by the so-called quality factor Q, defined by

$$Q = \frac{\kappa_\gamma}{\sqrt{\kappa_h}}, \tag{11}$$

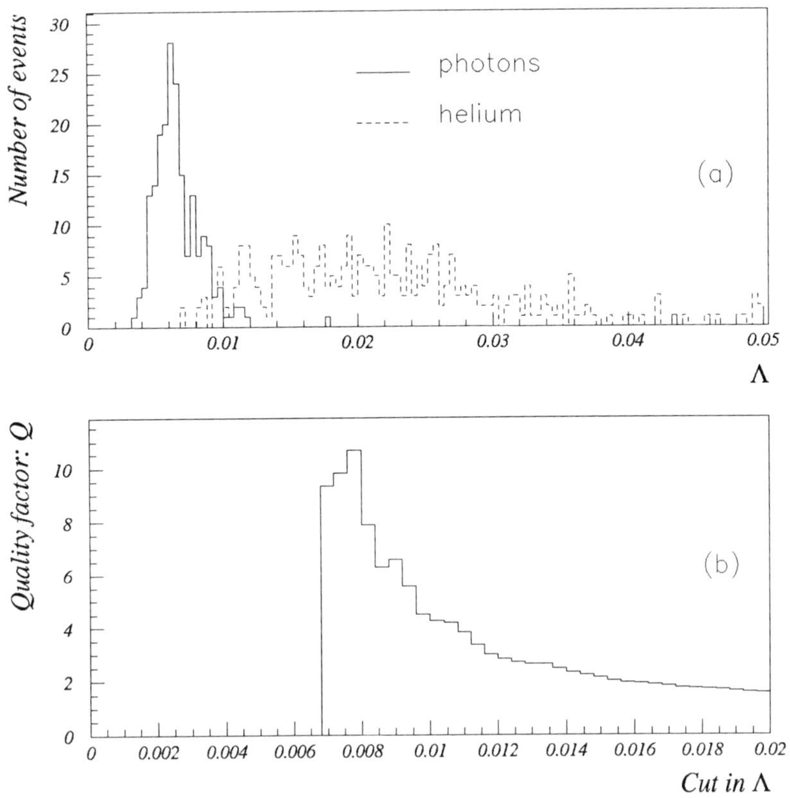

Figure 5. *a) The length* Λ *of the multifractal spectrum* $f(\alpha)$ *for accurately filtered photons and* γ*-filtered helium events belonging to the* $\Delta N_s = 5000 - 7500$ *bin. b) Quality factor as a function of the cut in the length* Λ.

where κ_γ and κ_h are the fractions (efficiencies) of photons and hadrons kept by the algorithm, namely $\kappa_\gamma = N_\gamma^{cut}/N_\gamma$ and $\kappa_h = N_h^{cut}/N_h$, where N^{cut} is the number of remaining particles after the cutting procedure.

In Fig. 5a we represent a histogram of this cutting parameter Λ for photon and helium samples. The quality factor obtained as a function of the applied cut in Λ is shown in Fig. 5b. It can reach values of up to $Q \approx 7$ for a cut in $\Lambda \approx 0.008$ units.

Similarly, the most convenient value of Λ for different databases can be calculated. Table 2 summarizes the cutting parameter and the quality factor obtained for the separation of different γ/hadron sub-samples. It is observed that, as the nuclear mass number increases, the separation procedure becomes more effective. For a given hadronic primary, the separation quality factor improves when the number of secondary particles increases, as a result of better statistics.

Table 2. Q-values for an optimal cut for the Λ parameter from γ/hadron sub-samples.

Separation	ΔN_s	Λ	Q
γ/proton	0-2500	$\leq$0.020	2.0
	2500-5000	$\leq$0.010	3.5
	5000-7500	$\leq$0.006	4.0
γ/helium	0-2500	$\leq$0.022	2.5
	2500-5000	$\leq$0.010	5.0
	5000-7500	$\leq$0.008	7.0
γ/oxygen	0-2500	$\leq$0.0250	3.5
	2500-5000	$\leq$0.0150	8.0
	5000-7500	$\leq$0.0083	10.0
γ/iron	0-2500	$\leq$0.035	10.0
	2500-5000	$\leq$0.015	15.0
	5000-7500	$\leq$0.013	∞

5 Conclusions

We have generated different Monte Carlo samples of extensive air showers initiated by protons and γ rays and also by helium, oxygen and iron nuclei, and we have studied the fluctuations of the density of secondary particles at detection level. Their structure is nontrivial and shows the typical characteristics of a 1/f noise, after removal of the white noise component. The scaling properties of the secondary particle density statistical moments have been studied. The moment scaling function $K(q)$ and also the associate spectrum $\tau(q)$ is calculated, and we obtained the multifractal spectrum $f(\alpha)$ for all the samples.

It is shown that neat differences exist in the pattern of $f(\alpha)$ for the different primary cosmic rays. The length Λ of the multifractal spectrum is much shorter for electromagnetic showers than for hadronic showers. Thus Λ can be considered as a cutting parameter that is able to separate γ-induced EAS from a whole sample of heterogeneous primary cosmic rays. This separation is important for high energy astronomy, since only electromagnetic EAS provide information on the direction of the source of the cosmic rays. The enrichment in primary cosmic γ rays is measured by the quality factor Q associated to a cutting in the separation parameter Λ. The separation Q-value increases quickly with the nuclear mass number of the cosmic ray. The separation quality is best when the number of secondary particles of the shower is large. We conclude that multifractal analysis of extensive air showers provides a new, effective γ/hadron separation method, which ca be combined with other methods currently used to identify high energy γ rays arriving from space.

Acknowledgments

This work is supported in part by Spanish Government grants for the research projects BFM2000-0600 and FTN2000-0963-C02.

References

1. P. Sokolsky, *Introduction to Ultra-high Energy Cosmic Ray Physics*, (Addison Wesley, Redwood City, CA, 1989).
2. T.K. Gaisser, *Cosmic Rays and Particle Physics*, (Cambridge University Press, 1990).
3. D.J. Fegan, J. Phys. G **23** , 1013 (1997).
4. J. Kempa, J. Phys. G **20** , 215 (1994).
5. E. Faleiro and J.L. Contreras, J. Phys. G **24**, 1795 (1998).
6. E. Faleiro and J.M.G. Gómez, in *Fractals and Beyond*, Ed. M.M. Novak (World Scientific, 1998).
7. E. Faleiro and J.M.G. Gómez, Europhys. Lett. **45**, 437 (1999).
8. E. Faleiro and J.M.G. Gómez, Fluctuations and Noise Lett., in press.
9. T.C. Halsey, M.H. Jensen, L.P. Kadanoff, I. Procaccia, and B.I. Shraiman, Phys. Rev. A **33**, 1141 (1986).
10. G. Paladin and A. Vulpiani, Phys. Rep. **156**, 148 (1987).
11. A. Saiz and V.J. Martinez, Phys. Rev. E **54**, 2431 (1996).
12. H.E. Stanley and P. Meakin, Nature **335**, 405 (1988).
13. F. Schmitt, D. Lavallée and D. Schertzer, Phys. Rev. Lett. **68**, 305 (1992).
14. P. Grassberger, Phys. Lett. A **97**, 227 (1983).
15. D. Schertzer and S. Lovejoy, in *Nonlinear Variability in Geophysics 3: Scaling and multifractal processes*, Lecture Notes NVAG3, EGS (1993).
16. J.N. Capdevielle et al., *KfK Report 4998*, Kernforschungzentrum, Karlsruhe (1992).
17. P.Z. Peebles, *Probability, Random Variables and Random Signal Principles* (McGraw-Hill (1987).
18. M.B. Weissman, Rev. Mod. Phys. **60**, 537 (1988).

A LATTICE GAS MODEL OF ELECTROCHEMICAL CELLS : MEAN-FIELD KINETIC APPROACH

M.-O. BERNARD, M. PLAPP, J.-F. GOUYET
Laboratoire de Physique de la Matière Condensée,
Ecole Polytechnique, F-91128 Palaiseau, France
E-mail: jean-francois.gouyet@polytechnique.fr

The present work is an attempt to simulate electrochemical cells and growth structures that form during electrodeposition. For that purpose we use a lattice gas model with charged particles, and build mean-field kinetic equations for their evolution, together with a Poisson equation for the electric potential, and oxido-reduction reactions on the electrode surfaces for the growth. In this preliminary study we confirm the viability of this approach by simulating the ion kinetics in front of planar electrodes during growth and dissolution.

1 Introduction

Electrochemical phenomena are ubiquitous in nature, and of particular industrial importance. They play a basic role in batteries, corrosion problems, electrodeposition of parts and circuitry, as well as in biochemical reactions. Electrochemical growth may lead to the formation of complex and highly branched structures that can be fractals or densely branched, depending on the experimental conditions[1,2,3,4]. Figure 1 shows an example of a fractal growth structure. While diffusion-limited aggregation (DLA)[5] and similar models produce structures that are strikingly similar to such experimental pictures, they greatly simplify the underlying phenomena and can hence not yield detailed information on the relation between growth conditions and characteristics of the growth structures such as growth speed, branch thickness and overall structure.

Figure 1. Electrodeposition of copper on a substrate (reprinted with permission from V. Fleury). The quasi-two-dimensional cell is composed of two copper electrodes and a $CuSO_4$ solution for the electrolyte (sample size $3 \times 2\,\mathrm{mm}^2$). Only a small part of the sample is shown.

To obtain such information, the motion of charged particles as well as their production/consumption by chemical reactions have to be taken into account. Despite the fact that charged systems are everywhere around and in us, many questions remain open. The main difficulty comes from the coexistence of competing short-range Van der Waals or chemical interactions, and long-range Coulomb interactions. Our aim is to simulate at least qualitatively in two dimensions electrochemical growth leading to branched structures such as observed experimentally[1,2,3,4].

For that purpose, we use a lattice gas model that includes charged particles and uses simple microscopic transformation rules to simulate the salient features of the electrochemical process, including both the diffusion kinetics of the charged and neutral species, and the oxido-reduction phenomena on the electrode interfaces. There exist, to our knowledge, no theoretical studies of the behavior of an entire electrochemical cell based on a microscopic model; lattice gas models have been used to simulate phenomena located on the electrode surfaces, such as adsorption or underpotential deposition[6], and for studies of ionic transport at liquid-liquid interfaces[7]. Marshall and Mocskos [8] have combined a lattice model for the electrodes and a continuum treatment for the electrolyte to simulate ramified growth; however, in this approach the detailed description of the interface is lost. While our model still contains strong simplifications, it provides a consistent description of the whole cell and is much closer to the basic microscopic physics and chemistry than the original DLA model and its generalizations that use, for example, a uniform drift to simulate the effect of the electric field [9].

To investigate the dynamics of the model, we will extend the formalism of Mean-Field-Kinetic-Equations (MFKE)[10,11,12] that has been used to study numerous transport and growth phenomena in alloys, including diffusion and ordering kinetics, spinodal decomposition and dendritic growth[?,14,15]. To describe electrodeposition, we need to include charged particles. This implies that, in addition to the kinetic equations driving the particle motion, we have to solve the Poisson equation that determines the local electrical potential. The chemical potentials present in the MFKE are then replaced by electrochemical potentials. In this way, the present method is able to establish a link between a microscopic lattice model and macroscopic phenomenological equations[16,17]. We present here some results on the growth and dissolution of planar electrodes in a binary electrolyte that are in good qualitative agreement with the macroscopic model of Chazalviel[17]. Finally, we report preliminary two-dimensional simulations that exhibit fingered growth structures.

2 Model

We consider an electrochemical cell, composed of a binary electrolyte and two metallic electrodes. The absence of supporting electrolyte, generally used to suppress ion migration, corresponds here to the experimental situation in Fig. 1. To take into account the crystalline structure of the electrodes, they are modeled by a lattice, whose sites are occupied by metallic atoms. It is then convenient to represent the electrolyte by the same lattice, occupied by a solvent, and by cations and anions. We consider a two-dimensional lattice gas on a square lattice with lattice spacing a. The cations M^+ give metallic atoms M^0 after reduction, and the anions A^- are considered to be non reactive. Their presence ensures the electroneutrality of the

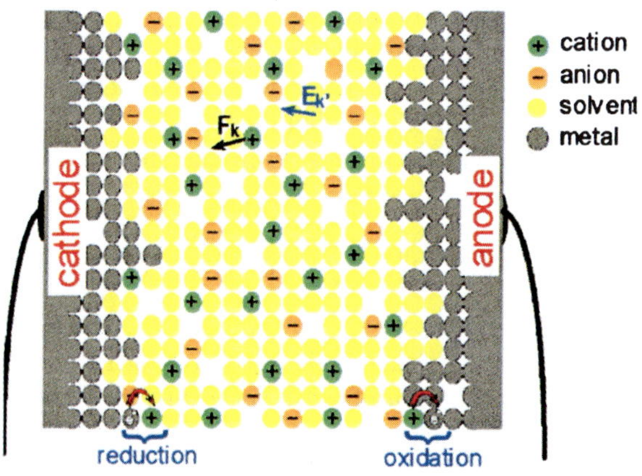

Figure 2. The lattice gas model used in the present study. A fixed potential difference is imposed across the electrodes. The ions in the electrolyte are submitted to an electric field $\mathbf{E_k}$ (and a force $\mathbf{F_k} = q\mathbf{E_k}$) at their lattice site position $\mathbf{k}$. The various species have short range interactions (in the present work, attractive interactions are considered between solvent and ions, solvent and solvent, and metal and metal). Oxido-reduction reactions appear on the electrode interfaces.

solution at equilibrium. The solvent S is neutral, but can interact through short range interactions with the other species. For simplicity, the two electrodes are supposed to consist of the same metal, as is the case in the experiment of Fig. 1. We suppose steric exclusion between the different species, i.e. a given site can be occupied by only one species or it can be empty. The diffusion processes are mediated by a small number of vacancies, i.e. a species is allowed to jump onto a neighboring empty site (using an exchange process between nearest neighbor species would also be possible, but is more complicated to implement). To model short-range interactions, we introduce nearest neighbor attractive interactions on the lattice. In addition, we have long range Coulomb interactions between the charged species. We must, in addition, consider electrons in the metallic electrodes: they have a particular status that will be discussed later.

The establishment of the Mean-Field-Kinetic-Equations (MFKE) follows the same procedure as for neutral particles[10,11,12]. We first write the Boolean kinetic equations on the lattice, starting from the general master equation,

$$\frac{\partial}{\partial t}P(\{n\},t) = \sum_{\{n'\}} \left[W(\{n'\} \to \{n\})P(\{n'\},t) - W(\{n\} \to \{n'\})P(\{n\},t)\right] , \quad (1)$$

which gives the probability P to find at time t a given configuration $\{n\}$, that is the occupation of each site by one of the species $\alpha = M^0, M^+, A^-, S$ or by a vacancy v. $\{n\}$ is the set of the occupation numbers $n_{\mathbf{k}}^{\alpha}$ on each site $\mathbf{k}$ of the lattice ($n_{\mathbf{k}}^{\alpha} = 1$ if $\mathbf{k}$ is occupied by species α and 0 otherwise). Double occupancy exclusion imposes

$$\sum_{\alpha} n_{\mathbf{k}}^{\alpha} + n_{\mathbf{k}}^{v} = 1 . \quad (2)$$

$W(\{n\} \to \{n'\})$ is the probability per unit time that configuration $\{n\}$ is changed into configuration $\{n'\}$. During the process $\{n\} \to \{n'\}$, one particle (metal, ion, or solvent) jumps to one of its vacant nearest neighbor sites. The jump probability depends not only on the local interactions with their neighborhood, but for the charged species also on the local electric field.

In the present work, we will be interested in the time evolution of the average concentrations of the various species ($\alpha = +, -, 0, S, v$),

$$p_{\mathbf{k}}^{\alpha} = \langle n_{\mathbf{k}}^{\alpha} \rangle = \sum_{\{n\}} n_{\mathbf{k}}^{\alpha} P(\{n\}, t). \tag{3}$$

In addition, we define $p_{\mathbf{k}}^{e}$ as the excess electron concentration on site $\mathbf{k}$ (that is, the difference between the actual number of electrons and the equilibrium number for a metal concentration $p_{\mathbf{k}}^{0}$, see below). The short range (attractive) interaction between two species α and β is limited to nearest neighbors and noted $\epsilon^{\alpha\beta}$. But in principle any longer range attractive or repulsive interaction can be considered. The temperature is fixed and constant in the whole system, and the external control parameter is the potential difference ΔV applied across the cell.

3 Mean-Field Kinetic and Poisson Equations

To establish the Mean-field-Kinetic-Equations, we follow the lines described in previous papers concerning spinodal decomposition, ordering or dendritic growth[12,13,14]. Two important new elements appear: *a*) there is an electric potential related to the charge distribution, and *b*) oxidation and reduction may take place at the electrodes.

3.1 Kinetic equations of the particles

The local concentration of the species M^0 and M^+ is modified by transport (diffusion and migration in the electric field) and by the chemical reaction; for the other species A^- and S, only transport is present. The MFKE derived from the microscopic model read

$$\frac{\partial p_{\mathbf{k}}^{+}}{\partial t} = -\sum_{\mathbf{a}} \widetilde{J}_{\mathbf{k},\mathbf{k}+\mathbf{a}}^{+} - \sum_{\mathbf{a}} \sigma_{\mathbf{k},\mathbf{k}+\mathbf{a}}, \tag{4}$$

$$\frac{\partial p_{\mathbf{k}}^{0}}{\partial t} = -\sum_{\mathbf{a}} \widetilde{J}_{\mathbf{k},\mathbf{k}+\mathbf{a}}^{0} + \sum_{\mathbf{a}} \sigma_{\mathbf{k},\mathbf{k}+\mathbf{a}}, \tag{5}$$

$$\frac{\partial p_{\mathbf{k}}^{-}}{\partial t} = -\sum_{\mathbf{a}} \widetilde{J}_{\mathbf{k},\mathbf{k}+\mathbf{a}}^{-}, \tag{6}$$

$$\frac{\partial p_{\mathbf{k}}^{s}}{\partial t} = -\sum_{\mathbf{a}} \widetilde{J}_{\mathbf{k},\mathbf{k}+\mathbf{a}}^{s}. \tag{7}$$

Here, $\widetilde{J}^{\alpha}_{\mathbf{k},\mathbf{k}+\mathbf{a}}$ is the diffusion current of species α on the bond linking site $\mathbf{k}$ to its nearest neighbor site $\mathbf{k}+\mathbf{a}$. Generalizing previous studies[10,11,12,15] to the electrochemical case, we can write this current as the product of a bond mobility $\widetilde{M}^{\alpha}_{\mathbf{ij}}$ times the (discrete) gradient of an electrochemical potential $\widetilde{\mu}^{\alpha}_{\mathbf{i}}$,

$$\widetilde{J}^{\alpha}_{\mathbf{k},\mathbf{k}+\mathbf{a}} = -\widetilde{M}^{\alpha}_{\mathbf{k},\mathbf{k}+\mathbf{a}} \mathfrak{D}_{\mathbf{a}} \widetilde{\mu}^{\alpha}_{\mathbf{k}}\,, \tag{8}$$

where $\mathfrak{D}_{\mathbf{a}}$ is a difference operator acting on the site coordinates, $\mathfrak{D}_{\mathbf{a}} f_{\mathbf{k}} = f_{\mathbf{k}+\mathbf{a}} - f_{\mathbf{k}}$. The electrochemical potential,

$$\widetilde{\mu}^{\alpha}_{\mathbf{k}} = \mu^{\alpha}_{\mathbf{k}} + q^{\alpha} V_{\mathbf{k}} = -\sum_{\beta}\sum_{\mathbf{a}} \varepsilon^{\alpha\beta} p^{\beta}_{\mathbf{k}+\mathbf{a}} + kT \ln\left(\frac{p^{\alpha}_{\mathbf{k}}}{p^{v}_{\mathbf{k}}}\right) + q^{\alpha} V_{\mathbf{k}}, \tag{9}$$

is the sum of three contributions: a local energy due to the interaction of species α with its local environment, an entropy term (these two constitute the chemical potential $\mu^{\alpha}_{\mathbf{k}}$), and an electrostatic energy ($V_{\mathbf{k}}$ is the electrostatic potential at site $\mathbf{k}$, and q^{α} is the electric charge of species α). The presence of the vacancy concentration in the denominator of the entropic contributions comes from the constraint of Eq. (2). The mobility along a bond $\mathbf{i}$ -$\mathbf{j}$ is given by (see an analogous case, in the absence of electric charges[20]),

$$\widetilde{M}^{\alpha}_{\mathbf{k},\mathbf{k}+\mathbf{a}} = \frac{w^{\alpha}}{kT} p^{v}_{\mathbf{k}} p^{v}_{\mathbf{k}+\mathbf{a}} \exp\frac{(\widetilde{\mu}^{\alpha}_{\mathbf{k}} + \widetilde{\mu}^{\alpha}_{\mathbf{k}+\mathbf{a}})}{2kT} \operatorname{shc}\frac{(\widetilde{\mu}^{\alpha}_{\mathbf{k}+\mathbf{a}} - \widetilde{\mu}^{\alpha}_{\mathbf{k}})}{2kT}\,, \tag{10}$$

where we have used the notation $\operatorname{shc} u = \sinh u/u$ (close to equilibrium, $\widetilde{\mu}^{\alpha}_{\mathbf{k}+\mathbf{a}} \cong \widetilde{\mu}^{\alpha}_{\mathbf{k}}$ and $\operatorname{shc}\left[(\widetilde{\mu}^{\alpha}{}_{\mathbf{k}+\mathbf{a}} - \widetilde{\mu}^{\alpha}_{\mathbf{k}})/kT\right] \cong 1$).

Finally, $\sigma_{\mathbf{k},\mathbf{k}+\mathbf{a}}$ is the current of electronic charges from $\mathbf{k}+\mathbf{a}$ to $\mathbf{k}$ (current of positive charges from $\mathbf{k}$ to $\mathbf{k}+\mathbf{a}$) reducing the cations on site $\mathbf{k}$ (resp. electronic current issued from the oxidation of the metal) via a corresponding elimination (resp. creation) of electrons on site $\mathbf{k}+\mathbf{a}$,

$$M^{+}(\mathbf{k}) + e^{-}(\mathbf{k}+\mathbf{a}) \rightleftharpoons M^{0}(\mathbf{k}). \tag{11}$$

The direction of the reaction depends on the relative magnitude of the electrochemical potentials of the involved species. Reduction of cations on a site $\mathbf{k}$ of the cathode appears when

$$\widetilde{\mu}^{+}_{\mathbf{k}} + \widetilde{\mu}^{e}_{\mathbf{k}+\mathbf{a}} > \widetilde{\mu}^{0}_{\mathbf{k}}; \tag{12}$$

otherwise, the metal is oxidized. The reduction rate on site $\mathbf{k}$ is the sum of all the reaction paths $\sum_{\mathbf{a}} \sigma_{\mathbf{k},\mathbf{k}+\mathbf{a}}$. Following[18], we can write the reaction rate,

$$\sigma_{\mathbf{k},\mathbf{k}+\mathbf{a}} = w^{*}_{\mathbf{k},\mathbf{k}+\mathbf{a}} \left(\exp\frac{\widetilde{\mu}^{+}_{\mathbf{k}} + \widetilde{\mu}^{e}_{\mathbf{k}+\mathbf{a}}}{kT} - \exp\frac{\widetilde{\mu}^{0}_{\mathbf{k}}}{kT}\right) \tag{13}$$

The coefficient $w^{*}_{\mathbf{k},\mathbf{k}+\mathbf{a}}$ is determined by comparison with the mesoscopic theory of Butler-Volmer and corresponds to the electronic tunneling from the metal surface to the nearest neighboring cation; $\widetilde{\mu}^{e}_{\mathbf{k}}$ is the local chemical potential of the electrons (for the definitions of both quantities, see below).

3.2 The Poisson problem

To determine the electrostatic potential for a given charge distribution, we solve a discrete version of the Poisson equation using the lowest order discretization for the Laplacian that involves only nearest neighbor sites,

$$-4V_{\mathbf{k}} + \sum_{\mathbf{a}} V_{\mathbf{k}+\mathbf{a}} = -\frac{a^{2-d}}{\epsilon} \sum_{\alpha=+,-,e} q^{\alpha} p^{\alpha}_{\mathbf{k}} \tag{14}$$

where ϵ is the permittivity (for simplicity, the same for all species), and q^{α} the charge of species α ($q^e = -e$, for the electrons); a is the lattice spacing and d the spatial dimension.

The boundary conditions at the metal-electrolyte interface have to be considered with special care. In a "macroscopic" picture where this interface is of arbitrary form but sharp (i.e. represented by a mathematical line), the electric field is zero in the metal, and the potential is constant and equal to the imposed boundary condition up to the sharp interface. If there is an electric field in the electrolyte, surface charges are created. In the mean-field representation outlined above, the interface is diffuse, i.e. "smeared out" over several lattice sites, and both the definition of the boundary condition and the creation of surface charges have to be consistently implemented. We solve these problems by the introduction of very mobile electrons diffusing from site to site in the metal, and solve the Poisson equation for all the charges, including electrons. More precisely, we denote by $p^e_{\mathbf{k}}$ the deviation from the neutral state expressed in electrons per site. Hence, $p^e_{\mathbf{k}} > 0$ corresponds to an excess of electrons, $p^e_{\mathbf{k}} < 0$ to an electron deficit. Their chemical potential is defined by

$$\widetilde{\mu}^e_{\mathbf{k}} = E_F + q^e V_{\mathbf{k}} + \frac{p^e_{\mathbf{k}}}{\mathcal{D}(E_F)} \tag{15}$$

where $q^e = -e$, E_F is the Fermi level of the metal, and $\mathcal{D}(E_F)$ is the density of electronic states at the Fermi level. This corresponds to screening in the Thomas-Fermi approximation[19]. The electronic current is then

$$\widetilde{J}^e_{\mathbf{k},\mathbf{k}+\mathbf{a}} = -\widetilde{M}^e_{\mathbf{k},\mathbf{k}+\mathbf{a}} \mathfrak{D}_{\mathbf{a}} \widetilde{\mu}^e_{\mathbf{k}} \tag{16}$$

and the time evolution of the excess electron concentration is

$$\frac{\partial p^e_{\mathbf{k}}}{\partial t} = -\sum_{\mathbf{a}} \widetilde{J}^e_{\mathbf{k},\mathbf{k}+\mathbf{a}} - \sum_{\mathbf{a}} \sigma_{\mathbf{k}+\mathbf{a},\mathbf{k}} \tag{17}$$

where the last term is the reaction term, active only in the solid-electrolyte interface. To force the electrons to stay in the metallic regions, we write the mobility in the form,

$$\widetilde{M}^e_{\mathbf{k},\mathbf{k}+\mathbf{a}} = \frac{w^e}{kT} f(p^0_{\mathbf{k}}) f(p^0_{\mathbf{k}+\mathbf{a}}) \tag{18}$$

where f is chosen important only if the metallic concentration is large enough. In particular, the rough surface, with a lower mean concentration, will present a low electronic mobility. We have used for f a monotonous function varying from 0 when

$p = 0$, to 1 for $p = 1$, with a rapid variation around some concentration p_c that is reminiscent of some percolation threshold. A convenient choice is,

$$f(p) = \frac{\tanh[(p - p_c)/\xi] + \tanh[p_c/\xi]}{\tanh[(1 - p_c)/\xi] + \tanh[p_c/\xi]}. \tag{19}$$

where ξ is of the order of the interface thickness.

The above expressions are chosen such that the excess of electrons will be localized in the interface. This model provides a fast way to calculate the surface charges on the electrodes. In practice, to save computational time we impose a fixed potential value inside both electrodes up to a given distance from the interfaces, which reduces the diffusion time of the electrons from the source to the surface where they are involved in the chemical reaction.

The interpolation function $f(p)$ is also used to specify the prefactor $w^*_{\mathbf{k},\mathbf{k}+\mathbf{a}}$ of the reaction rate (eq. 13),

$$w^*_{\mathbf{k},\mathbf{k}+\mathbf{a}} = w^* \left(1 - f(p^0_{\mathbf{k}})\right) f(p^0_{\mathbf{k}+\mathbf{a}}) \; . \tag{20}$$

In this way, the transfer is localized around the metal-electrolyte interface, where occupied metallic sites and electrolyte sites are neighbors. w^e in eq. 18 and w^* in eq. 20 are constant frequency factors.

4 Simulations

The first step, before considering two-dimensional cells in which electrodeposition creates dendritic patterns, is to check if one-dimensional systems work correctly. We have performed calculations on 200-site cells. The two electrodes are identical, with a thickness of 40 layers. The initial system is neutral everywhere: anion and cation concentrations are taken equal and there are no electrons in excess. We have chosen, in addition to the Coulomb interactions, attractive interactions between the solvent and anions, cations, and itself, as well as between metallic atoms: $\varepsilon^{s+} = \varepsilon^{s-} = \varepsilon^{ss} = \varepsilon^{00} = 1$, with the choice $kT = 1$, all the other interactions being zero[a]. The initial concentrations of each species are chosen such that the system is in equilibrium in the absence of an applied potential. This corresponds to a uniform chemical potential for each species. The various parameters that control the relative importance of the diffusion and reaction processes are chosen as follows:

- Initial concentrations :

$$\begin{aligned} p^0_{electrodes} &= 0.925, & p^0_{electrolyte} &= 0.0025 \\ p^{\pm}_{electrodes} &= 3 \times 10^{-4}, & p^{\pm}_{electrolyte} &= 0.01 \\ p^{s}_{electrodes} &= 0.0025, & p^{s}_{electrolyte} &= 0.907 \end{aligned}$$

- Jump frequencies : $w^\alpha = 1$, for $\alpha = +, -, 0, s$; and $w^e = 10^{-3}$, for the electrons (the electron mobility remains nevertheless faster than the mobility of the other species).

[a] The problem is oversimplified for the electrodes as it is known that quantum effects are important in the interaction between metallic atoms, and do not reduce to simple nearest neighbor interactions. This has no strong consequences for the present study.

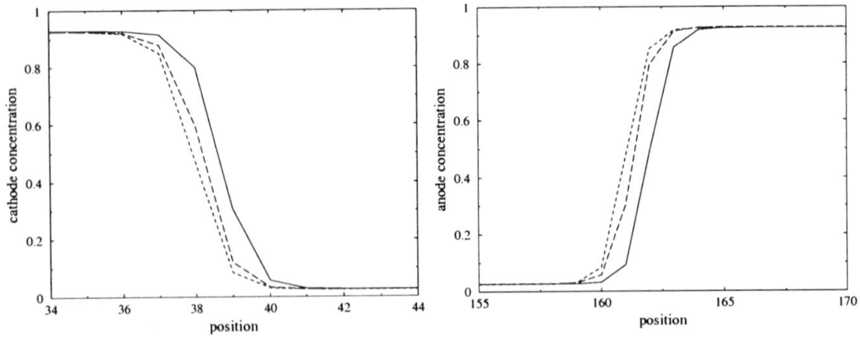

Figure 3. Picture showing the dissolution of metal at the anode (on the right), and the electrodeposition (reduction) at the cathode (on the left), at times $t = 0$ (dotted line), $t = 0.5 \times 10^5$ (dashed line), and $t = 5 \times 10^5$ (solid line). The last time corresponds to one monolayer of metal dissolved or deposited.

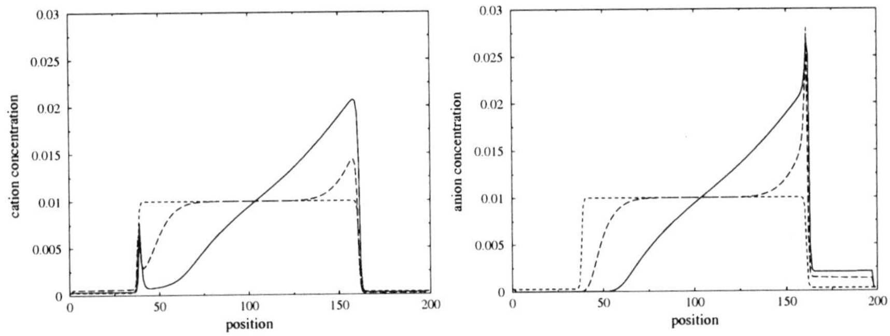

Figure 4. Cation (left) and anion (right) concentration profiles at times $t = 0$ (dotted line), $t = 0.5 \times 10^5$ (dashed line), $t = 5 \times 10^5$ (solid line).

- Transfer frequency : $w^* = 10^{-5}$; Fermi energy : $E_F/kT = 4.42$; density of states at the Fermi level: $\mathcal{D}(E_F) = 1000/kT$.
- Electrical parameters : $q^{\pm} = \pm e; q^e = -e; e = 1$; (dimensionless) permittivity: $\epsilon = 5 \times 10^{-2}$; applied potential difference : $10kT/e$.

When a potential is applied, the ionic species start to migrate, double layers appear on the interfaces, reduction processes take place on the cathode, and oxidation processes on the anode. Figure 3 shows the growth of the cathode and the dissolution of the anode. We notice that the shape of the metal concentration profile across the interface is essentially preserved, and that the deposited metal has the same concentration as the initial electrode.

The interesting results concern the kinetics of the ions. In Fig.4 that shows the evolution of the ion concentrations, we observe the progressive formation of a concentration gradient between the anode and the cathode with, for the cations, a concentration peak on the cathode (accompanied by an electronic layer on the

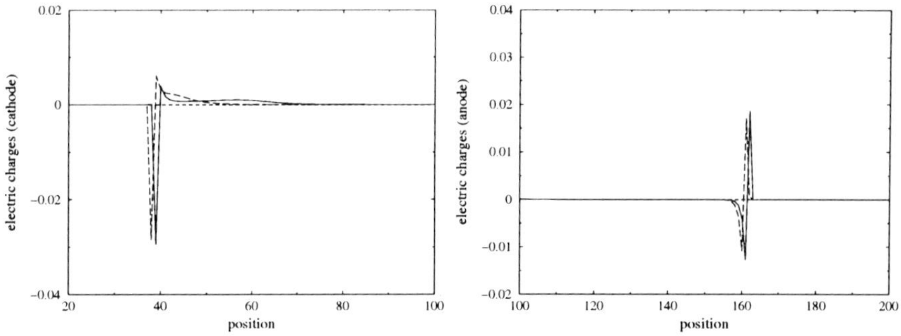

Figure 5. Charge distribution at times $t = 0.5 \times 10^5$ (dashed line) and $t = 5 \times 10^5$ (solid line). In addition to the double layers close to the interfaces, there is an extended space charge in front of the cathode as found by Chazalviel [17].

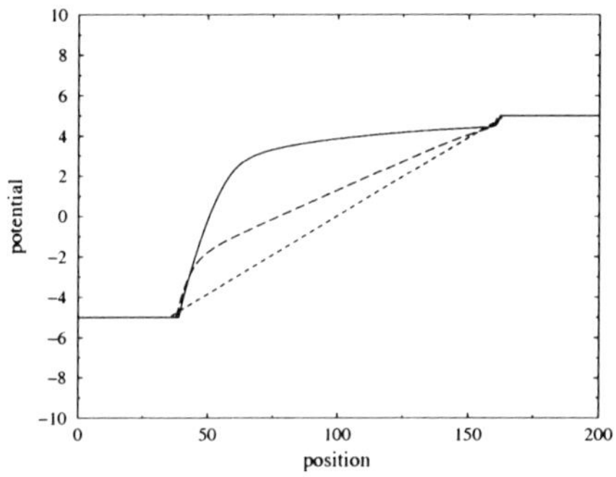

Figure 6. Potential profiles at times $t = 0$ (dotted line), $t = 0.5 \times 10^5$ (dashed line), $t = 5 \times 10^5$ (solid line).

metal surface). The anion concentration profile presents on the contrary an increase close to the anode.

Figure 5 shows the total charge distribution at two successive times, where we can see, in addition to the double-layers located around the interfaces, an excess of cations over the anions, on a range of about 30 lattice distances. This extended space charge leads to an important potential drop on the cathode side, as can be seen in Fig. 6 that shows the electric potential. In summary, the electrolyte domain can be divided into four regions that can be clearly distinguished at times larger than $t = 5 \times 10^5$ steps: the double layers very close to the interfaces, an almost electroneutral region where the concentration gradients are approximately constant, and the ion-depleted space charge region close to the cathode. According to Chazalviel[17], this extended space charge is crucial for the emergence of ramified growth: the strong electric field close to the surface leads to a Laplacian instability of a flat front, and the one-dimensional calculation becomes invalid.

5 Two-dimensional simulations: dendritic growth

We present now preliminary simulations of two-dimensional samples. We use a 100×40 sample with the same parameters as in the preceding one-dimensional simulations, except for the potential difference ($100kT/e$, corresponding to $\Delta V \approx 2.5$ V at room temperature), the transport frequency ($\omega^* = 6 \times 10^{-3}$) and the metal jump frequency ($\omega^0 = 10^{-3}$).

The geometry of the growth appears to be very sensitive to several of the model parameters, in particular to the applied voltage and the ratio of transfer frequency and metal jump frequency. In figure 7, we observe an increase of the cation concentration in front of the growing dendrite that induces a subsequent tip-splitting; for slightly different parameters, this event happens much later. A more detailed investigation is needed to elucidate the tip-splitting mechanism and the precise role of the various parameters. Also, in the present study, due to the small size of the sample, the gradients in front of the dendrites strongly depend on time. At the end, when the tips approach the anode and the gradients become very strong, the dendrite exhibits an anomalous behavior, characterized by a weakly connected growth and low metal concentration.

6 Conclusion

We have shown in this preliminary study that it is possible to build Electrochemical-Mean-Field-Kinetic Equations (EMFKE) that are able to reproduce qualitatively the behavior of electrochemical cells with planar electrodes. Our model leads naturally to the formation of the extended space charge, in addition to the Helmholtz double layers. This extended space charge is known to play a crucial role in the selection of the growth velocity and the dense branching structure of the deposit[4]. Furthermore, we have shown that our model leads to the emergence of dendritic growth in two-dimensional simulations.

The above EMFKE contain all the ingredients necessary to simulate dendritic growth by electrodeposition in two and three dimensions. We expect the crystalline anisotropy to be conveniently simulated by the intrinsic lattice anisotropy as in previous work on alloys[15,20]. It should be emphasized, however, that several limitations of the model have to be overcome to achieve a more quantitative modeling. Most seriously, lattice gas models cannot manage the very different length scales present in electrochemical cells, that range from the interatomic distance (Å), the capillary length that describes the effect of surface tension (of the order of the interface thickness, around 1nm), the Debye length (a few nanometers), the diffusion length (a few hundred μm), up to the cell size of a few millimeters. Since the lattice gas model explicitly contains the interatomic distance, the available computational resources do not allow to treat correctly all these scales at the same time, and the results must remain qualitative. Some of these problems could be resolved by adapting the more phenomenological phase-field method[21] to electrochemical systems and by using modern multi-scale algorithms; however, in such approaches, the direct microscopic interpretation is lost. Finally, we have ignored here the hydrodynamic convection currents that, at least in thick enough cells, play an important role in

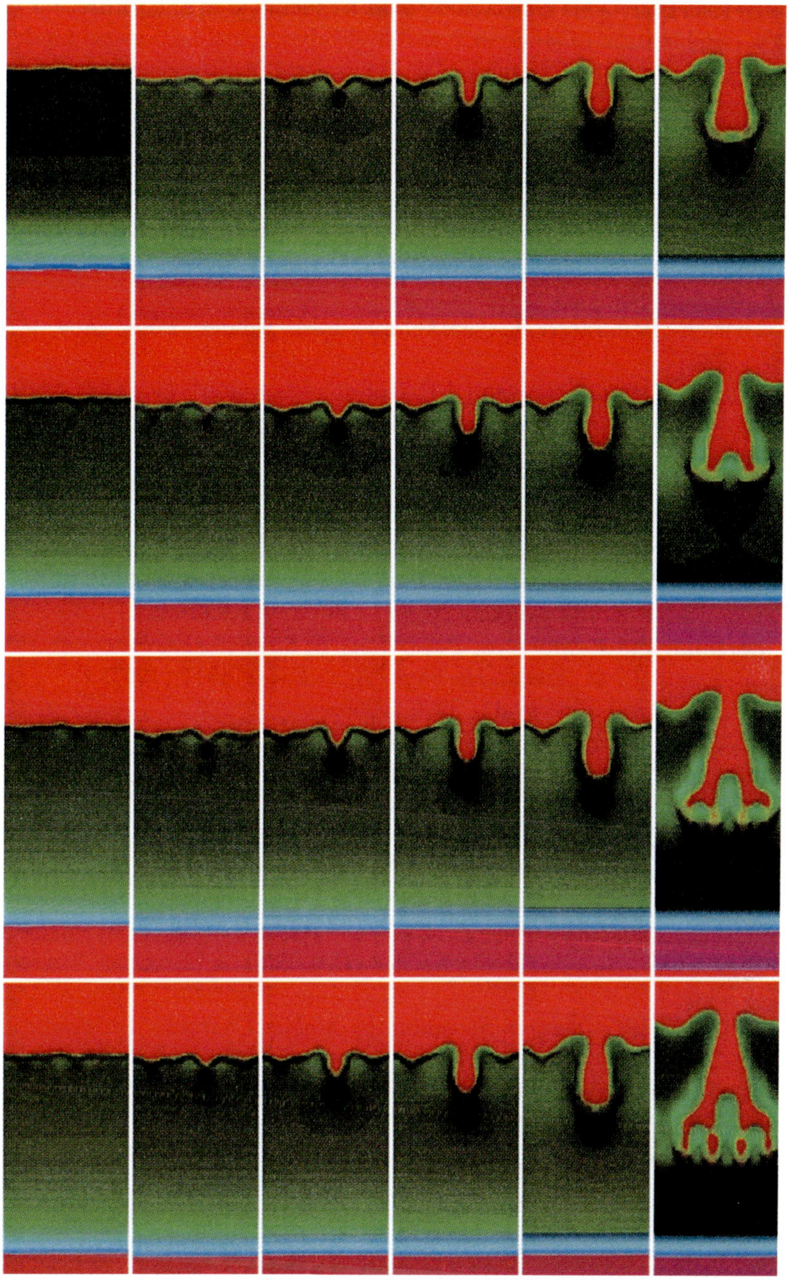

Figure 7. Evolution of a 100 × 40 electrochemical cell. The applied potential corresponds to $\Delta V = 2.5$ Volts. The cathode is on top, and growth precedes downwards. The concentrations of metal, anions, and cations are coded in red, green, and blue, respectively, and the cation and anion concentrations are re-scaled from approximately 8×10^{-2} to 1 to be visible. Time evolves from top left to bottom right, by columns. Dendrites grow on the cathode, while the anode is progressively dissolved (the images are re-centered during the evolution).

the pattern selection. This convection could be included in the future, for example by using a combination of stochastic and hydrodynamical lattice-gas models.

Discussions with J.-N. Chazalviel, V. Fleury, and M. Rosso are greatly acknowledged. Laboratoire de Physique de la Matière Condensée is Unité Mixte 7643 of CNRS and Ecole Polytechnique.

References

1. M. Matsushita, M. Sano, Y. Hayakawa, H. Honjo, and Y. Sawada, *Phys. Rev. Lett.*, **53**, 286 (1984).
2. D. Grier, E. Ben-Jacob, R. Clarke, and L.M. Sander, *Phys. Rev. Lett.*, **56**, 1264 (1986).
3. Y. Sawada, A. Dougherty, and J.P. Gollub, *Phys. Rev. Lett.*, **56**, 1260 (1986).
4. V. Fleury, M. Rosso, J.-N. Chazalviel, and B. Sapoval, *Phys. Rev. A*, **44**, 6693 (1991).
5. T. A. Witten and L. M. Sander, *Phys. Rev. Lett.*, **47**, 1400 (1981); *Phys. Rev. B*, **27**, 5686 (1983).
6. P.A. Rikvold, G. Brown, M.A. Novotny, and A. Wieckowski, *Colloids and Surfaces A: Physicochemical and Engineering Aspects* , **134**, 3 (1998).
7. W. Schmickler, *J. Electroanal. Chem.*, **460,** 144 (1999).
8. G. Marshall, and P. Mocskos, *Phys. Rev. E*, **55**, 549 (1997)
9. S.C. Hill, and J.I.D. Alexander *Phys. Rev. E*, **56**, 4317 (1997).
10. G. Martin, *Phys. Rev. B*, **41**, 2279 (1990).
11. J.-F. Gouyet, *Europhys. Lett.*, **21** , 335 (1993).
12. V. Vaks, and S.Beiden, *Sov. Phys. JETP*, **78**, 546 (1994).
13. J.-F. Gouyet, *Phys. Rev. E*, **51**, 1695 (1995).
14. V. Dobretsov, V. Vaks, and G. Martin, *Phys. Rev. B* , **54**, 3227 (1996).
15. M. Plapp and J.-F. Gouyet, *Phys. Rev. E*, **55**, 45 (1997).
16. J.S. Kirkaldy, *Can. J. Phys.*, **57**, 717 (1979).
17. J.-N. Chazalviel, *Phys. Rev. A*, **42**, 7355 (1990).
18. W. Schmickler, *Interfacial Electrochemistry*, Oxford University Press, (1996), pages 58-63.
19. J.-N. Chazalviel, *Coulomb Screening by Mobile Charges* , Birkhäuser (1999), pages 22-27.
20. M. Plapp and J.-F. Gouyet, *Phys. Rev. Lett.*, **78**, 4970 (1997); *Eur. Phys. J. B*, **9**, 267 (1999).
21. A. Karma and W.-J. Rappel, *Phys. Rev. E*, **57**, 4323 (1998).

FLAME FRONT INSTABILITIES AND DEVELOPMENT OF FRACTAL FLAMES

VITALIY BYCHKOV

Institute of Physics, Umeå University, S-901 87 Umeå, Sweden
E-mail: vitaliy.bychkov@physics.umu.se

Experimental observations indicate that the hydrodynamic flame instability result in development of a fractal structure at a flame front. We develop the theory of flame dynamics and stability and find estimates for the fractal dimension of a flame front. The obtained theoretical results are in a good agreement with experimental measurements.

Introduction

A premixed flame front is a region of chemical reactions with heat release propagating with subsonic velocity due to thermal conduction [1]. Velocity U_f of planar flames varies from 10 cm/s to 10 m/s depending on a particular fuel mixture with thickness of a flame front L_f about $10^{-4}-10^{-3}$cm. Both flame velocity and thickness are controlled by thermal diffusivity of the burning matter and by characteristic reaction time. Typically hydrodynamic motion of a flame occurs on length scales much larger than the flame thickness (for example, burning chamber of a car engine is about 10 cm) and therefore flame may be treated as a very thin front separating reactants from the products of burning. Because of heat release in the reaction the density of burning products ρ_b is usually much smaller than the density of the fuel mixture ρ_f with the expansion coefficient $\Theta=\rho_f/\rho_b$. Flame dynamics depends strongly on the expansion coefficient, which takes rather large values $\Theta=5-10$ for laboratory and industrial flames.

One of the most important problems in combustion science is to find velocity of flame propagation [1, 2]. While velocity of a planar flame U_f is determined by thermal and chemical fuel parameters, the resulting velocity of flame propagation U_w depends also on the flame shape: the more curved and wrinkled the flame front, the faster the flame propagates. In absence of external turbulence a curved flame shape results from intrinsic flame instabilities such as the Darrieus-Landau (DL) instability inherent to any premixed flame in a gaseous fuel [3]. Because of the DL instability small perturbations bend an initially planar flame front, if the perturbation wavelength exceeds the cut-off wavelength λ_c determined by thermal conduction and by finite flame thickness [4]. Typically the cut-off wavelength is considerably larger than the flame thickness and for most of the laboratory flames one has $\lambda_c>20L_f$.

While the linear stage of the DL instability has been well investigated, flame dynamics at the nonlinear stage remained a point of discussions for a long time. Particularly, it was unclear if the instability leads to spontaneous turbulization of a flame front suggested by Landau [3] or a flame front becomes stabilized at the nonlinear stage, as it was proposed by Zeldovich [1]. Recent experiments [5-7] demonstrated one more option of a fractal structure developing at a flame front. We have investigated dynamics and stability of curved flames and obtained that outcome of the flame instability depends on the characteristic hydrodynamic length scale of the flow (for example, on the tube radius) with smooth curved flames occurring in tubes of moderate width and fractal flames developing in very wide tubes. The estimates of fractal dimension of a flame front following from our theory are in a good agreement with experimental measurements.

Theory of curved stationary flames

The theory of curved stationary flames has been developed in [8,9] and compared to the results of numerical experiments [10-13]. As the amplitude of flame perturbations grows, "cusp"-points appear at the front stabilizing the DL instability at the nonlinear stage. Characteristic shape of a curved stationary flame may be illustrated by Figure 1(a) with the cusp pointing to the products of burning and a smooth hump directed to the reactants (the ideally slip and adiabatic walls at $x=0,R$ play the role of symmetry axes). Shape $z=F(\mathbf{x})$ and velocity U_w of a curved flame front may be described by the following nonlinear equation derived in [8]

$$1-U_w/U_f+\frac{\Theta}{2}(\nabla F)^2+$$
$$\frac{(\Theta-1)^3}{16\Theta}\left[(\nabla F)^2-\left(\hat{\Phi}F\right)^2\right]-\frac{\Theta-1}{2}\left(1-\frac{\lambda_c}{2\pi}\hat{\Phi}\right)\hat{\Phi}F=0 \qquad (1)$$

with the operator defined as multiplication by the absolute value of the wave-number in Fourier-space

$$\hat{\Phi}F=\frac{1}{4\pi^2}\int|\mathbf{k}|F_k\exp(i\mathbf{k}\cdot\mathbf{x})d\mathbf{k}. \qquad (2)$$

Analytical and numerical solution to the nonlinear equation demonstrates, that a curved shape of a flame front develops in sufficiently wide tubes exceeding some critical width $R>R_c$ with $R_c=\lambda_c/2$ for 2D flows and $R_c\approx 0.63\lambda_c$ in cylindrical tubes (the numerical factor is determined by properties of Bessel-functions). In narrow tubes thermal conduction damps the instability and makes curved flame shape impossible. Because of the curved shape the flame surface area increases, flame consumes more fuel per unit time and propagates faster. Characteristic velocity increase because of the curved flame shape has been found for 2D flows in [8]

$$U_{2D}/U_f=1+\frac{1}{2}\frac{\Theta(\Theta-1)^2}{\Theta^3+\Theta^2+3\Theta-1}, \qquad (3)$$

and a similar formula for 3D flows has been proposed in [9]

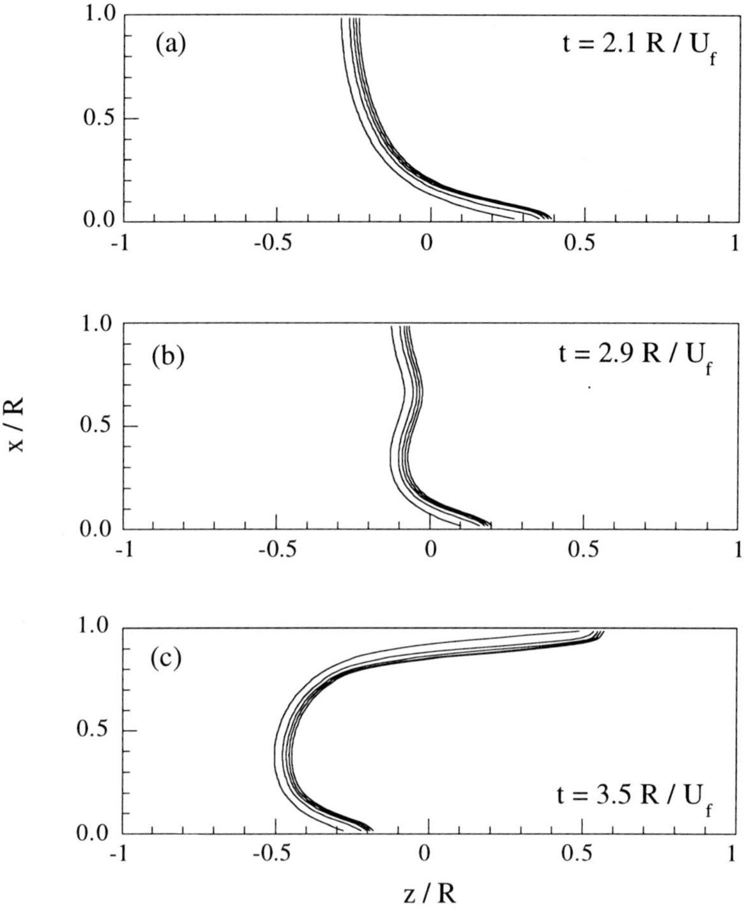

Figure 1. Evolution of a flame front isotherms with $\Theta = 8$ in a relatively wide tube of $R = 4.6R_c$. The flame propagates to the left.

$$U_{3D}/U_f = 1 + \frac{\Theta(\Theta-1)^2}{\Theta^3+\Theta^2+3\Theta-1}. \tag{4}$$

The obtained analytical formulas agree quite well with the results of numerical simulations [10-13] and experiments [6,7] predicting approximately twice-larger velocity for 3D curved stationary flames in comparison with planar ones.

Stability limits of curved stationary flames

In very wide tubes a flame front similar to that shown in Fig. 1 (a) becomes unstable again. Indeed, in wide tubes the flame radius of curvature tends to infinity in comparison with the flame thickness, the flame becomes locally planar and infinitely thin, which inevitably leads to the secondary DL instability of the flame front with respect to perturbations of a small scale. Such mechanism of the secondary DL instability has been proposed in [14], and the theory of the secondary instability has been developed in [13,15,16]. Particularly, stability limits of curved stationary flames have been found as a solution to an eigenvalue problem for the geometry of a 2D flow in [15] and for axisymmetric flames propagating in cylindrical tubes in [16]. For the case of a 2D flow the instability growth rate turns to be a purely real value, and by this reason the stability limits may be calculated on the basis of the linearised nonlinear equation (1)

$$\Theta\nabla F_s\cdot\nabla\tilde{F}+\frac{(\Theta-1)^3}{8\Theta}\left[\nabla F_s\cdot\nabla\tilde{F}-\hat{\Phi}F_s\cdot\hat{\Phi}\tilde{F}\right]-\frac{\Theta-1}{2}\left(1-\frac{\lambda_c}{2\pi}\hat{\Phi}\right)\hat{\Phi}\tilde{F}=0, \qquad (5)$$

where F_s and $\tilde{F}$ denote the stationary solution and the small perturbations, and the eigenvalue R_w is the critical tube width at which Eq. (5) has a solution with adiabatic boundary conditions at the tube walls. Re-scaling coordinates in Eq. (5) by the critical tube radius R_w one can place the eigenvalue inside the linearised equation. Solution to the eigenvalue problem demonstrates that curved stationary flames do become unstable in sufficiently wide tubes of $R>R_w$. The stability limits are presented in Fig. 2 versus the expansion coefficient Θ by the dashed line (2D flow) and the solid line (cylindrical tubes). As one can see, in the domain of realistic expansion coefficients $\Theta=5-10$ the secondary critical tube width R_w exceeds the primary one R_c by the factor about 4 both for the 2D flow and for the curved flames in cylindrical tubes, since two curves are practically undistinguishable. Solution to the eigenvalue problem agrees well with the results of numerical experiments [13] shown in Fig. 2 by markers. Evolution of the flame shape close to the stability limits is presented in Fig. 1 (a) – (c). The secondary DL instability close to the stability limits results in development of an extra cusp at the flame front accompanied by additional increase of the flame velocity. Theoretical results agree also with experiments on expanding flame balls [5-7]. When a flame ball ignited at a point grows, the ball radius characterising length scale of flame dynamics sweeps all values of interest from zero up to several meters. It allows one to observe all stages of the DL instability step by step: the period of stable propagation of an almost ideally spherical flame, development of the primary DL instability, onset of the secondary DL instability and so on. Indeed, in all experiments [5-7] more or less the same features of flame dynamics have been observed. At a certain ball radius the flame surface gets

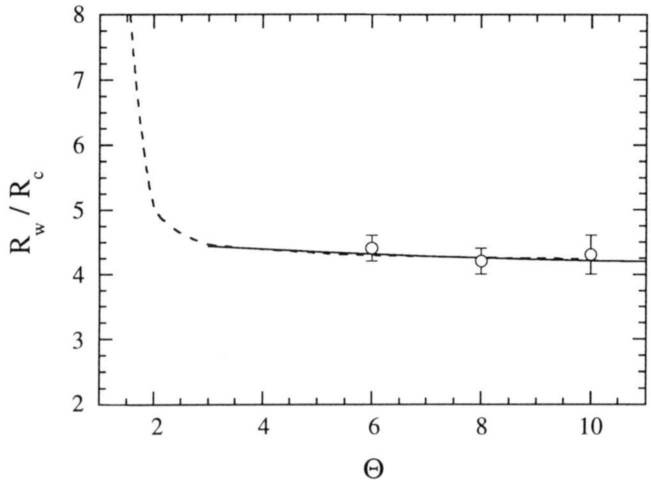

Figure 2. Stability limits (the ratio of the secondary critical tube width and the primary one R_w/R_c) versus the expansion coefficient. The solid line corresponds to the cylindrical tubes, the dashed line corresponds to the 2D flow, the markers show results of numerical experiments.

covered by cracks, which divide the spherical front into a number of primary cells of a rather large angular size. At this time velocity of flame propagation exceeds the planar flame velocity by a factor of 2 in agreement with the theory of the previous Section. As the flame ball grows the length scale of the primary cells increases and fine structure appears at the cells accompanied by noticeable flame acceleration. Keeping in mind the above results on flame stability one can identify the cracks as the primary DL instability, while the subsequent development of the fine structure may be interpreted as the secondary instability. In this sense it is interesting to study evolution of the characteristic cell size at an expanding flame front. According to the recent experiments [7] on propane and methane flames the typical initial cell size is about the cut-off wavelength of the DL instability λ_c, then it increases with maximal values reaching about $4\lambda_c$, then the fine structure develops and the observed cell size takes values between λ_c and $4\lambda_c$ in agreement with the above theory.

Development of fractal flame structure

For very wide tubes $R >> R_w$ development of the DL instability leads to a fractal flame structure as shown qualitatively in Fig. 3. The fractal structure is similar to that observed in [5-7] with the flame velocity depending on the largest possible length scale characterizing flame dynamics as

$$U_{fractal} \propto R^d \tag{6}$$

where $2+d$ is the fractal dimension of the flame front and d is the excess

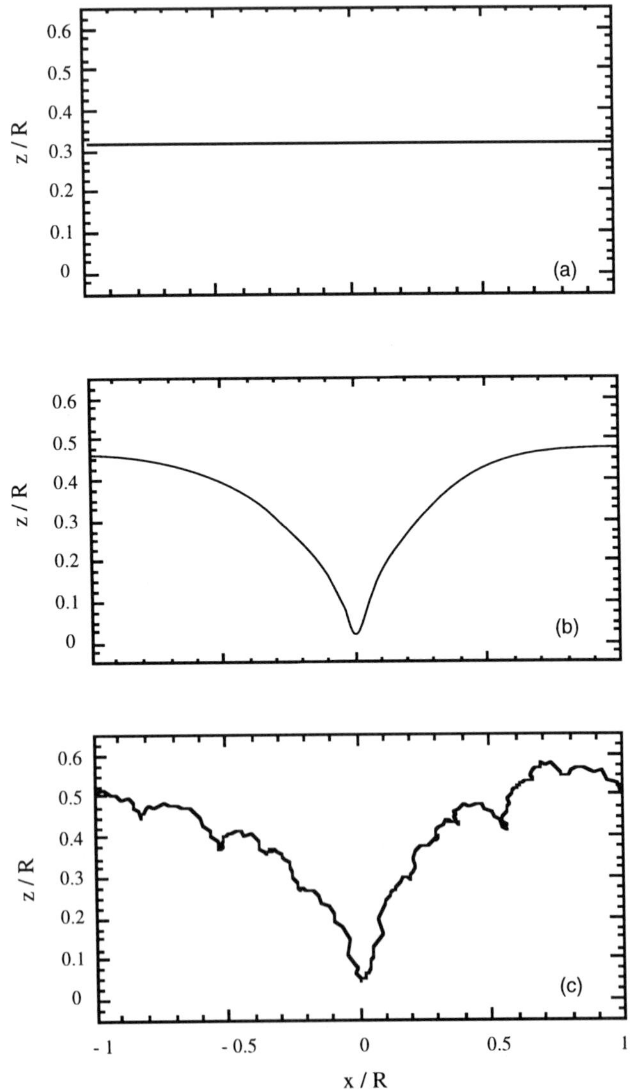

Figure 3. Qualitative picture of a flame structure for (a) $R < R_c$; (b) $R_c < R < R_w$; (c) $R >> R_w$. The flame propagates upwards.

of the flame front dimension over the embedding dimension. The fractal structure of a flame front implies cascades of humps and cusps of different sizes imposed one on another, so that the fractal excess may be estimated in the following way [17]. Let every step of the cascade decrease the cell size by factor $b = R_k / R_{k+1}$, and increase the front velocity (proportional to the flame surface area) by factor $\beta = U_{k+1} / U_k$. The cascading process is limited from below, since the cell size cannot be less than the cut-off wavelength λ_c, for which thermal conduction suppresses the instability. The cell size is limited from above too by the tube diameter $2R$ or by the radius of a flame ball. The fractal structure implies large number of cascades $N = \ln(2R/\lambda_c)/\ln b >> 1$ and the total flame surface is

$$U_{fractal} = U_N = U_f \beta^N = U_f (2R/\lambda_c)^d, \tag{7}$$

where the fractal excess is $d = \ln(\beta/b)$. Development of the primary cells and the subsequent secondary instability may be considered as a first step in the fractal flame structure. Then one can evaluate the factor β for a 3D flow as $\beta_{3D} = U_{3D}/U_f$ with U_{3D} given by Eq. (4) and a similar formula for a 2D flow. Evaluation for the factor b follows from the obtained stability limits of the secondary DL instability $b = R_w / R_c \approx 4.2$ both for 3D and 2D flows. The respective estimate for the fractal excess of 3D and 2D flames is shown in Fig. 4 with the typical experimental domain presented by the shaded region. As one can see, the developed theory predicts the fractal dimension

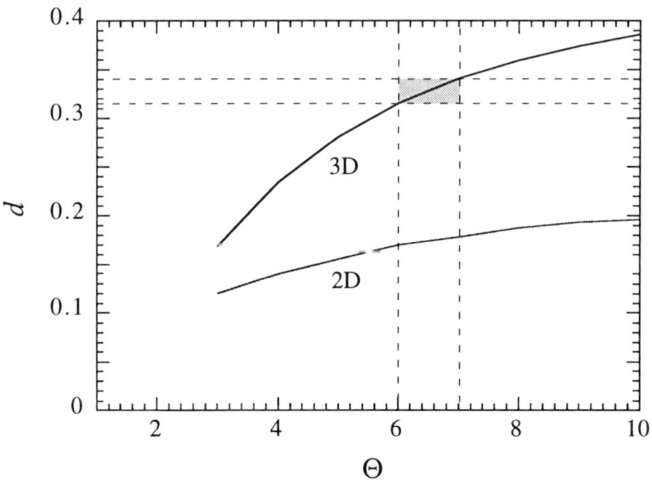

Figure 4. Fractal excess versus the expansion coefficient. The shaded region presents typical experimental domain.

2.32-2.34 for 3D flames with expansion coefficients $\Theta = 6-7$ typically used in combustion experiments. The fractal dimension depends on the expansion coefficient of the flame increasing with increase of the thermal expansion. The analytical estimates agree well with the experimental results on the fractal dimension of accelerating self-turbulized spherical flames, for which the fractal dimension 2.33 has been measured [5-7].

Acknowledgements

This work has been supported by the Swedish Research Council (VR) and by the Swedish Royal Academy of Sciences.

References

1. Zeldovich Ya.B., Barenblatt G.I., Librovich V.B., Makhviladze G.M., The Mathematical Theory of Combustion and Explosion (Consultants Bureau, NY, 1985).
2. Griffiths J.F. and Barnard J.A., Flame and Combustion (Blackie Academic and Professional, London, 1995).
3. Landau L.D., *Zh. Eksp.Teor. Fiz.* **14** (1944) 240.
4. Pelce P. and Clavin P., *J. Fluid Mech.* **124** (1982) 219.
5. Gostintsev Y.A., Istratov A.G., Shulenin Y.V., *Comb. Expl. Shock Waves* **24** (1988) 70.
6. Bradley D., Sheppard C., Woolley R., Greenhalgh D.A., Lockett R.D., *Comb. Flame* **122** (2000) 196.
7. Bradley D., Cresswell T.M., Puttock J.S., *Comb. Flame* **124** (2001) 551.
8. Bychkov V.V., *Phys.Fluids* **10** (1998) 2091.
9. Bychkov V.V. and Kleev A.I., *Phys. Fluids* **11** (1999) 1890.
10. Bychkov V.V., Golberg S.M., Liberman M.A., Eriksson L.E., *Phys. Rev.* **E 54** (1996) 3713.
11. Bychkov V., Golberg S., Liberman M., Kleev A.I., Eriksson L., *Comb. Sci. Tech.* **129** (1997) 217.
12. Kadowaki S., *Phys. Fluids* **11** (1999) 3426.
13. Travnikov O.Yu., Bychkov V.V., Liberman M. A., *Phys. Rev.* **E 61** (2000) 468.
14. Zeldovich Ya.B., Istratov A.G., Kidin N.I., Librovich V.B., *Comb. Sci. Tech.* **24** (1980) 1.
15. Bychkov V.V., Kovalev K.A., Liberman M.A., *Phys. Rev.* **E 60** (1999) 2897.
16. Senchenko S., Bychkov V.V., Liberman M.A., *Comb. Sci. Tech.* (2001) in press.
17. Bychkov V.V. and Liberman M.A., *Phys. Rev. Lett.* **76** (1996) 2814.

FRACTAL FUNCTIONS USING CONTRACTION METHOD IN PROBABILISTIC METRIC SPACES

JÓZSEF KOLUMBÁN

Babes-Bolyai University, Faculty of Mathematics and Computer Science, Cluj-Napoca,Romania

E-mail: kolumban@math.ubbcluj.ro

ANNA SOÓS

Babes-Bolyai University, Faculty of Mathematics and Computer Science, Cluj-Napoca,Romania

E-mail: asoos@math.ubbcluj.ro

In this paper, using probabilistic metric spaces techniques, we can weak the first moment condition for existence and uniqueness of selfsimilar fractal functions.

The most known fractals are invariant sets with respect to a system of contraction maps, especially the so called selfsimilar sets.

Recently Hutchinson and Rüschendorf gave a simple proof for the existence and uniqueness of invariant fractal sets and fractal functions using probability metrics defined by expectation. In these works a finite first moment condition is essential.

In this paper, using probabilistic metric spaces techniques, we can weak the first moment condition for existence and uniqueness of selfsimilar fractal functions.

The theory of probabilistic metric spaces, introduced in 1942 by K. Menger, was developed by numerous authors. The study of contraction mappings for probabilistic metric spaces was initiated by V. M. Sehgal, and H. Sherwood.

1 Selfsimilar fractal functions

Denote (X, d) a complete separable metric space Let $g : I \to X$, where $I \subset \mathbb{R}$ is a closed bounded interval, $N \in \mathbb{N}$ and let $I = I_1 \cup I_2 \cup \cdots \cup I_N$ be a partition of I into disjoint subintervals. Let $\Phi_i : I \to I_i$ be increasing Lipschitz maps with $p_i = Lip\Phi_i$. We have $\sum_{i=1}^{N} p_i \geq 1$ and if the Φ_i are affine then $\sum_{i=1}^{N} p_i = 1$. If $g_i : I_i \to \mathbf{X}$, for $i \in \{1, ..., N\}$ define $\sqcup_i g_i : I \to \mathbf{X}$ by

$$(\sqcup_i g_i)(x) = g_j(x) \text{ for } x \in I_j.$$

A *scaling law* $\mathbf{S}$ is an N-tuple $(S_1, ..., S_N)$, $N \geq 2$, of Lipschitz maps $S_i : \mathbf{X} \to \mathbf{X}$. Denote $r_i = LipS_i$. *A random scaling law* $\mathbf{S} = (S_1, S_2, ..., S_N)$ is a random variable whose values are scaling laws. We write $\mathcal{S} = dist\mathbf{S}$ for the probability distribution determined by $\mathbf{S}$ and $\stackrel{d}{=}$ for the equality in distribution.

Let $\mathbf{S} = (S_1, ..., S_N)$ be a scaling law. For the function $g : I \to X$ define the function $\mathbf{S}g : I \to X$ by

$$\mathbf{S}g = \sqcup_i S_i \circ g \circ \Phi_i^{-1}.$$

We say g *satisfies the scaling law* $\mathbf{S}$, or is a *selfsimilar fractal function*, if

$$\mathbf{S}g \stackrel{d}{=} g.$$

Fix $0 < p \leq \infty$. Let

$$L_\infty = \{g : I \to X \mid \; esssup_{x \in X} d(\,g(x), a) < \infty\},$$

$$L_p = \{g : I \to X \mid \int d(g(x), a)^p < \infty\}, \text{ if } 0 < p < \infty,$$

for some a$\in \mathbb{R}$.
The metric $d_p$on L_p is the complete metric defined by

$$d_\infty(f,g) = es \sup_x d(f(x), g(x)),$$

$$d_p(f,g) = \left(\int d(f(x), g(x))\right)^{\frac{1}{p} \wedge 1} \text{ if } 0 < p < \infty.$$

Let $\lambda_\infty = \max_i r_i$ and $\lambda_p = \sum_i p_i r_i^p$,for $0 < p < \infty$.
In Hutchinson and Rüschendorf prove the following:

Theorem 1 *([5]) If* $\mathbf{S} = (S_1, S_2, ..., S_N)$ *is a scaling law with* $\lambda_p < 1$ *for some* $0 < p \leq \infty$ *then there is a unique* $g^* \in L^p$ *such that* g^* *satisfies* $\mathbf{S}$.
Moreover, for any $g_0 \in L^p$,

$$esssupd_\infty(\mathbf{S}^k g_0, g^*) \leq \frac{\lambda_\infty^k}{1 - \lambda_\infty} esssupd_\infty(g_0, \mathbf{S}g_0) \to 0,$$

$$d_p(\mathbf{S}^k g_0, g^*) \leq \frac{\lambda_p^{k(\frac{1}{p} \wedge 1)}}{1 - \lambda_p^{\frac{1}{p} \wedge 1}} d_p(g_0, \mathbf{S}g_0) \to 0,\, 0 < p < 1$$

as $k \to \infty$.

For the random version we start with the random scaling law. Let $\mathbf{S} = (S_1, ..., S_N)$ be a random scaling law and let $G = (G_t)_{t \in I}$ be a stochastic process or a random function with state space $(X, \mathcal{X})$,where $\mathcal{X}$.is the Borel σ-algebra on X. The trajectory of the process G is the function $g : I \to$ X.The trajectory of the random function $\mathbf{S}g$ is defined up to probability distribution by

$$\mathbf{S}g = \sqcup_i S_i \circ g^{(i)} \circ \Phi_i^{-1},$$

where $\mathbf{S}, g^{(1)}, ..., g^{(N)}$ are independent of one another and $g^{(i)} \stackrel{d}{=} g$, for $i \in \{1, ..., N\}$.
If $\mathcal{G} = distg$ we define

$$\mathcal{S}\mathcal{G} = dist\mathbf{S}g.$$

We say g or $\mathcal{G}$ *satisfies the scaling law* $\mathbf{S}$, or is a *selfsimilar random fractal function*, if

$$\mathbf{S}g \stackrel{d}{=} g, \quad \text{or equivalently} \quad \mathcal{S}\mathcal{G} = \mathcal{G}.$$

Beginning from any $g_0 \in L_p$ Hutchinson and Rüschendorf define [5] a sequence of random functions

$$\mathbf{S}g_0 = \sqcup_i S_i \circ g_0 \circ \Phi_i^{-1},$$

$$\mathbf{S}^2 g_0 = \sqcup_{i,j} S_i \circ S_j^i \circ g_0 \circ \Phi_j^{-1} \circ \Phi_i^{-1},$$

$$\mathbf{S}^3 g_0 = \sqcup_{i,j,k} S_i \circ S_j^i \circ S_k^{ij} \circ g_0 \circ \Phi_k^{-1} \circ \Phi_j^{-1} \circ \Phi_i^{-1},$$

etc.; where $\mathbf{S^i} = (S_1^i, S_2^i, ..., S_N^i)$, for $i \in \{1, ..., N\}$, are independent of each other and of $\mathbf{S}$, the $\mathbf{S}^{ij} = (S_1^{ij}, S_2^{ij}, ..., S_N^{ij})$, for $i, j \in \{1, ..., N\}$ are independent of each other and of $\mathbf{S}$ and $\mathbf{S^i}$, etc.

Theorem 2 *(Hutchinson and Ruschendorf ([5])) If there exists a random function h such that*

$$esssup_\omega d_\infty(h^\omega, \delta_a^\omega) < \infty \qquad or \tag{1}$$

$$E_\omega^{\frac{1}{p}} d_p^p(h^\omega, \delta_h^\omega) < \infty \;\; for 1 \le p < \infty \quad or \tag{2}$$

$$E_\omega d_p(h^\omega, \delta_h^\omega) < \infty \;\; for \; 0 < p < 1, \tag{3}$$

and if $\mathbf{S} = (S_1, ..., S_N)$ is a random scaling law which satisfies either

$$\lambda_p := E \sum_{i=1}^{N} p_i r_i^p < 1 \; and \; E \sum_{i=1}^{N} p_i d^p(S_i(a), a) < \infty, \; or \tag{4}$$

$$\lambda_\infty := esssup_\omega \max_i r_i < 1 \; and \; esssup_\omega \max_i d^p(S_i(a), a) < \infty, \tag{5}$$

then there exists a unique g^* such that $\mathbf{S}g^* \stackrel{d}{=} g^*$ and for any $g_0 \in L_p$,

$$esssup d_\infty(\mathbf{S}^k g_0, g^*) \le \frac{\lambda_\infty^k}{1 - \lambda_\infty} esssup d_\infty(g_0, \mathbf{S}g_0) \to 0,$$

$$E^{\frac{1}{p}} d_p^p(\mathbf{S}^k g_0, g^*) \le \frac{\lambda_p^{\frac{k}{p}}}{1 - \lambda_p^{\frac{1}{p}}} E^{\frac{1}{p}} d_p^p(g_0, \mathbf{S}g_0) \to 0, \; 1 \le p < \infty$$

$$E d_p(\mathbf{S}^k g_0, g^*) \le \frac{\lambda_p^k}{1 - \lambda_p} E_p^d(g_0, \mathbf{S}g_0) \to 0, \; 0 < p < 1$$

as $k \to \infty$, where g^* does not depend on g_0. In particular, $\mathbf{S}^k g_0 \to g^*$ a.s. Moreover, up to probability distribution, g^* is the unique function such that $E \int log|g^*| < \infty$ and which satisfies $\mathbf{S}$.

However, using contraction method in probabilistic metric spaces,instead of (1) we can give weaker conditions for the existence and uniqueness of invariant random function.

Theorem 3 *Let $\mathcal{E}_p$ be the set of random functions $(G_t)_{t \in I}$ with state space X and let $\mathbf{S}$ be a random scaling law. Suppose there exists $h \in \mathcal{E}_p$ and a positive number γ such that either*

$$P(\{\omega \in \Omega \,|\, esssup_x d(h^\omega(x), \mathbf{S}h^\omega(x)) \ge t\}) \le \frac{\gamma}{t} \; for \; all \quad t > 0$$

and $\lambda_\infty := esssup_\omega \max_i r_i^\omega < 1$ or

$$P(\{\omega \in \Omega \,|\, \int_I d(h^\omega(x), \mathbf{S}h^\omega(x))^{\frac{1}{p} \wedge 1} \ge t\}) \le \frac{\gamma}{t} \; for \; all \quad t > 0$$

and $\lambda_p := E\sum_{i=1}^N p_i r_i^{\omega p} < 1$. *Then there exists* $G^* \in \mathcal{E}_p$ *such that* $\mathbf{S}g^* = g^*$. *Moreover, up to a probability distribution* g^* *is the unique function in* $\mathcal{E}_0 = \cup_{p>0}\mathcal{E}_p$.

Application 1 *The Brownian motion can be characterized as the fixed point of a scaling operator.*

For each $\alpha > 0$, *let* $B^\alpha : [0,1] \to \mathbb{R}$ *denote the constrained Brownian motion given by*

$$B^\alpha(t+h) - B^\alpha(t) \stackrel{d}{=} N(0,\alpha h), \quad for \quad t > 0\, and\, h > 0,$$

$$B^\alpha(0) = 0 \; a.s., \quad and \quad B^\alpha(1) = 1 \; a.s.,$$

where $N(0,\alpha h)$ *denotes the normal distribution with mean* 0 *and variance* αh. *For fix* $p \in \mathbb{R}$ *consider the Brownian motion* $B^\alpha|_{B^\alpha(\frac{1}{2})=p}$ *constrained by* $B^\alpha(\frac{1}{2}) = p$.

Let $S_1, S_2 : \mathbb{R} \to \mathbb{R}$ *the affine transformation characterized by*

$$S_1(0) = 0,\, S_1(1) = S_2(0) = p,\, S_2(1) = 1.$$

If

$$r_1 = LipS_1 = |p|, \quad r_2 = LipS_2 = |1-p|,$$

then

$$B^\alpha|_{B^\alpha(\frac{1}{2})=p}(t) \stackrel{d}{=} S_1 \circ B^{\frac{\alpha}{2r_1^2}}(2t), \quad t \in [0, \frac{1}{2}].$$

Similarly

$$B^\alpha|_{B^\alpha(\frac{1}{2})=p}(t) \stackrel{d}{=} S_2 \circ B^{\frac{\alpha}{2r_1^2}}(2t-1), \quad t \in [\frac{1}{2}, 1].$$

Now define

$$\Phi_1 : [0,1] \to [0, \frac{1}{2}], \quad \Phi_1(s) = \frac{s}{2},$$

$$\Phi_2 : [0,1] \to [\frac{1}{2}, 1], \quad \Phi_1(s) = \frac{s+1}{2}.$$

It follows that

$$B^\alpha|_{B^\alpha(\frac{1}{2})}(t) \stackrel{d}{=} \sqcup_{i=1}^2 S_i \circ B^{\frac{\alpha}{2r_i^2}} \circ \Phi_i^{-1}(t), \quad t \in [0,1].$$

Now let p^α *be random point with distribution* $N(0, \frac{\alpha}{2})$ *and let* $\mathbb{S}^\alpha = (S_1^\alpha, S_2^\alpha)$ *be the random scaling law obtained by defining* (S_1^α, S_2^α) *from the random point* p^α *in the same manner as* (S_1, S_2) *was previously defined from* p. *Let* $r_i^\alpha = Lip_i^\alpha$ *for* $i = 1, 2$ *and let* $r^\alpha = \max\{r_1^\alpha, r_2^\alpha\}$. *Denote* $\mathbb{S} = \{\mathbb{S}^\alpha | \alpha > 0.\}$

It follows for each $\alpha > 0$ *that*

$$B^\alpha \stackrel{d}{=} \sqcup_{i=1}^2 S_i^\alpha \circ B^{\frac{\alpha}{2r_i^2}(i)} \circ \Phi_i^{-1},$$

where $B^{\frac{\alpha}{2r_1^2}(1)} \stackrel{d}{=} B^{\frac{\alpha}{2r_1^2}}$ *and* $B^{\frac{\alpha}{2r_2^2}(2)} \stackrel{d}{=} B^{\frac{\alpha}{2r_2^2}}$ *are chosen independently of one another.*

Thus the family of constrained Brownian motion $\{B^\alpha | \alpha > 0\}$ *satisfies the family of scaling laws* $\mathbb{S}$.

2 Proof of Theorem 3

2.1 Menger spaces

Let $\mathbf{R}$ denote the set of real numbers and $\mathbf{R}_+ := \{x \in \mathbf{R} : x \geq 0\}$. A mapping $F : \mathbf{R} \to [0,1]$ is called a *distribution function* if it is non-decreasing, left continuous with $\inf_{t \in \mathbf{R}} F(t) = 0$ and $\sup_{t \in \mathbf{R}} F(t) = 1$ (see [1]). By Δ we shall denote the set of all distribution functions F. Let Δ be ordered by the relation "$\leq$", i.e. $F \leq G$ if and only if $F(t) \leq G(t)$ for all real t. Also $F < G$ if and only if $F \leq G$ but $F \neq G$. We set $\Delta^+ := \{F \in \Delta : F(0) = 0\}$.

Throughout this paper H will denote the Heaviside distribution function defined by

$$H(x) = \begin{cases} 0, & x \leq 0, \\ 1, & x > 0. \end{cases}$$

Let X be a nonempty set. For a mapping $\mathcal{F} : X \times X \to \Delta^+$ and $x, y \in X$ we shall denote $\mathcal{F}(x,y)$ by $F_{x,y}$, and the value of $F_{x,y}$ at $t \in \mathbf{R}$ by $F_{x,y}(t)$, respectively. The pair $(X, \mathcal{F})$ is a *probabilistic metric space* (briefly *PM space*) if X is a nonempty set and $\mathcal{F} : X \times X \to \Delta^+$ is a mapping satisfying the following conditions:

1^0. $F_{x,y}(t) = F_{y,x}(t)$ for all $x, y \in X$ and $t \in \mathbf{R}$;
2^0. $F_{x,y}(t) = 1$, for every $t > 0$, if and only if $x = y$;
3^0. if $F_{x,y}(s) = 1$ and $F_{y,z}(t) = 1$ then $F_{x,z}(s+t) = 1$.

A mapping $T : [0,1] \times [0,1] \to [0,1]$ is called a *t-norm* if the following conditions are satisfied:

4^0. $T(a,1) = a$ for every $a \in [0,1]$;
5^0. $T(a,b) = T(b,a)$ for every $a, b \in [0,1]$
6^0. if $a \geq c$ and $b \geq d$ then $T(a,b) \geq T(c,d)$;
7^0. $T(a, T(b,c)) = T(T(a,b), c)$ for every $a, b, c \in [0,1]$.

A *Menger space* is a triplet $(X, \mathcal{F}, T)$, where $(X, \mathcal{F})$ is a probabilistic metric space, T is a t-norm, and instead of 3^0 we have the stronger condition

8^0. $F_{x,y}(s+t) \geq T(F_{x,z}(s), F_{z,y}(t))$ for all $x, y, z \in X$ and $s, t \in \mathbf{R}_+$.

The (t, ϵ)-topology in a Menger space was introduced in 1960 by B. Schweizer and A. Sklar [9]. The base for the neighbourhoods of an element $x \in X$ is given by

$$\{U_x(t,\epsilon) \subseteq X : t > 0, \epsilon \in]0,1[\},$$

where

$$U_x(t,\epsilon) := \{y \in X : F_{x,y}(t) > 1 - \epsilon\}.$$

In 1966, V.M. Sehgal [11] introduced the notion of a contraction mapping in PM spaces. The mapping $f : X \to X$ is said to be a *contraction* if there exists $r \in]0,1[$ such that

$$F_{f(x),f(y)}(rt) \geq F_{x,y}(t)$$

for each $x, y \in X$ and $t \in \mathbf{R}_+$.

A sequence $(x_n)_{n\in\mathbf{N}}$ from X is said to be *fundamental* if

$$\lim_{n,m\to\infty} F_{x_m,x_n}(t) = 1$$

for all $t > 0$. The element $x \in X$ is called *limit* of the sequence $(x_n)_{n\in\mathbf{N}}$, and we write $\lim_{n\to\infty} x_n = x$ or $x_n \to x$, if $\lim_{n\to\infty} F_{x,x_n}(t) = 1$ for all $t > 0$. A probabilistic metric (Menger) space is said to be *complete* if every fundamental sequence in that space is convergent.

Let A and B nonempty subsets of X. The *probabilistic Hausdorff-Pompeiu distance* between A and B is the function $F_{A,B} : \mathbf{R} \to [\mathbf{0}, \mathbf{1}]$ defined by

$$F_{A,B}(t) := \sup_{s<t} T(\inf_{x\in A} \sup_{y\in B} F_{x,y}(s), \inf_{y\in B} \sup_{x\in A} F_{x,y}(s)).$$

In the sequel we remember some properties proved in [6,7]:

Proposition 1 *If $\mathcal{C}$ is a nonempty collection of nonempty closed bounded sets in a Menger space $(X, \mathcal{F}, T)$ with T continuous, then $(\mathcal{C}, \mathcal{F}_\mathcal{C}, T)$ is also Menger space, where $\mathcal{F}_\mathcal{C}$ is defined by $\mathcal{F}_\mathcal{C}(A, B) := F_{A,B}$ for all $A, B \in \mathcal{C}$.*

Proof. See [6,11]. □

Proposition 2 *Let $T_m(a, b) := \max\{a + b - 1, 0\}$. If $(X, \mathcal{F}, T_m)$ is a complete Menger space and $\mathcal{C}$ is the collection of all nonempty closed bounded subsets of X in $(t, \epsilon)-$ topology, then $(\mathcal{C}, \mathcal{F}_\mathcal{C}, T_m)$ is also a complete Menger space.*

Proof. See [7]. □

2.2 E-spaces

The notion of E-space was introduced by Sherwood [12] in 1969. Next we recall this definition. Let $(\Omega, \mathcal{K}, P)$ be a probability space and let (Y, ρ) be a metric space. The ordered pair $(\mathcal{E}, \mathcal{F})$ is an *E-space over the metric space* (Y, ρ) (briefly, an E-space) if the elements of $\mathcal{E}$ are random variables from Ω into Y and $\mathcal{F}$ is the mapping from $\mathcal{E} \times \mathcal{E}$ into Δ^+ defined via $\mathcal{F}(x, y) = F_{x,y}$, where

$$F_{x,y}(t) = P(\{\omega \in \Omega |\ d(x(\omega), y(\omega)) < t\})$$

for each $t \in \mathbf{R}$. Usually $(\Omega, \mathcal{K}, P)$ is called the base and (Y, ρ) the target space of the E-space. If $\mathcal{F}$ satisfies the condition

$$\mathcal{F}(x, y) \neq H, \ \ for \ \ x \neq y,$$

with H defined in paragraf 3.1., then $(\mathcal{E}, \mathcal{F})$ is said to be a *canonical E-space.* Sherwood [12] proved that every canonical E-space is a Menger space under $T = T_m$, where $T_m(a, b) = \max\{a + b - 1, 0\}$. In the sequel we suppose that $\mathcal{E}$ is a canonical E-space.

The convergence in an E-space is exactly the probability convergence. The E-space $(\mathcal{E}, \mathcal{F})$ is said to be complete if the Menger space $(\mathcal{E}, \mathcal{F}, T_m)$ is complete.

Proposition 3 *If (Y,ρ) is a complete metric space then the E-space $(\mathcal{E},\mathcal{F})$ is also complete.*

Prof.: See [7]

The next result was proved in [7]:

Theorem 4 *Let $(\mathcal{E},\mathcal{F})$ be a complete E- space, $N \in \mathbf{N}^*$, and let $f_1,...,f_N : \mathcal{E} \to \mathcal{E}$ be contractions with ratio $r_1,...r_N$, respectively. Suppose that there exists an element $z \in \mathcal{E}$ and a real number γ such that*

$$P(\{\omega \in \Omega | \rho(z(\omega), f_i(z(\omega)) \geq t\}) \leq \frac{\gamma}{t}, \tag{6}$$

for all $i \in \{1,..,N\}$ and for all $t > 0$. Then there exists a unique nonempty closed bounded and compact subset K of $\mathcal{E}$ such that

$$f_1(K) \cup ... \cup f_N(K) = K.$$

Corollary 1 *Let $(\mathcal{E},\mathcal{F})$ be a complete E- space, and let $f : \mathcal{E} \to \mathcal{E}$ be a contraction with ratio r. Suppose there exists $z \in \mathcal{E}$ and a real number γ such that*

$$P(\{\omega \in \Omega |\ \rho(z(\omega), f(z)(\omega)) \geq t\}) \leq \frac{\gamma}{t} \text{ for all } t > 0.$$

Then there exists a unique $x_0 \in \mathcal{E}$ such that $f(x_0) = x_0$.

Remark: If the mean values $\int_\Omega d(z(\omega), f_i(x(\omega)))dP$ for $i \in \{1,...,N\}$ are finite, then by the Chebysev inequality, condition (6) is satisfied. However, the condition (6) can also be satisfied for $\int_\Omega d(z(\omega), f(z(\omega)))dP = \infty$. For example, let $\Omega =]0,1]$ with the Lebesque measure and let $f(x) = \frac{x(\omega)}{3} + \frac{1}{\omega}$. Then for $z(\omega) \equiv 0$, the above expectation is ∞, but, for $\gamma = 1$, the condition (6) holds.

2.3 Proof of Theorem 3

Let $f : \mathcal{E}_p \to \mathcal{E}_p$,

$$f(g) = \mathbf{S}g = \sqcup_i S_i \circ g^{(i)} \circ \Phi_i^{-1},$$

where $\mathbf{S}, g^{(1)},...,g^{(N)}$ are independent of one another and $g^{(i)} \stackrel{d}{=} g$.

We first claim that, if $g \in \mathcal{E}_p$ then $f(g) \in \mathcal{E}_p$. For this, choose i.i.d. $g^{(\omega)} \stackrel{d}{=} g$. and $(S_1^\omega,...,S_N^\omega) \stackrel{d}{=} \mathbf{S}$ independent of $g^{(\omega)}$. For $p = \infty$ we have

$$\begin{aligned} ess \sup_x d(\mathbf{S}g^{(\omega)}(x), a) &= ess \sup_x d(\sqcup_i S_i^\omega \circ g_i^{(\omega)} \circ \Phi_i^{-1}(x), a) \leq \\ &\leq ess \sup_x \max_i r_i d(g_i^{(\omega)} \circ \Phi_i^{-1}(x), b) \leq \\ &\leq \max_i r_i ess \sup_x d(g_i^{(\omega)} \circ \Phi_i^{-1}(x), b) < \infty, \end{aligned}$$

where $b = \mathbf{S}(\delta_a)$. For $0 < p < \infty$ the proof is similar.

For $g_1, g_2 \in \mathcal{E}_p$ and $p = \infty$ we have

$$F_{f(g_1),f(g_2)}(t) = P(\{\omega \in \Omega \mid ess \sup_x d(\mathbf{S}g_1(x), \mathbf{S}g_2(x)) < t\}) =$$

$$= P(\{\omega \in \Omega \mid ess \sup_x d(S_i^\omega \circ g_1^{(i)} \circ \Phi_i^{-1}(x), S_i^\omega \circ g_2^{(i)} \circ \Phi_i^{-1}(x)) < t\}) \geq$$

$$\geq P(\{\omega \in \Omega \mid \lambda_\infty ess \ \sup_x d(g_1^{(i)} \circ \Phi_i^{-1}(x), g_2^{(i)} \circ \Phi_i^{-1}(x)) < t\}) = F_{g_1,g_2}(\frac{t}{\lambda_\infty})$$

for all $t > 0$.

Similarly if $0 < p < 1$. It follows that f is a contraction with ratio λ_∞ or λ_p and we may apply the Corollary 1 for $r = \lambda_\infty$ or $r = \lambda_p$ respectively.

For the uniqueness of $distg^*$ satisfying **S** let $\mathcal{G}$ be the set of probability distributions of members of $\mathcal{E}$. We define on $\mathcal{G}$ the probability metric by

$$F_{\mathcal{G}_1,\mathcal{G}_2}(t) := \sup_{s<t} \sup\{F_{g_1,g_2}(s) \mid g_1 \stackrel{d}{=} \mathcal{G}_1,\, g_2 \stackrel{d}{=} \mathcal{G}_2\}.$$

One checks that $\mathcal{S}$ is a contraction map with contraction constant λ_∞ or λ_p. Let $\mathcal{G}^*$ and $\mathcal{G}^{**}$ such that $\mathcal{S}\mathcal{G}^* = \mathcal{G}^*$ and $\mathcal{S}\mathcal{G}^{**} = \mathcal{G}^{**}$.

As in the proof of the Theorem 4, one can show that

$$F_{\mathcal{G}^*,\mathcal{G}^{**}}(t) = 1 \text{ for all } t > 0. \qquad \square$$

References

1. Gh.Constantin, I.Istraţescu, **Elements of Probabilistic Analysis,** *Kluwer Academic Publishers,* 1989.
2. J.E.Hutchinson, **Fractals and Self Similarity,** Indiana University Mathematics Journal **30**, 5 (1981), 713-747.
3. J.E.Hutchinson, L.Rüschendorf, **Random Fractal Measures via the Contraction Method,** Indiana Univ. Math. Jouranal **47**, 2 (1998), 471-489.
4. J.E.Hutchinson, L.Rüschendorf, **Random Fractal and probability metrics,** Research Report **MRR48**, (1998) Australian National University.
5. J.E.Hutchinson, L.Rüschendorf, **Selfsimilar Fractals and Selfsimilar Random Fractals,** Progress in Probability **46**, (2000), 109-123.
6. J.Kolumbán, A. Soós, **Invariant sets in Menger spaces,** Studia Univ. "Babes-Bolyai", Mathematica **43**, 2 (1998), 39-48.
7. J.Kolumbán, A. Soós, **Invariant sets of random variables in complete metric spaces,** Studia Univ. "Babes-Bolyai", Mathematica , (2001), accepted.
8. S.T.Rachev, **Probability Metrics and the Stability of Stochastic Models,** *Wiley,* 1991.
9. B.Schweizer, A.Sklar, **Statistical Mertic Spaces,** Pacific Journal of Mathematics, **10**, 1 (1960), 313-334.
10. B.Schweizer, A.Sklar, **Probabilistic Mertic Spaces,** *North Holland, New-York, Amsterdam, Oxford,* 1983.
11. V.M.Sehgal, **A Fixed Point Theorem for Mappings with a Contractive Iterate,** Proc. Amer. Math. Soc. **23**, (1969), 631-634.
12. H.Sherwood, **E-spaces and their relation to other classes of probabilistic metric spaces,** J.London Math. Soc. **44**, (1969), 441-448.

GROWTH DYNAMICS OF ROTATING DLA-CLUSTERS

ALEXANDER LOSKUTOV, DMITRY ANDRIEVSKY,
VICTOR IVANOV AND ALEXEI RYABOV

Physics Faculty, Moscow State University, Moscow, 119899 Russia

Phone: +7-095-939-5156, Fax: +7-095-939-2988

E-mail: loskutov@moldyn.phys.msu.su

Keywords: diffusion limited aggregation, cluster growth

Theoretical and computer simulation analysis of clusters growing by diffusion limited aggregation under rotation around a germ is presented. The theoretical model allows to study statistical properties of growing clusters in two different situations: in the static case (the cluster is fixed), and in the case when the growing structure has a nonzero rotation velocity around its germ. By the direct computer simulation the growth of rotating clusters is investigated. The fractal dimension of such clusters as a function of the rotation velocity is found. It is shown that for small enough velocities the fractal dimension is growing, but then, with increasing rotation velocity, it tends to unity.

1 Introduction

In this study we consider the problem of the fractal growth via adsorption in the plane and in the three-dimensional Euclidean space. The characteristic property of such processes is that the fractal growth is not described by known equations. *Statistical models* are usually used. The most known process of irreversible adsorption is the *diffusion limited aggregation* (DLA) [1–4]. In the simplest case this is a cluster formation when the randomly moving particles are sequentially and irreversibly deposited on the cluster surface. As a result of such clusterization a DLA-fractal is formed.

Depending on the embedding dimension D_e the fractal dimension D_f (the exponent in the dependence of the "mass" of the DLA-cluster on the characteristic linear size, $M \sim R^{D_f}$) takes the following values [5–6]:

Table 1. The values of the fractal dimension D_f of DLA-clusters for some embedding dimension.

D_e	D_f	D_e	D_f	D_e	D_f	D_e	D_f
2	1.71	3	2.49	4	3.40	5	4.33

These values are obtained by direct numerical simulations, and for $D_e>5$ there are no exact data. At the same time, analytical investigations are in a quite satisfactory agreement with numerical simulations, but also only for $D_e=3,...,5$.

During two last decades the fractal properties of growing DLA-clusters have been widely discussed in the scientific literature [1–2]. The enhanced interest to this model is connected with a variety of its applications to description of different physical phenomena. However, a very interesting extension of the problem, namely the study of DLA-fractals which are growing under rotation has not been investigated in detail (some aspects of this problem were discussed in [7], see also refs. therein).

We present here both the theoretical approach which enables us to calculate the fractal dimension of rotating DLA-clusters and the results of direct computer simulations. We propose a statistical annular model of the fractal growth in D_e=2 and D_e=3 Euclidean spaces (D_e>3 to be obtained). This model allows us to find the dimension of fixed DLA-clusters ("classical" DLA) and the dimension of clusters which have a rotation around the nucleating center. Therewith the convergence of numerical algorithms is two orders of magnitude faster than for usual case of DLA.

We have performed also direct computer simulations of DLA under rotation. Our preliminary results indicate the transition from fractal to non-fractal behavior of growing cluster on different length scales upon increasing the rotation velocity.

2 Theoretical Model

Suppose that particles have a diameter d. Let us divide the corresponding embedding space into $n_{\max}$ concentric rings with a middle radius $r_n = nd, n = 0,\ldots,n_{\max}$. Assume that the center of rings coincides with the center of a fractal, and the width of rings is equal to d. Let us consider models with the following properties:

a) particles can be situated only within some ring and can jump from one ring to another;
b) within a ring N_n particles can occupy $M_n = [2\pi r_n / d]$ cells, where [...] is an integer part;
c) particles can be arbitrarily distributed within the ring.

The case *b* corresponds to cellular model. The third case, (*c*), is the transient one between continuous and cellular models. When a particle jumps to another ring, it can be adsorbed here with probability corresponding to collision with one of *N* particles in the ring. It is obvious that in the case of cellular model this probability is

$$\tilde{p}_n = \frac{N_n}{M_n} \tag{1}$$

where N_n is the number of particles in an *n*-th layer (ring). In the case of equidistributed particles, the adsorption probability is the following:

$$\tilde{p}_n = 1 - \frac{(L_n - (N_n + 1)d)^{N_n - 1}}{(L_n - d)(L_n - (N_n + 1)d)^{N_n - 2}} \theta[L_n - (N_n + 1)d], \tag{2}$$

where $L_n = 2\pi r_n$ is the length of a circle along which particles are distributed, and $\theta[x]$ is a Heaviside function. Calculating this probability is quite a simple task of the known random sequential adsorption (RSA) theory: *N* particles of a length *d* are arbitrarily distributed on a segment of the length *L*, and we throw a new particle on this segment. What is the probability that this particle will not cover any other one?

Suppose that the particle is within *n*-th layer. Then the jump probability to (*n*+1)-th layer is $p_n^1 = (1 + d / r_n)/3$, to (*n*-1)-th layer is $p_n^{-1} = (1 - d / r_n)/3$, and the probability to be in the same layer is $p_n^0 = 1/3$. In general, $p_n^k = (1 + kd/r_n)/3$, $\sum_k p_n^k = 1$, k=-1, 0, 1. The term $\pm d/r_n$ follows from the assumption that the jump probability to a ring should be proportional to its length. Now we can obtain the probability of adsorption of the particle in the n-th layer:

$$P_n = \sum_{k=-1}^{1} \tilde{p}_{n+k} p_n^k. \tag{3}$$

Here $\tilde{p}_{n+k}$ is the probability of adsorption of the particles within $(n+k)$-th layer and p_n^k is the jump probability to this layer. If the adsorption does not take place, the particle moves to another layer with the probability $Q_n = 1 - P_n$. Corresponding jump probabilities under the condition that the adsorption did not take place, are the following:

$$q_n^k = \frac{p_n^k (1 - \tilde{p}_{n+k})}{Q_n}. \tag{4}$$

These expressions follow from Bayes formula of the full probability.

Thus, the constructed fractal is a set of layers which are characterized by a pair of numbers M_n and N_n. For the simplicity, let us assume that the particle mass is equal to unity. Then the fractal mass will be equal to $m = \sum_{n=1}^{n_{\max}} N_n$. The dimension of such a fractal can be obtained from the dependence $m(n_{\max})$. In the real DLA process we get $m(n_{\max}) \propto n_{\max}^{D_f}$, where D_f is the fractal dimension. In the double logarithmic scale this dependence is a straight line with the slope coefficient D_f. Additionally, we can define the local dimension as a function of $n_{\max}$,

$$D_f(n_{\max}) = \frac{d \log m(n_{\max})}{d \log n_{\max}}. \tag{5}$$

The constructed model can be easily generalized for a 3D fractal. In this case the jump probabilities are equal to

$$p_n^k = 1 + k\frac{d}{r_n^2}, \quad k = -1,0,1 .$$

The maximal number of particles in a layer can be estimated as follows: $M_n = \lfloor \pi r_n^2 / d^2 \rfloor$. The adsorption probability (see eq.(1)) in the cellular model is the same, but in the case of an arbitrary distribution of particles there is no exact expression. However, we can construct many approximations. At the same time, expressions (3) and (4) are the same as for the 3D fractal.

Now let us consider a rotating fractal. As it follows from direct numerical simulations (see below), in this case one or two growing branches of the fractal predominate, and particles are distributed within them in a quite compact way. So, the assumption that particles are equidistributed within a layer is not correct. Therefore, we should consider the cellular model.

Let us denote a frequency of the fractal rotation by ω. We propose the following model: rotation is a continuous process, but particles can make their jumps at a time interval τ. Between the jumps the particle is at rest and it can be adsorbed at any time. Let us pass to a rotational frame of reference. Then during the time τ, the particle is shifted to an angle $\varphi = \omega\tau$. In this case we have the effective particle size:

$$d_{ef} = d + \varphi r_n = d + \omega\tau r_n .$$

In addition, suppose that the layer already contains N_n particles. Then the adsorption probability is

$$\tilde{p}_n = \frac{N_n + \omega\tau r_n / d}{M_n} .$$

The other expressions remain the same.

3 Numerical Investigations of the Model

The model described above has been numerically investigated in two following cases: in 2D case as a cellular model of packing particles in layers with and without rotation, and in 3D case. In each case the logarithmic dependence of the total mass of particles inside a ring of the radius r on the value of r was obtained. The fractal dimensions were calculated by means of a linear approximation (as tangent of the slope angle). In our numerical simulations we have supposed that the diameter of particles is $d = 1$, the jump interval from one layer to another is $\tau = 1$, and the mass of a particle is equal to unity. So, the ring radius is $r_n = n$. The fractal dimension as a function of the fractal size was calculated by means of the numerical differentiation of curves shown in Fig.1 (see Fig.2

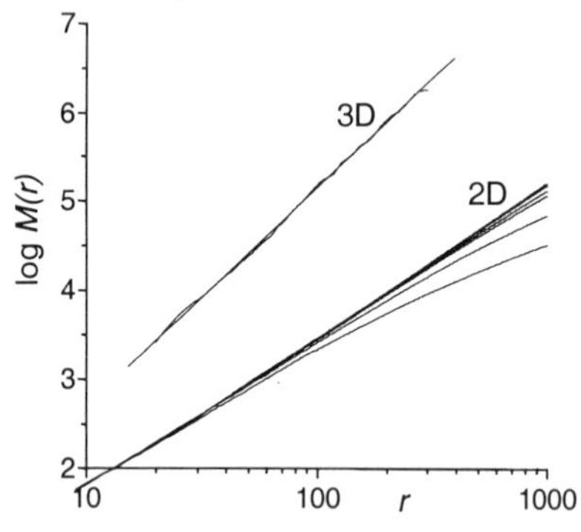

Fig. 1. The mass of the fractal as its function of the radius for 2D and 3D spaces.

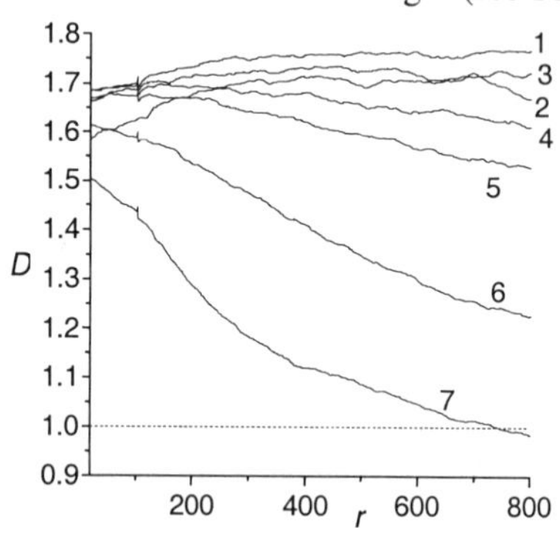

Fig. 2. Dependence of the fractal dimension on the fractal radius for different rotating frequencies.

and eq.(5)).

Fig.1 shows the dependence of logarithm of fractal mass as a function of radius for 2D and 3D fractals. In the 2D case the set of graphs correspond to different rotation frequencies (the values are shown below). The smallest slope of the graph corresponds to the maximal frequency. We have obtained 100 fractal clusters for every rotation frequency, and their mean value has been calculated. The visual distinction in graphs is not so radical, however the regression analysis allows us to find some differences. The fractal dimensions calculated by means of the linear approximation are shown in the following Table 2.

Table 2. The dependence of the fractal dimension on the rotating frequency.The numbers in the first line correspond to different curves shown in Fig.2.

	1	2	3	4	5	6	7
$\omega\cdot 10^4$	0	0.3	1	3	5	15	45
D_f	1.727 ± 0.002	1.722 ± 0.002	1.709 ± 0.002	1.675 ± 0.002	1.644 ± 0.002	1.504 ± 0.003	1.303 ± 0.005

As follows from numerical analysis, the fractal dimension decreases when the frequency growths. The results of numerical differentiation of graphs in Fig.1 are shown in Fig.2. This investigation is more detailed. The slope of curves (dimension) is varied in some limits. In the case of the rotating fractals the dimension tends to unity when the radius is growing. The numbers in Fig. 2 correspond to columns in Table 2. They mark curves in accordance with different frequencies. For example, number 1 corresponds to a fractal without rotation.

In the 3D case the linear approximation of graphs (see Fig.1) gives the following value of fractal dimension: $D_f = 2.471 \pm 0.005$. This result is in a quite good agreement with the direct numerical simulation.

Thus, it follows from the obtained results that for the fractals without rotation the statistical approach to the fractal growth is in agreement with the results of direct numerical simulations of DLA. Obviously, the differences in the dimension obtained in the case of *rotating* fractals are the results of different methods of the calculation. As in the case of any statistical approach, the proposed model of fractal growth allows us to get only *global* properties of the system, while the use of the sliding windows method supposes knowledge of information about *local* properties. Substantively, the method of sliding windows allows us to estimate the dimension of a fractal branch, whereas for the statistical approach there is no notion of the fractal branch, and the dimension is calculated for the whole fractal set. Moreover, application of the statistical approach assumes implicitly an isotropic particle distribution. Evidently, the appearance of the anisotropy in the rotating fractals (i.e. in the case of prevalence of some branch) can explain the increase of the fractal dimension.

The interesting and not yet investigated questions are the following: at what fractal radius r_{cr} a certain fractal branch begins to dominate? It is obvious that $r_{cr} = r(\omega)$. What is the growing rate of a fractal and what is the law of twisting of fractal branches?

4 Direct Computer Simulation of Rotating DLA-Clusters

We consider the DLA-fractal growth in the two-dimensional space on the square lattice. The size of a simulation box was equal to 4000×4000 lattice sites. A diffusing particle starts at the randomly chosen point on the circle of radius $1.5R_{cl}$ centered at the origin of the coordinate system where the cluster seed is placed. Here, R_{cl} is the maximal distance from the cluster origin to the occupied sites. The particle diffuses freely inside the square box with the side length of $2R_{cl}$ and with impenetrable walls until it is added to the growing cluster. This happens when the particle visits the lattice site which is one of the nearest neighbors to one of the occupied sites. At that moment a new diffusing particle begins its random walk. A snapshot of a usual DLA-cluster is presented in Fig. 3.

To model the rotation of the DLA cluster we have added a constant angular velocity to the diffusion motion of a particle. This means that it is necessary to add to a random displacement of the particle on the lattice also a constant displacement along the circle

Fig. 3. DLA cluster without rotation (M = 50000).

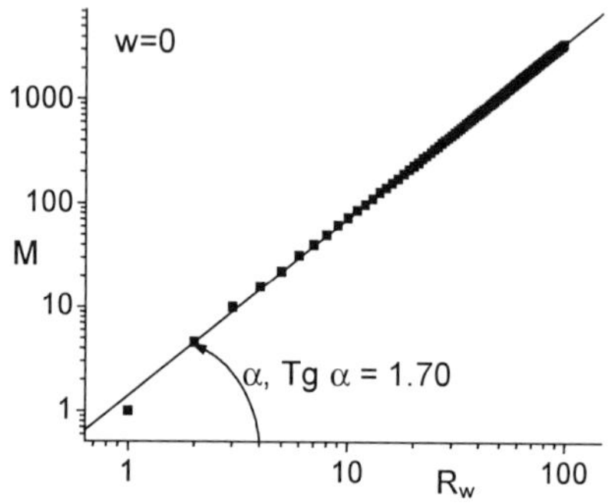

Fig. 4. Mass vs. Radius dependence for usual DLA cluster.

centered at the origin of the coordinate system and with the radius which is equal to the current distance of the diffusing particle from this origin. The particle has usually non-integer coordinates (x_f, y_f) which are transformed into the integer ones (x,y) by taking the integer part: $x = int(x_f)$, $y = int(y_f)$. The angular velocity ω was kept constant. The value $\omega=1$ corresponds to the rotation of the particle on the angle $d\varphi$ at each time step, where $d\varphi$ is chosen in such a way that the particle located at the maximal possible distance from the cluster origin could not jump over one lattice site. Each cluster grows until $R_{cl} < R_s$, where R_s is the maximal size of the lattice, $R_s=1000$, or until its mass is less than 50000 occupied sites. According to the maximal possible size of the cluster we have chosen $d\varphi=1/R_s=0.001$ *rad.* Our angular velocity ω is measured in radian per one Monte Carlo step. Our model is different from that one used by Lemke at al. in [7]. We use the original version of the Witten-Sander model on 2d lattice [3] while authors of [7] study the off-lattice realization of the ballistic version of the Witten-Sander model. The rotation of the cluster is also introduced in different way in [7] where the condition of linear and angular momentum conservation has been considered explicitly. Our model corresponds to the infinite initial mass situation in [7].

For each value of the angular velocity we sampled the statistics of 10 independent DLA-clusters. We have calculated the fractal dimension of DLA-aggregates by analyzing the cluster mass as a function of its radius $M \propto R^{D_f}$, where M is the number of occupied sites in the circle R, and D_f is the fractal dimension of DLA-cluster (unlike ref.[7] where the box-counting dimension was measured). This dependence is presented in Fig. 4 (log-log plot) for the case of the usual DLA without rotation. To improve the statistics, the method of sliding windows has been used. 100 points have been chosen randomly on the cluster. Each of them was treated as the center of a square with the side R_w from 1 to 100, and the number of occupied sites inside that square was calculated. The dependence $M(R_w)$ was averaged over these 100 different "windows" and over 10 independent clusters.

In Figs. 3–9 the DLA clusters for different angular velocities ω are shown. One can

Fig. 5. DLA cluster with rotation: $\omega = 0.03 \cdot 10^{-3}$, M = 50000.

Fig. 6. DLA cluster with rotation: $\omega = 0.1 \cdot 10^{-3}$, M = 50000.

Fig. 7. DLA cluster with rotation: $\omega = 0.5 \cdot 10^{-3}$, M = 50000.

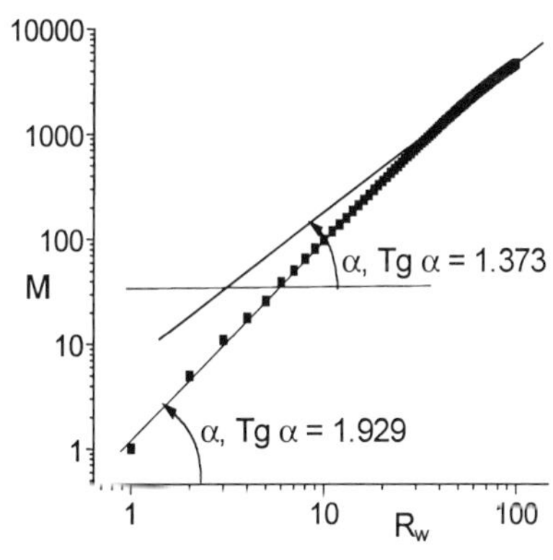

Fig. 8. Mass vs. Radius dependence for rotating DLA cluster. $\omega = 0.5 \cdot 10^{-3}$.

clearly see how clusters are changed upon increasing ω. At ω=0 the cluster is practically isotropic (Fig.3). First, upon increasing ω two arms become more pronounced, although many arms are still present (Fig.5). Then only one of these two arms begins to dominate, and finally, the cluster practically degenerates into a "tail" (Fig. 6). With the further increase in ω this "tail" becomes thinner (Fig. 7) and tends to a thin line (Fig. 9). It is possible to indicate several regimes in the cluster growth. If $R_{cl}<R_1(\omega)$, then several arms

are growing; after this, for $R_1<R_{cl}<R_2$, one arm starts to dominate, and finally, for large enough $R_{cl}>R_2$ this arm changes its curvature as follows from Fig.7. Possible physical reason for this behavior is the dependence of the tangential velocity of particles on the distance from the cluster seed: the larger cluster hits more faster particles in comparison to the smaller one. The spiral form appears to be slightly different here and in [7] for large

Fig. 9. DLA cluster with rotation: $\omega = 4.5\cdot10^{-3}$, M = 5000.

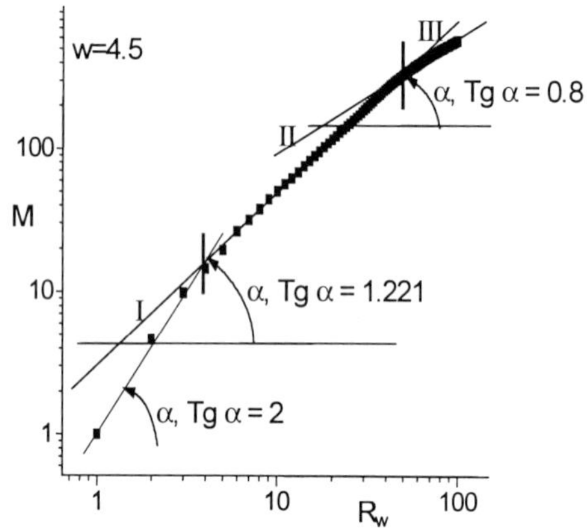

Fig. 10. Mass vs. Radius dependence for rotating DLA cluster: $\omega = 4.5\cdot10^{-3}$.

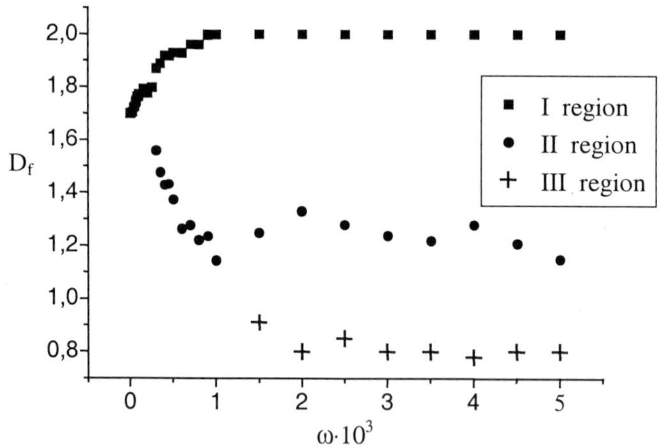

Fig. 11. Dependence of fractal dimension on rotation velocity.

angular velocities; this is probably due to the different models.

In Figs. 4, 8, 10 for different angular velocities of the DLA-cluster rotation from ω=0 to $4.5\cdot10^{-3}$ the "mass" of a DLA fractal as the function of the window size is presented (double logarithmic plots; the slope is equal to the fractal dimension D_f). Starting from some rotation velocity ($\omega \approx 0.3\cdot10^{-3}$) there is a clear bending point on the plot (Fig. 8; regions I and II). This can be explained by the fact that starting from this velocity the width of the "tail" becomes smaller than the maximal size of "sliding windows" which is equal to 100. Therefore by measuring the number of occupied sites, the empty space inside the window starts to play an important role. A position of the bending point indicates the width of the "tail". Upon increasing ω the slope of the line in the region I goes to 2, and the slope in the region II tends to unity. The cluster becomes more dense on small scale and degenerates into the thin stripe or even into a line. The bending point is shifted to the region of small radii (the width of the stripe is decreasing). At $\omega > 1.5 \cdot 10^{-3}$, one can find even three regions on the plot (Fig. 10; regions I, II and III). In Fig.11 the dependencies of the slopes in the regions I, II and III on the rotation velocity are presented. It turns out that the slope in the region III is smaller than unity. This is, of course, an artefact and it can be explained only as a finite size effect. Namely, if the size of the window is large enough, and if the center of the window is placed close to one of the ends of the thin spiral line, this window includes the cluster sites only "on one side" from its center another side being completely "empty". In other words, there is no line anymore in the window but only half a line, this leading to the fractal dimensionality less than the unity. We are studying now the finite size effects more attentively.

5 Concluding Remarks

As our main result, we report the fractal dimension D_f as a function of the angular velocity for two-dimensional DLA-cluster under rotation. This dependence has been obtained both by means of a new statistical annular model of the fractal growth and by means of direct computer simulations. The fractal dimension decreases when the rotation velocity is growing. We have found the clear evidence of a transition between fractal (at low rotation velocities) and non-fractal (at high velocities) regimes. The results of our simulation part correspond rather well to those obtained in [7] for infinite initial mass case. We did not perform here a detailed analysis of number of spiral arms, which was done very nicely in [7]. Instead, we concentrated in the present paper on studying the difference in fractal behavior on different scales what was not done in [7]. The DLA-cluster under rotation shows different fractal dimensions when analyzed on different length scales. To analyze the multifractal properties of DLA-cluster in more detail the growth-site probability density should be calculated. This is what we intend to do in the nearest future.

6 Acknowledgements

We would like to thank Prof. S. Nechaev for useful discussions and Prof. J.J.Arenzon for making us aware of Ref.7.

References

1. P. Meakin, In: *Phase Transitions and Critical Phenomena*, vol.12, C.Domb and J.L. Lebowitz (Eds.), Academic Press, 1988, p.335.
2. J.-F. Gouyet, *Physics and Fractal Structures*, Springer-Verlag Berlin, Masson, Paris, 1996.

3. T.A. Witten, J.M. Sander, *Phys. Rev. Lett.* 47, 1400 (1981), *Phys. Rev. A* 29, 2017 (1984).
4. E. Brener, H. Levine, Y.Tu, *Phys. Rev. Lett.* 66, 1978 (1991).
5. T.C. Halsey, M. Leibig, *Phys. Rev. A* 46, 7793 (1992).
6. P. Meakin, *Phys. Rev. A* 27, 1495 (1983).
7. N.Lemke, M.G.Malcum, R.M.C.de Almeida, P.M.Mors, and J.R.Iglesias, *Phys.Rev.E* 47, 3218 (1993)

THE PRESENCE OF THE SELF-SIMILARITY IN ARCHITECTURE: SOME EXAMPLES

NICOLETTA SALA
Academy of Architecture of Mendrisio, University of Italian Switzerland
Largo Bernasconi CH- 6850 Mendrisio
Switzerland
E-mail: nsala @ arch.unisi.ch

Benoit Mandelbrot showed how Fractals can occur in many different places in both Mathematics and elsewhere in Nature. A fractal object is self – similar if it has undergone a transformation whereby the dimensions of the structure were all modified by the same scaling factor. The new shape may be smaller, larger, translated, and/or rotated, but its shape remains similar. "Similar" means that the relative proportions of the shapes' sides and internal angles remain the same. Our sense, having evolved in nature's self-similar cascade, appreciate self-similar in designed objects. It is possible to demonstrate which the fractal shapes have known to artists and architects for centuries.
The aim of this paper is to present some examples of self-similarity in architectural and design projects. We will refer of the building's self-similarity in different cultures (e.g., Western and Oriental culture) and in different periods (e.g. in the Middle Ages until today).
We have organized our study using two different approaches: the unconscious building's self-similarity (as a result of an aesthetics sense), and the conscious building's self- similarity (as a result of a specific and conscious act of design).

1 Introduction

Fractal objects are irregular in shape, and they are generally self-similar and independent of scale. There are several mathematical structures that are fractals; for instance the Koch snowflake [26], the Julia set [21], and the Lorenz attractor [7]. For many centuries architecture has followed the Euclidean geometry and Euclidean shapes (bricks, boards) it is no surprise that buildings have Euclidean aspects. On the other hand, some architectural styles are informed by Nature, and much of Nature is manifestly fractal. So perhaps we should not be so surprised to find fractal architecture. As we shall see, fractals appear in architecture for reasons other than mimicking patterns in Nature. Our fractal analysis in architecture can be divided in two stages:

- on a little scale analysis (e. g, to analyse the single building shape);
- on a large scale analysis (e.g., to study the urban growth) [1, 2, 22, 24].

In the little scale analysis we can observe:

- the box-counting dimension (to determine the fractal dimension of a design to use this parameter as a critical tool) [4];
- the building's self-similarity (e.g., a building's component which repeats itself in different scales) [24].

In this paper we shall present some examples of the building self similarity.

2 Building's self-similarity

We can classify the building's self-similarity in two different ways: unconscious, when the fractal quality has been unintentional chosen for an aesthetic sense, and conscious, when the fractal quality is in every case the result of a specific and conscious act of design [24].

2.1 Unconscious building's self-similarity

Unconscious building's self-similarity is present in different cultures and different styles. For example, Indian and Southeast Asian temples and monuments exhibit a fractal structure. In fact, the towers are surrounded by smaller towers, surrounded by still smaller towers, and so on, for eight or more levels. In these cases the proliferation of towers represents various aspects of the Hindu pantheon [16]. Figure 1 represents a fractal Indian temple. Figure 2 shows an example of moghul art: the Taj Mahal (India) (1632 – 1648). Its name is the deformation of "Muntaz Mahal" which means "the favorite of the Harem". In this funeral mosque the self-similarity is present in the arches' shape repeated in four different scales (see figure 2). Humayun's Mausoleum at Dehli (India) (moghul art, 1557 – 1565) presents a descending fractal structure [10].

Figure 1. An Indian fractal (Tiruphati).

Figure 2. The Taj Mahal temple.

There are other examples of self-similarity in Oriental architecture. Figure 3a illustrates the Kaiyuan Si Pagoda, Chinese architecture (Song Dynasty, 1228 – 1250, Quanzhou,

Fuqian). Observing Figure 3b, which represents Kaiyuan Si Pagoda's plan, we can note the self similarity in the octagonal shape [18].

a) b)

Figure 3. Kaiyuan Si Pagoda, we can note the self-similarity in the shape (a) and in the plan (b).

Other example of Chinese fractal architecture is present in the Jinzan Monk Tomb (746, Song Shan Mountains, Henan) illustrated in the Figure 4a. We can note, another time, the presence of an octagonal fractal shape in the plan (Figure 4b).

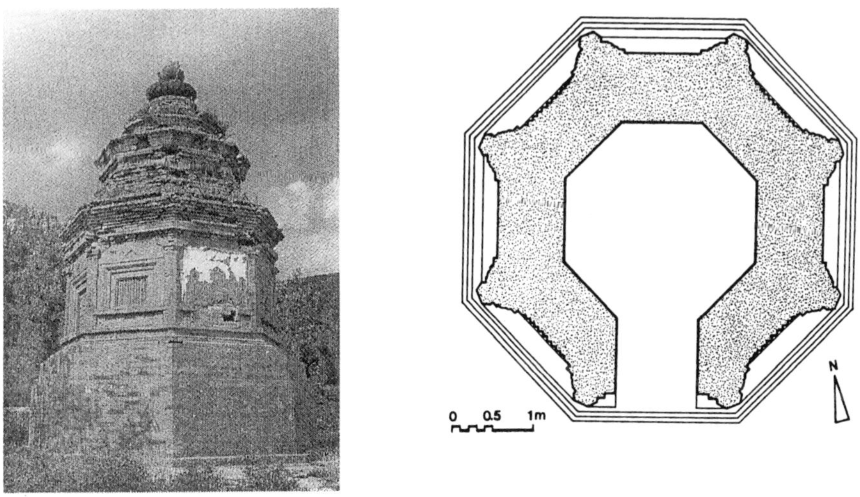

a) b)

Figure 4 Jinzan Monk Tomb, the self-similarity is present in the shape (a) and in the plan (b).

We can also find the presence of fractal geometry and the self-similarity in the Sacred Stupa Pha That Luang - Vientane (Laos), where the basic shape is repeated in different scales (see Figure 5) [24], and in the Royal Palace of Mandalay (Burma) (Figure 6).

Figure 5. The Sacred Stupa (Vientiane, Laos).

Figure 6 Royal Palace (Burma) an example of fractal architecture.

In the Western architecture we can find the oldest handmade fractal object in the cathedral of Anagni (Italy) [24]. Inside the cathedral, built in the year 1104, there is a floor, illustrated in Figure 7a, which is adorned with dozens of mosaics, each in the form of a Sierpinski gasket fractal (shown in the Figure 7 b).

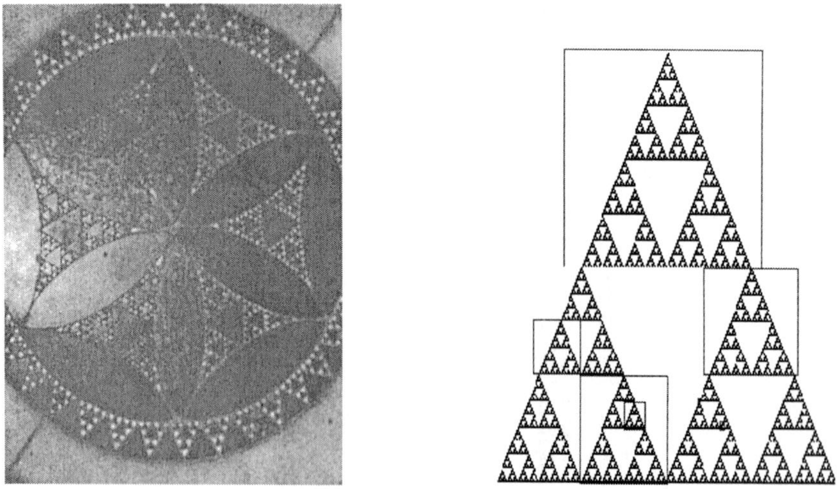

a) b)

Figure 7. The floor of the Cathedral of Anagni (a) and the Sierpinski gasket (b).

The intricate decoration of Renaissance and Baroque architecture, especially as expressed in cathedrals, frequently exhibited scaling over several levels.

The art historian George Hersey of Yale University points out fractal characteristics in Bramante's 1506 plan for the new St. Peter's "*Handbooks usually describe this design as a Greek cross with domed crossing and symmetrically placed subsidiary domes. We will briefly quote myself: Symmetrically clustered within the inside corners formed by the cross's arms are four miniature Greek crosses, that, together, make up the basic cube of the church's body. The arms of these smaller crosses consist of further miniatures. And their corners, in turn, are filled in with smaller chapels and niches. In other words, Bramante's plan may be called fractal: it repeats like units at different scales.*" [15].
Figure 8 shows the Bramante's 1506 plan for the new St. Peter.
Figure 9 illustrates the da Vinci (1452 – 1519) plan for a domed cathedral with four levels of domes.

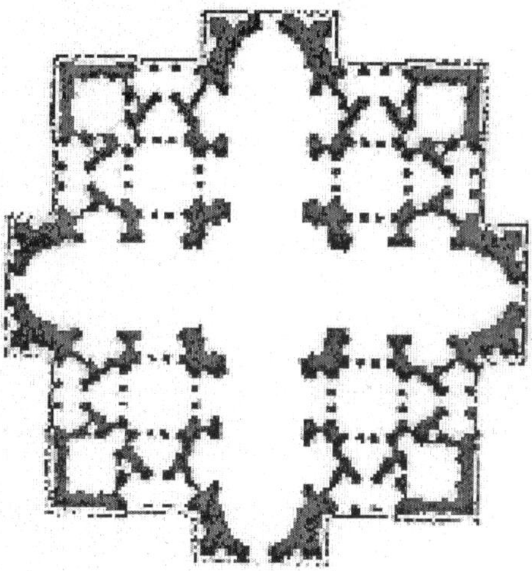

Figure 8. Bramante's 1506 plan for the new St. Peter.

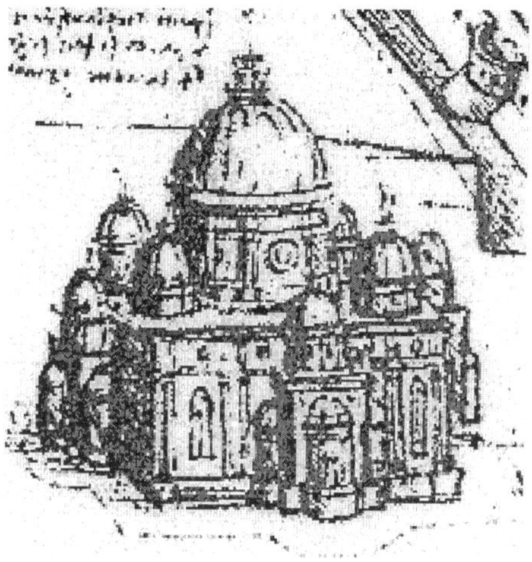

Figure 9. A da Vinci's plan for a cathedral with four levels of domes.

The fractal geometry is present in the "Castel del Monte" (Andria, Apulia, Southern Italy) built by the Holy Roman Emperor Friederich II of Hohenstaufen (1194 - 1250) in the last decade of his life. We can find a self-similar octagon in the plan (in analogy with the Figure 3b and 4b). In fact, the outer shape is an octagon, as is the inner courtyard. Even the eight small towers have octagonal symmetry. It is interesting to note which is possible to find a connection between the Castel del Monte's shape and the Mandelbrot set (Figure 10b). Castel del Monte has other interesting implications with geometry, in fact the planimetric aerial photo (Figure 10a) shows that the tangents of the octagon forming the inner courtyard intersect at the centers of the octagonal corner towers. This involves a geometric relationship between the towers and the inner courtyard, established by the similarity describe in a Gotze's work [12]. There is also the presence of the golden ratio [20, 24] .

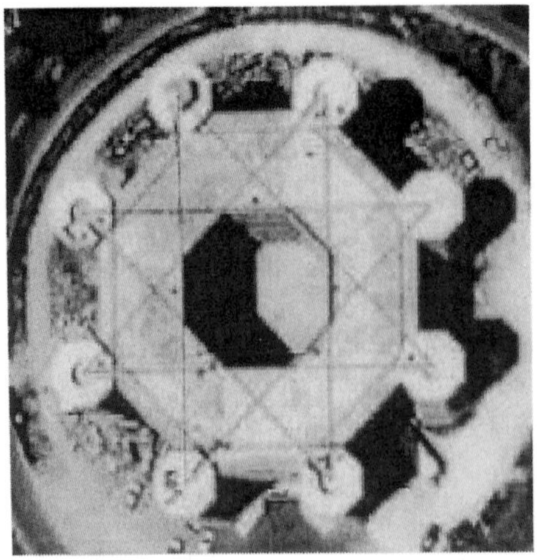

a) b)

Figure 10 The connection between Castel del Monte (a) and a Mandelbrot set (b).

2.2 Conscious building's self-similarity

The conscious building's self-similarity is a recent discovery by the twentieth century architects as a result of a specific and conscious act of design.

For example, Frank Lloyd Wright (1867-1959), in his late work ("Palmer house" in Ann Arbor, Michigan, 1950-1951) has used some self-similar equilateral triangles in the plan (Figure 11). A kind of "nesting" of fractal forms can be observed at two point in the Palmer house: the entry way and the fireplace. At these places one encounters not only actual triangles but also implied (truncated) triangles. At the entrance there are not only the triangles composing the ceramic ornament, there is also triangular light fixture atop of triangular pier. There is a triangle jutting forward overhead. The fireplace hearth is a triangular cavity enclosed between triangular piers. The concrete slab in which the grate rests is a triangle. The hassocks are truncated triangles [9].

Remembering the definition of the fractal as "*a geometrical figure in which an identical motif repeats itself on an ever diminishing scale*", the Palmer house is an excellent illustration of this concept.

It is interesting to note that the Wright's approach in the Palmer House, using some self-similar triangle, is in agreement with a viewpoint of an other important ninetieth century architect: Eugène Viollet-le-Duc (1814-1879). Viollet-le-Duc has built a methodological approach to the architecture using the triangle as a universal shape [14], but Viollet-le-Duc's self-similarity is unconscious and it is present because he was a perceptive observer of the Natural phenomena (for example the creation of the alpine glacier). Figure 12 shows a Viollet-le-Duc's draw dedicated to the geometric analysis of the old hospital of Compiègne (13rd century), we can note the self-similar triangle representation in the draw [24].

Other Wright's example of fractal architecture are the Robie house and the Marion County Civic Center, San Rafael (1957) where the self similarity is present in the external arches.

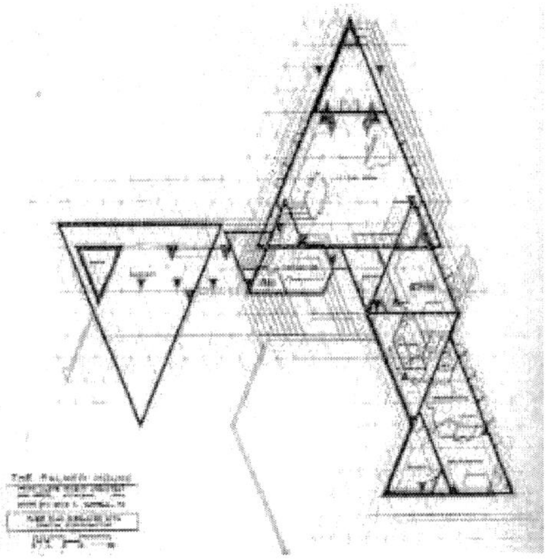

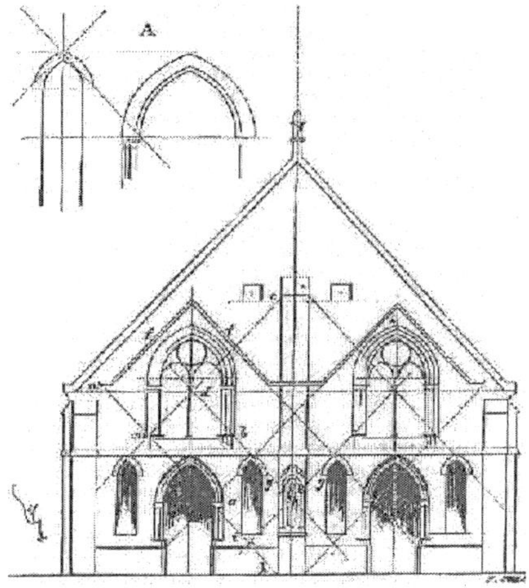

Figure 11 The plan of Palmer house.

Figure 12 Viollet-le-Duc's draw.

Italian architect Paolo Portoghesi has used the conscious self-similarity to realize the Chamber of Deputies (Rome, Italy, 1967), shown in Figure 13, where we can observe some self-similar spirals [23].

The group ARM (Ashton Raggatt McDougall), a Melbourne office, has realised the Storey Hall in Melbourne (Figure 14). The building, on one of the main downtown streets of Melbourne, is part of the highly urbanised campus of RMIT. It has fractals forms of windows, doorway, bronze surface and metal cornice which are based on the aperiodic tiling pattern derived by Roger Penrose [17].

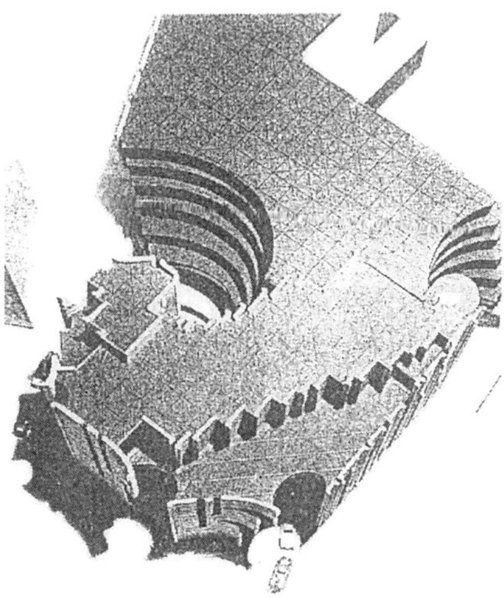

Figure 13 Portoghesi's Chamber of deputies (Rome, Italy)

Figure 14 ARM's Storey Hall (Melbourne, Australia)

We can find other interesting example of conscious self-similarity in the Malevich's buildings.

Kazimir Malevich (1878 – 1935) was an important figure in Russian and Soviet art and architecture in the early 20th century. Largely self-educated, Malevich was from the beginning of his artistic career, "*not concerned with nature or analyzing visual impressions, but with man and his relation to the cosmos.*" [12, p. 145]. Much of his work belongs to the Suprematist school. During the 20s, he began expressing architectural projects as 3-dimensional sculptures. Some examples of this Arkhitektonics are marvellous instances of fractals in architecture. Malevich creates buildings with ambiguous scales, erasing the difference in scale between buildings and people. This is achieved by surrounding the largest component of a building with a cascade of smaller and smaller copies, number and scale governed by an approximate 1/f relation (see Figure 15).

Figure 15. A Malevich's fractal building.

The Israeli architect Zvi Hecker, in the project of the Heinz-Galinski School (Berlin, 1993- 1995), has used spiral sunflower geometry (anticlockwise) plus concentric curves and self-similar curves and fish-shapes (see Figures 16a, and 16b) [17].

This small institution, built in the leafy suburbs of Charlottenberg, for 420 pupils is the first Jewish school to be constructed in Germany for 60 years. The Heinz-Galinski School creates a landform out of explicit metaphors. Mountain stairways, snake corridors, fish-shaped rooms are pulled together with an overall sunflower geometry.

The Heinz-Galinski School has to play an ethnic role in Berlin which is similar to Libeskind's Jewish Extension to the Berlin Museum: it must fit in and yet be unmistakably other. Hecker has underlined the paradox of *"a wild project"* that has *"very precise mathematical construction...Above all there is its cosmic relationship of spiral orbits, intersecting one another along precise mathematical trajectories "* [17, p. 25].

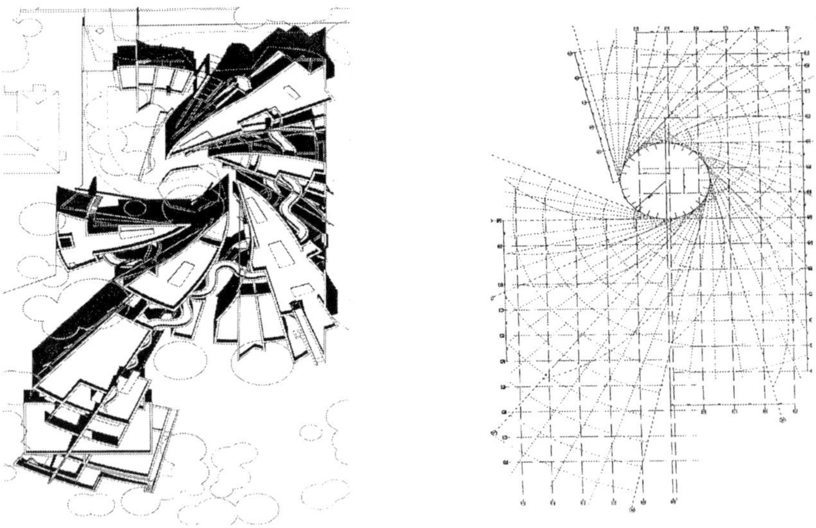

a) b)

Figure 16 Heinz-Galinski School, Berlin.

3 Conclusions

Fractal geometry and its connection between Chaos theory can help to introduce the new complexity paradigm in architecture. Fractal distributions can be used to generate complex rhythms for use in design. As an example, the fractal dimension of a mountain ridge behind an architectural project could be measured and used to guide the fractal rhythms of the project design [4]. The project design and the site background would then have similar rhythmic characteristics. In both criticism and design, fractal geometry provides a quantifiable calibration tool for the mixture of order and surprise.

This paper has presented only some particular aspect of the self-similarity in the buildings, but we can also apply the fractal geometry in the urban growth [1, 2]. In fact, the skeletal structure of the industrial city is tree-like with radial street systems converging on the historic core. When these tree-like structure are embedded into urban development we begin to see typical patterns emerge which are clearly fractal [3]. Fractal cities presented in Batty and Longley's book are generated by means of cellular automata models and interpret them in the context of self-organised critically [22, 27].

We are sure that the fractal geometry is helping to newly define a new architectural models where the broken symmetry, the self-similarity and the Chaos theory play a central role [5, 6].

References

1. Batty M., Cities as Fractals: Simulating Growth and Form. In Crilly A. J., Earnshaw R. A. and Jones H. *Fractals and Chaos* (Springer - Verlag, New York 1991), pp. 43 – 69.
2. Batty M. and Longley P., *Fractal Cities* (London Academic Press, London, 1994).

3. Batty M., Longley P., The Fractal City. *Architectural Design, New Science = New Architecture,* Academy Group, London, n. 129 (1997) pp. 74 – 83.
4. Bovill C., *Fractal Geometry in Architecture and Design* (Birkhäuser, Boston, 1995).
5. Briggs J., *Fractals The Patterns of Chaos* (Thames and Hudson, London, 1992).
6. Briggs J., *Estetica del caos* (Red Edizioni, Como 1993).
7. Crilly J. and Earnshaw R. A., Jones H., *Fractals and Chaos.* (Springer - Verlag, New York, 1991) pp. 43 - 69.
8. Donato F. and Lucchi Basili L., *L'ordine nascosto dell'organizzazione urbana.* (Franco Angeli Editore, Milano 1996).
9. Eaton L. K., Fractal Geometry in the Late Work of Frank Llyod Wright: the Palmer House. In Williams K. (edited by*), Nexus II: Architecture and Mathematics,* (Edizioni Dell'Erba, Fucecchio, 1998) pp. 23 - 38.
10. Fivaz R., *L'ordre et la volupté* (Press Polytechniques Romandes, Lausanne, 1988).
11. Gotze H., Friedrich II and the Love of Geometry. In Williams K. (edited by*), Nexus: Architecture and Mathematics*, (Edizioni Dell'Erba, Fucecchio, 1996) pp. 67 - 80
12. Gotze H., Die Baugeometrie von Castel del Monte, *Sitzungsberichte del Heidelberger Akademie der Wissenscahften, Phil. Hist. Klasse*, Jahrg. 1991, Bericht 4.
13. Gray C., *The Russian Experiment in Art 1863 – 1922* (Thames and Hudson, London, 1962).
14. Gubler, J. La metafora della montagna nell'architettura di Viollet-le-Duc. *Alpi Gotiche – L'alta montagna sfondo del revival medievale,* (Edizione Museo Nazionale della Montagna "Duca deli Abruzzi" Club Alpino Italiano – Sezione di Torino, 1998) pp. 53-54.
15. Hersey G., *The Monumental Impulse* (MIT Press 1993).
16. JacksonW. , Other shore fractals: Hindu trascendence symbols and the modelling of wholeness.
17. JencksC. , Landform Architecture Emergent in the Nineties. *Architectural Design, New Science = New Architecture,* Academy Group, London, n. 129, (1997) pp. 15 – 31.
18. Liu L. G., *Chinese Architecture* (Academy Press, London, 1989).
19. Mandelbrot B., *The Fractal Geometry of Nature* (W.H. Freeman, New York 1982).
20. Patruno L., Castel del Monte il mistero di Federico. *Specchio n. 90*, (1997) pp. 132 - 149.
21. Peitgen H-O. and Richter P.H., *The Beauty of Fractals: Images of Complex Dynamical Systems* (Springer-Verlag, Berlin, 1986).
22. Portugali J., *Self-Organization and the City* (Springer, Berlin, 1999).
23. Prioni G. , *Paolo Portoghesi, opere e progetti* (Electa, Milano, 1996).
24. Sala N., Fractal Models In Architecture: A Case Of Study *Proceedings International Conference on "Mathematics for Living"* Amman, Jordan, November 18-23, (2000), pp. 266 – 272.
25. Viollet-le-Duc E., *L'architettura ragionata,* (Jaka Book, Milano, 1982).
26. von Koch H., Sur une courbe continue sans tangente, obtenue par une construction géometrique élémentaire. *Arkiv för Matematik* 1, (1904) pp. 681-704.
27. Wolfram, S., Universality and complexity in cellular automata. *Physica D* 10, (1984), pp. 1-35.

AN APPROACH TO RAY TRACING AFFINE IFS FRACTALS

TOMEK MARTYN

Computer Graphics Laboratory, Institute of Computer Science, Warsaw University of Technology, ul. Nowowiejska 15/19, 00-665 Warsaw, Poland

E-mail: martyn@ii.pw.edu.pl

In this paper a method for ray tracing affine IFS fractals is presented. Like the previous approaches, the ray-fractal intersection algorithm presented here uses the object instancing principles. However, in contrast to the former ones, our method allows ray-tracing fractals at pixel size accuracy in an extremely efficient way. Moreover, our approach makes it possible to solve the spatial aliasing problem in a trivial and efficient way by appropriate corrections of sizes of instancing volumes. To take advantage of coherency and decrease computational redundancy, which is caused by searching for an intersection with each ray separately, the idea of tracing ray bundles is introduced.

Keywords: Fractals; IFS attractors; Ray tracing; Object instancing; Realistic image synthesis.

1 Introduction

The application of fractals to modeling and investigation of natural phenomena has a long history. Thanks to the observations by B. B. Mandelbrot [5], many people nowadays regard fractal geometry as a "geometry of nature". Anyway, one is certain – it provides a powerful tool for description of irregularity and complexity. Fractal geometry makes it possible to describe various shapes which very often resemble natural creations. It does so in a concise and natural way as classical geometry when the latter defines, for example, the shape of a sphere or a cone. In terms of computer graphics, the conciseness of description should imply low memory requirements for visualization of fractals. Moreover, fractal formulas specify objects with the infinite level of detail, i.e. each part of the object is made up of the infinite number of parts. These properties of fractals make them very useful for computer graphics, especially as models for the realistic image synthesis. However, the advantages of fractal descriptions have not often been completely exploited in ray tracing.

First of all, visualization of fractal objects has often been done on the basis of indirect models, which consist of a large number of Euclidean figures. The models were generated from fractal formulas in a preprocessing stage and then they were stored in memory during the whole time of visualization. Consequently, memory requirements were high and the conciseness of description got lost during the ray tracing process.

Secondly, in contrast to fractal descriptions, indirect models represent a finite level of detail. Thus, if the aim is visualization at pixel size accuracy then in the case of change of visualization parameters (for example, while generating animation frames), the models should be actualized in a preprocessing stage. In practice, such an approach usually involves the repetition of the model generation.

The aim of this paper is to demonstrate a method for ray tracing of fractals which are described by affine iterated function systems (IFSes); these fractals are usually called affine IFS attractors. The results of the previous works on the subject were algorithms exploiting the object instancing technique. Their main disadvantages are that they either required a lot of computation [3, 4] or they did not take into consideration visualization at pixel size accuracy [8]. All the methods were also sensitive to spatial-aliasing and limited to IFSes consisting of invertible affine mappings only (see Sec. 3.1). Like the previous ones, our method uses object instancing principles but, in contrast, it is general in the sense of mapping invertibility, efficient, aliasing-resistant and allows ray tracing affine IFS attractors at pixel size accuracy and low memory requirements.

2 Affine IFS Attractor Anatomy and the Object Instancing Technique

An *affine iterated function system* (IFS for short) on a normed space $(\mathbf{R}^3, \|\,.\,\|)$ is a finite set $\{w_i\}_{i=1}^N$ of N affine mappings $w_i : \mathbf{R}^3 \to \mathbf{R}^3$ that are contractions with respect to the norm $\|\,.\,\|$. Given an IFS, we can construct the so-called *Hutchinson operator* $W : H \to H$ which acts on the collection H of all compact and nonempty subsets of $\mathbf{R}^3$ in the following way:

$$W(B) = \bigcup_{i=1}^{N} w_i(B), \quad B \in H. \tag{1}$$

The operator W is a contractive transformation on H regarded as the metric space (H, h), where h is the Hausdorff metric induced by the metric $d(\mathbf{x},\mathbf{y}) = \|\mathbf{x}-\mathbf{y}\|$, $\mathbf{x},\mathbf{y} \in \mathbf{R}^3$ [1]. Moreover, (H, h) is a complete metric space. Thus, as a one of the consequences of the Banach fixed point theorem, we have that W possesses exactly one fixed point $A \in H$:

$$W(A) = \bigcup_{i=1}^{N} w_i(A) = A. \tag{2}$$

The set A is called the *attractor* of the IFS that the operator W is constructed with.

The above formula reveals an important feature of the IFS attractors, namely, they are self-tiled sets, i.e. every attractor consists of N subsets that are images of the attractor under the IFS mappings. Since the IFS mappings are contractive, the component subsets are smaller than the attractor, for

$$\begin{aligned} \operatorname{diam}(w_i(A)) &= \max\{\|w_i(\mathbf{x}) - w_i(\mathbf{y})\| : w_i(\mathbf{x}), w_i(\mathbf{y}) \in w_i(A)\} \\ &\le \lambda_i \max\{\|\mathbf{x}-\mathbf{y}\| : \mathbf{x},\mathbf{y} \in A\} = \lambda_i \operatorname{diam}(A), \end{aligned} \tag{3}$$

where $\lambda_i \in [0,1)$ denotes the contractivity factor of the mapping w_i. Like the whole attractor, each its subset $w_i(A)$ can be decomposed into N smaller affine copies $w_i \circ w_k(A)$, $k = 1, \ldots, N$, of the attractor, because

$$w_i(A) = w_i(W(A)) = w_i(\bigcup_{k=1}^{N} w_k(A)) = \bigcup_{k=1}^{N} w_i \circ w_k(A) \tag{4}$$

and

$$\operatorname{diam}(w_i \circ w_k(A)) \le \lambda_i \lambda_k \operatorname{diam}(A). \tag{5}$$

Proceeding in that way, we can decompose the attractor into smaller and smaller subsets that reach the attractor's points in the limit.

The self-affine geometry of the IFS attractors makes it possible to approximate these sets by means of the object instancing technique. Starting with the volume enclosing the attractor, we construct recurrently an approximating set, which includes the attractor, by exchanging each volume with the union of N smaller ones, until the diameter of every volume satisfies an approximation accuracy. The union of volumes that are substituted for a volume bounding the attractor's subset $w_i \circ \ldots \circ w_k(A)$, $i,\ldots, k = 1,\ldots,N$, is determined on the basis of the component affine mappings of the operator $W_{i\ldots k} : H \to H$ specified as

$$W_{i\ldots k}(.) = w_i \circ \ldots \circ w_k \circ W(.) = \bigcup_{l=1}^{N} w_i \circ \ldots \circ w_k \circ w_l(.) \tag{6}$$

The new bounding volumes are calculated either by the direct application of the operator $W_{i\ldots k}$ to generate N affine images of the initial volume that bounds the attractor [3, 8] or they can be determined as volumes surrounding those strict affine images, by a smart analysing of the operators component mappings $w_i \circ \ldots \circ w_k \circ w_l$, $l = 1,\ldots, N$. The approach presented in the next section can be classified to that latter category.

3 The ray-attractor intersection problem

A ray-attractor intersection is determined by exploiting the object instancing method as a routine for constructing a hierarchy of bounding volumes. A ray hits the attractor at a point represented by a terminal node of the hierarchy if the ray pierces every ancestor of that node. Obviously, explicit storing the hierarchy tree is highly memory-consumptive, because the number of volumes depends exponentially on approximation accuracy. In order to guarantee linear memory requirements, a sensible solution is to create and remove volumes of single paths of the tree while searching for an intersection proceeds. A volume is created if its direct ancestor is too "big" (in the sense of approximation accuracy) and intersected with the ray. A volume is removed if its direct descendants have already been removed or they will not be created.

The key factor that influences the computational cost of the above process is the geometry of applied bounding volumes. If approximation accuracy is identified with that specified by pixel sizes, then the shape of bounding volumes used does not affect the final result of the attractor's visualization. In this case the choice of the type of bounding volumes should be done taking into account the computational costs related to

- the calculation of a ray-volume intersection,
- the procedure of determining descendant volumes in the object instancing process,
- the determination of a volume extent to check whether it is of pixel size and can be regarded as a primitive for shading.

3.1 Previous work

In [4] balls were used as bounding volumes. However, despite a relatively cheap computational cost of the ball-ray intersection procedure [2], in order to determine new instances, the eigenvalues of symmetric matrices had to be calculated. This, in turn, involves the use of an expensive iterative method of *Jacobi Transformations* [7] in each step of the exchanging process. As a result, such a ray-attractor intersection routine required a lot of computation (see Sec. 3.2.1), which made the whole visualization process low efficient.

The same drawback appeared in [3], where, instead of balls, bounding ellipsoids were used. This time, the Jacobi Transformations method was applied to test if a current ellipsoid was of a pixel size (and could be identified with an attractor's point) or it had to be replaced with other "smaller" ones during the object instancing process.

The method presented in [8] supplies a little bit different approach to exploiting the object instancing idea. Rather than testing the intersections of a ray with affine images of the initial bounding volume, the intersections of affine images of the ray with the volume were determined. In the case of IFS attractors, a ray was transformed with inverted IFS mappings, and to enclose an attractor axis-aligned bounding box was used. Thus, from the point of view of the previous methods, such an approach can be considered as a variant of that by Hart et al. [3], in which instead of ellipsoids, bounding parallelepipeds are used. Although the idea of transforming rays makes ray tracing IFS attractors potentially very efficient, however it has at least two shortcomings. First of all, the authors did not take into consideration visualization at pixel size accuracy, which, if they did so, would make such an approach more complicated and much less efficient [6]. Secondly, the method is spatial aliasing sensitive, since the direct application of contractive mappings to determine bounding parallelepipeds can result, in general case, in arbitrary small (or thin) volumes; too small to detect their intersections with scanning rays. Another disadvantage of that approach can be seen in the restriction to invertible affine mappings only (for example the original definition of Barnsley's fern uses a non-invertible mapping to encode the fern's stem).

3.2 *New Approach*

In view of the shortcomings of the previous methods, a new approach to object instancing for ray tracing IFS attractors is demonstrated below. The general idea used in our method is similar to that the previous works were based on. However, our approach using in a smart way axis-aligned boxes as instancing volumes avoids the necessity of eigenvalues calculation, which makes the ray-attractor intersection routine extremely efficient. Moreover, the method is aliasing-resistant and not restricted to invertible IFS mappings only.

3.2.1 Axis-aligned bounding box determination

Let C be the axis-aligned cube with edges of the unit length and centred at the origin of the coordinate system in the real space $\mathbf{R}^3$ equipped with the Euclidean norm. The vertices of the cube are the vectors of coordinates $\mathbf{c}_k = [c_{1k}, c_{2k}, c_{3k}]^{\mathrm{T}}$, $c_{ik} = \pm\frac{1}{2}$, $k = 1,\ldots, 8$, so the diameter of C is equal to $\sqrt{3}$. Since any affine mapping $w(\mathbf{x}) = \mathbf{A}\mathbf{x} + \mathbf{t}$, $\mathbf{A} = [a_{ij}]_{i,j=1,2,3}$, $\mathbf{t} = [t_1, t_2, t_3]^{\mathrm{T}}$, preserves the relation of parallelism, the image set $w(C)$ is, in general, a parallelepiped centred at $\mathbf{t}$ and spanned on the vectors $w(\mathbf{c}_k) = [p_{1k}, p_{2k}, p_{3k}]^{\mathrm{T}}$, $k = 1,\ldots, 8$, where

$$p_{ik} = t_i + \sum_{j=1,2,3} a_{ij} c_{jk}, \qquad i = 1, 2, 3. \tag{7}$$

The coordinates of the vectors $\mathbf{c}_k$ represent all triplets of the elements from $\{-\frac{1}{2}, \frac{1}{2}\}$; thus the maximum and respectively minimum of the ith coordinates of the vertices $w(\mathbf{c}_k)$ satisfy

$$\max_{k=1,\ldots,8} \{p_{ik}\} = t_i + \frac{1}{2} \sum_{j=1,2,3} |a_{ij}| \tag{8.a}$$

$$\min_{k=1,\ldots,8} \{p_{ik}\} = t_i - \frac{1}{2} \sum_{j=1,2,3} |a_{ij}| . \tag{8.b}$$

Hence, we immediately have that the smallest axis-aligned box B_w, which bounds the parallelepiped $w(C)$ is the one centred at $\mathbf{t}$ and with the vertices of coordinates $\mathbf{b}_k = [b_{1k}, b_{2k}, b_{3k}]^{\mathrm{T}}$, where $b_{ik} = t_i \pm \frac{1}{2} \sum_{j=1,2,3} |a_{ij}|$. The diameter of the box holds

$$\mathrm{diam}(B_w) = \sqrt{\sum_{i=1,2,3} \left(\sum_{k=1,2,3} |a_{ik}| \right)^2} . \tag{9}$$

Since the value $\sum_{k=1,2,3} |a_{ik}|$ represents the extent of $w(C)$ with respect to the ith axis of the coordinate system, so $\mathrm{diam}(w(C)) \geq \max_{i=1,2,3} \{ \sum_{k=1,2,3} |a_{ik}| \}$. (Note that the right side is the value of the maximum norm of the matrix $\mathbf{A}$.) Hence, on the basis of (9), $\mathrm{diam}(B_w) \leq \sqrt{3\,\mathrm{diam}(w(C))^2}$ and as a consequence the inequality

$$\frac{1}{\sqrt{3}} \mathrm{diam}(B_w) \leq \mathrm{diam}(w(C)) \leq \mathrm{diam}(B_w) \tag{10}$$

is always satisfied.

If $A \subset \mathbf{R}^3$ is the attractor of an IFS $\{w_i\}_{i=1}^N$, then for any invertible mapping $g: \mathbf{R}^3 \to \mathbf{R}^3$ the set $g(A)$ is the attractor of the IFS $\{g \circ w_i \circ g^{-1}\}_{i=1}^N$, for $A = \bigcup_{i=1}^N w_i(A)$ implies $g(A) = \bigcup_{i=1}^N g \circ w_i \circ g^{-1}(g(A))$. Hence, if B_0 is an axis-aligned box that bounds the attractor A, and $g: \mathbf{R}^3 \to \mathbf{R}^3$ is the mapping that takes the cube C to B_0, then $C \supset g^{-1}(A)$. Consequently, if we transform the IFS $\{w_i\}_{i=1}^N$ with the mapping $g^{-1}: \mathbf{R}^3 \to \mathbf{R}^3$, obtaining the IFS $\{f_i\}_{i=1}^N$, $f_i = g^{-1} \circ w_i \circ g$, we can determine an appropriate axis-aligned box $B_{i_1 \ldots i_n}$ enclosing a given attractor subset $w_{i_1} \circ \ldots \circ w_{i_n}(A)$, $i_k \in \{1, \ldots, N\}$, on the basis of the composition mapping $g \circ f_{i_1} \circ \ldots \circ f_{i_n}$ in the way described above. The trasformation g can be specified as the affine mapping $g(\mathbf{x}) = \mathrm{diag}(l_1, l_2, l_3)\mathbf{x} + \mathbf{t}_0$, where l_i denotes the B_0 extent with respect to the ith axis of the coordinate system, and $\mathbf{t}_0$ is the center of the box. In turn, the box B_0 can be determined from a finite sequence of points generated by the Chaos Game with the starting point $\mathbf{x}_0$ being the fixed point of any of the IFS mappings, i.e. $\mathbf{x}_0 = w_i(\mathbf{x}_0)$.

The inequality (10) guarantees that the diameter $\mathrm{diam}(B_{i_1 \ldots i_n})$ is at most $\sqrt{3}$ times greater than the diameter of the image set $g \circ f_{i_1} \circ \ldots \circ f_{i_n}(C) = w_{i_1} \circ \ldots \circ w_{i_n}(B_0)$. Since the determination of a box takes place in each step of the object instancing process, our method provides, in the worst case, $\sqrt{3}$ times worse convergence to primitives than the ones using eigenvalues calculation. However, if we compare constant time, which is supplied by our approach for bounding box determination, with the computational cost of Jacobi Transformations, then the advantages of our method are clearly seen. Our approach requires only $12a + 3m$ operations to determine a bounding box, where a and respectively m denote the floating point addition/subtraction and multiplication/division respectively. In contrast, a single step of Jacobi Transformations for a 3×3 symmetric matrix requires $12a + 11m$ floating-point operations plus calculation of two square roots. In order to achieve results at accuracy necessary for the aim of visualization, on average from 4 to 14 algorithm's steps are required, and before they proceed, the 3×3 symmetric matrix $\mathbf{A}^{\mathrm{T}}\mathbf{A}$ must be determined. As a result, it would be rather difficult to find a fractal counterpart for which the previous approaches turned out to be more efficient than the one presented in this paper.

Another important consequence of the inequality (10) is that our approach can be used for ray tracing attractors of IFSes made of mappings that are contractive with respect to any norm on $\mathbf{R}^3$, (e.g. fractal interpolation functions are often described by IFSes including mappings that are contractions in a norm other than the Euclidean; see [1]). This follows from the fact that all the norms on $\mathbf{R}^n$ are equivalent, i.e. for any norms $\|\cdot\|_1$ and $\|\cdot\|_2$, there is a positive constant $c \in \mathbf{R}$, such that $\|\mathbf{x}\|_1 \le c\|\mathbf{x}\|_2$, $\forall \mathbf{x} \in \mathbf{R}^n$.

Additional advantages which make our method better than the former ones are that the IFS mappings do not have to be invertible, and the computation of an intersection of a ray with an axis-aligned box is more efficient than the one related to an ellipsoid or even a ball; see e.g. [2].

3.2.2 Pixel size accuracy and antialiasing

To visualize an attractor at pixel size accuracy we need to decide whether the extent of a bounding box exceeds the value implied by the size of a pixel or not. In other words, we have to know when the box can be identified with an attractor point and considered as a leaf of the hierarchy tree.

Let $R(t) = \mathbf{o} + t\mathbf{d}$ be a ray and S be the minimum ball enclosing a given box, where the ball's center is $\mathbf{c}$ and the radius is r. The ball S is a leaf (in the context of the ray) if

- $r \le \frac{1}{2}u\|\mathbf{c}-\mathbf{o}\|$, and $R(t)$ is an eye ray, where u is the extent of the pixel (intersected by R) at the unit distance from $\mathbf{o}$;
- $r \le \frac{1}{2}\|\mathbf{c}-\mathbf{o}\|\frac{s}{\|\mathbf{s}-\mathbf{o}\|}$, and $R(t)$ is a shadow ray cast from a point source of light, where s is the radius of a ball that can be treated as a leaf with the center at the shadowed point $\mathbf{s}$.

(For more details see [3].)

If a scanning ray does not intersect a bounding box it may be caused by individual lengths of respective edges of the box regardless of its diameter. Moreover neither the super-sampling nor the covers idea adapted to ray tracing IFS attractors in [3] are an appropriate solution to that spatial-aliasing problem, because each box's extent relevant to the coordinate system axis can be arbitrarily small.

The cube with the edges of the length $\sqrt{3}^{-1}p$ is the box of the greatest volume that can be enclosed by a ball of the diameter p; thus to avoid the problem, we claim the length of the edges not to be less than the value $\sqrt{3}^{-1}p$, where p is the pixel size at the box's center. Consequently, we calculate the length l_i, i = 1, 2, 3, of the box's edges parallel to the ith axis of the coordinate system as

$$l_i = \begin{cases} \sum\limits_{k=1,2,3} |a_{ik}|, \text{ if } \sum\limits_{k=1,2,3} |a_{ik}| > \sqrt{3}^{-1}p \\ \sqrt{3}^{-1}p, \text{ otherwise} \end{cases} . \tag{11}$$

If for every i = 1, 2, 3, $\sum_{k=1,2,3} |a_{ik}| \le \sqrt{3}^{-1}p$, then the diameter of the box determined according to the rule (11) is equal to the pixel size p, and the box is regarded as a primitive for shading.

4 Efficiency improvements

In order to decrease the computational redundancy related to the determination of mapping compositions and axis-aligned boxes, it is recommended to take advantage of coherency and searching the hierarchy tree with a bundle of rays in a single routine call, rather than with each ray separately. Such an intersection routine proceeds as a recurrent filter that allows only rays hitting the current box to pass a lower level of the hierarchy; then a sub-bundle of the current bundle is specified.

Tracing of ray bundles involves greater memory requirements than the standard method because the data related to the result of the tracing has to be stored in memory for every ray of the bundle at the same time. Because of this, bundles are recommended only for eye rays and their derivative shadow rays rather than rays of the next levels of recurrent ray tracing (i.e. reflected/refracted rays and their derivative shadow rays), that should be treated in the standard manner. Besides, for reflected/refracted rays which are derivatives of eye rays "close" to each other, the spatial coherency in the sense of the structure of the hierarchy tree is generally lost as the rays are reflected/refracted by curved surfaces.

In the case of eye rays a bundle is a collection of rays that scan a rectangular part of the screen and on the basis of their intersection with objects the shadow bundle is created. Shadow rays should be cast from a light source rather than from shadowed surfaces, because in the former case only one point specifying the origin of all the rays have to be stored.

An additional improvement is to specify the minimum rectangular screen extents both for a box and a bundle in each step of the recurrent intersection searching. If do so, the efficiency of the intersection routine can be increased by applying the simple clipping of the bundle in the screen coordinates. Only those rays which hit the common part of the rectangular extents are tested against an intersection with the box. The extent for a bundle is determined during the calculation of ray-box intersections as the maximum

Figure 1. Shadows of the Plato World

Figure 2. Somewhere else.

and minimum of screen coordinates of the pixels pierced by the rays with acceptable intersections with the box of the previous level.

5 Results

From a number of images generated to test our approach, four have been chosen and displayed here to demonstrate the ability of IFS descriptions to model complex scenes for ray tracing.

Figure 3. Forest scene.

Figure 4. Call from the future past.

Fig.1 shows a group of self-similar fractal sets; amongst others Sierpinski's Tetrahedron and Menger's Spong are presented.

Fig.2 exemplifies the application of IFS for modeling a terrain; three IFS surfaces were used, each of which specified by nine affine mappings. The deformed variations of Sierpinski's Tetrahederon were obtained by parameterization the original IFS code for this classical fractal set.

Fig.3 is another example of modeling complex scenes by means of IFSes. The scene includes about 30 models of trees encoded with parameterized hierarchical IFS. To model a field of grass a hierarchical IFS equipped with a stochastic automaton was used.

Fig.4 demonstrates the application of our approach to render parameterized 3D versions of Barnsley's Fern. One of the important things about this fractals is that the stem, like in the original IFS, is specified by a noninvertible affine mapping.

The images were ray-traced with 1280×960 screen resolution using super-sampling and 1280 rays per eye bundle (40×32 block) on a PC equipped with Intel Pentium 466 MHz Processor and 128 MB RAM from which only about 4 MB was used during the rendering process.

6 Summary

In this paper a new approach to object instancing for affine IFS attractors has been presented. Thanks to a smart use of axis-aligned bounding boxes as instancing volumes, ray tracing the IFS fractals can be done at pixel size accuracy in an extremely efficient manner. Moreover the geometry of boxes makes it possible to solve the spatial aliasing problem in a simple way by controlling and correction of the box sizes. Another advantage of our method is that it allows an IFS to include both invertible and noninvertible affine mappings. To minimize computational redundancy, which is common when object instancing is used in the ray-tracing context, the idea of tracing bundles of rays has been introduced. Consequently, even scenes of high level of complexity can be ray-traced using machines less powerful than graphics workstations.

Acknowledgements: The author is grateful to Prof. Jan Zabrodzki at CG Lab of Warsaw University of Technology for his help and valuable comments on the manuscript.

References

1. Barnsley M. F., Fractals Everywhere, 2nd ed. (Academic Press, Boston, 1993).
2. Glassner A. S. (ed.), An Introduction to Ray Tracing (Academic Press, London, 1989).
3. Hart J. C. and DeFanti T. A., Efficient antialiased rendering of 3-D linear fractals, *Computer Graphics 25* (1991) pp. 91–100.
4. Hepting D., Prusinkiewicz P. and Saupe D., Rendering methods for iterated function systems. In *Proceedings of Fractals '90* (1990), IFIP.
5. Mandelbrot B. B., The Fractal Geometry of Nature (W. H. Freeman and Co., San Francisco, 1982).
6. Martyn T., Ray Tracing Affine IFS Attractors, Ph.D. Thesis, Warsaw University of Technology, Poland, 1999, (in Polish).
7. Press W. H., Flannery B. P., Teukolsky S. A. and Vetterling W. T., Numerical Recipes in C (Cambridge University Press, 1988).
8. Traxler C. and Gervautz M., Efficient ray tracing of complex natural scenes. In *Fractal Frontiers*, ed. by M. M. Novak, and T. G. Dewey (World Scientific, Singapore, 1997).

MODELING AND APPROXIMATION OF FRACTAL SURFACES WITH PROJECTED IFS ATTRACTORS

E. GUÉRIN, E. TOSAN AND A. BASKURT

LIGIM - EA 1899 - Computer Graphics, Image and Modeling Laboratory
Claude Bernard University, Lyon, France
Bât. 710 - 43, bd du 11 novembre 1918 - 69622 Villeurbanne Cedex
Tel.: (33) 4.72.43.26.10 - Fax : (33) 4.72.43.13.12
E-mail: [eguerin|et|abaskurt]@ligim.univ-lyon1.fr

A method for modeling and approximating rough surfaces is introduced. A fractal model based on projected IFS attractors allows the definition of free form fractal shapes controled with a set of points. This flexible model has good fitting properties for recovering surfaces. The approximation is formulated as a non-linear fitting problem and resolved using a modified LEVENBERG-MARQUARDT minimisation method. The main applications are surface modeling, shape description and geometric surface compression.

1 Introduction

Basically, the problem of approximating the surface of 3D objects consists in finding a model that represents a set of data points:

$$(x_i, y_i, z_i) \in \mathbb{R}^3,\ \forall i = 0, \ldots, n$$

A wide variety of representation methods have been proposed for modeling these surfaces[1]. Unfortunately, these models do not recover rough surfaces, i.e. surfaces defined by continuous functions that are nowhere differentiable. Models that are able to produce rough surfaces are mostly based on random processes. This is the reason why these models are not suitable for approximation. In order to propose an efficient solution to the problem of rough surface approximation, the current study proposes a parametric model based on a deterministic fractal approach.

In [2] and [3], we have proposed a model for fractal curves and surfaces. This model combines two classical models: a fractal model (IFS attractors) and a CAGD model (free form shapes). This model is called projected IFS attractors. A set of control points allows an easy and flexible control of the fractal shape generated by the IFS model and provide a high quality fitting, even for surfaces with sharp transitions. In [4] and [5], we have proposed an approximation method for curves based on this model. In [6], we give the extension of this method to surfaces. In this paper, we present a general formulation of surface modeling and some approximation results.

2 Model

2.1 IFS

Introduced by BARNSLEY[7] in 1988, the IFS (Iterated Function Systems) model generates a geometrical shape or an image [8] with an iterative process. An IFS-based modeling system is defined by a triple $(\mathcal{X}, d, \mathcal{S})$ where:

- $(\mathcal{X}, d)$ is a complete metric space, $\mathcal{X}$ is called *iteration space*;

- $\mathcal{S}$ is a semigroup acting on points of $\mathcal{X}$ such that: $\lambda \in \mathcal{X} \mapsto T\lambda \in \mathcal{X}$ where T is a contractive operator, $\mathcal{S}$ is called *iteration semigroup*.

An *IFS* $\mathbb{T}$ *(Iterative Function System)* is a finite subset of $\mathcal{S} : \mathbb{T} = \{T_0, ..., T_{N-1}\}$ with operators $T_i \in \mathcal{S}$. We note $\mathcal{H}(\mathcal{X})$ the set of non-empty compacts of $\mathcal{X}$. The associated HUTCHINSON operator is:

$$K \in \mathcal{H}(\mathcal{X}) \mapsto \mathbb{T}K = T_0K \cup ... \cup T_{N-1}K \ .$$

This operator is contractive in the new complete metric space $\mathcal{H}(\mathcal{X})$ and admits a fixed point, called *attractor* [7]:

$$\mathcal{A}(\mathbb{T}) = \lim_{n\to\infty} \mathbb{T}^n K \ with \ K \in \mathcal{H}(\mathcal{X})\,.$$

2.2 Parameterisation of attractors

By introducing a finite set Σ, the IFS can be indexed $\mathrm{T} = (T_i)_{i\in\Sigma}$ and the attractor $\mathcal{A}(\mathbb{T})$ has an *address function* [7] defined on Σ^ω, the set of infinite words of Σ:

$$\sigma \in \Sigma^\omega \mapsto \phi(\sigma) = \lim_{n\to\infty} T_{\sigma_1}...T_{\sigma_n}\lambda \in \mathcal{X} \text{ with } \lambda \in \mathcal{X}\,. \tag{1}$$

When operators match joining condition [2,3,9], this function defines parameterised curves or surfaces. For curves, a single indexing $\tilde{\Sigma} = \{0, ..., N-1\}$ is sufficient [10]:

$$\Phi(s) = \phi(\sigma) \text{ with } s = \sum_{i=1}^{\infty} \frac{1}{N^i}\sigma_i$$

where $\sigma = \sigma_1 \ldots \sigma_n \ldots$ corresponds to the development of the scalar s in base N. For surfaces, it is more convenient to use a double indexing $\Sigma = \{0, \ldots, N-1\} \times \{0, \ldots, N-1\}$ [11]:

$$\Phi(s,t) = \phi(\rho) \text{ with } \rho = (\sigma_1, \tau_1) \ldots (\sigma_n, \tau_n) \ldots \in \Sigma^\omega$$

where $\sigma = \sigma_1 \ldots \sigma_n \ldots$ and $\tau = \tau_1 \ldots \tau_n \ldots$ are respectively the development of s and t in base N.

2.3 Projected attractors

The main idea of our model is drawn from the formula of free form surfaces used in CAGD:

$$F(s,t) = \sum_{j\in J} \Phi_j(s,t) p_j$$

where p_j constitutes a grid of control points (see Fig. 1), and Φ_j are blending functions. These blending functions have the following property:

$$\forall (s,t) \in [0,1]^2 \ \sum_{j\in J} \Phi_j(s,t) = 1$$

The way to obtain the same property for IFS attractors is to use a barycentric metric space.

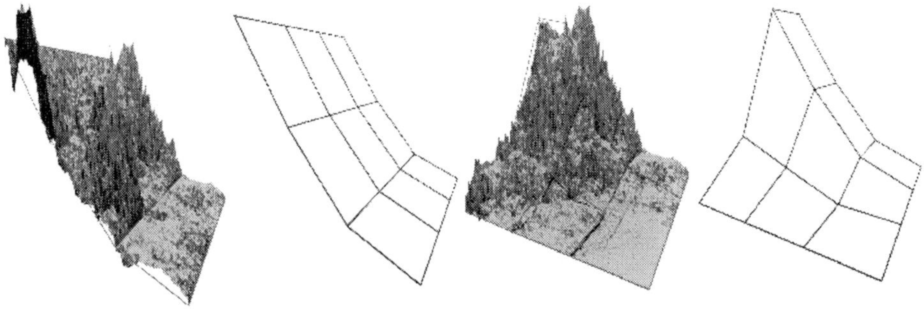

Figure 1. Deformation of a free form surface using the control grid.

In classical fractal interpolation [7,12] or fractal compression [8], the complete metric space $\mathcal{X}$ used is $\mathbb{R}^2$ or $\mathbb{R}^3$, and the iteration semigroup is constituted of contractive affine operators. Our work consists in enlarging iteration spaces [2,3]. This model uses a barycentric space $\mathcal{X} = \mathcal{B}^J$:

$$\mathcal{B}^J = \{(\lambda_j)_{j \in J} \mid \sum_{j \in J} \lambda_j = 1\}\,.$$

For curves, this barycentric space is used with $\tilde{J} = \{0, \dots, m\}$, for surfaces with $J = \{0, \dots, m\} \times \{0, \dots, m\}$. Then, the iteration semigroup is constituted of matrices with barycentric columns:

$$S_J = \{T \mid \sum_{j \in J} T_{ij} = 1, \, \forall i \in J\}\,.$$

This choice leads to the generalization of IFS attractors named *projected IFS attractors*:

$$P\mathcal{A}(\mathbb{T}) = \{P\lambda \mid \lambda \in \mathcal{A}(\mathbb{T})\}$$

where P is a grid of control points $P = (p_j)_{j \in J}$ and $P\lambda = \sum_{j \in J} \lambda_j p_j$. In this way, we can construct a fractal function [2,3,9] using the projection:

$$F(s,t) = P\Phi(s,t) = \sum_{j \in J} \Phi_j(s,t) p_j$$

where $\Phi(s,t)$ is a vector of functions $\Phi(s,t) = (\Phi_j(s,t))_{j \in J}$ and J is the double index set $J = \{0, \dots, m\} \times \{0, \dots, m\}$. Fig. 1 shows the action of the control grid on the *free form surface* defined by the function.

2.4 Tabulation of parametric surfaces

With a tabulation process[5,6], considering only the values of s and t multiple of $\frac{1}{N^p}$ leads to a simplification in the computing without any loss of information. The surface tabulation is a grid defined by:

$$F(\frac{i}{N^p}, \frac{j}{N^p}) = P\Phi(\frac{i}{N^p}, \frac{j}{N^p}) \text{ with } (i,j) \in \{0, \dots, N^p - 1\} \times \{0, \dots, N^p - 1\}$$

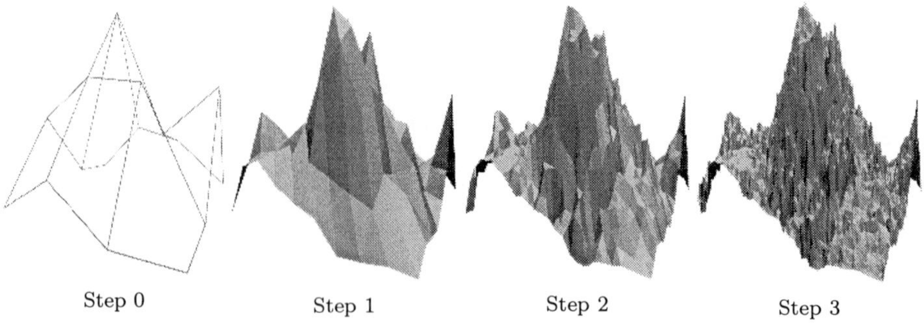

Figure 2. Three first iterations of the construction process beginning with the initial 4×4 control grid (Step 0).

In this special case, developments of $\frac{i}{N^p}$ are ended by a infinite sequences of 0:

$$\begin{cases} \sigma = \sigma_1 \dots \sigma_p \, 00 \dots \\ \tau = \tau_1 \dots \tau_p \, 00 \dots \end{cases}$$

Then, $F(\frac{i}{N^p}, \frac{j}{N^p})$ simplifies in:

$$\begin{aligned} F(\tfrac{i}{N^p}, \tfrac{j}{N^p}) &= P\phi\,((\sigma_1, \tau_1) \dots (\sigma_p, \tau_p) \dots (0,0) \dots (0,0) \dots) \\ &= PT_{\sigma_1\tau_1} \dots T_{\sigma_p\tau_p}\phi\,((0,0) \dots (0,0) \dots) = PT_{\sigma_1\tau_1} \dots T_{\sigma_p\tau_p}\Phi(0,0) \end{aligned}$$

By choosing simplifications (but no restrictions) such as $\Phi(0,0) = e_{00}$, the surface tabulation can be generated computing only p iterations without any loss of information:

$$F(\frac{i}{N^p}, \frac{j}{N^p}) = PT_{\sigma_1\tau_1} \dots T_{\sigma_p\tau_p} e_{00}$$

Fig. 2 shows the three first iterations of the construction process.

3 Surface representation

Any couple $(P, \mathbb{T})$ does not describe a parametric surface $F(s,t)$. A joining condition must apply to the IFS to be sure the generated object is a surface. Classically, this joining condition is obtained by tensor products of curves. This method, very used in CAGD, can be extended to surfaces defined by IFS. Therefore, it can be applied to the modeling of fractal surfaces [2,3,11,12]. The function $\Phi(s,t) = \tilde{\Phi}^u(s) \otimes \tilde{\Phi}^v(t)$ is defined by the IFS $T_{i,j} = \tilde{T}_i^u \otimes \tilde{T}_j^v$.

However, the fractal famiily of surfaces defined by IFS is much larger. This section explains in details how it is possible to provide a joining condition such that the function $(s,t) \in [0,1]^2 \mapsto \Phi(s,t)$ is well defined [9].

3.1 *Joining condition*

The joining condition of the quadrangular patch is drawn from the structure of $[0,1]^2$ (see fig. 3).

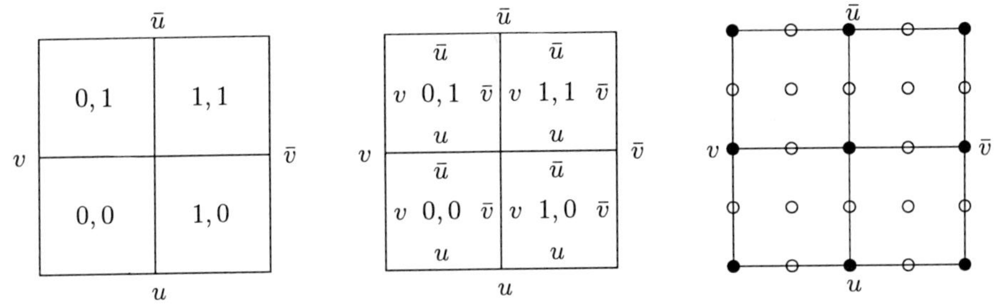

Figure 3. Quadrangular subdivision scheme: boundaries, joinings and grid.

Boundary curves $\Phi_u(s) = \Phi(s, 0)$, $\Phi_{\bar{u}}(s) = \Phi(s, 1)$, $\Phi_v(s) = \Phi(0, s)$, $\Phi_{\bar{v}}(s) = \Phi(1, s)$ are defined by IFS T^γ with indices in $\tilde{\Sigma} = \{0, ..., N-1\}$ extracted from patch IFS (see Fig. 3):

$$T_i^\gamma = T_{\xi_i^\gamma} \text{ with } \xi_i^u = (i, 0),\ \xi_i^{\bar{u}} = (i, N-1),\ \xi_i^v = (0, i),\ \xi_i^{\bar{v}} = (N-1, i)$$

for $i = 0, ..., N-1$. These boundary curves are drawn from the corresponding address functions:

$$\Phi_\gamma(s) = \phi^\gamma(\sigma) = \phi(\xi^\gamma(\sigma)) \text{ with } \xi^\gamma : \sigma \in \tilde{\Sigma}^\omega \mapsto \xi_{\sigma_1}^\gamma ... \xi_{\sigma_n}^\gamma ... \in \Sigma^\omega$$

A necessary and sufficient condition for the function $\Phi : (s, t) \mapsto \phi(\rho)$ to be defined and to be continuous is that the composants $T_{i,j}\phi$ of the address function ϕ join along their borders $T_{i,j}\phi^\gamma$ (see proof in [9]). This condition is an equation system:

$$\begin{cases} T_{i,j}\phi^{\bar{u}}(\sigma) = T_{i,j+1}\phi^u(\sigma) & \text{for } i = 0, ..., N-1 \text{ and } j = 0, ..., N-2 \\ T_{i,j}\phi^{\bar{v}}(\sigma) = T_{i+1,j}\phi^v(\sigma) & \text{for } i = 0, ..., N-2 \text{ and } j = 0, ..., N-1 \end{cases}$$

This equation system can be expressed as linear constraints on the $T_{i,j}$ matrices.

3.2 *Constraints*

Now, consider that each patch boundary is the embedding of a curve defined on a grid $\tilde{J} = \{0, ..., m\}$ in the corresponding boundary of the patch control grid (see Fig. 4):

$$\phi^\gamma(\sigma) = \amalg_\gamma \tilde{\phi}^\gamma(\sigma) = \amalg_\gamma \sum_{k=0}^{m} \tilde{\phi}_k^\gamma(\sigma)\tilde{e}_k = \sum_{k=0}^{m} \tilde{\phi}_k^\gamma(\sigma)\, \amalg_\gamma\, \tilde{e}_k \tag{2}$$

The embeddings associated with the grid boundaries are defined by:

$$\amalg_u \tilde{e}_k = e_{k,0},\ \amalg_{\bar{u}} \tilde{e}_k = e_{k,m},\ \amalg_v \tilde{e}_k = e_{0,k},\ \amalg_{\bar{v}} \tilde{e}_k = e_{m,k} \text{ for } k = 0, .., m.$$

Let us denote $\tilde{\mathrm{T}}^\gamma = (\tilde{T}_i^\gamma)_{i=0,...,N-1}$ the IFS that defines $\tilde{\phi}^\gamma$, $T_{i,j}$ matrices must satisfy the following embedding constraints [9]:

$$\phi^\gamma = \amalg_\gamma \tilde{\phi}^\gamma \Leftrightarrow T_{\xi_i^\gamma} \amalg_\gamma = \amalg_\gamma \tilde{T}_i^\gamma \text{ pour } i = 0, .., N-1$$

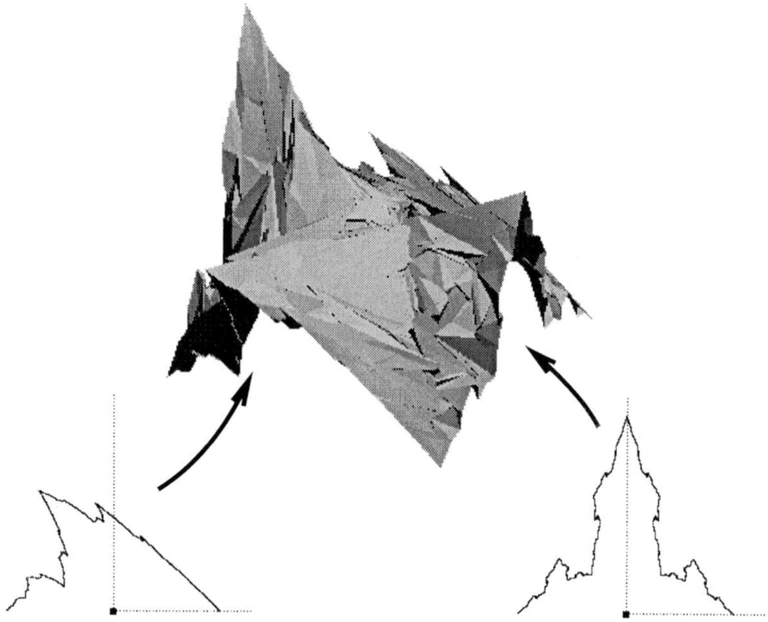

Figure 4. Quadrangular patch and the corresponding boundary curves.

Then, the joining equations can be expressed as constraints on $T_{i,j}$ and on boundary IFS $\tilde{\mathrm{T}}^{\gamma}$ [9]:

$$T_{i,j}\phi^{\bar{v}} = T_{i+1,j}\phi^{v} \Leftrightarrow T_{i,j}\,\amalg_{\bar{v}}\,\tilde{\phi}_{\bar{v}} = T_{i+1,j}\,\amalg_{v}\,\tilde{\phi}^{v}$$
$$\Leftrightarrow T_{i,j}\amalg_{\bar{v}} = T_{i+1,j}\,\amalg_{v} \text{ et } \tilde{\mathrm{T}}^{\bar{v}} = \tilde{\mathrm{T}}^{v}.$$

The argument is also valid with $\tilde{\mathrm{T}}^{\bar{u}} = \tilde{\mathrm{T}}^{u}$ (see Fig. 4).

3.3 Description

An IFS defined by N matrices $T_{i,j}$ that satisfy the previous constraints is equivalent to a subdivision scheme characterised by a grid of points (see Fig. 3).

In this way, the classical process of point interpolation to build a fractal curve in $\mathbb{R}^2$ [7] has been extended to fractal surfaces in $\mathbb{R}^3$ [12,13,14].

The proposed grid is composed of $(mN+1) \times (mN+1)$ points of $\mathcal{B}^J$ corresponding to the matrices columns of $T_{i,j}$:

$$T_{i,j}e_{kl} = \lambda_{mi+k,mj+l}$$

The joining equations are satisfied:

$$T_{i,j}\amalg_{\bar{v}} = T_{i+1,j}\amalg_{v} \Leftrightarrow \forall l = 0,..,m \;\; T_{i,j}\,\amalg_{\bar{u}}\,\tilde{e}_l = T_{i+1,j}\,\amalg_{u}\,\tilde{e}_l$$
$$\Leftrightarrow \forall l = 0,..,m \;\; T_{i,j}e_{m,l} = T_{i+1,j}e_{0,l}$$
$$\Leftrightarrow \forall l = 0,..,m \;\; \lambda_{mi+m,mj+l} = \lambda_{m(i+1),mj+l}$$

However, embedding equations are expressed as restrictions on grid boundaries:

$$T_{\xi_j^v}\amalg_v = \amalg_v \tilde{T}_j^v \Leftrightarrow \forall l = 0, ..., m \;\; T_{\xi_j^v} \amalg_v \tilde{e}_l = \amalg_v \tilde{T}_j^v \tilde{e}_l$$
$$\Leftrightarrow \forall l = 0, ..., m \;\; T_{0,j} e_{0l} = \amalg_v \tilde{T}_j^v \tilde{e}_l$$
$$\Leftrightarrow \forall l = 0, ..., m \;\; \lambda_{0,mj+l} = \amalg_v \tilde{\lambda}_{mj+l}^v$$

where $(\tilde{\lambda}_k^v)_{k=0,..,mb}$ is a set of points of $\mathcal{B}^{\tilde{J}}$ that characterise the curve $\tilde{\Phi}^v$.

The points $\lambda_{mi,mj}$ are transformed patch corners $T_{i,j}e_{00} = \lambda_{mi,mj}$, $T_{i,j}e_{m0} = \lambda_{mi+m,mj}$, $T_{i,j}e_{mm} = \lambda_{mi+m,mj+m}$, $T_{i,j}e_{0m} = \lambda_{mi,mj+m}$. If one takes $N = m$, a pretty good choice would be $\lambda_{mi,mj} = e_{ij}$. Changing of system coordinates R with $Re_{kl} = r_{kl}$ allows to consider the general case.

The choice of the description grids $(mN+1) \times (mN+1)$ is not any. Practically, the points are taken in the subspaces corresponding to parts of the control grid [6].

Considering the coefficients of the operators T_i and the coordinates of the control grid P as elements of a parameter vector $\mathbf{a}$ allows us to construct a family of functions $F_\mathbf{a}$ defined by couples $(P_\mathbf{a}, \mathbb{T}_\mathbf{a})$.

4 Approximation

Given a sampled surface $(s_i, t_j, \mathbf{Q}_{ij}) \in \mathbb{R}^3$, the challenge is to determine the projected IFS model which provides a good quality approximation of this surface. The approach proposed in the current study is similar to the one we introduced in [4,5] for curves. It is based on a non-linear fitting formalism. Then, we show how the approximation problem can be seen as a standard non-linear fitting problem.

4.1 Non linear fitting

Let $\mathbf{Q}_{ij\,(i=0,\dots,N^p\; j=0,\dots,N^p)}$ be a given surface to approximate. Let $F_\mathbf{a}$ be the function associated with the parameter vector $\mathbf{a}$. The approximation problem consists in determining the parameter vector $\mathbf{a}$ that minimizes the distance between the sampled surface $\mathbf{Q} = \{(\frac{i}{N^p}, \frac{j}{N^p}, \mathbf{Q}_{ij})\}$ and the function $F_\mathbf{a}$:

$$\mathbf{a}_{opt} = \underset{\mathbf{a}}{\text{argmin}}\; d(\mathbf{Q}, F_\mathbf{a})$$

wherc:

$$d(\mathbf{Q}, F_\mathbf{a}) = \sum_{ij} ||\mathbf{Q}_{ij} - F_\mathbf{a}(\frac{i}{N^p}, \frac{j}{N^p})||^2$$

4.2 Solving the non-linear fitting problem

Our resolution method is based on the LEVENBERG-MARQUARDT algorithm [15]. This algorithm is a numerical resolution of the following fitting problem:

$$\mathbf{a}_{opt} = \underset{\mathbf{a}}{\text{argmin}} \sum_{i=0}^{M} (v_i - f(\mathbf{a}, u_i))^2$$

where vectors $\mathbf{v}$ and $\mathbf{u}$ are the fitting data and f is the fitting model.

In order to resolve our approximation problem using this algorithm, we have to consider the following data:

$$\mathbf{v} = (v_0, \ldots, v_M) = (0, \ldots, 0)\ ;\ \mathbf{u} = (u_0, \ldots, u_M) = (0, \ldots, M)$$

where $M = (N^p + 1)^2$. Then, the fitting model is:

$$f(\mathbf{a}, k) = ||\mathbf{Q}_{i_k j_k} - F_{\mathbf{a}}(\frac{i_k}{N^p}, \frac{j_k}{N^p})||$$

where $i_k = k \mod N^p$ and $j_k = k/N^p\ \forall k = 0, \ldots, (N^p + 1)^2$.

The LEVENBERG-MARQUARDT method combines two types of approximation for minimizing the square distance. The first consists in a quadratic approximation. When this fails, the method tries a simple linear approximation. These approximations are computed with the provided partial derivatives of the fitting model. In our case, these partial derivatives are numerically computed by a perturbation vector [4,5]:

$$\delta\mathbf{a}_i = (\underbrace{0, \ldots, 0, \varepsilon}_{i}, 0, \ldots, 0)$$

Then, the computation of partial derivatives is approximated by:

$$\frac{\partial f}{\partial a_i}(\mathbf{a}, u) \simeq \frac{f(\mathbf{a} + \delta\mathbf{a}_i, u) - f(\mathbf{a}, u)}{\varepsilon}$$

4.3 Results

We show three examples of approximation. The model used is composed of a 3×3 patch IFS and a 4×4 control grid. A total of 232 parameters is needed to code the entire model (subdivision schema and control grid). Fig. 5 shows the original elliptic paraboloid shape and the approximated surface reconstructed with our method. The equation of this shape is:

$$z = z_0 - (\frac{x - x_0}{\sigma_x})^2 - (\frac{y - y_0}{\sigma_y})^2$$

Note that the original and the approximated shapes are very similar. Our model allows to reconstruct smooth surfaces[2,3], and not only rough surfaces. Fig. 6 shows our first experiments on a natural surface. The original surface has been extracted from a geological database (found at the United States Geological Survey Home page `http://www.usgs.org`). The general aspects of the approximated surfaces are similar to the original ones.

A grey-level image can be seen as a surface. Fig. 7 shows the approximation of a grey-level image with both image and surface point of view. This image is an astronomy picture (a white dwarf). Fig. 8 shows the comparison between our approximation method and JPEG algorithm. Table 1 gives numerical results related to this comparison in terms of error (PSNR) and compression ratio. To obtain the same error, the JPEG algorithm provide a compression ratio that is lower than the one obtained with our method. We can also see that the performance is not very altered when each model parameter is quantified to 8 bits.

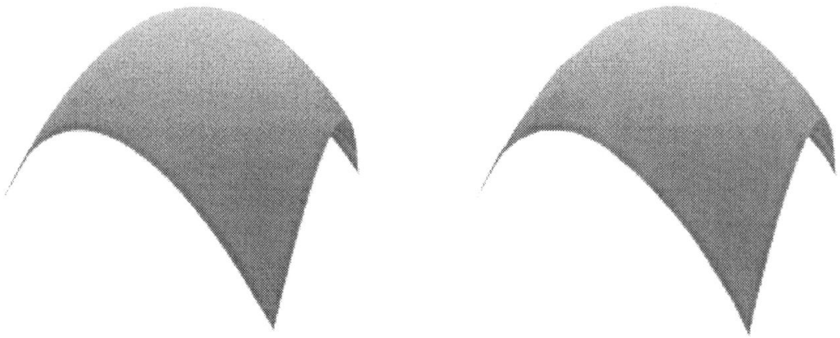

Original Approximation

Figure 5. Smooth surface.

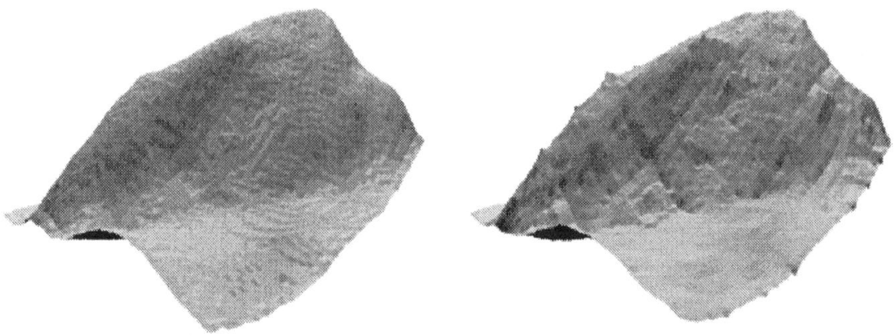

Original Approximation

Figure 6. Geological surface.

5 Conclusion

We presented a new approach for modeling and approximating both rough or smooth objects. This method is based on a fractal model named projected IFS attractors. This model is a parametric description which has the advantage of compactly describing the surface shape making it useful for geometric modeling and image synthesis. First results show that our approximation method is an interesting approach for rough object reconstruction and grey-level image description.

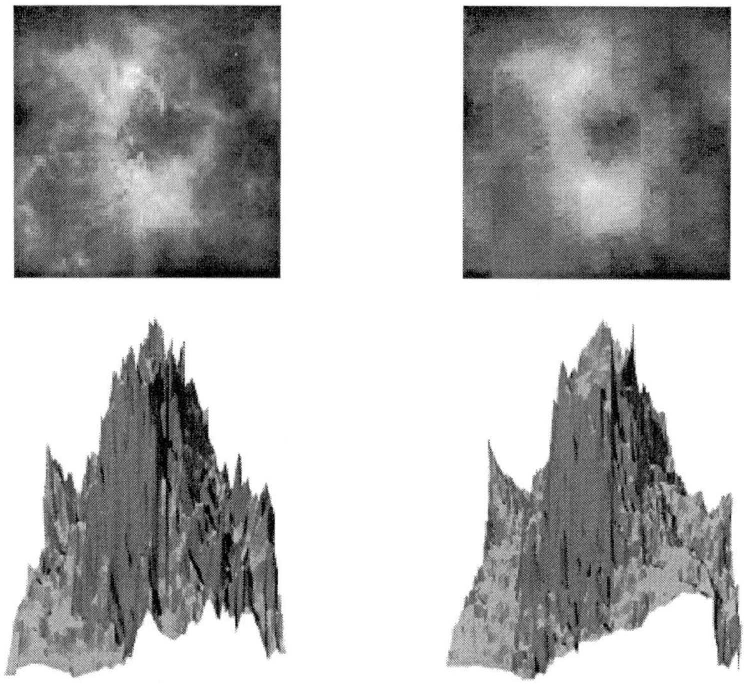

Original　　　　Approximation

Figure 7. Grey-level image.

a)　　b)　　c)　　d)　　e)

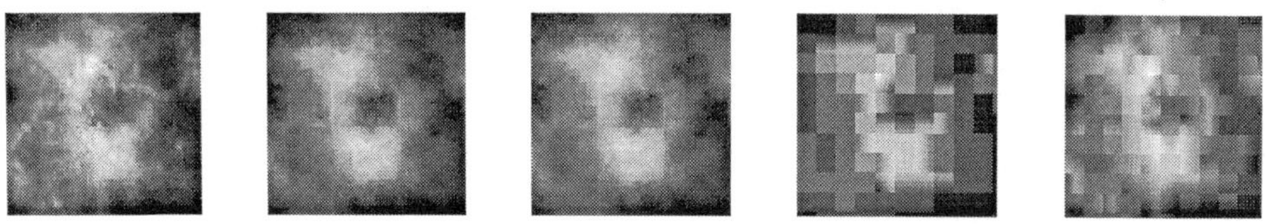

Figure 8. Compression of the original image (a) with our model (b) and (c), and with JPEG algorithm (d) and (e).

Compression type	PSNR (dB)	Code length	Compression ratios
Projected IFS model (b) 16 bits / scalar	26.6	464 bytes	14.5
Projected IFS model (c) 8 bits / scalar	26.4	232 bytes	29
JPEG (d)	23.6	545 bytes	12.3
JPEG (e)	26.5	592 bytes	11.4

Table 1. PSNR and compression ratios

References

1. R. M. Bolle and B. C. Vemuri, *On Three-Dimensional Surface Reconstruction Methods*, *IEEE Transactions on Pattern Analysis and Machine Intelligence* **13**, 1, 1 (1991).
2. C. E. Zair and E. Tosan, *Fractal modeling using free form techniques*, *Computer Graphics Forum* **15**, 3, 269 (1996), EUROGRAPHICS'96 Conference issue.
3. C. E. Zair and E. Tosan, *Computer Aided Geometric Design with IFS techniques*, in M. M. Novak and T. G. Dewey, eds., *Fractals Frontiers*, 443–452 (World Scientific Publishing, 1997).
4. E. Guérin, E. Tosan and A. Baskurt, *Fractal coding of shapes based on a projected IFS model*, in *ICIP 2000*, volume II, 203–206 (2000).
5. E. Guérin, E. Tosan and A. Baskurt, *A fractal approximation of curves*, *Fractals* **9**, 1, 95 (2001).
6. E. Guérin, E. Tosan and A. Baskurt, *Fractal Approximation of Surfaces based on projected IFS attractors*, in *Proceedings of EUROGRAPHICS'2001, short presentations* (2001).
7. M. Barnsley, *Fractals everywhere* (Academic Press, 1988).
8. A. E. Jacquin, *Image coding based on a fractal theory of iterated contractive image transformations*, *IEEE Trans. on Image Processing* **1**, 18 (1992).
9. E. Tosan, *Surfaces fractales définies par leurs bords*, in L. Briard, N. Szafran and B.Lacolle, eds., *Journées "Courbes, surfaces et algorithmes", Grenoble* (1999).
10. H. Prautzsch and C. A. Micchelli, *Computing curves invariant under halving*, *Computer Aided Geometric Design* , 4, 133 (1987).
11. C. A. Micchelli and H. Prautzsch, *Computing surfaces invariant under subdivision*, *Computer Aided Geometric Design* , 4, 321 (1987).
12. P. Massopust, *Fractal Functions, Fractal Surfaces and Wavelets* (Academic Press, 1994).
13. P. Massopust, *Fractal surfaces*, *Journal of Mathematical Analysis and Applications* , 151, 275 (1990).
14. H. Xie and H. Sun, *The study of Bivariate Fractal Interpolation Functions and Creation of Fractal Interpolated Surfaces*, *Fractals* **5**, 4, 625 (1997).
15. W. H. Press, B. P. Flannery, S. A. Teukolsky and W. T. Vetterling, *Numerical Recipes in C : The Art of Scientific Computing* (Cambridge University Press, 1993).

RESCALED RANGE ANALYSIS OF THE FREQUENCY OF OCCURRENCE OF MODERATE-STRONG EARTHQUAKES IN THE MEDITERRANEAN AREA

YEBANG XU AND PAUL W. BURTON

School of Environmental Sciences, University of East Anglia, Norwich NR4 7TJ, UK

E-mail: y.b.xu@uea.ac.uk and p.burton@uea.ac.uk

Rescaled range analysis of the natural contemporary earthquake history in the Mediterranean area produces a Hurst exponent, H, of 0.803 (±0.022). An H of 0.5 for Brownian motion is the dividing line between persistent and antipersistent time series, and so this indicates a seismic process that is persistent with long seismic "memory". These earthquake occurrences are consistent with a non-independent, biased random, process. When aftershocks of strong earthquakes are removed from the natural earthquake history the new H is 0.763 (±0.023), confirming a general process with long seismic memory. Comparing either rescaled range curve with that generated from an independent, random process, confirms the non-independent and biased randomness of both earthquake histories. The V statistic analysis of the natural earthquake history shows scaling changes at the 40, 55 and 120 month time index, points at which growth of the seismic process pauses, and which are relevant to identification of seismic cycles. The 55-month scaling break is clear in the V statistic analysis of the aftershock-removed earthquake history, although the other two scaling breaks are relatively subdued. The long "period" scaling break lends itself to tentative seismic hazard forecasts for the up-coming period in the area, whereas that for 55 months may represent a non-periodic cycle length.

Keywords: earthquakes, rescaled range analysis, persistence, V statistic and Mediterranean.

1 Introduction

Do earthquake sequences have a long "memory"? What are the characteristics of an individual earthquake time series? These questions are very basic, and should be unavoidable if earthquake forecasting is to more effectively evaluate expectations of moderate-strong earthquakes. Such questions are inevitably linked to man's efforts to mitigate against damage or long-term economic loss. In view of this Mediterranean Europe will be the example region used herein, a region where many studies based on more traditional comparisons and statistics have already been made e.g. mapping seismic hazard throughout Europe using an extreme value approach. However, detailed research based on the above problems requires new fractal methodology leading to fresh geophysical implications.

In addition, one might ask, is there a seismic cycle? Romanowicz [12] suggested that the great earthquakes of the past 80 years alternate between strike-slip and dip-slip mechanism on a 20-30 years cycle. Johnson and Sheridan [6] doubt this suggestion because of the difficulty of detecting random and non-random patterns in their computer based random simulations. However, epochs of relative seismic activity and quiescence appear to alternate for earthquakes (M≥ 6) in China during a period of about 500 years (1400-1970 AD), and this provides some evidence - suggesting seismic periodic cycles on this broad time scale [3]. However, it is usually very difficult to identify a seismic cycle in practice. Especially difficult when a complete record of earthquakes is available only for a time shorter than a complete seismic cycle. How may random or non-random process be detected and distinguished? How may a seismic cycle be identified? Is a seismic cycle periodic or non-periodic? Our aim herein is to make a step further into this subject.

The first step is not to regard the seismic time series as independent and identically distributed, and this immediately differs from most traditional approaches. It follows that those statistical analyses for independent distributions are not suitable. A non-parametric approach is appropriate. Rescaled range (R/S) analysis is a very robust tool for investigating non-linear or fractal time series. This method was discovered by Hurst [5] and developed in Mandelbrot's studies on biased random walks (e.g. [9]). This analysis technique is used to resolve long memory effects and fractional Brownian motion. This analysis technique was originally invented and used by Hurst to analyse reservoir capacity associated with the seasonally flooding Nile and has more recently been mentioned by Lomnitz [7] in relation to earthquake slip. The technique has been successful in a variety of applications: river discharges, rainfall, capital markets, tree-rings, sunspot numbers etc. (e.g. [10, 11]. Obviously, this method is very valuable, and appropriate to our purpose, and so it will be introduced into this study of earthquake time series.

The Mediterranean area is a region of high seismicity, Greece alone accounting for 2% global seismicity. Many studies of seismic hazard, seismicity and related topics have been made in this area, such as temporal scaling ranges, fractal dimensions, lacunarity and the different hierarchies for spatial fractal evolution of seismicity (Xu and Burton [14, 15]). However, rescaled range analysis of the frequency of moderate-strong earthquakes in the Mediterranean area is in its infancy [2]. Results on the existence of earthquake memory would be of value to the sciences of earthquake forecasting and long-term considerations of economic loss mitigation, particularly if accompanied by evidence of cyclicity.

2 Seismic Rescaled Range Analytical Method and Data

The steps for the application of the rescaled range analytical method (R/S) [5, 9] to the history of moderate-strong earthquakes in the Mediterranean area are similar to those for capital markets suggested by Peters [10, 11]. These steps are as follows.

(A) Calculate the monthly frequency of earthquakes in the earthquake time series. Then create a new series consisting of these values of earthquake monthly frequencies in chronological order, i.e. a new time series. Let N be the total length of the new series, that is the total number of monthly observations.

(B) Choose lengths ni for appropriate sub-periods. In this study: ni = 4, 5, 6, ..., N/2 (months); with i = 1,2,3, ..., (N/2)-3.

(C) Divide this series for earthquake frequency versus time into M contiguous sub-periods of length ni, where M = N/ni.

(a) In each sub-period labelled J_m (m = 1,2, 3, ..., M), each monthly frequency value is denoted by $F_{k,m}$ (k = 1,2, ..., ni) and the average frequency value is

$$av_m = (1/ni) \sum^{ni}_{k=1} F_{k,m} \quad (1)$$

(b) Calculate the following series:

$$Y_{k,m} = \sum^{k}_{i=1} (F_{i,m} - av_m) \quad (2)$$

with k = 1,2, ..., ni.

$Y_{k,m}$ is the accumulated departure from av_m far each sub-period J_m.

(c) For each sub-period J_m, calculate the range

$$R_{Jm} = \max(Y_{k,m}) - \min(Y_{k,m}) \quad (3)$$

Where $1 \leq k \leq ni$.

and the standard deviation

$S_{Jm}= [(l/ni) \Sigma^{ni}_{k=l} (F_{k,m} - av_m)^2]^{1/2}$ (4)

Now, calculate R_{Jm} / S_{Jm} for each sub-period J_m. These are conventionally "R/S" values.

(d) For M contiguous sub-periods of length ni, the average R/S value is

$(R/S)_{ni} = (l/M) \Sigma^m_{k=l}(R_{Jm}/ S_{Jm})$. (5)

(D) We use Log $(R/S)_{ni}$ as the dependent variable and Log (ni) or Log (Time) as the independent variable for a graph. The slope of the ensuing linear regression, if the regression exists, is the Hurst exponent, H.

Peters [10] developed a significance test for R/S analysis. Based on the formulae suggested by Anis and Lloyd [1], he gave the following equation to circumscribe systematic deviation of the R/S statistic:

$E (R/S_{ni}) = [(ni -0.5) / ni] (ni \pi /2)^{-0.5} \Sigma^{ni-1}_{r=l}[(ni- r) / r]^{0.5}$. (6)

In this paper, we use equation 6 to calculate the expected R/S values for independent random variables - Gaussian random variables.

The statistic V_{ni} [5, 10] is defined as follows:

$V_{ni} = (R/S)_{ni} / (ni)^{0.5}$. (7)

In this analysis equation 7 is used to attempt to discern seismic cycles.

3 History of Moderate-strong Earthquakes in the Mediterranean area: Persistence and Memory

The study region is the Mediterranean area within latitudes 30.00° N - 50.00° N and longitudes 10.00° W – 40.00° E. The time period is from January 1973 to October 1997. The total number of earthquakes in the catalogue used for this area and period is 7,318. These data come from the USGS earthquake database. According to the International Seismological Centre, the earthquake magnitude threshold for recent earthquake bulletins (e.g. 1995) for the Mediterranean area is about 3.9 M. Comparing the USGS data with the ISC reveals no large differences, and the USGS data can be used until the most recent years in this research. Generally these data are of high quality and a reasonable magnitude threshold range for the earthquakes is $Ms \geq 4.0$ for completeness. These data form a solid base for the study and the epicentres are illustrated in Fig. 1. The high seismicity illustrated in Fig. 1 spread over 298 months. Monthly frequencies are expressed in Fig. 2. The dashed line in Fig. 2 represents the distribution of monthly frequency of earthquakes versus time in the Mediterranean area for the whole study period. This graph fluctuates rapidly and appears very complex. These are the primary data for the following rescaled range analysis.

Log (R/S) of the earthquake frequency is plotted versus time (in months) in Fig. 3. This figure also contains logarithmic values of E(R/S), calculated using equation 6, to check against the null hypothesis that the system is an independent process. The distribution of Log (R/S) versus Log (Time) deviates systematically from that for E(R/S). This means that the temporal distribution of moderate-strong earthquakes in the Mediterranean area is not an independent process. The statistical results generate a Hurst exponent, H, equal to 0.8031 (±0.022) with correlation coefficient, r_c, of 0.9862 for this seismic process. The importance of the Hurst exponent is, for instance, that it is a measure of the bias in the fractional Brownian motion. A Hurst exponent in the range H > 0.5 is for persistent or trend reinforcing series. Here, H =0.8031, considerably more than 0.5. This means that this seismic system has increased in the previous seismic period, and the chances are that it will continue to increase in the next seismic period. In other words, the interpretation is that this is a persistent seismic time series and it has a long memory. This suggests that long-term correlation will exist between current earthquake events and future events. On this evidence, these are therefore the fundamental characteristics for this historical process of moderate-strong earthquake occurrence in the Mediterranean area. An obvious practical implication is that estimations of earthquake hazard and risk using statistical studies based on the assumption of independence should be modified. New studies taking into account a long "memory" in the seismic process should be implemented.

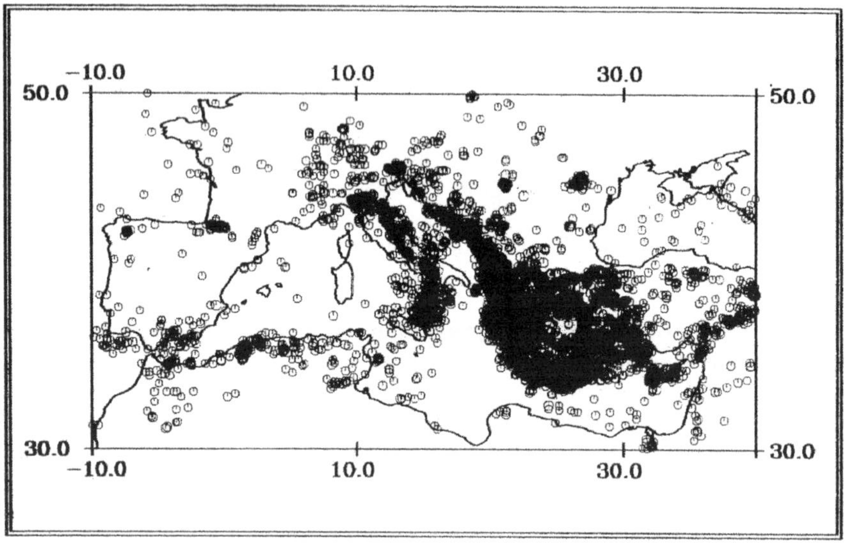

Scale 1: 45000000

Figure 1 Epicentres of moderate-strong earthquakes (Ms ≥ 4.0) in the Mediterranean area during January 1973 to October 1997.

Aftershocks of several large earthquakes (e.g. Ms≥6) exist in the time series of moderate-strong earthquakes in the Mediterranean area. These are known dependent events; they have an influence on the temporal seismic distribution and traditionally a seismologist might remove such events prior to any conventional statistical analysis. We shall now also adopt a procedure designed to remove aftershocks to deepen our investigation of the characteristics of this particular time series. We apply a commonly accepted procedure suggested by [4] to remove these aftershocks in main shock dependent tempero-spatial windows [T(M), L(M)]. Here, M is the main shock magnitude, T[M] is the temporal span and L(M) is the spatial radius for aftershocks associated with the parent event (Ms≥6). Any earthquake which is contained in this window is regarded as an aftershock of this parent earthquake. Such associated earthquakes are removed. Our window algorithm for aftershocks shown in Table 1 is consistent with the principle suggested by [4].

The remvals of aftershocks associated with parent events (Ms≥6.0) are made from application of the Gardner-Knopoff window algorithm [4]. It should be noted that the removal of aftershocks associated with a parent event probably has some influence on the temporally and spatially adjacent parent events. However, there is no influence if aftershocks of all parent events are removed from the moderate-strong earthquake time series as a whole. The solid line in Fig. 2 represents the distribution of earthquake monthly frequency versus time after the removal of aftershocks.

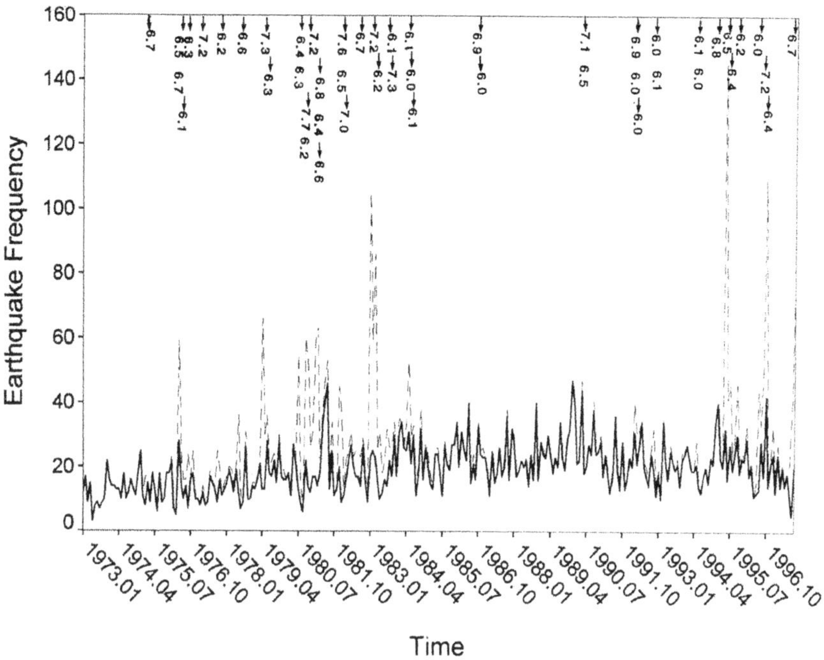

Figure 2 Earthquake monthly frequency during January 1973 to October 1997 for all moderate-strong earthquakes in the Mediterranean area with Ms ≥ 4 (dashed line) and with aftershocks of strong earthquakes (Ms ≥ 6) removed (solid line).

Comparison between the solid and the dashed lines in Fig. 2 suggests that most frequency peaks are caused by concentrations of aftershocks associated with strong earthquakes (Ms≥6). Removal of all the associated aftershocks removes the main peaks from the dashed line in Fig.2. After aftershock removal, the distribution represented by the solid line appears more complicated. The graph of Log (R/S) versus Log (Time) for the temporal distribution of moderate-strong earthquakes after the removal of the aftershocks is achieved; E(R/S) for an independent random variable is superimposed in the same graph. Log (R/S) is still seen to deviate from the trend of E(R/S), although the amount of deviation apparent in this graph is less than that in Fig. 3. And the statistical results for this graph still demonstrate a well formed Hurst exponent at H = 0.7627 (±0.0233) (note > 0.5) with correlation coefficient r_c = 0.9829. In other words, this still is a persistent earthquake time series which has long memory.

The dashed line in Fig. 2 illustrates earthquake monthly frequency versus time for the regional seismicity with aftershocks. It is worth emphasising that this is the real and natural distribution of the seismicity. Fig.2 reinforces the observation that most peaks of earthquake frequency are caused by dense, temporary gatherings of aftershocks associated with main events (Ms≥6.0). This natural process causes a large, visual divergence swinging upwards from the expected R/S values of an independent variable in Fig. 3. The Hurst exponent for the natural process is 0.8031, which is much larger than 0.5. It is this large divergence which means that this natural earthquake time series is a persistent time series with long memory. This essential characteristic probably reflects a degree of temporal self-organisation is in harmony with the ordered spatial seismic distribution already observed in the Mediterranean area, both linked through an organic but complex tectonic stress field.

Table 1. Temporo-spatial windows for removal of aftershocks associated with main shock magnitude M (after Gardner and Knopoff, [4])

Main shock magnitude	L(km)	T(months)
$6.0 \leq M < 6.3$	54	17
$6.3 \leq M < 6.8$	61	27
$6.8 \leq M < 7.3$	70	31
$7.3 \leq M < 7.8$	81	32
$7.8 \leq M < 8.3$	94	33

The main peaks of the dashed line (with aftershocks) are removed when the Gardner-Knopoff [4] window is applied to remove aftershocks producing the solid line in Fig. 2. However, the rescaled range analysis still suggests that this earthquake time series without aftershocks is still persistent with long memory. This is a vital result indicating that the main shocks pruned of associated aftershocks also exhibit a degree of temporal organisation within a dominating tectonic stress field.

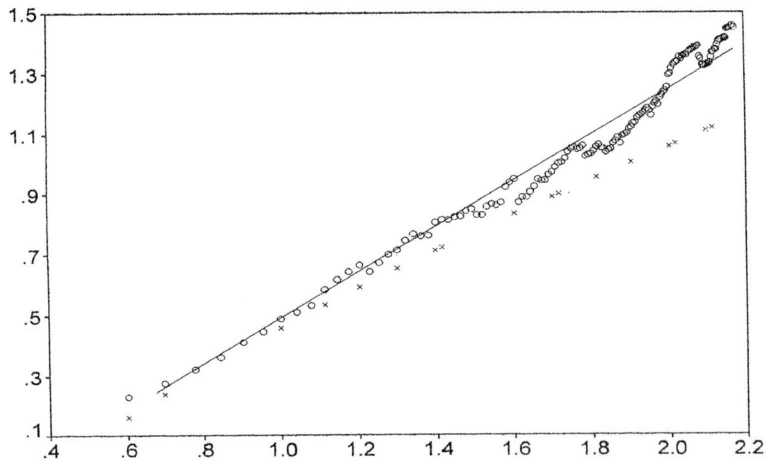

Figure 3. Rescaled range analysis of the natural earthquake history for earthquakes with $Ms \geq 4$. Log R/S of earthquake monthly frequency is plotted on the y-axis versus log time (months) and represented by the circle plotting points. The linear regression fitted to these data yields a Hurst exponent of 0.803 (±0.022). For comparison values of Peters [10] formula for E(R/S) provides expected values of R/S for an independent random or Gaussian variable; Log E(R/S) are represented by the cross plotting points.

The methodology of aftershock removal from earthquake series is problematic and may be debated. Is the method suggested by Gardner and Knopoff [4] ideal or are there more appropriate methods? Indeed any method of removing aftershocks is arbitrary to some degree and may to some extent destroy the natural seismic distribution by direct intervention of the seismologist. Specific methods may serve as a tool to study some detailed characteristics of temporal textures. Here it is observed for the Mediterranean area that both the natural and "artificial" (without aftershocks) distributions of moderate-strong earthquakes are persistent time series with long memory. It follows that practical applications incorporating earthquake forecasts should seek means to modify analyses and forecasts based only on an independent random model of seismicity, as these do not reflect the entire objective situation, and seek improved means to reflect a deepening understanding of seismicity.

4 History of Moderate-strong Earthquakes in the Mediterranean area: "V" Statistic and Recurrence

V values are calculated using equation 7 and Fig. 4 is an example showing the relation between the V statistic and time. Peters [10] pointed out that if the V statistic versus time graph is upwardly sloping then R/S is scaling at a faster rate than the square root of time: the process is persistent. Conversely, if the graph is sloping downward then the process is anti-persistent. For an independent, random process the graph will be flat. The graph in Fig. 4 is upwardly sloping as a whole and so this R/S earthquake process is scaling at a rate faster than the square root of time: it is persistent. This provides further support for the inferences obtained from analysis related to Fig. 3.

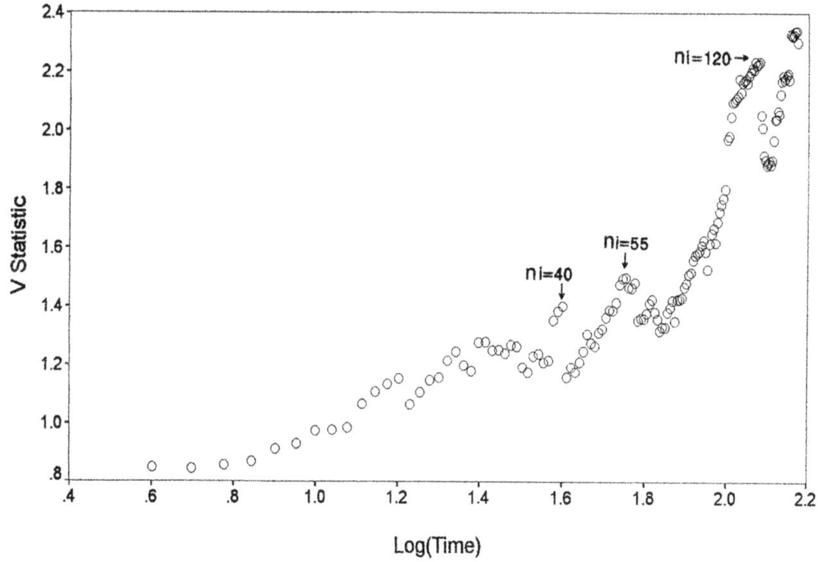

Figure 4. V statistic analysis of the natural earthquake history for earthquakes with Ms $\geq$ 4.0. Time is in months. See equation 7 and text. Note that the curve trends upwards but with obvious breaks in scaling at 40, 55 and 120 months.

The V statistic graphed in Fig. 4 has scaling changes visible at ni = 40, 55, and 120 months. The most obvious scaling break is at ni = 120 months where the V statistic seems to stop growing. It appears that the long-memory process has dissipated at this point, or, crucially, this indicates the existence of a cycle which may be periodic or non-periodic [10]. This inference should be borne in mind when reconsidering the dashed line in Fig. 2. The dashed line in Fig. 2 suggests two groups of peak values of earthquake monthly frequency (≥80): the first is contained during January, 1983 to March, 1983; and the second during May, 1995 to October, 1996. These two groups of peaks divide the whole earthquake temporal distribution (the dashed line in Fig. 2) into two periods which have an average duration, or period, of approximately 11 years. This conforms reasonably to the cycle indicated at ni = 120 months or 10 years in the V statistic of Fig. 4.

A *forward look*: A further inference based on the above results using all earthquakes, including aftershocks, is that the temporal distribution can be divided into three periods (Fig. 2). Period 1 is March 1975 to February 1985 with the first strong earthquake (March 27th, 1975; 6.7 Ms) being regarded as the start point. Period 2 is March 1985 to February 1995 and Period 3 then follows after February 1995. Period 1 is seismically active with 3024 events of Ms≥4.0 and 29 strong earthquakes with Ms≥6.0. The total seismic moment release for the strong earthquakes (Ms≥6.0) in Period 1 is 1.75×10^{21} N m, calculated using the equation Log Mo =1.5 Log Ms + 9.198 (±0.317) [8]. Period 2, although far from quiescent, is quieter with a large drop in the number of strong earthquakes: there are 2862 events with Ms≥4.0 including only 11 strong earthquakes with Ms≥6.0. The total seismic moment release for strong earthquakes in Period 2 is calculated as 1.62×10^{20} N m, and this only amounts to about 9% of that in Period 1, that is, about an order of magnitude less seismic energy release. The inferred Period 3 commenced in March 1995, continuing for only two years, to the ending date of this study in October 1997. Yet this embryonic Period 3 contains 1097 events (Ms≥4.0), including eight strong earthquakes (Ms≥6.0) distributed through two projecting peaks of the dashed line for earthquake monthly frequency (Fig. 2) in May 1995 and October 1996. This suggests that Period 3 may be an active one, and, if it has seismicity characteristics similar to Period 1, then one could forecast that Period 3 would continue until 2005. The forecast would be for about 1927 earthquakes (Ms≥4.0), including 21 strong earthquakes (Ms≥6.0), to occur after October 1997 and until 2005. The total seismic moment release could be about 1.75×10^{21} N m for strong earthquakes (Ms≥6.0) in Period 3, and allowing for the 1.68×10^{20} N m released during March 1995 to October 1997 forecasts 1.58×10^{21} N m to be released after October 1997. This is course estimation; one that should be confirmed or corrected in future, and viewed with interest as a speculative estimation based on earthquake history spanning a limited human time-scale and an exceptionally short seismotectonic time-scale.

There also seem to be at least two further scaling breaks (Fig. 4), at ni = 55 months and ni =40 months, although such periods can not be found simply by inspection of the dashed line in Fig. 2. These two breaks are non-typical. It may be that ni = 55 reflects a non-periodic cycle length averaging somewhere between four and five years between every two peaks of earthquake monthly frequency (frequency≥55). The scaling break at ni = 40 months might reflect a quasi-period of about three years within Period 1, also for earthquake monthly frequency≥55. The observations must limit such interpretations to speculation requiring future investigation.

When aftershocks are removed, using the Gardner-Knopoff [4] window, the earthquake monthly frequency (solid line in Fig. 2) shows a more complex fluctuation with most high valued peaks obviously removed. The corresponding V statistic graph for this aftershock-removed distribution is also obtained (not illustrated). The scaling break at ni = 40 months is now much less obvious and the longer time break appears slightly shifted to about ni = 105 months (compared to ni = 120 for the total earthquake distribution in Fig. 4), although the logarithmic shift is only from about 2.02 to 2.07. So reduced evidence remains for "stop growth" near these points. Yet the scaling break at ni = 55 months remains clear, confirming that cycles that are periodic or non-periodic, may exist on this time scale.

5 Conclusions

R/S analysis of the moderate-strong earthquake history in the Mediterranean produced a Hurst exponent, H, of 0.803 (±0.022). The Hurst exponent greater than 0.5 is indicative of long memory in a temporal process, a persistent time series. The significance test for the R/S analysis also confirms that this seismic process is not independent and purely random. When aftershocks of the strong earthquakes (Ms≥6.0) are removed then the highest peaks of earthquake monthly frequency disappear, leaving a complex fluctuation in the distribution (Fig. 2). Nevertheless a trial R/S analysis of this earthquake history with aftershocks removed still indicated a non-independent, biased random, process, because the R/S curve drifted away from the null-hypothesis random variable E(R/S) curve. Indeed the Hurst exponent only reduced from its previous value of H =0.803 (±0.022) to a new value of H = 0.763(±0.023), still substantially in excess of 0.5 (Burton and Xu [2]). But do these two earthquake histories with long memory contain evidence of cycles within them?

The V statistic now provides further evidence of biased randomness in the earthquake history through detection of cycles that can be periodic or non-periodic. The V statistic versus time, for the earthquake history with aftershocks is upwardly sloping as a whole (Fig. 4), reinforcing the R/S result that scaling is at a rate faster than the square root of time and hence indicative of a seismic process with long memory. The obvious break at ni = 120 months in the V statistic means that growth has stopped at this point and the long-memory process has dissipated. Interpreting this break as indicative of a periodic cycle in the earthquake history (Fig. 2) could be useful to attempt earthquake forecasting into an upcoming Period 3 (March 1995 -2005). On this model one may expect in the upcoming period that there will be about 1927 earthquakes (Ms≥4.0). This includes 21 strong earthquakes with Ms≥6.0 producing a cumulative seismic moment of 1.58×10^{21} N m to be released during October 1997 to 2005. The V statistic features two other scaling changes, at 40 and 55 months respectively, which could reflect non-periodic or quasi-periodic cycles. The V statistic versus time for the earthquake history without aftershocks is also upwardly sloping as a whole, again supportive of a seismic process with long memory. The scaling break at 40 months is now very subdued and the long-period break is shifted slightly left to 105 months but also is subdued. The scaling break in the V statistic at 55 months remains very obvious in and may represent a non-periodic cycle length. The trend of the V statistic versus time for both earthquake histories, with and without aftershocks, confirms a seismic process with long memory.

An holistic view of geodynamics and spatial and temporal variations of texture in seismicity is a desirable goal. The Mediterranean area spans the plate boundary between the African and Eurasian plates. This area responds to an organic but complicated tectonic stress field caused by dynamics generating east-west opening of the Atlantic ocean and north-south closing of the Thetas sea [13]. Spatial organisation of the seismicity in the Mediterranean area (Fig. 1) reflects this opening of the Atlantic Ocean, closing of the Thetys sea and the relative movement of secondary tectonic blocks. A persistent seismic texture with long memory is indicated by the results of the R/S analyses. This should not be a surprise: it is reasonable that the temporal distribution of the moderate-strong earthquake history in this area is non-random and includes an element of ordering temporal texture. It should not be a surprise if this temporal texture is co-ordinated with the spatial seismic order and an organic but complicated tectonic stress field. These three constitute a holistic picture of seismicity. The relationships among these different facets need refining by further study.

Aftershocks are non-independent events. Seismologists remove aftershocks before statistical analysis believing that they will then analyse independent events. We know for the Mediterranean area that the earthquake history is one of non-independent events even without aftershocks. We are now also able to extend these analyses to preliminary identification of seismic cycles within the non-independent process in time. “Memory” in the seismic process should be a consideration in seismic hazard analysis.

Acknowledgment

This research was supported by grant CEC (NNE5/1999/381: SHIELDS).

References

1. Anis, A. A. and Lloyd, E. H., The expected value of the adjusted rescaled Hurst range of independent normal summands, *Biometrika*, **63**. (1976)
2. Burton, P.W. and Xu, Y., Rescaled range analysis providing preliminary evidence for time-varying seismicity in Greece and Europe, in *Proc. XXVII Gen. Ass. European Seismological Commission* (2000) pp. 197-202.
3. Fu, S. and Liu, B., *Seismology (in Chinese)*, (Seismological Press. Beijing 1991)
4. Gardner, J. K. and Knopoff, L., Is the sequence of earthquakes in southern California with aftershocks removed, Poissonnian? *Bull. Seism. Soc. Am.*, **64** (1991) pp. 1363-1367.
5. Hurst, H. E., Long-term storage capacity of reservoirs, *Transactions of the American Society of Civil Engineers*, **116** (1951).
6. Johnson, S. M. and Sheridan, J. M., Distinguishing between random and nonrandom pattern in the energy release of great earthquakes, *J. Geophy. Res*. **102** (1997) pp. 2853-2855.
7. Lomnitz. C., *Fundamentals of earthquake prediction* (John Wiley & Son. Inc., New York 1994)
8. Main. I. G. and Burton. P. W., Moment-magnitude scaling in the Aegean area, *Tectonophysics* **179** (1990) pp. 273-285.
9. Mandelbrot. B., Statistical methodology for non-periodic cycles: from the covariance to R/S analysis, *Annals of Economics Social Measurement* **1** (1972).
10. Peters, E., *Fractal market analysis* (John Wiley & Sons, Inc. New York, 1994).
11. Peters, E., *Chaos and order in the capital markets* (John Wiley & Sons. Inc., New York, 1996).
12. Romanowicz, B., The spatiotemporal pattern in the energy release of great earthquake, *Science* **260** (1993) pp. 1923-1926.
13. Udias. A. and Buform, E., Regional stresses in the Mediterranean region derived from focal mechanism of earthquakes, in *Recent evolution and seismicity of the Mediterranean region*, edited by Boschi. E.. Mantovani, E. and Morelli, A. (Kluwer Academic Publisher, London, 1993) pp. 261-268.
14. Xu, Y. and Burton, P. W., Microearthquake swarms: scaling and lacunarity, *Geophys. J. Int*., **131** (1997) pp. F1-F8.
15. Xu, Y. and Burton, P. W., Spatial fractal evolution and hierarchies for microearthquakes in central Greece, *Pure & Applied Geophys*., **154** (1999) pp. 73-99.

LOCALIZED PRINCIPAL COMPONENTS

ANTOINE SAUCIER
École Polytechnique de Montréal, C.P. 6079, Succ. Centre-ville, Montréal (Québec), Canada, H3C 3A7
E-mail: antoine@cerca.umontreal.ca

We introduce a new class of principal components that are *localized* in space. Like classical principal components, these functions correlate best with the signal, within the limits of a localization constraint. These *localized principal components* can be regarded as *optimal interpolators* for a given signal. In that sense, they have something in common with some of the wavelets presented in the literature.

1 Introduction

Principal components analysis (i.e. Karhunen-Loève expansions) is a powerful tool in signal processing [1] [2] and statistics [3] [4]. Principal components (PCs) form a basis of functions for which the convergence rate is highest (on average) for a given collection of functions. It follows that PCs are often used for compression purposes.

At the question "Why do wavelets work?", Wim Sweldens [5] answers that "building blocks that have space and frequency localization ... will be able to reveal the internal correlation structure of the data sets." Conversely, we will attempt to define functions that correlate best with a signal, but that also have space localization. The resulting functions would then be *adapted* to the signal of interest.

In this paper, we introduce a new class of principal components that are *localized* in space. Like classical PCs, these functions are *derived* from the correlation matrix of the signal. They also *correlate best* with the signal, within the limits imposed by a localization constraint.

This paper is structured as follows. In section 2, we revisit and illustrate Karhunen-Loève expansions. Our presentation based on Lagrange multipliers is inspired from Karson [3]. In section 3, we introduce LPCs and show how they can be computed. In the last part, we examine the problem of building complete bases formed of LPCs.

2 Principal Component Analysis Revisited

2.1 Basics and Notations

Principal component analysis (PCA) can be regarded as a way to compress the representation of a collection of vectors. If these *random vectors* $\mathbf{V}$ belong to a vector space of finite dimension N, then they can be expanded on an orthonormal basis $(\mathbf{e}_1, \mathbf{e}_2, ..., \mathbf{e}_N)$ and be written in the form

$$\mathbf{V} = \sum_{i=1}^{N} X_i \, \mathbf{e}_i \tag{1}$$

where the *random coordinate* $X_i = \mathbf{e}_i \cdot \mathbf{V}$ is the projection of $\mathbf{V}$ on $\mathbf{e}_i$, and where "$\cdot$" denotes a scalar product. There is an infinity of orthonormal bases, but one of them has a convergence rate which is highest (on average) *for a given collection of vectors*: The Karhunen-Loève basis. This basis is composed of the *eigenvectors* $\mathbf{\Psi}_k, k = 1, 2, ..., N$ of the correlation matrix defined by $C_{i,j} = \langle X_i \ X_j \rangle$, where $\langle ... \rangle$ denotes an expectation value. The eigenvectors are sorted according to decreasing eigenvalues, i.e. $\mathbf{\Psi}_1$ has the largest eigenvalue. The $\mathbf{\Psi}_k$s are also called the *principal components* of $\mathcal{S}$. It will be useful to review briefly the origin of this result.

In this paper, the vectors of interest are digital *samples* taken from a *stationary* 1D signal. The signal $S(z)$ is sampled from a *window* at discrete equidistant spatial coordinates $z_i, i = 1, 2, ..., N$ and the *sampled function* $\mathbf{V} = (S(z_1), S(z_2), ..., S(z_N))$ is a N-dimensional vector. The collection of samples obtained from all possible windows of a given size N defines the ensemble $\mathcal{S}(N)$ (the window size is related to N by $L_w = (N-1)\Delta z$, where Δz is the spacing between consecutive coordinates). The vectors can always be *centered*, i.e. the mean vector $\langle \mathbf{V} \rangle$ will be subtracted from each element of $\mathcal{S}(N)$. We can therefore assume in the following that $\langle \mathbf{V} \rangle = 0$. In practice, the expectation values are estimated by averaging over $\mathcal{S}(N)$, i.e. by *spatial averaging*.

The *canonical basis* of $\mathbb{R}^N$ is formed of the orthonormal vectors $(1, 0, 0, ..., 0)$, $(0, 1, 0, ..., 0), ..., (0, 0, 0, ..., 1)$ that will be denoted by $\mathbf{u}_1, \mathbf{u}_2, ..., \mathbf{u}_N$ respectively. We can always expand $\mathbf{V}$ on this basis, i.e.

$$\mathbf{V} = \sum_{i=1}^{N} S(z_i) \ \mathbf{u}_i \tag{2}$$

We search for another orthonormal basis $\mathbf{e}_i$, $i = 1, 2, ..., N$ that converges as fast as possible in $\mathcal{S}(N)$. Since $\mathbf{V} = \sum_{i=1}^{N} X_i \ \mathbf{e}_i$, where $X_i = S(z_i)$, then $||\mathbf{V}||^2 = \mathbf{V} \cdot \mathbf{V} = \sum_{i=1}^{N} X_i^2$. The contribution of $\mathbf{e}_1$ to $||\mathbf{V}||^2$ is X_1^2, and therefore its mean contribution is $\langle X_1^2 \rangle$. For a fast convergence, a suitably chosen $\mathbf{e}_1$ should therefore maximize $\langle X_1^2 \rangle$.

Since $X_1 = \mathbf{e}_1 \cdot \mathbf{V}$, then the optimization problem consists in finding an $\mathbf{e}_1$ that *maximizes* $\langle (\mathbf{e}_1 \cdot \mathbf{V})^2 \rangle$ with the normalization constraint $\mathbf{e}_1 \cdot \mathbf{e}_1 = 1$. This is a classic case for a Lagrange multipliers method. We form the function

$$U = \langle (\mathbf{e}_1 \cdot \mathbf{V})^2 \rangle - \lambda (\mathbf{e}_1 \cdot \mathbf{e}_1 - 1) \tag{3}$$

where λ is a Lagrange multiplier. If we denote the components of $\mathbf{e}_1$ in the canonical basis $\{\mathbf{u}_j, j = 1, ..., N\}$ by α_i, $i = 1, 2, ..., N$, then (3) takes the explicit form

$$U = \langle \left(\sum_{i=1}^{n} X_i \ \alpha_i \right)^2 \rangle - \lambda \left(\sum_{i=1}^{n} \alpha_i^2 - 1 \right) \tag{4}$$

Expanding and averaging (4) then yields

$$U = \sum_{i=1}^{N} \sum_{j=1}^{N} \langle X_i X_j \rangle \ \alpha_i \alpha_j - \lambda \left(\sum_{i=1}^{n} \alpha_i^2 - 1 \right) \tag{5}$$

The last expression reveals the essential role of the positive definite *correlation matrix* $C_{i,j} \equiv \langle X_i \ X_j \rangle$ in this optimization problem. This last equation can be rewritten in the equivalent vector form

$$U = \mathbf{e}_1 \cdot C \ \mathbf{e}_1 - \lambda (\mathbf{e}_1 \cdot \mathbf{e}_1 - 1) \tag{6}$$

The extremum conditions $\partial U/\partial \alpha_k = 0,\ k = 1, 2, ..., N$ lead to the system of equations

$$\mathrm{C}\ \mathbf{e}_1 = \lambda\ \mathbf{e}_1 \tag{7}$$

and therefore $\mathbf{e}_1$ must be one of the eigenvectors of C. The one to choose is the one for which $\mathbf{e}_1 \cdot \mathrm{C}\ \mathbf{e}_1$ is largest, but according to (7) we have $\mathbf{e}_1 \cdot \mathrm{C}\ \mathbf{e}_1 = \mathbf{e}_1 \cdot \lambda\ \mathbf{e}_1 = \lambda$ because $\|\mathbf{e}_1\|=1$. Hence *it is the eigenvector with the largest eigenvalue that should be selected as the first basis vector.* This vector is also called the *first principal component* of the set $\mathcal{S}(N)$.

Next, we may search for a *second* vector $\mathbf{e}_2$ that is normed, orthogonal to $\mathbf{e}_1$, and that maximizes $\mathbf{e}_2 \cdot \mathrm{C}\ \mathbf{e}_2$. The Lagrange function of this problem is

$$U = \mathbf{e}_2 \cdot \mathrm{C}\ \mathbf{e}_2 - \lambda(\mathbf{e}_2 \cdot \mathbf{e}_2 - 1) - \gamma\ \mathbf{e}_1 \cdot \mathbf{e}_2 \tag{8}$$

where λ and γ are two Lagrange multipliers. The extremum conditions $\partial U/\partial \alpha_k = 0,\ k = 1, 2, ..., N$ then lead to the system

$$\mathrm{C}\ \mathbf{e}_2 = \lambda\ \mathbf{e}_2 + \frac{\gamma}{2}\ \mathbf{e}_1 \tag{9}$$

Taking the scalar product of (9) by $\mathbf{e}_1$ and taking into account the orthogonality of $\mathbf{e}_1$ and $\mathbf{e}_2$ yields $\frac{\gamma}{2} = \mathbf{e}_1 \cdot \mathrm{C}\ \mathbf{e}_2$. C is symmetric and Hermitian, therefore $\frac{\gamma}{2} = \mathbf{e}_1 \cdot \mathrm{C}\ \mathbf{e}_2 = \mathrm{C}\ \mathbf{e}_1 \cdot \mathbf{e}_2 = \lambda_1 \mathbf{e}_1 \cdot \mathbf{e}_2 = 0$. Replacing $\gamma = 0$ in (9) leads to $\mathrm{C}\ \mathbf{e}_2 = \lambda\ \mathbf{e}_2$ and therefore $\mathbf{e}_2$ is also an eigenvector of C. The one selected must be orthogonal to $\mathbf{e}_1$ and have the largest eigenvalue possible, hence $\mathbf{e}_2$ *is the eigenvector with the second largest eigenvalue*, also called the *second principal component of S*. The other basis vectors can be derived by pursuing the same constrained optimization procedure. The final result is that the optimal basis of vectors is composed of the eigenvectors of C sorted according to decreasing eigenvalues.

2.2 Example

We consider a white noise with a uniform probability distribution in the range $[-1/2, 1/2]$. This noise, that will be denoted by $\mathrm{W}(-1/2, 1/2)$, is produced by drawing successive iterates from a pseudorandom number generator. We took a 10000 points sample of this noise and smoothed it over five points (using sliding window averaging), which resulted in a signal denoted by $S(i)$ (figure 1). Firstly, we estimated the autocorrelation function $R(n) = \langle S(i)S(i+n)\rangle_{\mathrm{S}},\ n = 0, 2, ..., 40$, where $\langle ... \rangle_{\mathrm{S}}$ denotes a spatial average. Secondly, a symmetric correlation matrix $\mathrm{C}(41\times41)$ was obtained from $R(n)$ with $\mathrm{C}_{i,j} \equiv R(|i-j|+1)$. Finally, the 41 normalized eigenvectors of this matrix were computed and sorted according to decreasing eigenvalues, and plotted in figure 1.

The first principal component, which in this case is $\cap$-shaped (figure 1), is the one that correlates best with the signal. A 41 points sample of this signal is shown in upper left plot of figure 1. It is interesting to observe that the principal components are *either odd or even functions with respect to their center.* This symmetry must be inherited from the properties of the correlation matrix, which is symmetric, non-negative definite and constant along diagonal lines (i.e. banded).

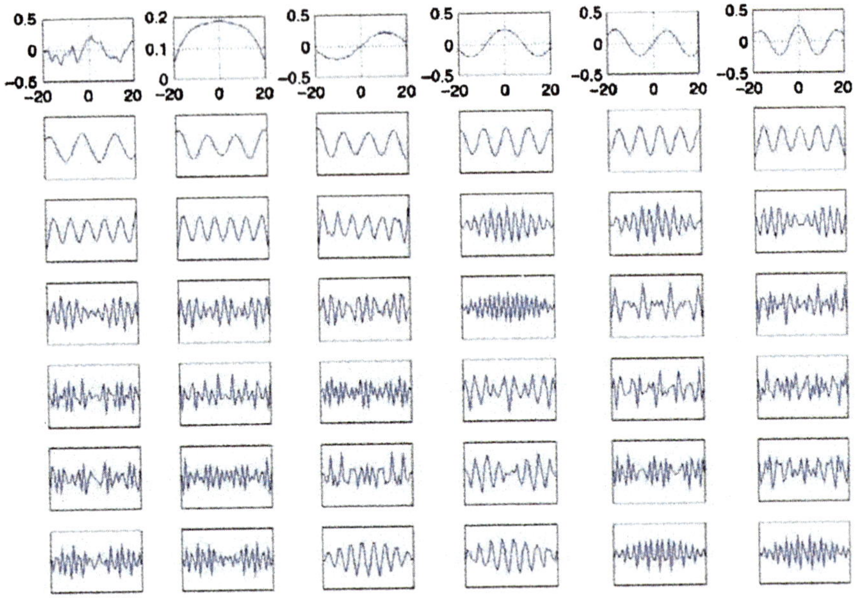

Figure 1. Principal components of a white noise W$(-1/2, 1/2)$ smoothed over 5 points. The upper left plot is a 41 points sample of this signal, while the other plots are the principal components sorted according to decreasing eigenvalues. The first principal component is therefore located on top, second plot from the left. When left blank, the horizontal axis graduation ranges from -20 to 20, while the vertical axis ranges from -0.5 to 0.5.

3 First Principal Component with a Localization Constraint

3.1 Theory

We will now search for a function $\mathbf{e}_1 = (\alpha_1, ..., \alpha_N)$ that maximizes $\mathbf{e}_1 \cdot \mathrm{C}\ \mathbf{e}_1$, as previously, but which in addition is *localized spatially*. In other words, we look for a function that describes the data best but that has a *limited spatial extent* (which is typically not the case in classical PCA). We assume that the function is localized around the *window center*, and that the localization constraint takes the form

$$\sum_{i=1}^{N} p(\alpha_i)\ \ell_i^2 = \delta_1^2 \tag{10}$$

where $\ell_i = i - i_c$, $i_c = (1 + N)/2$ is the window center, δ_1 is the *width* of $\mathbf{e}_1$ and $p(\alpha) \geq 0$ is a *probability function* (yet unspecified) that satisfies the normalization constraint

$$\sum_{i=1}^{N} p(\alpha_i) = 1 \tag{11}$$

The definition (10) is identical to the variance for a discrete random variable α. The constraint

$$\|\mathbf{e}_1\| = \sum_{i=1}^{N} \alpha_i^2 = 1 \tag{12}$$

on the norm of $\mathbf{e}_1$ must also be satisfied. The Lagrange function of this problem is

$$U = \sum_{i=1}^{N}\sum_{j=1}^{N} \mathrm{C}_{i,j}\alpha_i\alpha_j - \lambda\left(\sum_{i=1}^{N}\alpha_i^2 - 1\right) - \mu\left(\sum_{i=1}^{N} p(\alpha_i) - 1\right) - \gamma\left(\sum_{i=1}^{N} p(\alpha_i)\,\ell_i^2 - \delta_1^2\right) \tag{13}$$

where λ, μ and γ are Lagrange multipliers. Many choices are *a priori* possible for $p(\alpha)$. However, if we choose $p(\alpha) = \alpha^2$, then the constraints (11) and (12) become identical, which allows us to use the simpler Lagrange function

$$U = \sum_{i=1}^{N}\sum_{j=1}^{N} \mathrm{C}_{i,j}\alpha_i\alpha_j - \lambda\left(\sum_{i=1}^{N}\alpha_i^2 - 1\right) - \mu\left(\sum_{i=1}^{N}\alpha_i^2\,\ell_i^2 - \delta_1^2\right) \tag{14}$$

The extremum condition $\partial U/\partial\alpha_n = 0$ with $n = 1, 2, ..., N$ leads to

$$\sum_{j=1}^{N} \mathrm{C}_{n,j}\;\alpha_j = (\lambda + \mu\;\ell_n^2)\;\alpha_n, \qquad n = 1, 2, ..., N. \tag{15}$$

which can be rewritten in the equivalent form

$$\sum_{j=1}^{N} \mathrm{C}_{n,j}\;\alpha_j = \lambda(1 + r\;\ell_n^2)\;\alpha_n, \qquad n = 1, 2, ..., N. \tag{16}$$

where $r \equiv \mu/\lambda$. As long as $r \neq -\ell_n^2$, $n = 1, 2, ..., N$, we may divide both sides by $(1 + r\;\ell_n^2)$, which yields

$$\sum_{j=1}^{N} \frac{\mathrm{C}_{n,j}}{1 + r\;\ell_n^2}\;\alpha_j = \lambda\;\alpha_n, \qquad n = 1, 2, ..., N. \tag{17}$$

Equation (17) shows that $\mathbf{e}_1$ is an eigenvector of a new operator $\tilde{\mathrm{C}}(r)$ defined by

$$\tilde{\mathrm{C}}_{i,j}(r) \equiv \frac{\mathrm{C}_{i,j}}{1 + r\;\ell_i^2} \tag{18}$$

In the following, $\tilde{\mathrm{C}}_{i,j}(r)$ will be called the *localized correlation operator*. Is is emphasized that $\tilde{\mathrm{C}}(r)$ is *not symmetric* unless $r = 0$, whereas C is always symmetric. (17) can be rewritten in the equivalent vectorial form

$$\tilde{\mathrm{C}}(r)\;\mathbf{e}_1(r) = \lambda(r)\;\mathbf{e}_1(r) \tag{19}$$

The problem is therefore to solve the system (19) for $\mathbf{e}_1(r)$, λ and r, while satisfying the two constraints (12) and (10). For any given value of r, we can use standard methods to find the N eigenvectors $\mathbf{\Psi}_i(r), i = 1, 2, ..., N$. The eigenvector of interest, denoted by $\mathbf{\Psi}^*(r)$, is the one for which $\mathbf{\Psi}_k(r) \cdot \mathrm{C}\;\mathbf{\Psi}_k(r)$ is maximum, which is not necessarily the one with the largest eigenvalue. Indeed, the $\mathbf{\Psi}_i(r)$s are eigenvectors of $\tilde{\mathrm{C}}(r)$, *not* of C. The *energy function* $E(r) = \mathbf{\Psi}^*(r) \cdot \mathrm{C}\ \mathbf{\Psi}^*(r)$ measures the maximum correlation that can be achieved between $\mathbf{e}_1$ and the signal for given r. $\mathbf{\Psi}^*(r)$ can always be normalized, which yields a vector that satisfies both (19) and the normalization (12). However, r must be chosen to satisfy the localization constraint $\delta^2(r) = \sum_{i=1}^{N} (\alpha_i(r))^2\ \ell_i^2 = \delta_1^2$, where δ_1 is a user-given width parameter. In the next section, we will examine on a few examples how this problem can be solved in practice.

3.2 Computation of Localized Principal Components

The first step is to estimate the correlation matrix C, which is done as in section (2.2). The $\mathbf{\Psi}^k(r)$s are eigenvectors of $\tilde{\mathrm{C}}(r)$. According to definition (18), $\tilde{\mathrm{C}}(r)$ is *singular* for $r = r_i \equiv -1/\ell_i^2, i = 1, 2, ..., N$. Since $\ell_i = i - i_c$, where $i_c = (N+1)/2$, then $\min\{\ell_i\} = 1$ and $\ell_{\max} \equiv \max\{\ell_i\} = (N-1)/2$ and therefore the singularities belong to the interval $r_i \in [-1, -1/\ell_{\max}]$.

The second step is to compute $\tilde{\mathrm{C}}(r)$ for $-2 \leq r \leq 3$, using discrete values of r, and taking care of avoiding the N singular values $r_i = -1/\ell_i^2, i = 1, 2, ..., N$. For each r, we then compute the eigenvectors $\mathbf{\Psi}_k(r), k = 1, ..., N$ of $\tilde{\mathrm{C}}(r)$, and select the one for which the energy $E(r) = \mathbf{\Psi}_k(r) \cdot \mathrm{C}\ \mathbf{\Psi}_k(r)$ is maximum. Next, the width of this vector is computed with $\delta^2(r) = \sum_{i=1}^{N} (\alpha_i(r))^2\ \ell_i^2$. The LPC problem is solved if we can find a value of r that gives a user-defined value of $\delta(r)$, while maximizing the energy.

As an example, we applied the above procedure to a 50000 points white noise $\mathrm{W}(-1/2, 1/2)$ smoothed on 10 points, and obtained the functions $\delta(r)$ and $E(r)$ which are shown in figure 2. The peak corresponds to the rightmost singularity $r_{\max} = -1/\ell_{\max}$, and is close to zero. Experiments on various signals showed that the energy function is *always* significantly larger on the right side of this peak. The plot of $\delta(r)$ (figure 2) reveals that the equation $\delta(r) = \delta_1$ has *at least* two roots, i.e. one on each side of the peak. However, the one of interest maximizes $E(r)$ and is therefore on the right side of the peak. To get $\delta_1 \geq 1/2$, it is usually sufficient to consider only the range $r_{\max} < r \leq 3$.

3.3 Localized Principal Components for Smoothed White Noise

The LPCs were computed from a 10000 points white noise $\mathrm{W}(-1/2, 1/2)$ smoothed on 2 and 5 points. A window size $N = 41$ was used in both cases. The LPCs are shown in the figures 3 and 4, using 20 values of δ_1 equally spaced between $\delta_{\min} \equiv 1/2$ and $\delta_{\max} \equiv 1/2\sqrt{(N^2-1)/3}$ ($\delta_{\max}$ correspond to the variance of a discrete uniform probability distribution on N points). With $N = 41$, we get $\delta_{\max} \approx 11.8$.

Let us first consider the signal smoothed on 2 points (figure 3). For small δs, the LPC is bell shaped. This shape is obtained for most signals at small δs. Indeed, a strong localization constraint does not leave room for anything but a highly peaked function. The plots 2-3-4 of figure 3 exhibit LPCs with a slight dip in the negative on each side of the peak. These functions resemble several wavelets found in the literature, e.g. the Mexican hat wavelet [6], the Deslaurier-Dubuc

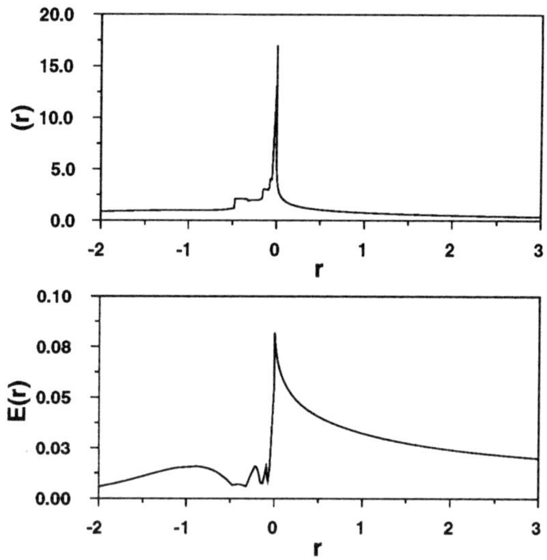

Figure 2. $\delta(r)$ and E(r) obtained for a white noise $W(-1/2, 1/2)$ smoothed on 10 points.

interpolation function of degree three [7], or biorthogonal wavelets [8]. As δ increases, a wave-shaped LPC appears (from the 6th plot in figure 3), and progressively evolves toward a nearly sinusoidal function for the largest δs.

Let us consider next the signal smoothed on 5 points (figure 4). For small δs, the results are similar, i.e. the LPC is bell shaped. For larger δs, the LPC tends to keep a bell shape but develops a positive background value. At the largest δs, the LPC becomes symmetric and bimodal. Similar results were obtained with a white noise $W(-1/2, 1/2)$ smoothed on 10 points.

4 Toward the Construction of a Basis of Localized Principal Components

Having obtained the first LPC $\mathbf{e}_1$, it would be interesting to construct other vectors that would form an orthonormal basis. Let us search for a *second* normalized vector $\mathbf{e}_2 = (\beta_1, ..., \beta_N)$ that maximizes $\mathbf{e}_2 \cdot \mathrm{C}\, \mathbf{e}_2$, which is orthogonal to $\mathbf{e}_1$ and localized in the same way (i.e. around the window center), yet *broader* than $\mathbf{e}_1$. The Lagrange function of this problem takes the form

$$U = \mathbf{e}_2 \cdot \mathrm{C}\ \mathbf{e}_2 - \lambda\left(\mathbf{e}_2 \cdot \mathbf{e}_2 - 1\right) - \mu\left(\sum_{i=1}^{N} \beta_i^2 \ell_i^2 - \delta_2^2\right) - 2\gamma\ \mathbf{e}_2 \cdot \mathbf{e}_1 \qquad (20)$$

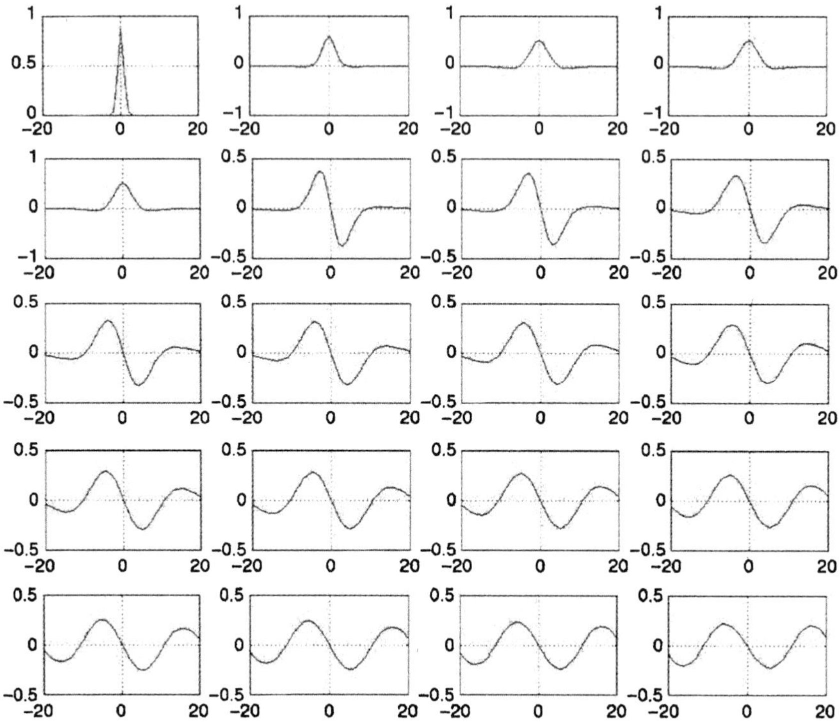

Figure 3. Localized principal components for a white noise W(−1/2, 1/2) smoothed on 2 points. δ_1 is minimum ($\delta_1 = 1/2$) for the upper-left figure, maximum for the lower-right figure ($\delta_1 = 11.8$), and increases linearly for the intermediate figures.

where $\delta_2 > \delta_1$ is the width of $\mathbf{e}_2$. (20) takes the explicit form

$$U = \sum_{i=1}^{N}\sum_{j=1}^{N} \mathrm{C}_{i,j}\beta_i\beta_j - \lambda\left(\sum_{i=1}^{N}\beta_i^2 - 1\right) - \mu\left(\sum_{i=1}^{N}\beta_i^2\,\ell_i^2 - \delta_2^2\right) - 2\gamma\sum_{i=1}^{N}\beta_i\alpha_i \quad (21)$$

The extremum condition $\partial U/\partial\beta_n = 0$ with $n = 1, 2, ..., N$ leads to

$$\sum_{j=1}^{N} \mathrm{C}_{n,j}\beta_j = \lambda\left(1 + r\,\ell_n^2\right)\beta_n + \gamma\,\alpha_n, \quad n = 1, 2, ..., N \quad (22)$$

where $r = \mu/\lambda$, as previously. If $r \neq -1/\ell_n^2$ with $n = 1, 2, ..., N$, then (22) can be rewritten in the form

$$\sum_{j=1}^{N} \frac{\mathrm{C}_{n,j}}{1 + r\,\ell_n^2}\,\beta_j = \lambda\,\beta_n + \frac{\gamma}{1 + r\,\ell_n^2}\,\alpha_n, \quad n = 1, 2, ..., N \quad (23)$$

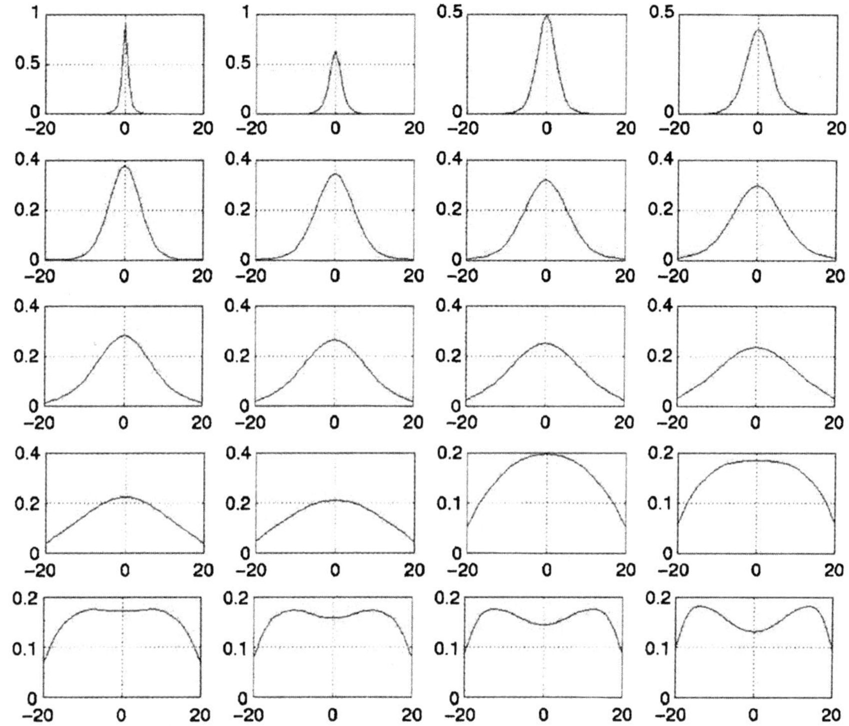

Figure 4. Localized principal components for a white noise W(−1/2, 1/2) smoothed on 5 points. δ_1 is minimum ($\delta_1 = 1/2$) for the upper-left figure, maximum for the lower-right figure ($\delta_1 = 11.8$), and increases linearly for the intermediate figures.

Introducing the diagonal operator $\Lambda_{i,j}(r) \equiv \delta_{i,j}/(1 + r\ \ell_i^2)$, (23) can be rewritten in the equivalent vectorial form

$$\tilde{\mathrm{C}}(r)\ \mathbf{e}_2 = \lambda\ \mathbf{e}_2 + \gamma\ \Lambda(r)\ \mathbf{e}_1 \tag{24}$$

with

$$||\mathbf{e}_2|| = 1\ \ ,\ \ \mathbf{e}_1 \cdot \mathbf{e}_2 = 0\ \text{ and } \sum_{i=1}^{N} \beta_i^2 \ell_i^2 = \delta_2^2 \tag{25}$$

At this stage in section 2.1, it was shown that for principal components (i.e. *non* localized) the Lagrange multiplier γ vanishes, and consequently the other vectors are simply the eigenvectors of C. Unfortunately, this simplification does not occur here because $\tilde{\mathrm{C}}(r)$ is *not* symmetric (hermitian). Indeed, taking the scalar product of (24) by $\mathbf{e}_1$ yields

$$\mathbf{e}_1 \cdot \tilde{\mathrm{C}}(r)\ \mathbf{e}_2 = \lambda\ \mathbf{e}_1 \cdot \mathbf{e}_2 + \gamma\ \mathbf{e}_1 \cdot \Lambda(r)\ \mathbf{e}_1 \tag{26}$$

Since $\tilde{C}(r)$ is not hermitian, we *cannot use* the chain of identities $\mathbf{e}_1 \cdot \tilde{C}(r)\,\mathbf{e}_2 = \tilde{C}(r)\,\mathbf{e}_1 \cdot \mathbf{e}_2 = \lambda\,\mathbf{e}_1 \cdot \mathbf{e}_2 = 0$, and consequently $\gamma \neq 0$. This result implies that the solution of the optimization problem (24)-(25) is no longer equivalent to an eigenvector problem, but rather leads to a more complex system of equations. Its solution will be left for future work.

5 Conclusions

We proposed a new approach to define *localized functions* that are *adapted* to any given type of stationary 1D signal. These *Localized principal components*, or LPCs, have the best possible correlation with the signal within the limits imposed by their limited width. LPCs can be regarded as *adaptive optimal interpolators* for a given signal. In that sense, they are similar to some of the wavelets found in the literature. If a complete orthonormal basis of multiscale LPCs could be constructed, then such a basis could be useful for the detection and filtration of outliers or *localized perturbations* within a signal.

6 Acknowledgments

Antoine Saucier is grateful to Christian Turgeon for his help with the matlab programming in this project. This project was supported financially by cerca (Centre de Recherche en Calcul Appliqué) and École Polytechnique de Montréal.

References

1. Stéphane Mallat. *A wavelet tour of signal processing.* Academic Press, San diego, London, Boston, New York, Sydney, Tokyo, Toronto, 1998. Section 9.1.3. for principal components.
2. Antoine Saucier and Jiri Muller. A generalization of multifractal analysis based on polynomial expansions of the generating function. In Dekking, Lévy Véhel, Lutton, and Tricot, editors, *Fractals: Theory and Applications in Engineering*, pages 81–91. Springer-Verlag, London, 1999.
3. Marvin J. Karson. *Multivariate statistical methods.* The Iowa state university press, Ames, Iowa, U.S.A, 1982. Chapter 8 for principal components analysis.
4. Jean-Paul Chilè and pierre Delfiner. *Geostatistics - Modeling Spatial Uncertainty.* John Wiley & sons, inc, New York, Chichester, Weinheim, Brisbane, Singapore, Toronto, 1999. Section 5.6.6 for principal component analysis.
5. W. Sweldens. Wavelets: What next? *Proc. IEEE*, 84(4):680–685, 1996.
6. Stéphane Mallat. *A wavelet tour of signal processing.* Academic Press, London, New York, Toronto, 1998. Section 4.3.1 : Mexican hat wavelet.
7. Stéphane Mallat. *A wavelet tour of signal processing.* Academic Press, San diego, London, Boston, New York, Sydney, Tokyo, Toronto, 1998. Section 7.6 for the Deslaurier-Dubuc interpolation function of degree 3.
8. Stéphane Mallat. *A wavelet tour of signal processing.* Academic Press, San diego, London, Boston, New York, Sydney, Tokyo, Toronto, 1998. Section 7.4.3 for biorthogonal wavelets.

CLUSTER FORMATION AND CLUSTER SPLITTING IN A SYSTEM OF GLOBALLY COUPLED MAPS

O. POPOVYCH

Department of Physics, University of Potsdam, PF 601553, Potsdam, Germany
E-mail: sascha@stat.physik.uni-potsdam.de

YU. MAISTRENKO

Institute of Mathematics, National Academy of Sciences of Ukraine, 01601 Kyiv, Ukraine
E-mail: maistrenko@imath.kiev.ua

E. MOSEKILDE

Department of Physics, Technical University of Denmark, 2800 Kgs. Lyngby, Denmark
E-mail: Erik.Mosekilde@fysik.dtu.dk

We study the transition from full synchronization (coherent motion) to two-cluster dynamics for a system of N globally coupled logistic maps. As the nonlinearity parameter is increased for the individual map, new periodic and strongly asymmetric two-cluster states appear in the same order as the periodic windows arise in the logistic maps. General expressions for the stability of K-cluster states in the full N-dimensional phase space are derived, and we show how transverse period-doubling bifurcations can lead to cluster splitting, e.g., to the transition from period-3 two-cluster dynamics to period-6 three-cluster dynamics. Bifurcations in the directions of the cluster subspace, on the other hand, only lead to more complex temporal behavior.

Keywords: Coupled chaotic maps; clustering; stability; cluster splitting

1 Introduction

The purpose of this paper is to study cluster formation in the ensemble of N globally coupled one-dimensional maps

$$x_i(n+1) = (1-\varepsilon) f\left(x_i(n)\right) + \frac{\varepsilon}{N} \sum_{j=1}^{N} f\left(x_j(n)\right), \quad i = 1, 2, \ldots, N, \tag{1}$$

where $\mathbf{x} = \{x_i(n)\}_{i=1}^{N}$ is the state vector at time $n = 0, 1, \ldots$, and $\varepsilon \in \mathbb{R}$ is the coupling parameter. $f : \mathbb{R} \to \mathbb{R}$ is the logistic map $f(x) = ax(1-x)$, and a is the nonlinearity parameter of this map. We shall also examine the processes by which a given cluster structure, via loss of synchronization for the oscillators in one of the clusters, can break up and form a new cluster structure.

Systems of coupled maps were introduced by Kaneko[1] and Waller and Kapral[2] in order to study spatio-temporal dynamics of extended systems and pattern formation[3]. Kaneko[4] has also considered globally coupled maps of the form (1) as a model of large systems of identical chaotic oscillators interacting via a mean field. Since then, globally coupled systems have attracted a rapidly growing interest in the scientific community, see, e.g., Ref. 5 and references therein. Examples of such systems are typically found in the biological sciences, where groups of cells or functional units are controlled by feedback signals that depend on their own aggregate behavior[6,7]. Systems of globally coupled chaotic oscillators may also arise in studies of Josephson junction arrays[8] and of multimode lasers[9]. Recently, Wang et al.[10] have provided experimental evidence of clustering in a system of globally coupled electrochemical reactors.

The simplest form of asymptotic dynamics that can arise in the globally coupled map system (1) is the fully synchronized (or coherent) state in which all elements display the same temporal behavior. In this state the motion is restricted to a one-dimensional invariant manifold $D_N = \{(x_1, x_2, \ldots, x_N) \mid x_1 = x_2 = \ldots = x_N\}$, the main diagonal in N-dimensional phase space, and along this manifold the dynamics is governed by the one-dimensional logistic map f. The coherent state is generally stable in the presence of a sufficiently strong global coupling.

Another form of asymptotic dynamics in system (1) takes the form of clustering (or partial synchronization[11]) where the population of oscillators splits into $K < N$ subgroups (clusters) such that all oscillators within a given cluster asymptotically move in synchrony. In the case, the coordinates of the state vector $\mathbf{x} = \{x_i\}_{i=1}^{N}$ can be represented as

$$\begin{array}{rcl} x_{i_1} = x_{i_2} = \ldots = x_{i_{N_1}} & \stackrel{def}{=} & y_1 \\ x_{i_{N_1+1}} = x_{i_{N_1+2}} = \ldots = x_{i_{N_1+N_2}} & \stackrel{def}{=} & y_2 \\ \cdots\cdots\cdots\cdots\cdots\cdots & & \\ x_{i_{N_1+N_2+\ldots+N_{K-1}+1}} = x_{i_{N_1+N_2+\ldots+N_{K-1}+2}} = \ldots = x_{i_N} & \stackrel{def}{=} & y_K. \end{array} \tag{2}$$

The positive integer N_j is the number of variables x_i belonging to the jth cluster, $j = 1, 2, \ldots, K$ so that $N_1 + N_2 + \ldots + N_K = N$. By virtue of the complete symmetry of the considered system, for any set $\{N_j\}$ the K-dimensional subspace defined by Eqs. (2) remains invariant for system (1).

Within the cluster subspace defined by (2) the dynamics is governed by the system of K coupled maps

$$y_i(n+1) = (1-\varepsilon) f(y_i(n)) + \varepsilon \sum_{j=1}^{K} p_j^{(K)} f(y_j(n)), \quad i = 1, \ldots, K. \tag{3}$$

This system is also a globally coupled map system, but with different weights $p_j^{(K)} = N_j/N, j = 1, 2, \ldots, K$, associated with the contribution of the jth cluster to the global coupling. By varying the parameters $p_j^{(K)}$ in (3) we can obtain the governing system for any possible K-cluster dynamics of the original system (1). We shall refer to the K-dimensional system (3) as F_K.

A necessary condition for the presence of stable K-cluster behavior in system (1) is that the system (3), with the assumed values of the parameters $p_j^{(K)}$, has an attractor $A^{(K)}$ that does not belong to any cluster subspace of lower dimension. For example, with even number of cites N system (1) may demonstrate symmetric two-cluster dynamics ($K = 2$ and $N_1 = N_2 = N/2$ in (2)) if the symmetric system F_2 (system (3) with $K = 2$ and with parameters $p_1^{(2)} = p_2^{(2)} = 0.5$) has a stable invariant set $A^{(2)} \not\subset D_2 = \{(y_1, y_2) \mid y_1 = y_2\}$.

Provided that the set $A^{(K)}$ is stable in the cluster subspace, the conditions for an attractor $A^{(K)}$ of system (3) to be stable in the whole N-dimensional phase space of system (1) are that it is also stable in the directions transverse to the corresponding cluster subspace. The transverse stability of $A^{(K)}$ may be asymptotic when the state attracts all trajectories from its neighborhood, or weak when $A^{(K)}$ is stable in the Milnor sense, i.e., it attracts a positive Lebesgue measure set of initial conditions[12].

2 Transverse stability of the partially synchronized state

Let us start by examining the conditions for transverse stability of the two-cluster state, $K = 2$ in (2). The Jacobian matrix of the N-dimensional map Φ defined by Eq. (1) has the following form:

$$\begin{bmatrix} f'(x_1)\left(1-\frac{N-1}{N}\varepsilon\right) & \frac{\varepsilon}{N}f'(x_2) & \cdots & \frac{\varepsilon}{N}f'(x_N) \\ \frac{\varepsilon}{N}f'(x_1) & f'(x_2)\left(1-\frac{N-1}{N}\varepsilon\right) & \cdots & \frac{\varepsilon}{N}f'(x_N) \\ \vdots & \vdots & \ddots & \vdots \\ \frac{\varepsilon}{N}f'(x_1) & \frac{\varepsilon}{N}f'(x_2) & \cdots & f'(x_N)\left(1-\frac{N-1}{N}\varepsilon\right) \end{bmatrix}. \tag{4}$$

Without loss of generality, we may renumerate the variables x_i in (1) with respect to the space index in such a way that state variables with subsequent indices belong to the same cluster

$$\begin{aligned} x_1 = x_2 = \ldots = x_{N_1} &\stackrel{def}{=} y_1 \\ x_{N_1+1} = x_{N_1+2} = \ldots = x_N &\stackrel{def}{=} y_2. \end{aligned} \tag{5}$$

Reduced on the subspace defined by Eqs. (5), the matrix (4) can be represented as

$$D\Phi = \begin{bmatrix} M(y_1) & L(y_2) \\ L^T(y_1) & N(y_2) \end{bmatrix},$$

where $M(y_1)$ and $N(y_2)$ are symmetric matrices of dimensions $N_1 \times N_1$ and $N_2 \times N_2$, respectively. Both matrices have the form of (4), depending on the variables y_1 and y_2, respectively. The matrix L (L^T is its transposed) is an $N_1 \times N_2$ rectangular matrix composed by elements of the form $\frac{\varepsilon}{N}f'(\cdot)$, i.e., $L(y) = [l_{ij}(y)], l_{ij}(y) = \frac{\varepsilon}{N}f'(y), i = 1, \ldots, N_1, j = 1, \ldots, N_2$.

The matrix $D\Phi$ has two distinct eigenvalues $\nu_1 = f'(y_1)(1-\varepsilon)$ and $\nu_2 = f'(y_2)(1-\varepsilon)$ of the multiplicities $N_1 - 1$ and $N_2 - 1$, respectively. The corresponding eigenvectors are

$$\begin{aligned} \mathbf{v}_{1,1} &= (\overbrace{1,-1,0,0,\ldots,0}^{N_1},\overbrace{0,0,0,0,\ldots,0}^{N_2}) \\ \mathbf{v}_{1,2} &= (1,0,-1,0,\ldots,0,0,0,0,0,\ldots,0) \\ &\cdots\cdots\cdots\cdots\cdots\cdots\cdots \\ \mathbf{v}_{1,N_1-1} &= (1,0,0,0,\ldots,-1,0,0,0,0,\ldots,0) \end{aligned}$$

and

$$\begin{aligned}\mathbf{v}_{2,1} &= (0,0,0,0,\ldots,0,1,-1,0,0,\ldots,0)\\ \mathbf{v}_{2,2} &= (0,0,0,0,\ldots,0,1,0,-1,0,\ldots,0)\\ &\ldots\ldots\ldots\ldots\ldots\ldots\ldots\ldots\\ \mathbf{v}_{2,N_2-1} &= (\underbrace{0,0,0,0,\ldots,0}_{N_1},\underbrace{1,0,0,0,\ldots,-1}_{N_2}).\end{aligned}$$

The two eigen-subspaces $V_1 = \{\mathbf{v}_{1,i}\}_{i=1}^{N_1-1}$ and $V_2 = \{\mathbf{v}_{2,j}\}_{j=1}^{N_2-1}$ are transverse to the synchronizing subspace defined by Eqs. (5) and $\dim V_1 \oplus V_2 = N - 2$.

Let now the two-dimensional system F_2 have an attractor $A^{(2)}$ that does not belong to the diagonal $D_2 = \{(y_1, y_2) \mid y_1 = y_2\}$. Stability of such a two-dimensional attractor (in the whole phase space $\mathbb{R}^N$ of system (1)) is controlled by the transverse Lyapunov exponents of $A^{(2)}$, which measure the average rate of growth of perturbations transverse to the two-cluster subspace. By virtue of the form of the transverse eigenvalues ν_1 and ν_2 and the fact that the eigenvectors do not depend on the phase coordinates, the transverse Lyapunov exponents of the two-cluster state are found to be

$$\begin{aligned}\lambda_{\perp,1}^{(2)} &= \lim_{k\to\infty}\frac{1}{k}\sum_{n=0}^{k-1}\ln\mid f'(y_1(n))(1-\varepsilon)\mid = \lim_{k\to\infty}\frac{1}{k}\sum_{n=0}^{k-1}\ln\mid f'(y_1(n))\mid + \ln|1-\varepsilon|\,,\\ \lambda_{\perp,2}^{(2)} &= \lim_{k\to\infty}\frac{1}{k}\sum_{n=0}^{k-1}\ln\mid f'(y_2(n))(1-\varepsilon)\mid = \lim_{k\to\infty}\frac{1}{k}\sum_{n=0}^{k-1}\ln\mid f'(y_2(n))\mid + \ln|1-\varepsilon|\,,\end{aligned} \tag{6}$$

evaluated for a typical trajectory $\{(y_1(n), y_2(n))\}_{n=0}^{\infty}$ in the attractor $A^{(2)}$. If both $\lambda_{\perp,i}^{(2)}$, $i = 1,2$ are negative, the attractor $A^{(2)}$ of the two-dimensional system F_2 is also an attractor, at least in the Milnor sense,[12] of the N-dimensional system (1)[13].

Hence, the procedure for finding stable *two-cluster states* of the N-dimensional system (1) is reduced to evaluating the two-dimensional system F_2 of the form (3) with $K = 2$ for different values of the parameter $p_1^{(2)}$ ($p_2^{(2)} = 1 - p_1^{(2)}$). First, we examine the existence of attractors $A^{(2)}$ of the system of two coupled maps F_2. Then two Lyapunov exponents $\lambda_{\perp,i}, i = 1,2$ of the form (6) are calculated to control the stability of $A^{(2)}$ in $\mathbb{R}^N$. For the parameter regions where $A^{(2)}$ exists and both Lyapunov exponents are negative, the system of N globally coupled maps (1) has an attracting two-cluster states. Its dynamics is given by the two-cluster attractor $A^{(2)}$ with a distribution of the maps between the clusters $Np_1^{(2)} : Np_2^{(2)}$. Note that this procedure does not depend on the number N of coupled oscillators in Eq. (1), but only on the ratio of the numbers of oscillators in the clusters. For example, if the two-dimensional system F_2 with $p_1^{(2)} = 1/3$ and $p_2^{(2)} = 2/3$ has an attractor $A^{(2)}$ and both the transverse Lyapunov exponents calculated by (6) are negative, then the N-dimensional system (1) will have corresponding stable two-cluster states for $N = 3\,(N_1 = 1, N_2 = 2), N = 6\,(N_1 = 2, N_2 = 4), N = 9\,(N_1 = 3, N_2 = 6), N = 12\,(N_1 = 4, N_2 = 8)$, etc.

By analogy, stability of a K-cluster state (2) with temporal dynamics given by an attractor $A^{(K)}$ of system (3) is controlled by K transverse Lyapunov exponents of the form

$$\lambda_{\perp,j}^{(K)} = \lim_{k\to\infty} \frac{1}{k} \sum_{n=0}^{k-1} \ln \mid f'(y_j(n)) \mid + \ln|1-\varepsilon|, \quad j=1,2,\ldots,K. \tag{6'}$$

When all the Lyapunov exponents are negative, $A^{(K)}$ provides the existence of stable K-cluster states for system (1).[a]

3 Periodic two-cluster states

By applying the above procedure we can determine the parameter combinations for which stable K-cluster dynamics exists. Figure 1a displays the regions (shown in black) in the (p,ε)-parameter plane (we denote $p = p_1^{(2)}$) where the system of N globally coupled maps displays stable stable two-cluster dynamics for $a = 4$. Figure 1b shows a similar diagram for $a = 3.84$ where the logistic map exhibits period-3 dynamics. The horizontal axis defines the distribution of oscillators between the two clusters $Np : N(1-p)$.

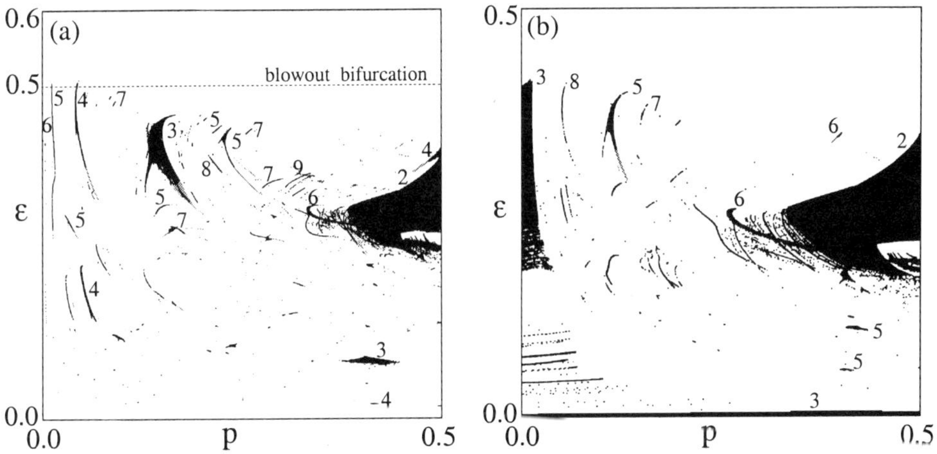

Figure 1. Regions of parameter plane where system (1) has attracting periodic two-cluster states with partition $\{p, 1-p\}$ are shown in black. Gray regions correspond to parameter values where a system F_2 still has attracting periodic cycles out of the diagonal D_2. The numbers indicate the temporal periods of the clustered dynamics. (a) $a = 4$ and (b) $a = 3.84$. For $a = 3.84$, regions of period-3 two-cluster dynamics are being formed both along the $\varepsilon = 0$ axis and along the $p = 0$ axis (b).

[a]Note, that the transverse Lyapunov exponent $\lambda_{\perp,j}^{(K)}$ enters the analysis with the multiplicity $N_j - 1$. For $N_j = 1$, the corresponding Lyapunov exponent is not involved in determining the stability.

The horizontal dashed line in Fig. 1a corresponds to the parameter-ε value ($\varepsilon_{bl} = 0.5$) at which the coherent (fully synchronized) chaotic state losses its stability via a blowout bifurcation[14]. As one can see from Fig. 1, the first two-cluster states to appear, when ε decreases, are highly asymmetric, i.e., states that correspond to small values of parameter p and for which the distributions of the oscillators between the clusters are strongly unequal.

The formation of stable two-cluster states in system (1) is associated with the appearance of stable off-diagonal periodic orbits in the two-dimensional cluster subspaces governed by system F_2. These periodic orbits appear to be stable in the transverse directions too. The phase diagram in Fig. 1 displays a surprising organization of the periodic regions (windows) to the right of the $p = 0$ axis: They follow the well-known sequence of windows of periodicity of the logistic map. Indeed, as one can see in Fig. 1 the widest asymmetric window is of period 3. To the right, the next relatively large window is of period 5, followed by windows of period 7 and 9. In between the period-3 and -5 windows, one can find a window of period 8. For smaller values of p (see Fig. 1a), one can find a period-adding sequence of windows of periods 4, 5, 6, etc. These periodic windows emerge in the (p, ε)-parameter plane through the $p = 0$ axis when, increasing parameter a, the logistic map passes through the windows of periodicity. For example, in Fig. 1b, the period-3 asymmetric window is shown when it emerges through the $p = 0$ axis at $a = 3.84$, i.e., at the same value when the logistic map develops a stable period-3 cycle.

To explain the mechanism for the emergence of the asymmetric periodic windows in the (p, ε)-parameter plane, let us examine the system F_2 of the form (3) with $K = 2$ and $p_1^{(2)} = 0$

$$\begin{aligned} y_1(n+1) &= (1-\varepsilon)f(y_1(n)) + \varepsilon f(y_2(n)) \\ y_2(n+1) &= f(y_2(n)). \end{aligned} \tag{7}$$

Choose a parameter-a value for which the logistic map f has a stable period-3 cycle γ_3 and consider the third iterate of the map $(x, y) \mapsto (\varphi(x, y), f(y))$, where $\varphi(x, y) = (1-\varepsilon)f(x) + \varepsilon f(y)$, assuming y to be one of the points of the cycle γ_3. We have $(x, y) \mapsto (\varphi_\varepsilon^3(x, y), y)$, where $\varphi_\varepsilon^3(x, y) = (1-\varepsilon)f\Big((1-\varepsilon)f\big[(1-\varepsilon)f(x) + \varepsilon f(y)\big] + \varepsilon f^2(y)\Big) + \varepsilon y$. Clearly, if $x = y$ then $\varphi_\varepsilon^3(y, y) = y$, which guarantees the existence of a symmetric (i.e., located on the diagonal D_2) period-3 cycle $\gamma_3^{(1)}$ for system (7). We are interested now in the existence of a stable period-3 cycle $\gamma_3^{(2)}$ out of the diagonal. To obtain it, let us find a value $x \neq y$ such that $\varphi_\varepsilon^3(x, y) = x$.

Consider deformations of the graph of $x \mapsto \varphi_\varepsilon^3(x, y)$ with decreasing $\varepsilon \in [0, 1]$. From the above expression for $\varphi_\varepsilon^3(x, y)$ it follows that $\varphi_1^3(x, y) = y$ and $\varphi_0^3(x, y) = f^3(x)$ for any x. Since φ_ε^3 smoothly depends on ε, its graph undergoes a smooth deformation from the horizontal line ($\varepsilon = 1$) up to the graph of the third iteration of the logistic map $f^3(x)$ ($\varepsilon = 0$). In Fig. 2a, the graphs of $\varphi_\varepsilon^3(x, y)$ as a function of x are plotted for different values of ε, when y is fixed to be one of the points of the cycle γ_3. With decreasing ε, the graph touches the line $\varphi_\varepsilon^3(x, y) = x$ and then starts to intersect it in two points A and B, see Fig. 2a. This leads to the formation of the stable period-3 cycle $\gamma_3^{(2)}$ for system (7). Point A in Fig. 2a corresponds to

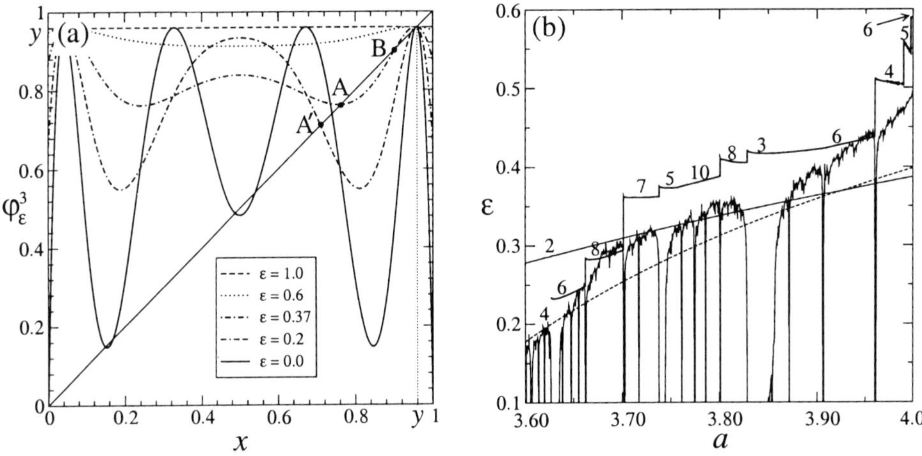

Figure 2. (a) Graphs of φ^3_ε as a function of x are plotted for different values of ε. y is fixed to be one of the points of the stable period-3 cycle γ_3 of the logistic map. Parameter $a = 3.84$. (b) Bifurcation curves of emergence of stable two-cluster dynamics in system (1) with strongly asymmetric (solid stepped curve) and symmetric (smooth solid and dashed curves) clusters. Fractal curve corresponds to the destabilization of the coherent chaotic state (blowout bifurcation).

the x-coordinate of a point of $\gamma_3^{(2)}$. This cycle is born via a saddle-node bifurcation and is placed out of the diagonal D_2. With further reduction of ε, the cycle $\gamma_3^{(2)}$ becomes unstable via a period-doubling bifurcation (the slop of the graph of $\varphi^3_\varepsilon(x, y)$ becomes less than -1, see point A' in Fig. 2a).

The non-diagonal stable cycles for system (7) produce the asymmetric periodic windows shown in Fig. 1. The windows split from the axis $p = 0$ and move to the right in the (p, ε)-parameter plane as parameter a increases. They explain the emergence of the highly asymmetric two-cluster states in the system of N globally coupled maps (1).

Note, that an analogous cascade of periodic windows also emerge through the axis $\varepsilon = 0$ for p close to 0.5, see Fig. 1. These windows cause manifestations of periodicity in the turbulent regime as recently described by Shimada and Kikuchi[15].

We summarize our findings in Fig. 2b. Here, bifurcation curves for the emergence of asymmetric two-cluster states (fully drawn stepwise curve) and symmetric two-cluster states (smooth solid and dashed curves) in system (1) are shown together with a curve of blowout bifurcation for the coherent chaotic state (fractal curve). Numbers correspond to the periods m of stable cycles $\gamma_m^{(2)}$ responsible for stable in-cluster dynamics.

4 Cluster splitting via period-doubling

Besides the regions of parameter space where periodic two-cluster dynamics is stable in the whole N-dimensional phase space, one can also find regions where more complex temporal dynamics occurs as well as regions where the periodic two-cluster dynamics although stable in the cluster subspace of F_2 is unstable in the full phase space of the original system (1). Some of the latter regions are indicated by a gray tone in Figs. 1a and 1b (e.g., the regions attached from below to period-3 and period-5 windows). A more detailed picture of the structure of the period-3 window is provided in Fig. 3a. Here, the region of stability for the period-3 cycle $\gamma_3^{(2)}$ in the two-dimensional system F_2 is delineated by solid curves. This region consists of two parts, namely, the region $C2P3^S$, where the cycle $\gamma_3^{(2)}$ is stable in the whole phase space of system (1), and the region $C2P3^U$, where it is stable only in the two-dimensional cluster subspace. The bold dashed curve provides transverse stability boundaries of two-cluster attractors developed from $\gamma_3^{(2)}$.

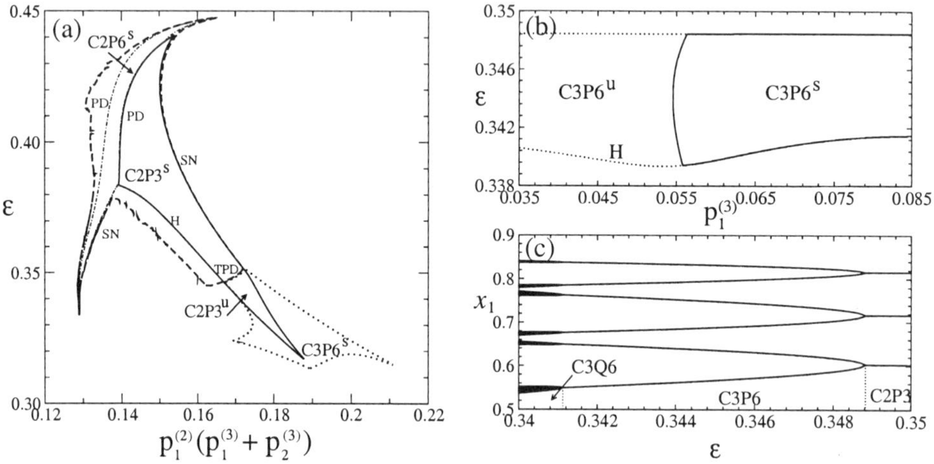

Figure 3. (a) Detailed structure of asymmetric period-3 window. $CkPm^S$ denotes a region where the system of globally coupled maps (1) has stable period-m k-cluster states. PD stands for period-doubling, SN for saddle-node, H for Hopf, and TPD for transverse period-doubling bifurcations. (b) The region, where system F_3 with parameter $p_i^{(3)}$ such that $p_1^{(3)} + p_2^{(3)} = p_1^{(2)} = 0.17$ has stable period-6 cycle $\gamma_6^{(3)}$, is bounded by dotted curves. This cycle is also stable in the whole N-dimensional phase space of system (1) in the region bounded by solid curves. (c) Bifurcation diagram at the transition from two-cluster period-3 state with partition $\{0.17, 0.83\}$ to three-cluster period-6 or quasiperiodic state with partition $\{0.08, 0.09, 0.83\}$. Parameter $a = 4$.

Two transverse multiplicators $\mu_{\perp,1}^{(2)}$ and $\mu_{\perp,2}^{(2)}$ of the cycle $\gamma_3^{(2)}$ can be found in accordance with (6). When the parameter point $(p_1^{(2)}, \varepsilon)$ crosses the TPD bifurcation curve with decreasing ε, the transverse multiplicator $\mu_{\perp,1}^{(2)}$ becomes less than -1. The cycle $\gamma_3^{(2)}$ loses its stability via a transverse period-doubling bifurcation giving rise to a period-6 cycle $\gamma_6^{(3)}$. After this bifurcation, the period-3 two-cluster state with partition $\{p_1^{(2)}, p_2^{(2)}\}$ is no longer stable. Instead, three-cluster states originated by $\gamma_6^{(3)}$ attain a stability.

To illustrate, let us consider the three-dimensional system F_3 of the form (3) with $K = 3$. Choose parameters $p_i^{(3)}$, $i = 1, 2, 3$, such that $p_1^{(3)} = p_2^{(3)} = \frac{1}{2} p_1^{(2)}$ and

$p_3^{(3)} = p_2^{(2)}$. In Fig. 3a, we have plotted a region depicted by dotted curve, where system F_3 has a stable period-6 cycle $\gamma_6^{(3)}$. The cycle appears also to be stable in the whole N-dimensional phase space of system (1): all three transverse multiplicators $\mu_{\perp,i}^{(3)}$, $i = 1, 2, 3$ of $\gamma_6^{(3)}$ are found to be less than 1 in absolute value. Inspection of Fig. 3a clearly shows that the stability regions $C2P3^S$ and $C3P6^S$ are separated by the TPD bifurcation curve. Therefore, period-doubling loss of transverse stability by the the two-cluster period-3 cycle $\gamma_3^{(2)}$ results in the appearance of stable three-cluster states governed by period-6 cycle $\gamma_6^{(3)}$. The two-cluster state with partition $\{p_1^{(2)}, p_2^{(2)}\}$ splits into a three-cluster state with partition $\{p_1^{(3)}, p_2^{(3)}, p_3^{(3)}\}$, where $p_3^{(3)} = p_2^{(2)}$. The cluster splitting is symmetric if $p_1^{(3)} = p_2^{(3)} = \frac{1}{2}p_1^{(2)}$.

The bifurcation described also leads to asymmetric splitting, as shown in Fig. 3b. Here, only the sum $p_1^{(3)} + p_2^{(3)} = p_1^{(2)} = 0.17$ is fixed ($p_3^{(3)} = p_2^{(2)} = 0.83$). A wide variety of stable three-cluster states with different partitions arises: $p_1^{(3)}$ can vary from 0.085 (symmetric splitting) to 0.056. For example, the two cluster state with the oscillator distribution 170 : 830 ($N = 1000$) may split into tree-cluster state with any distribution from 85 : 85 : 830 up to 56 : 114 : 830.

With further decrease of ε, the period-6 three-cluster cycle $\gamma_6^{(3)}$ bifurcates via a supercritical Hopf bifurcation (within the cluster subspace) transforming the in-cluster dynamics from periodic to quasiperiodic. The bifurcation diagram for the transition from period-3 two-cluster to quasiperiodic three-cluster dynamics is plotted in Fig. 3c.

The bifurcation scenarios described also take place for other asymmetric windows of period 4, 5, etc. presented in Fig. 1a to the left of the considered period-3 window. Again, the smaller-sized cluster splits via transverse period-doubling. For another period-4 window presented in Fig. 1a close to symmetric (p is close to 0.5, top, with gray region), the cluster splitting bifurcation is also caused by the transverse period-doubling. But in this case, the larger-sized cluster splits giving birth to three-cluster states with period-8 temporal dynamics.

Conclusion

One can think of a number of main scenarios for the break up of the coherent state in an ensemble of globally coupled chaotic oscillators. In one scenario the coherent state splits into a couple of more or less equally populated clusters with different dynamics, while in another scenario only a single or a few oscillators lose synchrony with the rest of the ensemble. We have previously considered the first case in significant detail[16]. In this connection we have shown that the loss of synchronization for the coherent state and the formation of clusters of synchronized behavior are two distinct processes, and that the type of behavior that arises after the loss of total synchronization depends sensitively on the dynamics of the individual map. Symmetric two-cluster formation was found to proceed through the stabilization of an asymmetric period-2 cycle or of an asymmetric period-4 cycle.

In the present work we considered the alternative desynchronization scenario, and again the outcome of the process was found to depend sensitively on the parameter of the individual map. With increasing nonlinearity parameter, new strongly asymmetric two-cluster states with periodic dynamics were found to arise in the same order as the periodic windows occur in the logistic map. As they first appear these two-cluster states only allow a vanishingly small fraction of the oscillators in one of the clusters. With further increase of the nonlinearity parameter, however, the region of stability shifts in the direction of a more even distribution of oscillators between the clusters. Our investigations also revealed that an analogous sequence of periodic two-cluster states emerges in the region of very low coupling where the dynamics is generally assumed to be totally incoherent (turbulent regime).

Another interesting result of our work is the observation of cluster splitting processes in connection with transverse period-doubling bifurcations. We have shown, for instance, how the period-3 two-cluster state can transform into a three-cluster state of period-6.

Acknowledgments

We thank A. Pikovsky for a number of illuminating discussions. O.P. acknowledges support from the Alexander-von-Humboldt Foundation.

References

1. K. Kaneko, *Prog. Theor. Phys.* **72** 480 (1984).
2. I. Waller and R. Kapral *Phys. Rev.* A **30** 2047 (1984).
3. R. Kapral Phys. Rev. A **31** 3868 (1985).
4. K. Kaneko, *Physica* D **41**, 137 (1990).
5. K. Kaneko, *Physica* D **103**, 505 (1997); *Phys. Rev. Lett.* **78**, 2736 (1997); *Physica* D **124**, 322 (1998); F. Xie and G. Hu, *Phys. Rev.* E **56**, 1567 (1997); N.J. Balmforth, A. Jacobson, and A. Provenzale, *Chaos* **9**, 738 (1999); P. Glendinning, *Phys. Lett.* A **259**, 129 (1999), **264**, 303 (1999); S.C. Manrubia and A.S. Mikhailov, *Europhys. Lett.* **50**, 580 (2000), **54**, 451 (2001).
6. K. Kaneko, *Physica* D **75**, 55 (1994).
7. E. Mosekilde, *Topics in Nonlinear Dynamics, Applications to Physics, Biology and Economic Systems*, (World Scientific, Singapore, 1996).
8. K. Wiesenfeld and P. Hadley, *Phys. Rev. Lett.* **62**, 1335 (1989).
9. K. Wiesenfeld, C. Bracikowski, G. James, and R. Roy, *Phys. Rev. Lett.* **65**, 1749 (1990).
10. W. Wang, I.Z. Kiss, and J.L. Hudson, *Chaos* **10**, 248 (2000).
11. M. Hasler, Yu. Maistrenko, and O. Popovych, *Phys. Rev.* E **58**, 6843 (1998); Yu. Maistrenko, O. Popovych, and M. Hasler, *Int. J. Bif. Chaos* **10**, 179 (2000).
12. J. Milnor, *Comm. Math. Phys.* **99**, 117 (1985).
13. P. Ashwin, J. Buescu, and I. Stewart, *Nonlinearity* **9**,703 (1996).
14. J.C. Sommerer and E. Ott, *Nature* (London) **365**, 136 (1993); E. Ott and J.C. Sommerer, *Phys. Lett.* A **188**, 39 (1994).
15. T. Shimada and K. Kikuchi, *Phys. Rev.* E **62**, 3489 (2000).
16. O. Popovych, Yu. Maistrenko, and E. Moseskilde, *Phys. Rev.* E **64**, 026205 (2001).

THE HIERARCHY STRUCTURES OF THE JULIA SETS

SY-SANG LIAW

Department of Physics, National Chung-Hsing University, 250, Guo-Kwang Road, Taichung, Taiwan

E-mail: liaw@phys.nchu.edu.tw

Julia sets of the complex mappings are normally fractals and complicated. Nevertheless, their structures can be decomposed into a series of hierarchy structures that are found in the simplest quadratic mapping. We show that the local structure of any Julia set of one-parameter mappings can be understood as a combination of quadratic Julia structures.

1 Introduction

The Julia sets of the mappings can be defined in several equivalent ways[1]. In this article we will consider only one-parameter complex mappings and define a Julia set of a mapping at a given particular value of parameter as the boundary of the set of the initial complex values that have finite absolute value after infinite times of iteration of the mapping. Julia sets are fractals except for some particular cases. Even in the simplest quadratic mapping $z \to z^2 + p$ the Julia sets show various complex structures[2]. We denote the Julia sets of the mapping $z \to z^2 + p$ as J_p. Fig. 1, 2 and 3 are three examples of connected cases of J_p with p = -0.5+0.5i, -1.754878 and $-0.156520 + 1.032247i$ respectively.

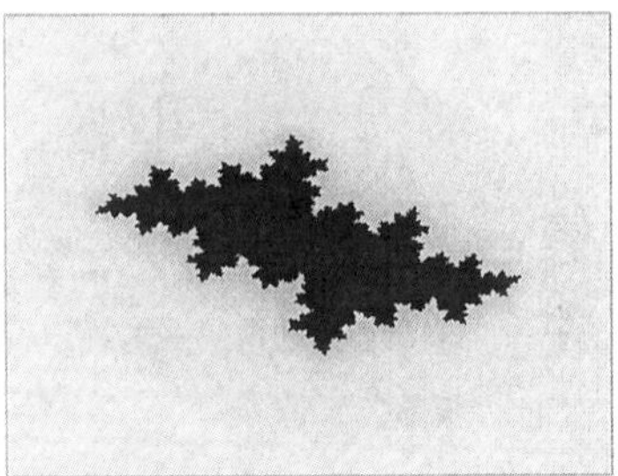

Fig. 1 The Julia set of the quadratic mapping $z \to z^2 + p$ with $p = -0.5 + 0.5i$

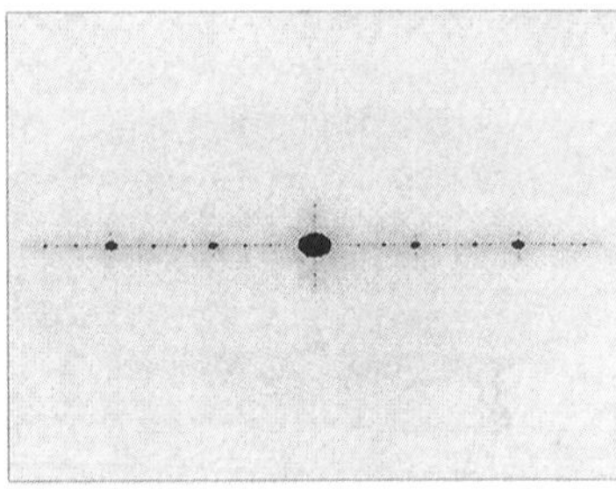

Fig. 2 The Julia set of the quadratic mapping $z \to z^2 + p$ with $p = -1.754878$

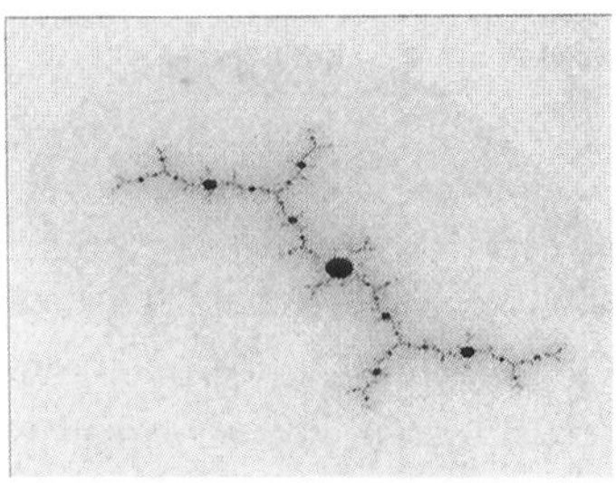

Fig. 3 The Julia set of the quadratic mapping $z \to z^2 + p$ with $p = -0.156520 + 1.032247i$

The set of the values p's that have connected Julia sets is known as the Mandelbrot set[3]. The Mandelbrot set has been proven to be connected itself[4] and there are infinitely many small Mandelbrot-like sets on the boundary of it[5]. In Figs. 4 and 5 we enlarge parts of the Mandelbrot set to reveal some of these Mandelbrot-like sets. The values of mapping eventually converge to k values for parameters within a particular Mandelbrot-like set. These values of periods are indicated in Figs. 4 and 5 for some Mandelbrot-like sets. A Mandlebrot-like set on the boundary of the Mandlebrot set is s small copy of the Mandelbrot set so that it is easy to see that there is a hierarchy structure for the Mandelbrot-like sets. Therefore, the period of a Mandlebrot-like set that is on the boundary of another Mandelbrot-like set of period k is multiple of k, denoted by mk.
period of a Mandlebrot-like set that is on the boundary of another Mandelbrot-like set of period k is multiple of k, denoted by mk.

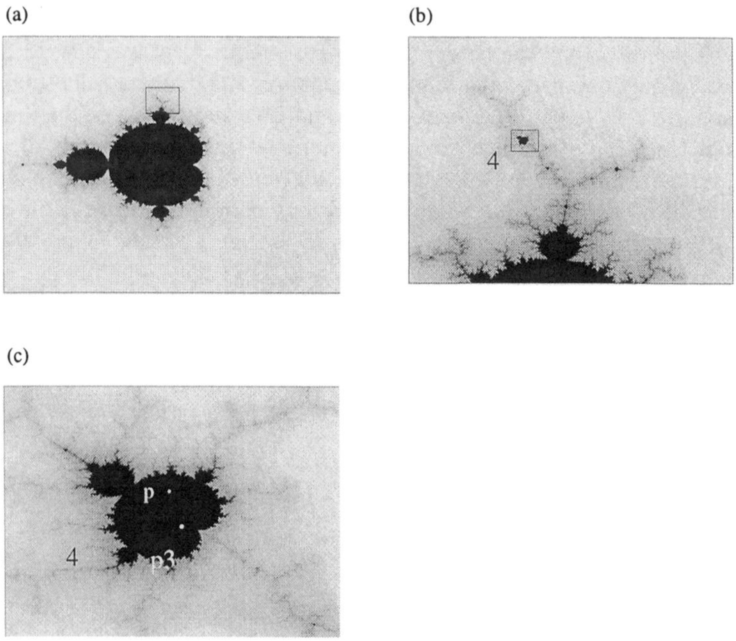

Fig. 4 (a)The Mandelbrot set and (b,c)the enlargement of a Mandelbrot-like set of period 4 on its boundary. $p_2 = -0.156520 + 1.032247i$ is the center of a Mandelbrot-like set of period 4. p is located at $p_2 + \varepsilon$. $\varepsilon = (-0.5 + 0.5i)/\alpha\gamma = -0.001041 + 0.005901i$, where α, γ are given in Eq. (2-3) evaluated at p_2.

In Figs. 6 and 7 we plot the filled-in Julia sets for parameters within some Mandelbrot-like sets shown in Fig. 4 and 5 respectively. We see that these Julia sets have hierarchy structures too. That is, distinct structures of J_p shown in Figs 1, 2, and 3 of different size are clearly seen in these Julia sets. In Sec. II we will give an explanation for the appearance of the hierarchy structure of these Julia sets. The explanation can be used for not only quadratic mappings but also for any one-parameter complex mappings. We give a further example in Sec. III. Sec. IV is conclusion and further possible work.

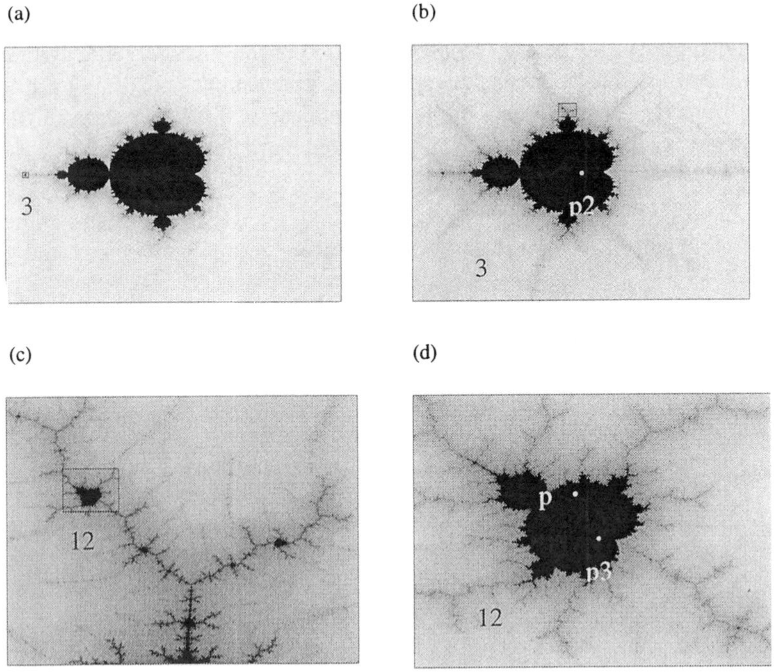

Fig. 5 (a) The Mandelbrot set and (b,c,d) the enlargements of two Mandelbrot-like set of period 3 and 12 on its boundary. $p_2 = -1.754878$ is the center of a Mandelbrot-like set of period 3. $p_3 = -1.7588113 + 0.01899490i$ is the center of a Mandelbrot-like set of period 12. p is located at $p_3 + \varepsilon$ $\varepsilon = (-0.5+0.5i)/\alpha\gamma = -0.00002316 + 0.00011017i$, where α, γ are given in Eq. (2-10) evaluated at p_3.

2 Determination of the local structures

Given a one-parameter complex mapping $f_p(z)$, Douady and Hubbard[5] have shown that there are infinite many Mandelbrot-like sets in the parameter space. The centers p_0 of these Mandelbrot-like sets[6] can be determined by solutions of the equation

$$f_p^k(z_c) = z_c \tag{2-1}$$

The size and orientation of the Mandelbrot-like set of period k centered at p_0 can then calculated[7] by using an effective quadratic mapping of $f_p(z)$ for p near p_0 and z near z_c :

$$\begin{aligned} f_{p_0+\varepsilon}^k(z) &= f_{p_0}^k(z_c) + \frac{1}{2}\frac{d^2}{dz^2} f_{p_0}^k(z_c)(z-z_c)^2 + \varepsilon \frac{d}{dp} f_{p_0}^k(z_c) + O(\varepsilon^2) \\ &\approx z_c + \alpha(z-z_c)^2 + \varepsilon\gamma \end{aligned} \tag{2-2}$$

Using the conjugate transformation[8], we have found[7] that the Mandelbrot-like set centered at p_0 is similar to the Mandelbrot set modified by the factor $1/\alpha\gamma$. Details of the calculations for a few one-parameter mappings are given in Ref.[9]. Using Eq.(2-2), we can also determine the local structure of the Julia set near z_c [10]. It is shown[10] that the Julia set within a disk of radius $2/|\alpha|$ centered at z_c is similar to J_p, with $p = \varepsilon\alpha\gamma$, modified by $1/\alpha$.

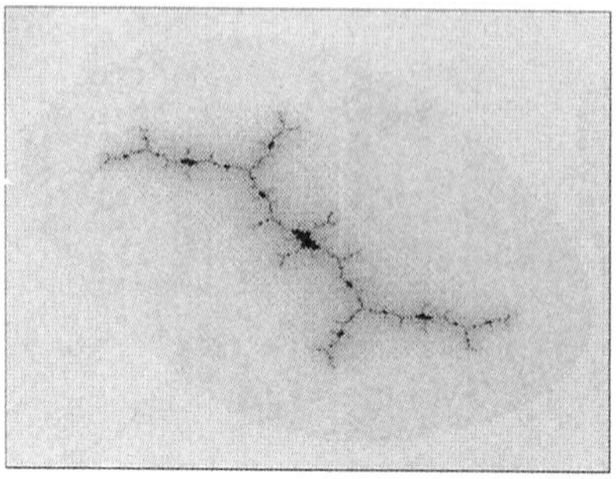

Fig.6 The Julia set of the quadratic mapping $z \to z^2 + p$ with $p = p_2 + \varepsilon = -0.157561 + 1.038148i$ as indicated in Fig. 4. Two levels of structures similar to Figs. 1 and 3 can be seen.

(a) (b)

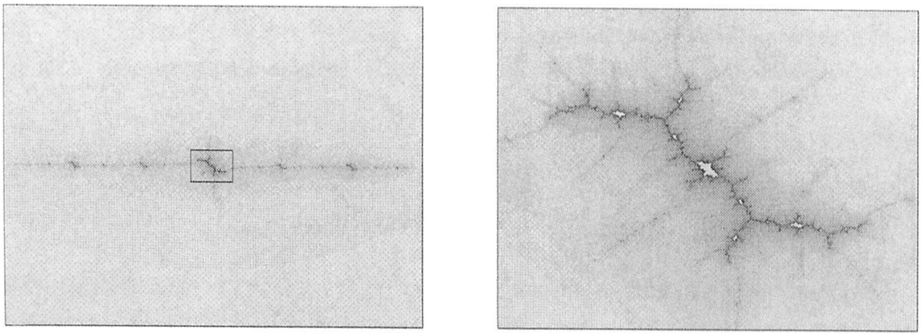

Fig. 7 (a) The Julia set and (b) an enlargement of its central region of the quadratic mapping $z \to z^2 + p$ with $p = p_2 + \xi = p_3 + \varepsilon = -1.7588345 + 0.0191051i$ as indicated in Fig. 5. Three levels of structures similar to Figs. 1, 2, and 3 are plotted in gray, black, and white respectively.

We have plotted some examples of the Julia sets revealing their hierarchy structures in Figs. 6 and 7. We will first analyze the one with two levels of structures for the simplest quadratic mapping shown in Fig. 6. Let $p = p_2 + \varepsilon$, where p_2 is the center of a period-4 Mandelbrot-like set. $p_2 = -0.156520 + 1.032247i$. $\varepsilon = -0.001041 + 0.005901i$. The first level structure of Julia set of $f_p(z)$ is simply of course J_p (Fig. 1). On the other hand, according to Eq. (2-2), we have

$$f^4_{p_2+\varepsilon}(z) = \alpha z^2 + \varepsilon\gamma,$$
$$\alpha = \frac{1}{2}\frac{d^2}{dz^2} f^4_{p_2}(z_c) = -10.55 - 5.45i, \quad \gamma = \frac{d}{dp} f^4_{p_2}(z_c) = -9.83 - 1.39i \tag{2-3}$$

Thus near $z_c = 0$ within a disk of radius $\rho = 2/|\alpha| = 0.168$, the second level structure of the Julia set is given by $\frac{1}{\alpha}J_p$ with $p = \varepsilon\alpha\gamma = -0.5 + 0.5i$, which is similar to Fig. 1. In previous work[10], we have shown that this set of second level of structure also appears at other location. Their positions z_0 are determined by the solution of the equation:

$$f^n_{p_2+\varepsilon}(z_0) = z_c, \quad n \in I \tag{2-4}$$

That is, at z_0 which is a solution of Eq.(2-4), there is a similar set of the second level structure to the one at z_c, only to modified[10] by a factor δ:

$$\delta = \frac{d}{dz} f^n_{p_{2+\varepsilon}}(z_0) \tag{2-5}$$

In fact, the set of all solutions z_0 of Eq.(2-4) determines the first level structure of the Julia set.

Now let us consider a three level hierarchy structure. Let p_1, p_2, and p_3 be the centers of the first, second and third level of the hierarchy structures. They are determined by the following equations respectively:

$$f_{p_1}(z_c) = z_c, \tag{2-6}$$
$$f^k_{p_2}(z_c) = z_c \tag{2-7}$$
$$f^{mk}_{p_3}(z_c) = z_c \tag{2-8}$$

For Fig. 5, $f_p(z) = z^2 + p$, $z_c = 0$, $k = 3$, $m = 4$, we obtain $p_1 = 0$, $p_2 = -1.754878$, $p_3 = -1.7588113 + 0.01899490i$. Consider the Julia set for parameter in the neighborhood of p_3, $p = p_3 + \varepsilon$, with $\varepsilon = -0.00002316 + 0.00011017i$.

Because of Eq. (2-8), using Eq. (2-2) we have for z near z_c:

$$\begin{aligned} f^{mk}_{p_3+\varepsilon}(z) &= f^{mk}_{p_3}(z_c) + \frac{1}{2}\frac{d^2}{dz^2} f^{mk}_{p_3}(z_c)(z - z_c)^2 + \varepsilon\frac{d}{dp} f^{mk}_{p_3}(z_c) + O(\varepsilon^2) \\ &\approx z_c + \alpha(z - z_c)^2 + \varepsilon\gamma \end{aligned} \tag{2-9}$$

where

$$\alpha = \frac{1}{2}\frac{d^2}{dz^2} f_{p_3}^{mk}(z_c) = 98.1 + 50.3i$$

(2-10)

$$\gamma = \frac{d}{dp} f_{p_3}^{mk}(z_c) = 56.7 + 5.95i$$

On the other hand, $p = p_3 + \varepsilon$ can be written as

$$p_3 + \varepsilon = p_2 + (p_3 - p_2 + \varepsilon) = p_2 + \xi \tag{2-11}$$
$$\xi = p_3 - p_2 + \varepsilon = -0.003934 + 0.018995i$$

Because that ξ is within the small copy of the Mandelbrot-like set centered at p_2, and because of Eq. (2-7), we can also expand the kth order iteration of $f_p(z)$ at p_2:

$$\begin{aligned} f_{p_2+\xi}^{k}(z) &= f_{p_2}^{k}(z_c) + \frac{1}{2}\frac{d^2}{dz^2} f_{p_2}^{k}(z_c)(z - z_c)^2 + \xi \frac{d}{dp} f_{p_2}^{k}(z_c) + O(\xi^2) \\ &\approx z_c + \alpha'(z - z_c)^2 + \xi\gamma' \end{aligned} \tag{2-12}$$

$$\alpha' = \frac{1}{2}\frac{d^2}{dz^2} f_{p_2}^{k}(z_c) = -9.30$$

(2-13)

$$\gamma' = \frac{d}{dp} f_{p_2}^{k}(z_c) = -5.65$$

(The approximation Eq.(2-12) is not as good as Eq. (2-9) because that $|\xi|$ is not that small.) Thus for parameter p in the neighborhood of p_2 within a disk of radius $2/|\alpha'\gamma'|$, that is $|\xi| < 2/|\alpha'\gamma'|$, the Julia set near z_c within a disk of radius $2/|\alpha'|$ is approximately similar to J_p/α with $p = \xi\alpha'\gamma'$. On the other hand, since p is also within a disk of $2/|\alpha\gamma|$ centered at p_3, because of Eq. (2-9), in the vicinity of z_c within a disk of radius $2/|\alpha|$ the Julia set is similar to J_p/α with $p = \varepsilon\alpha\gamma = -0.5 + 0.5i$. This J_p is shown in white in Fig. 7 near $z_c = 0$. Other copies of this J_p shown in the figure can also be calculated according to the equation similar to Eq.(2-5).

Notice that the range of the third level structure is smaller than that of the second level structure, and the second is smaller than the first. Intuitively, this will be always true. Mathematically, one has to prove the relation

$$|\alpha| > |\alpha'| > 1 \tag{2-14}$$

The proof of Eq. (2-14) is under investigation.

The argument presented here can be easily generalized to more levels of hierarchy structures of the Julia sets.

3. Further examples

In previous two sections we demonstrate the hierarchy structure of the Julia sets for the simplest quadratic mapping only. However, our argument presented in Sec. II is not confined to the quadratic mappings. We give an example here of the mapping with an entire function: $z \to p\sin(z/2)$.

Consider the parameter p within a Mandelbrot-like set of period $9 = 3 \cdot 3$, as shown in Fig. 8. $p_1 = 0$, $p_2 = 0.310955 + 3.041250i$ is the center of the Mandelbrot-like set of period 3, and $p_3 = 0.3427027 + 3.0119123i$ is the center of the Mandelbrot-like set of period 9 located at the antenna of the period-3 Mandelbrot-like set. The filled-in Julia set of this mapping with $p = p_3 + \varepsilon = 0.3426879 + 3.0115733i$ with $\varepsilon = -0.0000148 - 0.0003400i$ is plotted in Fig. 9. Three levels of structures are seen. The global first level structure is particular for this sinusoidal mapping with p relative to p_1. The second level structure, plotted with darker dots, is determined by effective quadratic mapping with the value p relative to p_2. Thus it is similar to J_p shown in Fig. 2. The third level structure, plotted in white, is also determined by effective quadratic mapping with the value p relative to p_3. Thus it is similar to J_p with $p = \varepsilon\alpha\gamma = -0.5 + 0.5i$ shown in Fig. 1 (The values α and γ are calculated according to Eq. (2-10) to be $\alpha = 10.4 + 26.0i$ and $\gamma = -69.5 + 26.4i$ respectively.)

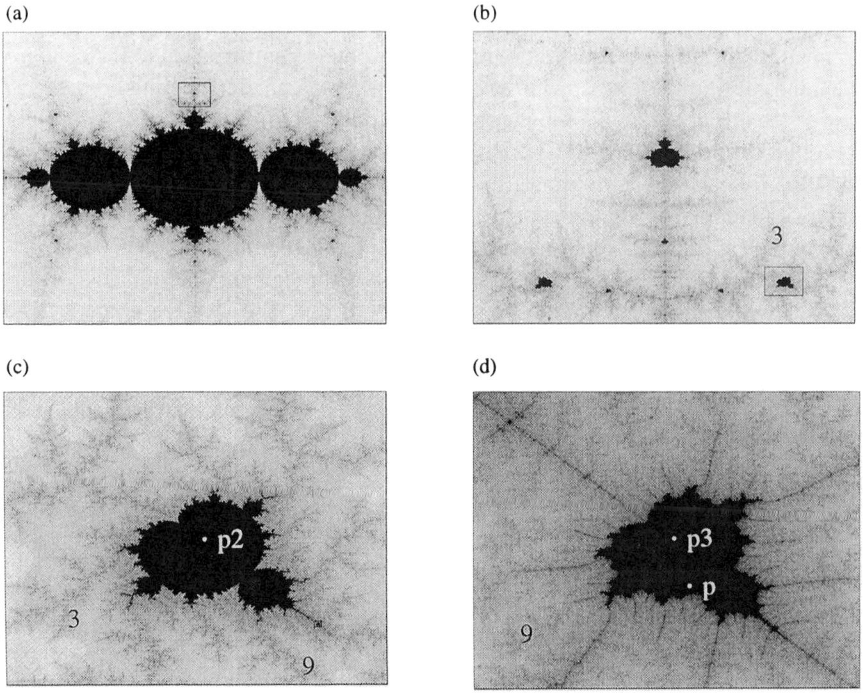

Fig. 8 Three levels of structures are plotted in gray, black, and white respectively. (a) The parameter space of the mapping $z \to p\sin(z/2)$. (b,c,d) The Mandelbrot-like set of period 3 and 9 are indicated in the magnified windows. $p_2 = 0.310955 + 3.041250i$ is the center of a Mandelbrot-like set of period 3. $p_3 = 0.3427027 + 3.0119123i$ is the center of a Mandelbrot-like set of period 9. p is located at $p_3 + \varepsilon$. $\varepsilon = (-0.5 + 0.5i)/\alpha\gamma = -0.0000148 - 0.0003390i$, where α, γ are given in Eq. (2-10) evaluated at p_3.

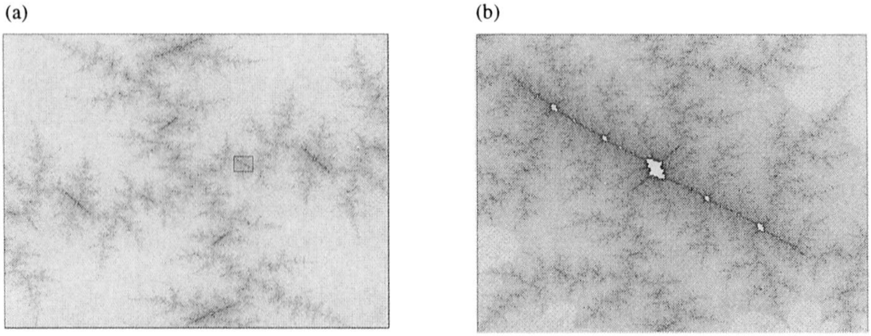

Fig. 9 (a) The Julia set and (b) an enlargement of a small region of the mapping $z \to p \sin(z/2)$ with $p = p_2 + \xi = p_3 + \varepsilon = 0.3426879 + 3.0115733i$ as indicated in Fig. 8. Three levels of structures are plotted in gray, black and white, respectively.

4 Conclusion

We have shown that the Julia set of any one-parameter complex mapping contains hierarchy structures. Except for the global first level structure, which is dependent on the form of the mapping, the local structure, including all higher levels of structures, can be determined by the effective quadratic mappings expanded with respect to the centers of the Mandelbrot-like sets that appear on in the set of all parameters having bounded orbits for one of the critical points of the mapping.

For the case when the critical points are degenerate, which we have not discussed in this article, one can still approximate the local structure of the Julia sets by simple polynomial mappings using the approximation proposed in Ref. [7].

Having understood the hierarchy structures of the Julia sets, we can then construct very complicated and very beautiful patterns very easily with the help of computers. This would be very useful in application such as designing and modeling. Note that besides the 2-dimensional structures, one can also use the quaternion mappings to construct 3-dimensional hierarchy structures of the Julia sets.

References

1. P. Blanchard, Bull. Amer. Math. Soc. **11**, 85-139 (1984).
2. See, for example, H.-O. Peitgen and P.H. Richter, *The Beauty of Fractals*, Springer-Verlag (1986).
3. B.B. Mandelbrot, Annals New York Academy of Sciences 357, 249-259 (1980).
4. A. Douady and J.H. Hubbard, CRAS Paris **294**, 123-126; Publications Mathematiques d'Orsay 84-02, Universite de Paris-Sud (1984).
5. A. Douady and J.H. Hubbard, Ann. Ecole Norm. Sup. (4)**8**, 287-343 (1985).
6. B. Branner, *Chaos and Fractals*, proceedings of symposia in applied mathematics 39, 75-105 (1989).
7. S.S. Liaw, *Find the Mandelbrot-like sets in any mapping*, submitted to Fractals (2001).

8. S.S. Liaw, Fractals **6**, 181-189 (1998).
9. S.S. Liaw, *Parameter space of one-parameter complex mappings*, Chaos, Solitons, and Fractals **13**, 761 (2002).
10. S.S. Liaw, *Structure of the cubic mappings*, Fractals **9**, 231(2001).

SOME NEW FEATURES OF INTERFACE ROUGHENING DYNAMICS IN PAPER WETTING, BURNING, AND RUPTURING EXPERIMENTS

ALEXANDER S. BALANKIN

Sección de Estudios de Posgrado e Investigación, Escuela Superior de Ingeniería Mecánica y Eléctrica, Ed. 5, 3er. Piso, Instituto Politécnico Nacional, México, D.F., México,07738
E-mail: balankin@hotmail.com, Tel.: (525) 729-6000+54589

DANIEL MORALES MATAMOROS

Instituto Mexicano de Petróleo, México, D.F.
E-mail: dmorales@imp.mx

The dynamics of interfaces growing in paper wetting, fracturing and burning processes is studied with use the same kinds of papers for different experiments. We were able to study five different types of kinetic roughening. Some new observations concerning the spatial-temporal dynamics of rough interfaces are reported. Specifically, we found that the types of kinetic roughening, as well as the scaling exponents, are dependent on the paper structure and the mechanism of interface formation. Moreover, we have observed that the local roughness exponent of moving wet front logarithmically increases from 0.5 at initial stage up to its stationary value, achieved before front saturation. We also found that the stress-strain fracture behaviour of some kinds of paper exhibits a statistical self-affine invariance with scaling exponent, which is equal to the local roughness exponent of rupture line. The same exponent governs the changes in the stress-strain curve as the strain rate increases. This gives rise to the facture-energy – time-to-fracture uncertainty relation, similar to the time-energy uncertainty relation in quantum mechanics. The physical implications of these points are discussed.

1 Introduction

Kinetic roughening of interfaces in nonequilibrium conditions has attached much attention in recent years [1-8]. This is mainly due to the many important applications of the theory of interface dynamics including fluid flow in porous media, self-affine crack mechanics, forest fire, and molecular-beam epitaxy among others. From a phenomenological point of view, there are many common aspects between interface dynamics in these systems, despite their different nature and huge scale differences. Different growth processes have very often been shown to exhibit scaling properties that allow one to divide the growth models into universality classes characterised by different sets of critical exponents. In this sense, experiments with paper can be used to understand some essential features of kinetic roughening in systems of different nature.

Although fractal growth models have provided a useful insight into a vast array of problems concerning interface roughening dynamics, many important questions are still open. In fact, it is particularly important to reach more complete and predictive theoretical understanding of interface growth in presence of correlated spatial-temporal noise. Clearly, such a problem can only be investigated by a comprehensive effort that involves computer simulations, analytical tools and suitably designed experiments.

The purpose of this work is to understand the nature of the quenched and temporal noises and their effect on the scaling dynamics of interfaces of different nature, growing in a given disordered medium. The main questions to be answered are: (1) What properties of the system determine the type of scaling behaviour and the values of scaling exponents? (2) Do the scaling exponents take only universal values or they change continuously with the order parameter of system? (3) What is the reason for difference in the values of scaling exponents obtained in similar experiments in different works? (4) How does the spatial roughness affect the temporal behaviour of growing interface? To gain an insight to these problems we perform a detailed study of the spatial-temporal behaviour of interfaces growing in papers with different structures by three different mechanisms: imbibition, combustion, and fracturing.

2 Experimental details

Paper is a porous composite material with anisotropic structure associated with an asymmetric orientation distribution of fibres. Despite the stochastic nature of the fibre distribution, the fibre and the pore structures of papers are not random but possess long-range correlations characterised by power-law behaviour of the space-density autocorrelation function [5-8]. Different kinds of paper have different structures, giving rise to different pore distributions, characterised by different scaling exponents of the autocorrelation function [8]. This allows us to study the effects of correlated quenched noise and medium anisotropy on spatial-temporal dynamics of interfaces formed by different processes, characterised by different types of temporal (thermal) noise.

In this work, we study kinetic roughening dynamics of interfaces growing in paper wetting, burning, and fracturing experiments. These experiments were performed on rectangular sheets of three different kinds of commercial paper: *"Secante"*(thickness = 0.34±0.07 mm, areal density = 210±18 g/mm^2), *"Filtro"* (thickness = 0.25±0.04 mm, areal density = 110±10 g/mm^2), and *"Toilet"* (thickness = 0.11±0.06 mm, areal density = 36.8±0.3 g/mm^2). These papers have well different structures characterised by different scaling exponents of space-density autocorrelation function (ϕ = 0.12±0.05, ϕ = 0.27±0.08, and ϕ = 0.43±0.13 for *"Secante"*, *"Filtro"*, and *"Toilet"* papers, respectively). Detailed multifractal analysis of paper structures will be published elsewhere.

The papers have a well-defined structural anisotropy associated with the preferred orientation of fibres in the so-called machine direction. Accordingly, we studied interfaces formed along and across the machine direction of a paper. The length of paper sheets (in the direction of the interface propagation) used in all experiments was 250 mm, whereas the sheet width was varied from $\mathbf{W_0}$ = 10 mm to $\mathbf{W_M}$ = 100 mm, with the following relation $\mathbf{W} = \lambda \mathbf{W_0}$ for scaling factor λ = 1, 1.5, 2, 3, 4, 5, 8 and 10. At least 30 experiments of each type were carried out for each sheet width. Statistical data were analysed with the help of @RISK 4.05 software [9]

To study the spatial-temporal dynamics of the wet interface, a vertical sheet of paper is wetted with black Chinese-ink solution. The evolution of the interface between the wet (black) and the dry (white) regions is recorded with a digital camcorder at 24 frames/sec (see [7]). We allow the interface to rise until it stops and no change in either the height or the shape of the interface is observed.

To obtain the flame front we take a sheet of paper and ignite it at one end by a linear electric heating wire. The sheets were burned in the horizontal position and the combustion rate was rather slow (in the range from 2 mm/sec to 10 mm/sec for different kinds of paper). Unfortunately, we were able to satisfactory record the fire propagation only in experiments with *"Toilet"* paper, since the combustion of other papers used in this work is accompanied by high heat radiation and smoke, owing to theirs larger thickness and high density. Accordingly, we have studied the temporal scaling of fire front only in *"Toilet"* paper, whereas in other papers we have analysed only the spatial scaling properties of the "post-mortem" burning and flame fronts (see figure 1 *a*), formed when the fire has been quenched, after it reaches a middle line marked o the specimen.

Mechanical tests were carried out on a 4505 INSTRON testing machine under carefully controlled temperature $\mathbf{T} = 20 \pm 3$°C and humidity 38 ± 6%. The deformation rate was controlled by grip displacement speeds of 0.5, 1.0, 5.0, and 10 mm/min. The paper failure process was recorded with a digital camcorder at 60 frames/sec.

The failure of these papers occur as the culmination of progressive damage, involving complex interactions between multiple growing micro-cracks (see figure 2 *a*). As a result, the descending parts of stress-strain curves display a stochastic behaviour (see figure 2 *b*). Note the difference between these experiments and the studies in references [4-6], in which the single-notched sheets were tested.

We note that *"Secante"* and *"Filtro"* papers have an elasto-plastic behaviour, whereas the *"Toilet"* paper displays the linear elastic behaviour up to the tensile stress σ_M (see figure 2 *b*).

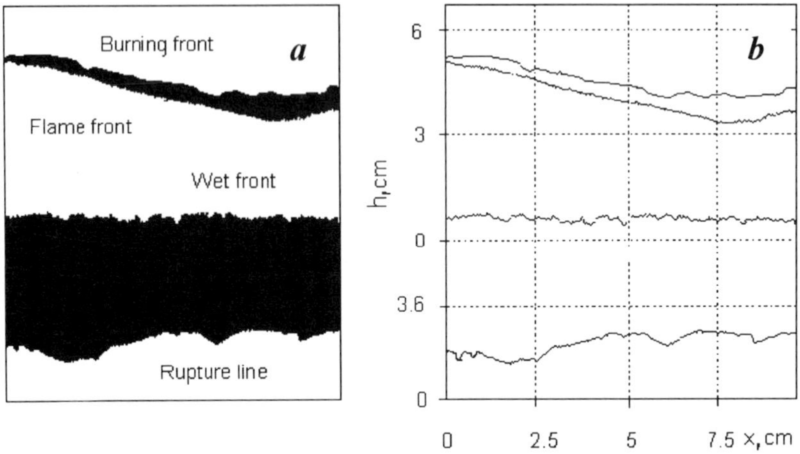

Figure 1. (*a*) Black-and-white image of wetting, burning, and flame fronts and rupture line in *"Secante"* paper and (*b*) the corresponding graphs **h(x)**.

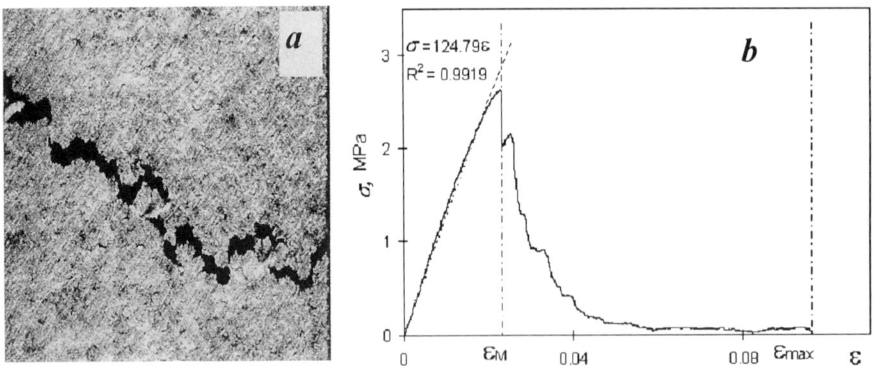

Figure 2. (*a*) Rupture line formation by cumulative damage and (*b*) stress-strain curve for the *"Toilet"* paper.

The interface images obtained from video frames have relatively low resolution, limited by the maximum frame size 600x600 pixels. Therefore, to fine study of the roughness of "post-mortem" burning, flame, and wet fronts and rupture lines, the tested specimens were scanned with HP-61000

Scan-Jet in black-and-white BMP-format with 600x600 dpi resolution. The profiles of studied interface from video frames, as well as from scanned images, were plotted using the *Scion Image* software [10] as single-valued functions $\mathbf{h(x, t)}$, in the XLS-format (see figure 1 *b*). Notice that in paper fracturing experiments, $\mathbf{h(x)}$ is a continuous function of $\mathbf{x}$ only after a paper sheet is divided in two parts.

A rough interface is characterised by the height fluctuations around the mean position of interface $\mathbf{h^*(t)} = \langle \mathbf{h(x, t)} \rangle_W$, where $\langle \ldots \rangle_W$ denotes an average over $\mathbf{x}$. Therefore, a basic quantity to look at is the global interface width $\mathbf{w(t,W)} = \mathbf{max}_{x\in W}[\mathbf{h(x,t)}] - \mathbf{min}_{x\in W}[\mathbf{h(x,t)}]$. The spatial correlations of rough interface are characterized by the height-height correlation function $G(\Delta,t) = \sqrt{\left\langle [h(x+\Delta,t) - h(x,t)]^2 \right\rangle_\Delta}$, where $\langle \ldots \rangle_\Delta$ denotes an average over $\mathbf{x}$ in widows of size $\Delta < \mathbf{W}$. Another useful characteristic function is the structure factor or power-spectrum of interface, defined as $S(q,t) = \langle \hat{h}(q,t)\hat{h}(-q,t) \rangle$, where $\hat{h}(q,t)$ is the Fourier transform of the interface height in a system of size $\mathbf{W}$.

Numerous observations have shown that at initial stage $\mathbf{t} < \mathbf{t_S}$, where $\mathbf{t_S}$ is the saturation time, the mean plane of interface moves as $\mathbf{h^*(t)} \propto \mathbf{t}^\delta$, where δ is the diffusion exponent [1,2]. At the same time, the global interface width scales as $\mathbf{w(t} \ll \mathbf{t_S,W)} \propto \mathbf{t}^\beta$, where β is the growth exponent, while for saturated interfaces $\mathbf{w(t} \geq \mathbf{t_S,W)} \propto \mathbf{W}^\alpha$, where α is the global roughness exponent [1-8].

Accordingly, $\mathbf{G^{1/2}(\Delta<\xi,\ t<<t_S)} \propto \mathbf{\Delta w(\Delta,t)} \propto \mathbf{t}^\beta$, where $\Delta w(\Delta,t) = \left\langle \left\langle [z(x,t) - \langle z(t) \rangle_\Delta]^2 \right\rangle_\Delta \right\rangle^{1/2}$ is the local width of interface and $\boldsymbol{\xi}\mathbf{(t)} \propto \mathbf{t}^{1/z}$ is the horizontal correlation length ($\mathbf{z}$ is the so called dynamic exponent). For the saturated interface $\mathbf{\Delta w(\Delta)} \propto \mathbf{G^{1/2}(\Delta)} \propto \Delta^\zeta$, where ζ is the local roughness exponent [1-8].

In this work, the scaling properties of interfaces were studied using five methods adopted in the commercial software "BENOIT 1.2" [11]: the variogram ($\mathbf{V} \propto \mathbf{G} \propto \Delta^{2\zeta}$), roughness-length (root-mean-square roughness $\propto \mathbf{\Delta w} \propto \Delta^\zeta$), power-spectrum (structure factor, $\mathbf{S} \propto \mathbf{q}^{-(2\alpha_S + 1)}$), and wavelets methods (the mother wavelet in BENOIT 1.2 is a step function) and the *R/S*-analysis ($\mathbf{R/S} \propto \Delta^\zeta$). These methods permit to determine the local roughness exponent ζ. Furthermore, the analysis of power-spectrum of interfaces in sheets of different width permits to identify the type of kinetic roughening [12,13]. In the case of anomalous roughening, the global roughness exponent α was determined from the scaling of sample averaged interface width [5,6]: $\mathbf{w_N} \propto \mathbf{W}^\alpha$.

3 Experimental results and discussion

3.1 Generic dynamic scaling in kinetic roughening

The generic scaling behaviour of structure factor was defied in [13] as $\mathbf{S(q,t)} \propto \mathbf{q}^{-\vartheta}\mathbf{s(qt^{1/z})}$, where function $\mathbf{s(qt^{1/z})}$ has the general form

$$\mathbf{s(y)} \propto \mathbf{y}^{2(\alpha - \alpha_S)},\ \text{if } \mathbf{y} >> \mathbf{1},\ \text{or } \mathbf{s(y)} \propto \mathbf{y}^{\vartheta},\ \text{if } \mathbf{y} << \mathbf{1}; \qquad (1)$$

here $\vartheta = 2\alpha_S + 1$, α_S is the spectral roughness exponent, and α is the global roughness exponent. Notice that if $\alpha \neq \alpha_S$ the power spectrum scales with the system width as $\mathbf{S(q,W)} \propto \mathbf{W}^{2\theta}$, where $\theta = \alpha - \alpha_S$.

In the absence of any characteristic length in the system, the interface growth are expected to show a power-law behaviour of the correlation function in space and time, and the Family-Vicsek dynamic scaling ansatz [1,2], $\mathbf{w(t,W)} = \mathbf{t}^{\alpha/z}\mathbf{f(W/\xi(t))}$, ought to hold. The scaling function $\mathbf{f(y)}$ behaves as $\mathbf{f} \propto \mathbf{y}^\alpha$, if $\mathbf{y} << \mathbf{1}$ ($\xi >> \mathbf{W}$), or $\mathbf{f} \propto$ **constant**, if $\mathbf{y} >> \mathbf{1}$. In such a case, the local interface width also scales as $\Delta\mathbf{w}(\Delta,\mathbf{t}) \propto \mathbf{G}^{1/2}(\Delta,\mathbf{t}) = \mathbf{t}^\beta\mathbf{g}(\Delta/\xi)$, where $\mathbf{g(y)}$ behaves as $\mathbf{g} \propto$ **constant**, if $\mathbf{y} >> \mathbf{1}$, and $\mathbf{g} \propto \mathbf{y}^\zeta$, if $\mathbf{y} << \mathbf{1}$, and the local roughness exponent ζ is equal to the global roughness exponent α, because of there is no characteristic length scale besides the system size. Therefore, the short time interface dynamic exponent is equal to $\mathbf{z} = \alpha/\beta$ [1,2]. The Family-Vicsek scaling of the structure factor reads as [2]: $\mathbf{S(q,t)} \propto \mathbf{q}^{-\vartheta}\mathbf{s(qt^{1/z})}$, where $\vartheta = 2\alpha + 1$ and $\mathbf{s(y)} \propto$ **constant**, when $\mathbf{y} >> \mathbf{1}$, and $\mathbf{s(y)} \propto \mathbf{y}^\vartheta$, when $\mathbf{y} << \mathbf{1}$, *i.e.* $\zeta = \alpha = \alpha_S$ [13].

Generally, however, $\zeta \leq \alpha$ [12]. Kinetic roughening with $\zeta < \alpha$ is called an anomalous roughening [1,2]. Different scaling forms can be treated as subclasses of the generic scaling ansatz (1). Authors of [13] have identified four different dynamic scaling regimes, associated with the bounds of scaling exponents (see table 1). In this work, we have observed all these regimes, as well as a new unconventional anomalous roughening (see table 1). Specifically:

1. Wet fronts in *"Secante"* and *"Filtro"* papers possess a statistical self-affine invariance with $\zeta = \alpha = \alpha_S = 0.63\pm0.02$ and $\zeta = \alpha = \alpha_S = 0.75\pm0.03$, respectively. Rupture lines along the machine direction in these papers also exhibit a statistical self-affine invariance. However, the crack roughness exponents differ from the wet front roughness exponents. Namely, $\zeta = \alpha = \alpha_S = 0.55\pm0.04$ for *"Secante"* and $\zeta = \alpha = \alpha_S = 0.42\pm0.04$ for *"Filtro"* papers.
2. The rupture lines across the machine direction in these papers exhibit an intrinsically anomalous roughness with the local exponent found to be equal to the crack roughness exponent for rupture lines in the perpendicular direction and the global roughness exponent equal to $\alpha = 0.71\pm0.05$ for *"Secante"* and $\alpha = 0.67\pm0.07$ for *"Filtro"* papers, respectively. Rupture lines across and along the machine direction in the *"Toilet"* paper are characterised by the same local roughness exponent $\zeta = \alpha_S = 0.75\pm0.05$, but different global exponents: $\alpha = 1.45\pm0.15$ along and $\alpha = 1.85\pm0.35$ across the paper machine direction.
3. The super-roughening regime was observed in wetting and burning experiments with the *"Toilet"* paper. We find that the wet front across the machine direction in this paper is characterised by $\zeta = 0.92\pm0.04$ and $\alpha = \alpha_S = 1.5\pm0.2$, whereas the burning front across and along the machine direction in this paper is characterised by $\zeta = 1.0\pm0.1$ and $\alpha = \alpha_S = 1.9\pm0.9$.
4. The wet front along the machine direction in the *"Toilet"* paper displays a anomalous roughening with $\alpha_S = 1.5\pm0.2 > 1$, $\zeta = 0.92\pm0.06$, and $\alpha = 1.1\pm0.2 < \alpha_S$.
5. Unconventional regime of anomalous kinetic roughening (see table 1) of flame fronts was observed in slow combustion of *"Secante"* (see [14]) and *"Filtro"* paper experiments, whereas the burning front roughness in these papers possesses statistical self-affine invariance.

Unconventional anomalous roughening is characterised by $\alpha_S < 1$. Furthermore, we find that the local roughness exponent of flame front, as well as the crossover length, ξ_C, separating the local and the global scaling intervals, are scaled with the paper sheet width (see also [14]) as

$$\zeta \propto \mathbf{W}^{\nu} \text{ and } \xi_C \propto \mathbf{W}^{\omega}, \tag{2}$$

where the exponent $\nu < 0$ was found to be equal to the exponent 2θ, which governs the structure factor scaling (1), while the exponent ω was found to be equal to the global roughness exponent, that is $\omega = \alpha > \alpha_S$. For flame fronts in the *"Secante"* paper we find $\alpha_S = 0.5(\vartheta - 1) = 0.53 \pm 0.02$, $\alpha = 0.83 \pm 0.03 = \omega$, and $2\theta = \nu = -0.40 \pm 0.05 < 0$ [14]. Flame fronts in the *"Filtro"* paper are characterised by $\alpha_S = 0.5 \pm 0.1$, $\alpha = 0.72 \pm 0.06 = \omega$, and $2\theta = \nu = -0.37 \pm 0.07$. This type of roughness behaviour may be attributed to the effect of turbulent airflow accompanying the combustion of the paper.

It is intriguing that in both papers the global roughness exponent of the flame front is found to be equal to the local roughness exponent of the corresponding burning front.

Table 1. Scaling exponents for different kinetic roughening regimes.

Roughening regime	α_S	ζ	α	ϑ	θ	ω	ν
Family-Vicsek ansatz	< 1	α_S	α_S	$2\alpha + 1$	0	-	0
Intrinsically anomalous roughening	< 1	α_S	$> \alpha_S$	$2\zeta + 1$	$\alpha - \zeta > 0$	0	0
Super roughening	> 1	≤ 1	α_S	$2\alpha + 1$	0	0	0
New class of anomalous roughening [8]	> 1	1	$\neq \alpha_S$	$2\alpha_S + 1$	$\alpha - \alpha_S$	?	0
Unconventional anomalous roughening [14]	< 1	$\propto \mathbf{W}^{-\nu} < \alpha$	$> \alpha_S$	$2\alpha_S + 1$	< 0	α	θ

An important point is that none mentioned above exponents are constants for a given system; rather they change from sample to sample in ranges that are lager than the corresponding statistical error within a sample. This can be attributed to the statistical variations in paper thickness, areal density, and fractal dimensions of fibre and pore structures. We note that the statistical data for all exponents satisfy a normal distribution with well different means for different kinds of paper, as well as for interfaces formed by different mechanisms in the same paper.

At the same time, it should be emphasized that despite a well-pronounced anisotropy of paper structures, in all cases a mean value of the local roughness exponents is independent on the interface orientation.

3.2 The fine temporal dynamic of growing interfaces

We found that the motion of the wet front in a paper has a stepwise nature. The height of the wetted area, as a function of time, displays a Devil's-staircase-like behaviour with scaling exponent δ, whereas the front width oscillates erratically with time. These erratic oscillations possess a statistical self-affine invariance in time with the scaling exponent χ, which is found to be equal to the growth exponent β (see, for details, [7]).

We also note that in all experiments, the values of χ, β, and δ, as well as the interface roughness exponents ζ, α_S, and α, vary from one experiment to another in wide ranges, and that their distributions obey a normal distribution.

Moreover, we found that in the paper wetting experiments the mean value of $\chi = \beta$ depends on the interface front orientation with respect to the fibre direction in the paper, whereas the mean values of other exponents do not depend on it. Data for *"Secante"* and *"Filtro"* papers are reported in [7], for the *"Toilet"* paper we find that wetting fronts across the machine direction is characterised by $\beta = \chi = 0.97 \pm 0.15$, whereas fronts along the machine direction scale with $\beta = \chi = 0.53 \pm 0.09$.

Furthermore, we found that the local roughness exponent of moving wet front increases from $\zeta(\mathbf{t} < \mathbf{t_0}) = 0.50 \pm 0.05$ up to $\zeta(\mathbf{t} \geq \mathbf{t_\xi}) = 0.75 \pm 0.03$ (see figure 3), where $\mathbf{t_0} = 10 \pm 7$ sec, $\mathbf{t_\xi} = 960 \pm 90$ sec $< \mathbf{t_S} = 1800 \pm 200$ sec. This indicates that at the initial stage ($\mathbf{t} < \mathbf{t_0}$) the wet front is random ($\zeta = \mathbf{0.5}$). Later, however, the long-range correlations from the pore structure induce the persistent correlations along the wet front, such that the local roughness exponent increases with time as

$$\zeta = \mathbf{0.5} + \rho \mathbf{Ln(t)}, \tag{3}$$

where $\rho = 0.46 \pm 0.12$ is a constant. Therefore, the structure induced correlations display power law behaviour within the time interval $(\mathbf{t_0}, \mathbf{t_\xi})$. An important point is that the moving front achieves a nonequilibrium stationary state ($\zeta = \mathbf{constant}$) before its saturations ($\mathbf{t_\xi} < \mathbf{t_S}$).

The mean values of all exponents obtained in paper wetting, burning, and rupturing experiments are dependent on the paper structure, which gives rise to a correlated quenched noise, as well as on the mechanism of interface formation, which controls a temporal noise in the system.

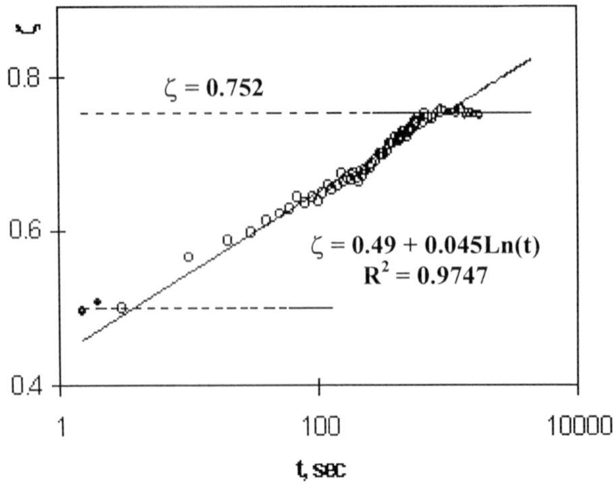

Figure 3. Log-log plot of the local roughness exponent versus time for the wet front in *"Filtro"* paper .

3.3 Self-affine invariance of descending parts of stress-strain behaviour of paper

We find that the descending parts of stress-strain curves (see figure 2 *b*) possess a self-affine invariance with the scaling exponent found to be equal to the local roughness exponent of rupture line, $\mathbf{h}(\lambda\mathbf{x}) \cong \lambda^{\zeta}\mathbf{h}(\mathbf{x})$. That is $\sigma(\lambda\varepsilon) = \lambda^{-\zeta}\,\sigma(\varepsilon)$.

The same exponent governs the changes in the stress-strain curve as the strain rate increases. Namely, we find that the strain interval $\Delta\varepsilon$ of descending parts of stress-strain curves and its standard deviation, $\mathbf{S_D}(\Delta\varepsilon)$, scale with the strain rate, $\dot{\varepsilon}$, as $\Delta\varepsilon \propto \dot{\varepsilon}^{1-\zeta}$ and $\mathbf{S_D}(\Delta\varepsilon) \propto \dot{\varepsilon}^{\zeta}$. Moreover, we find that $\zeta = \mathbf{D} - \mathbf{1}$, where $1 \leq \mathbf{D} \leq 2$ is the fractal dimension of fibre network measured by the box-counting method (see figure 4).

Furthermore, we find that

$$\mathbf{S_D}(\tau_F)\mathbf{U_F} = \mathbf{C_1} \quad \text{and} \quad \tau_F\, \mathbf{S_D}(\mathbf{U_F}) = \mathbf{C_2}, \tag{4}$$

where τ_F is the time-to-fracture, $U_F = \int_{\varepsilon_M}^{\varepsilon_{max}} \sigma d\varepsilon$ is the fracture energy, and $\mathbf{C_1}$ and $\mathbf{C_2}$ are constants. Notice that these relations are similar to the energy-time uncertainty relation in quantum mechanics. For *"Toilet"* paper ($\mathbf{D}$ = 1.75±0.08, ζ = 0.75±0.10) we find $\mathbf{C_1}$ = **85±35 mJ×min** and $\mathbf{C_2}$ = **38±15 mJ×min**.

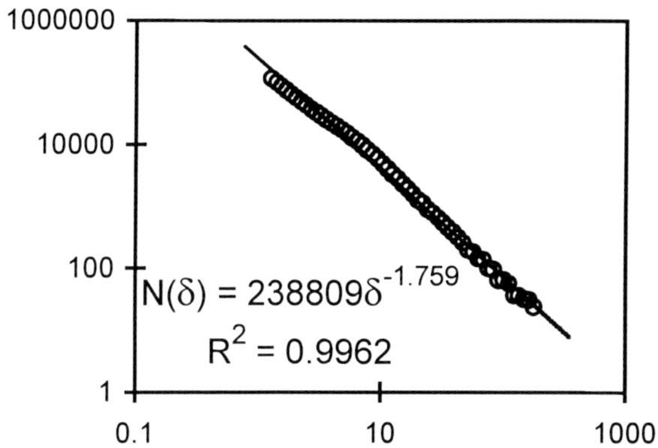

Figure 4. Log-log plot of the number of boxes covered the micrograph of *"Toilet"* paper versus the box size, δ, obtained with the help of BENOIT 1.2 software.

Relations (4) fail when the mechanical behaviour of paper is elasto-plastic. Detailed discussion of Eq. (4), as well as the fractal damage model, will be published elsewhere.

4 Conclusions

Results of present research show that the type of kinetic roughening of interfaces in nonequilibrium conditions, as well as the values of scaling exponents are dependent on the quenched correlations in the disordered medium, as well as on the temporal noise, determined by the interface growth mechanisms. Furthermore, the correlated spatio-temporal noise gives rise to the fractal behaviour of the system far from equilibrium.

Our data suggest that the scaling exponents change continuously with the parameters that controls spatial and temporal correlations in the system. Moreover, we found that the interface roughness exponents changes with time and achieves its stationary value before the interface saturation. This stationary value depends on the correlations in the quenched noise, as well as on the temporal noise in the system.

Acknowledgements

This work was supported by the National Council for Science and Technology (CONACYT) under the research grant 34951-U, and by the Mexican Petroleum Institute (IMP) under the research program D.00005.

References

1. Meakin P., *Fractals, scaling and growth far from equilibrium* (Cambridge University Press, New York, 1998).
2. Barabási A.-L., and Stanley H. E., *Fractal concepts in surface growth* (Cambridge University Press, New York, 1995).
3. Balankin A. S., Physics of fracture and mechanics of self-affine cracks. *Eng. Fracture Mech.* **57** (1997) pp. 135-204.
4. Balankin A. S., Hernandez L. H.., Urriolagoitia-Calderon, G., Probabilistic mechanics of self-affine cracks in paper sheets. *Proc. R. Soc. (London)* **455** (1999) pp. 2565-2575.
5. Balankin A. S. and Susarrey O., A new statistical distribution function for self-affine crack roughness parameters. *Phil. Mag. Lett.* **79** (1999) pp. 629–637.
6. Balankin A. S., Matamoros D. and Campos I., Intrinsic nature of anomalous crack roughening in an anisotropic brittle composite. *Phil. Mag. Lett.* **80** (2000) pp. 165–172.
7. Balankin A. S., Bravo A. and Matamoros D., Some new features of interface roughening dynamics in paper wetting experiments. *Phil. Mag. Lett.* **80** (2000) pp. 503–509.
8. Myllys M., Maunuksela J., Alava M., *et al.*, Kinetic roughening in slow combustion of paper, *http://cond-mat/0105234* (2001).
9. @RISK 4.05, *http://www.palisade.com* (Palisade Corporation, New York, 2000).
10. Scion Image, *http://www.scioncorp.com* (Scion Corporation, Maryland, 1998).
11. BENOIT 1.2, *http://trusoft-interational.com* (TruSoft-International, St. Petersburg, 1999).
12. López J. M., Scaling approach to calculate critical exponents in anomalous surface roughening. *Phys. Rev. Lett.* **83** (1999) pp. 4594-4597.
13. Ramasco J. J., López J. M. and Rodríguez M. A., Generic dynamic scaling in kinetic roughening. *Phys. Rev. Lett.* **84** (2000) pp. 2199–2202.
14. Balankin A. S. and Matamoros D., Unconventional anomalous roughening in the slow combustion of paper. *Phil. Mag. Lett.* **81** (2001) pp 495-503.

SIDEBRANCHING IN THE NONLINEAR ZONE: A SELF–SIMILAR REGION IN DENDRITIC CRYSTAL GROWTH

R. GONZÁLEZ-CINCA

Groupe de Physique des Solides, CNRS UMR 7588, Universités Denis Diderot and Pierre et Marie Curie, Tour 23, 2 place Jussieu, 75251 Paris Cedex 05, FRANCE

Permanent address: Departament de Física Aplicada, Universitat Politècnica de Catalunya, Campus Nord UPC, Ed. B5, J.Girona Salgado s/n, E-08034 Barcelona, SPAIN

E-mail: ricard@fa.upc.es

We have studied sidebranching induced by fluctuations in dendritic growth by means of a phase–field model. We have considered a region where the linear theories are not valid and we have computed the contour length and the area of the dendrite at different distances from the tip. The dependence of the ratio of both magnitudes with the undercooling shows a behaviour in agreement with previous experiments. The derived scaling relation implies that dendrites are self–similar in the considered region.

Keywords: solidification, self-similar, sidebranching, phase–field

1 Introduction

The study of dendritic patterns in nonequilibrium systems is a common point of interest of different disciplines.[1,2,3,4,5,6,7] In the case of solidification phenomena, increased attention is focused on sidebranching, which is the secondary branches that appear at both sides of a growing solid dendrite. Many theoretical[8,9,10,11,12,13] and experimental[14,15,16,17] studies of the region close to the tip in which linear theories are valid have been carried out. In particular, it was found that the qualitative behaviour of sidebranching in this region is independent of the origin of the noise that induces it.[13]

The region further down from the tip (the nonlinear zone) was studied in experiments with xenon dendrites by Hürlimann *et al.*[18] and with succinonitrile dendrites by Li and Beckermann.[19,20] They proposed an alternative set of integral parameters in order to describe the complex shape of a dendrite and the nonlinearities of dendritic solidification. Parameters characterizing independent parts of the dendrite (*e.g.* amplitude and wavelength of the sidebranching) do not take into account the interaction of the side branches through the diffusion field. Nonlinear effects such as *e.g.* coarsening make unclear which side branches should be included in the measurement of the wavelength and which others should not. However, the contour length U, the projection area F and the volume of a dendrite appear to be more appropriate. It was found that F varied linearly with U in a specific range of undercoolings. The average value of the slope of $F(U)$ for different data sets, M, had a very similar dependence with the undercooling than that of the tip radius. Thus, it could be concluded from the experiments that not only near the tip but also further down from it, lengths of a dendrite can be scaled with the tip radius.

In this paper we present a study of sidebranching by means of a phase–field model[21,22,23,24,25,26] which permits the simulation of moving solid–liquid interfaces. In particular, we focus on the nonlinear zone, where branches do not behave as free growing dendrites yet. We study the behaviour of the integral parameters and the role played by the tip radius.

In next Section we present the classical sharp–interface model that characterizes a solidification system, the phase–field model and the numerical procedure used in this work. In Sec. 3, results on the behaviour of integral parameters are presented. Finally, concluding remarks are presented in Sec. 4.

2 Model and numerical procedure

The standard sharp–interface macroscopic model[27] for free solidification of a pure substance considers the solid–liquid interface as a microscopically thin moving surface. This model relies on the heat diffusion equation together with two boundary conditions at the interface, which is assumed to be sharp. The diffusion equation for the temperature field T is given by

$$\frac{\partial T}{\partial t} = D\nabla^2 T, \tag{1}$$

where t is time and D is the diffusion coefficient ($D = k/c_p$, being k the heat conductivity and c_p the specific heat per unit volume).

One boundary condition is obtained from the conservation of the released latent heat at the moving interface:

$$Lv_n = Dc_p[(\nabla_n T)_S - (\nabla_n T)_L], \tag{2}$$

where L is the latent heat per unit volume, v_n is the normal velocity of the interface, ∇_n is the normal derivative at the interface and S and L refer to solid and liquid respectively. The left–hand–side corresponds to the rate at which heat is produced at the interface per unit area while the right–hand–side is the total energy flux away from the interface.

The sharp interface model is completed with a thermodynamic boundary condition, the Gibbs–Thomson relation:

$$T_{int} = T_M - \frac{T_M}{L}[\sigma(\theta) + \sigma''(\theta)]\kappa - v_n\beta_k(\theta). \tag{3}$$

where T_{int} is the temperature at the interface, T_M is the melting temperature, $\sigma(\theta)$ is the anisotropic surface tension (where θ is the angle between the normal to the interface and some crystallographic axis), κ is the local curvature of the interface and $\beta_k(\theta)$ is an anisotropic kinetic term.

The set of equations (1), (2) and (3) defines a full time–dependent free–boundary problem.

In recent years, phase–field models have received increased attention. In these models an additional non–conserved scalar order parameter or phase field ϕ is introduced, whose time evolution equation is coupled with the heat diffusion equation through a source term in order to take into account the boundary conditions at the interface. When the equations are integrated the system is treated as a whole and no distinction is made between the interface and the bulk. The phase field takes constant values in each of the bulk phases (for example, $\phi = 0$ in the solid and $\phi = 1$ in the liquid) changing continuously between them over a transition layer, the interfacial thickness ϵ. This is precisely the parameter that controls the convergence to the sharp–interface limit. The main computational advantage of phase–field models is that the location of the interface does not have to be explicitly determined, but it is given by those points where $\phi = 0.5$.

The corresponding equations for the time evolution of the phase field and the dimensionless temperature can be written in the following form:[21]

$$\epsilon^2\tau(\theta)\frac{\partial\phi}{\partial t} = \phi(1-\phi)\left(\phi - \frac{1}{2} + 30\epsilon\beta\Delta u\phi(1-\phi)\right)$$

$$-\epsilon^2\frac{\partial}{\partial x}\left[\eta(\theta)\eta'(\theta)\frac{\partial\phi}{\partial y}\right] + \epsilon^2\frac{\partial}{\partial y}\left[\eta(\theta)\eta'(\theta)\frac{\partial\phi}{\partial x}\right] + \epsilon^2\nabla\left[\eta^2(\theta)\nabla\phi\right] \tag{4}$$

$$\frac{\partial u}{\partial t} + \frac{1}{\Delta}\left(30\phi^2 - 60\phi^3 + 30\phi^4\right)\frac{\partial\phi}{\partial t} = \nabla^2 u + \psi(x,y,t) \tag{5}$$

where $u(\mathbf{r},t)$ is the diffusion field and Δ is the dimensionless undercooling. Lengths are scaled in some arbitrary reference length ω, while times are scaled by ω^2/D. In these equations θ is the angle between the x-axis and the gradient of the phase field. $\eta(\theta)$ is the anisotropy of the surface tension. The anisotropy of the kinetic term is then given by $\tau(\theta)/\eta(\theta)$. β is equal to $\frac{\sqrt{2}\omega}{12d_o}$ and d_o is the capillary length.

An external noise is introduced through the additive term ψ in the heat equation. It has been demonstrated[13] that sidebranching induced by this kind of noise is qualitatively similar to that of internal noise. Thus, one can consider it appropriate to the study of the nonlinear zone of sidebranching. In our two-dimensional simulations the noise term is evaluated at each uncorrelated cell of lateral size Δx simply as $I \cdot r$, where I denotes the amplitude of the noise, and r is a uniform random number in the interval $[-0.5, 0.5]$. The phase-field model equations have been solved on rectangular lattices using first-order finite differences on a uniform grid with mesh spacing Δx. An explicit time-differencing scheme has been used to solve the equation for ϕ, whereas for the u equation the alternating-direction implicit (ADI) method was chosen, which is unconditionally stable.[21] The kinetic term has been taken as isotropic, which leads to $\tau(\theta) = m\eta(\theta)$ with constant m. A four-fold surface tension anisotropy $\sigma = \sigma(0)(1 + \gamma cos(4\theta))$ has been considered. This gives rise to dendrites growing with perpendicular side branches perpendicular to them.

The growth morphologies were obtained by setting a small vertical seed ($\phi = 0$, $u = 0$) in the center of either of the two shortest sides of the system and imposing $\phi = 1$ and $u = -1$ on the rest of the system. We have considered symmetric boundary conditions for ϕ and u on the four sides of the system, although it was checked that they did not influence on the results presented in this paper.

We have used a set of phase–field model parameters that gives rise to a growing needle without sidebranching when no noise ($I = 0$) is added to the simulations. This assures us that the sidebranching observed when $I \neq 0$ is not due to computational roundings. The fixed parameters for all the simulations have been $\beta = 400$,

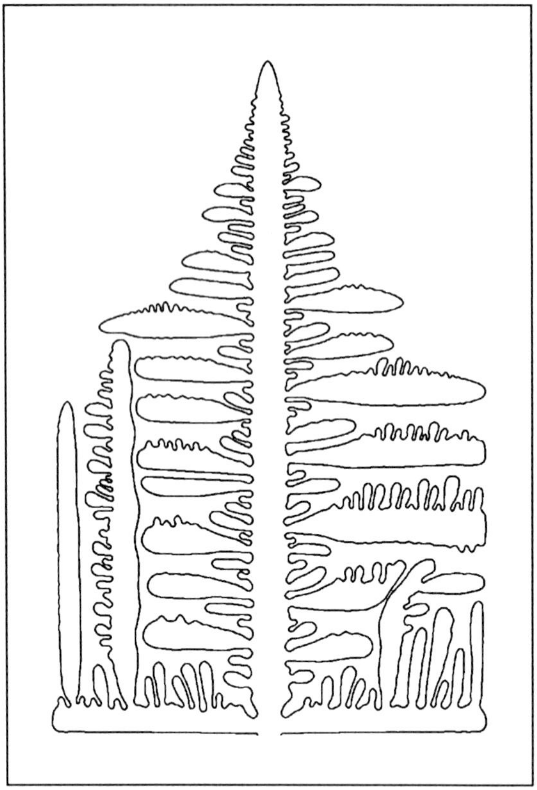

Figure 1. Dendrite obtained with a phase–field model.

$\gamma = 0.045$, $m = 20$ and $\epsilon = 0.003$. The value of Δ has been varied in the range $0.4 - 0.5$. The noise amplitude was kept constant ($I = 19$) in all the simulations and the time and spatial discretizations used were $\Delta t = 10^{-4}$ and $\Delta x = 0.01$. In Fig. 1 it is shown a typical growing dendrite obtained with the phase–field model.

The velocity and the radius of the tip are very weakly affected when noise is introduced. However, side branches appear at both sides of the main dendrite, yielding approximately a 90° angle with it. Further down the tip one can clearly observe competition between branches which gives rise to a coarsening effect. When branches arrive very close to the vertical boundaries of the system, they are stop by the effect of the symmetrical boundary conditions (no heat flux), as can be observed in the figure.

3 Integral parameters

We have measured the contour length U and the area F of the dendrite. U is the length of the contour of the dendrite measured from the tip to a distance y

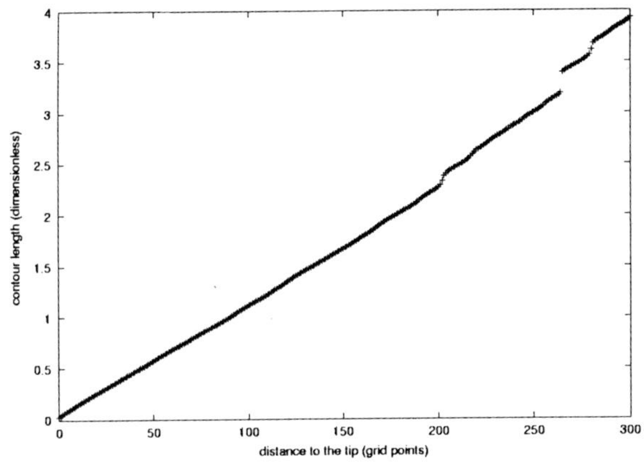

Figure 2. Contour length of the dendrite *vs.* distance to the tip.

further back while F is one half of the area of the dendrite. Both magnitudes have been measured for different values of the dimensionless undercooling. Considering the origin of coordinates at the tip, the contour length and the area have been calculated from the coordinates of the dendrite contour by

$$U = \sum_{i=1}^{n} [(y_{i+1} - y_i)^2 + (x_{i+1} - x_i)^2]^{1/2} \tag{6}$$

and

$$F = \sum_{i=1}^{n} \frac{(x_{i+1} + x_i)}{2} (y_{i+1} - y_i), \tag{7}$$

where n corresponds to each distance to the tip for which we calculated U and F. It is shown in Fig. 2 and Fig. 3 the contour length and the area, respectively, as functions of the distance to the tip for $\Delta = 0.475$. We have observed that F/ρ^2, ρ being the tip radius, approximately varies as $(y/\rho)^{1.6}$ far from the tip. The value of the exponent coincides with the analytical calculation for a nonaxisymmetric needle crystal,[10] and is similar to the value obtained in experiments.[20]

The values of $U(y)$ and $F(y)$ have been computed for the same dendrite (same undercooling) at different times. The dependence of the area on the contour length has been studied for all of these data sets. We have considered both magnitudes measured between the tip and the positions where x becomes a local minimum. We found the same linear relation between F and U for all the data sets and we called M the slope of the graph $F(U)$.

We have also measured the tip radius of the dendrites obtained by computing $\rho = \phi_x/\phi_{yy}$ at the tip. The tip radius and the slope M were found to have a very similar dependence on the undercooling. This is reflected in Fig. 4, where the normalized slope M/ρ is represented as a function of the undercooling. The value of M/ρ is found to be independent of the undercooling. This result is in agreement with experimental observations.[18]

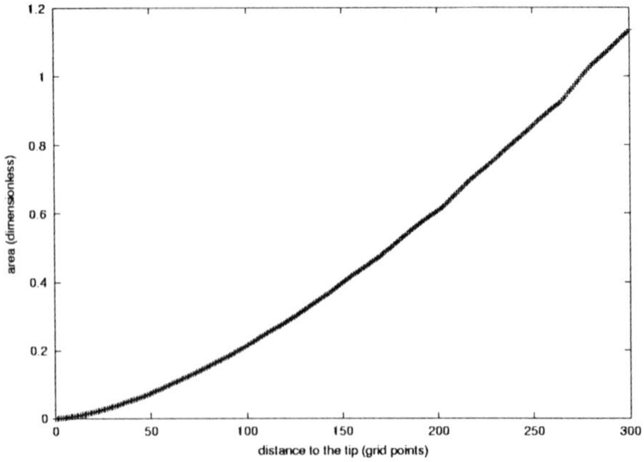

Figure 3. Area of a half of the dendrite *vs.* distance to the tip.

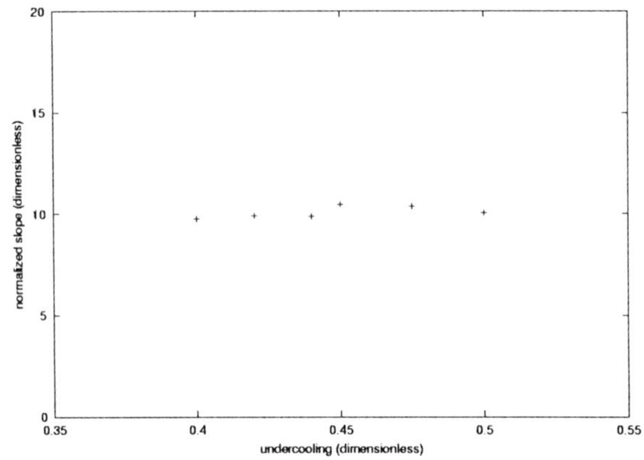

Figure 4. Normalized slope of the graph $F(U)$ *vs.* undercooling.

4 Concluding remarks

We have studied sidebranching induced by fluctuations in dendritic growth by means of a phase–field model, focusing on the nonlinear zone of the dendrite. Our contribution to a better knowledge of sidebranching under these conditions has been done through the study of the nonlinear regime by means of the integral parameters. We have measured the contour length and the area of the dendrite as functions of the distance to the tip and we have found a linear relationship between both integral parameters. The corresponding slope is proportional to the tip radius in the considered range of undercoolings. These results confirm previous experiments and give evidence of the importance of the tip radius as a length scale in the nonlinear regime. The fact that the scaling relation obtained is undercooling independent implies that the dendrites are self–similar in the sidebranching region corresponding to the nonlinear zone and for the considered range of undercoolings. These results offer some new insight into the understanding of a fully developed dendrite. Further work increasing the statistics and the range of undercoolings is currently being carried out.

Acknowledgments

The author thanks L. Ramírez-Piscina for valuable discussions. This research is supported by the Dirección General de Investigación (Spain) (Project BFM2000-0624-C03-02) Comissionat per a Universitats i Recerca (Spain) (Projects 1999SGR00145 and 2000XT00005), and European Commission (TMR Network Project ERBFMRXCT96-0085).

References

1. D'Arcy W. Thompson, *On Growth and Form*, Cambridge University Press (1952).
2. *Dynamics of Curved Fronts.* Ed. P. Pelcé. Perspectives in Physics. Academic Press (1988).
3. J.D. Murray, *Mathematical Biology*, Springer–Verlag (1989).
4. *Growth and Form. Nonlinear Aspects.* Eds. M. Ben Amar, P. Pelcé and P. Tabeling, vol. 276 of NATO ASI series B, Plenum Press (1988).
5. M.C. Cross and P.C. Hohenberg, *Rev. Mod. Phys.*, **65**, 851 (1993).
6. *Handbook of Crystal Growth.* Ed. D.T.J. Hurle, vol. 1B, North-Holland (1993).
7. *Spatio Temporal Patterns.* Eds. P.E. Cladis and P. Palffy–Muhoray, vol. XXI of Santa Fe Institute Studies in the Science of Complexity, Addison-Wesley (1994).
8. M.N. Barber, A. Barbieri and J.S. Langer, *Phys. Rev. A*, **36**, 3340 (1987).
9. J.S. Langer, *Phys. Rev. A*, **36**, 3350 (1987).
10. E. Brener and D. Temkin, *Phys. Rev. E*, **51**, 351 (1995).
11. A. Karma and W.-J. Rappel, *Phys. Rev. E*, **60**, 3614 (1999).
12. S.G. Pavlik and R.F. Sekerka, *Physica A*, **277**, 415 (2000).
13. R. González-Cinca, L. Ramírez-Piscina, J. Casademunt and A. Hernández-Machado, *Phy. Rev. E*, **63**, 051602 (2001).
14. A. Dougherty, P.D. Kaplan and J.P. Gollub, *Phys. Rev. Lett.*, **58**, 1652 (1987).
15. X. W. Qian and H. Z. Cummins, *Phys. Rev. Lett.*, **64**, 3038 (1990).
16. U. Bisang and J. H. Bilgram, *Phys. Rev. Lett.*, **75**, 3898 (1995).
17. U. Bisang and J. H. Bilgram, *Phys. Rev. E*, **54**, 5309 (1996).
18. E. Hürlimann, R. Trittibach, U. Bisang and J. H. Bilgram, *Phys. Rev. A*, **46**, 6579 (1992).
19. Q. Li and C. Beckermann, *Phys. Rev. E*, **57**, 3176 (1998).
20. Q. Li and C. Beckermann, *Acta Mater.*, **47**, 2345 (1999).
21. A. A. Wheeler, B. T. Murray and R. J. Schaefer, *Physica D*, **66**, 243 (1993).
22. G. B. McFadden, A. A. Wheeler, R. J. Braun, S. R. Coriell and R. F. Sekerka, *Phys. Rev. E*, **48**, 2016 (1993).
23. S.-L. Wang, R. F. Sekerka, A. A. Wheeler, B. T. Murray, S. R. Coriell, R. J.

Braun and G. B. McFadden, *Physica D*, **69**, 189 (1993).

24. R. González-Cinca, L. Ramírez-Piscina, J. Casademunt, A. Hernández-Machado, L. Kramer, T. Tóth Katona, T. Börzsönyi and Á. Buka, *Physica D*, **99**, 359 (1996).
25. T. Tóth-Katona, T. Börzsönyi, Z. Váradi, J. Szabon, Á. Buka, R. González-Cinca, L. Ramírez-Piscina, J.Casademunt and A. Hernández-Machado, *Phys. Rev. E*, **54**, 1574 (1996).
26. R. González-Cinca, L. Ramírez-Piscina, J. Casademunt, A. Hernández-Machado, T. Tóth Katona, T. Börzsönyi and Á. Buka, *J. Cryst. Growth*, **193**, 712 (1998).
27. J.S. Langer, *Lectures in the Theory of Pattern Formation*, Chapter 10 in *Chance and Matter* (1986 Les Houches Lectures) edited by J. Souletie, J. Vannimenus and R. Storra, North Holland 1987, pp. 629-711.
28. E.A. Brener and V.I. Melnikov, *Adv. in Phys.*, **40**, 53 (1991).

SCALING LAWS AND FREQUENCY DECOMPOSITION FROM WAVELET TRANSFORM MAXIMA LINES AND RIDGES

M. HAASE, J. WIDJAJAKUSUMA [1], R. BADER [2]
Institut für Computeranwendungen (ICA II), Stuttgart University, Pfaffenwaldring 27, 70569 Stuttgart, Germany, E-mail: mh@ica.uni-stuttgart.de

[1] *Institut für Mechanik (Bauwesen), Stuttgart University, Pfaffenwaldring 7, 70569 Stuttgart, Germany, E-mail: jack@mechbau.uni-stuttgart.de*

[2] *Institut für Musikwissenschaften, Hamburg University, Neue Rabenstr.13, 20354 Hamburg, Germany, E-mail: R.Bader@t-online.de*

Wavelet techniques have now become well established for various applications. They are especially attractive for a reliable characterization of the scaling behaviour of functions and measures with non-oscillating singularities. Another important feature of wavelets is their ability to decompose vibrations into components according to their instantaneous frequencies. The essential information about scaling and instantaneous frequencies is contained in a small subset of the redundant continuous wavelet transform, namely in the maxima lines and ridges, which can be considered as a *fingerprint* of the signal. We show that even oscillating singularities can be easily characterized using complex progessive wavelets. We derive differential equations for two families of wavelets which allow a direct numerical integration of maxima lines and ridges. The applications presented range from fractal basin boundaries, oscillating singularities, system identification in engineering structures and design to a problem in musicology. Wavelets are used to characterize the timbre of instruments. The fine structure of transients allow an identification of instruments and instrument classes.
Keywords: Wavelet transform, maxima lines and ridges, scaling, instantaneous frequencies, musicology

1 Introduction

Large classes of signals arising in mathematics, natural and engineering sciences as well as in musical sounds exhibit a very complex, irregular behaviour. Wavelet techniques provide new solutions for the analysis of a large range of functions. One of the most valuable features of the continuous wavelet transform (CWT) is that it allows a very precise analysis of the regularity properties of a signal. As long as a function does not contain oscillating singularities, the standard wavelet transform modulus maxima (WTMM) method is a useful technique which has now become well established [1,2]. It allows us to extract the local scaling behaviour of singularities and provides one of the most reliable methods for the determination of singularity spectra [5]. This is possible by analysing the scaling behaviour along the *maxima lines* where the modulus of the CWT is concentrated. In order to include oscillating singularities, we propose a modification of the WTMM method which uses complex progressive wavelets as the kernel.

Another main feature of the wavelet transform is its capacity to decouple vibrations according to their instantaneous frequencies. Choosing an analysing complex progressive wavelet as the kernel for the CWT, which is well localized in Fourier space, we see that the modulus of the CWT concentrates near a series of curves called *ridges*[3].

Commonly, the maxima lines and ridges are calculated as follows. First, the CWT is evaluated by a Fourier domain product at discrete points (a_i, b_j). Local maxima are next determined for either each scale a_i or constant time b_j for maxima lines and ridges respectively. An additional chaining stage is then necessary to obtain connected lines, which might be difficult in the case of bifurcating lines.

In contrast with the above method we calculate the CWT *continuously* along the relevant lines. The advantage is, that unnecessary calculations and difficult chaining algorithms are avoided. In the case of the Gaussian and Morlet family of wavelets a set of ordinary differential equations for the parameterized lines $a(s), b(s)$ can be derived.

The outline of the paper is as follows. In section 2, we briefly summarize some basic properties of the CWT which are needed for our approach. Two families of wavelets with an increasing number of vanishing moments are presented in section 3. In section 4, differential equations are derived which allow a direct tracing of maxima lines and ridges. Some examples are presented illustrating the capability of the WTMM method. We show that the information contained in the ridges can be used for an approximative analysis of the time-frequency components in asymptotic signals. In addition to an application in engineering science we emphasize the relevance of the method to musicology. The fine structure of transients allows an identification of instruments and instrument classes. This is done in section 5. Conclusions are drawn in section 6 and future work is discussed.

2 Continuous wavelet transform - some basic features

The continuous wavelet transform (CWT)

$$Wf(a,b) = \frac{1}{a}\int_{-\infty}^{+\infty} f(t)\,\overline{\psi\left(\frac{t-b}{a}\right)}\,\mathrm{d}t \qquad (a, b \in \mathbf{R},\, a > 0) \tag{1}$$

decomposes the function $f(t) \in L^2(\mathbf{R})$ hierarchically in terms of elementary components $\psi\left(\frac{t-b}{a}\right)$. They are obtained from a single *analysing wavelet* $\psi(t)$ by means of dilations and translations. Here $\overline{\psi}(t)$ denotes the complex conjugate of $\psi(t)$, a the scale and b the shift parameter. $\psi(t)$ has to be chosen so that it is well localized both in physical and Fourier space.
The signal $f(t)$ can be uniquely recovered by the inverse wavelet transform

$$f(t) = \frac{1}{C_\psi}\int_{-\infty}^{+\infty}\int_{0}^{\infty} Wf(a,b)\,\psi\left(\frac{t-b}{a}\right)\frac{\mathrm{d}a}{a}\,\mathrm{d}b, \tag{2}$$

if $\psi(t)$ (resp. its Fourier transform $\hat{\psi}(\omega)$) satisfies the admissibility condition

$$C_\psi = \int_{0}^{+\infty} \frac{|\hat{\psi}(\omega)|^2}{\omega}\,\mathrm{d}\omega < \infty, \tag{3}$$

which reduces to $\int\limits_{-\infty}^{\infty} \psi(t)\,\mathrm{d}t = 0$ for $\psi(t) \in L^1(\mathbf{R})$. In order to make the entire range of singularities accessible we require $\psi(t)$ to be orthogonal to polynomials up to order N

$$\int\limits_{-\infty}^{+\infty} t^k \psi(t)\,\mathrm{d}t = 0 \qquad\qquad 0 \leq k \leq N. \tag{4}$$

Meanwhile, it is a standard technique to extract microscopic information about the scaling properties of cusp singularities from the scaling behaviour along wavelet transform modulus maxima lines [2,8]. Assuming a cusp singularity with Hölder exponent $h(t_0) \in (n, n+1)$ at t_0, the CWT scales like

$$|Wf(a,t_0)| \leq C a^{h(t_0)} \tag{5}$$

for $a \to 0$, provided the analysing wavelet chosen has $n_\psi > n$ vanishing moments. However, in the presence of oscillating singularities like

$$f(t) = |t - t_0|^{h(t_0)} \sin(|t - t_0|^{-\beta(t_0)}), \tag{6}$$

the standard WTMM method gives irrelevant information on the Hölder regularity of the function [7,6]. In general, *two* exponents h, β are necessary to describe the singular behaviour of a function $f(t)$, namely the Hölder exponent h and the oscillation exponent β describing the local power law divergence of the instantaneous frequency. For this case, we propose to use *complex progressive* wavelets to analyse the scaling behaviour, see section 3. In this case, the modulus of the CWT scales like

$$|Wf(a,t_0)| \leq C a^{h(t_0)/(1+\beta(t_0))} \tag{7}$$

as long as $\psi(t)$ has enough vanishing moments.

3 Two families of wavelets

Two families of wavelets, each of specific use for different purposes, are introduced. The first is the Gaussian family of real wavelets which are obtained as derivatives of the Gaussian function. They are very efficient in detecting cusp singularities and sharp signal transitions and in characterizing their scaling properties. The second is the Morlet family which consists of complex progressive wavelets. They allow us to determine the Hölder and oscillating exponents of oscillating singularities. Morlet wavelets allow vibration modes to be decoupled and the temporal evolution of frequency transients and damping coefficients to be measured.

We define the family of real Gaussian wavelets of n-th order $\psi_n(t)$ as

$$\psi_0(t) = \mathrm{e}^{-t^2/2}, \qquad \psi_n(t) = \frac{\mathrm{d}}{\mathrm{d}t}\psi_{n-1}(t) \qquad (n \in \mathbf{N}). \tag{8}$$

For $n > 0$, the functions $\psi_n(t)$ fulfil the admissibility condition, eq.(3), and can thus be used as analysing wavelets. Although $\psi_n(t)$ has an infinite support, the function as well as its Fourier transform decay rapidly to zero. For all practical calculations, it can therefore be considered to be well localized in time and frequency.

In contrast with real wavelets, complex wavelets can separate amplitude and phase, enabling the measurement of instantaneous frequencies and their temporal evolution. Let us define a family of complex wavelets [9], which are obtained as derivatives of the classical Morlet wavelet $\Psi_0(t)$

$$\Psi_0(t) = \mathrm{e}^{-t^2/2}\,\mathrm{e}^{i\omega_0 t}\,, \qquad \Psi_n(t) = \frac{\mathrm{d}}{\mathrm{d}t}\Psi_{n-1}(t) \qquad (n \in \mathbf{N}). \tag{9}$$

$\Psi_0(t)$ does not fulfil the admissibility condition eq.(3) in a strict sense. However, for practical purposes, because of the fast decay of its envelope towards zero, we can consider the Morlet wavelet $\Psi_0(t)$ to be admissible for $\omega_0 \geq 5$. On the other hand, all derivatives of $\Psi_0(t)$ are wavelets in a strict sense independently on ω_0.

By induction, it can be shown that the following relation holds [9]

$$(n+1)\Psi_n(t) + (t - i\omega_0)\,\Psi_{n+1}(t) + \Psi_{n+2}(t) = 0. \tag{10}$$

Eq.(10) serves as a basis for the derivation of a partial differential equation for the wavelet transform [9] with Ψ_n as the kernel

$$\left(a\frac{\partial^2}{\partial b^2} - \frac{\partial}{\partial a} - i\omega_0\frac{\partial}{\partial b} + \frac{n}{a}\right) W^n f = 0. \tag{11}$$

Here, we used the notation $Wf(a,b) =: W^n f$. The key to the usefulness of this family of wavelets is the fact that all members are progressive (or analytic) meaning

$$\hat{\Psi}_n(\omega) = 0 \qquad \text{for} \qquad \omega < 0. \tag{12}$$

For the simple case of $f(t) = \cos(\omega t)$ and using $\Psi_0(t)$ as kernel this leads to a localization of the modulus of the CWT around the line $a = \omega_0/\omega$.
Inserting $\omega_0 = 0$ into eq.(11), one obtains an analogous partial differential equation for the Gaussian family of wavelets

$$\left(a\frac{\partial^2}{\partial b^2} - \frac{\partial}{\partial a} + \frac{n}{a}\right) W^n f = 0. \tag{13}$$

4 Direct tracing of maxima lines and ridges

Exploring the special properties of the family of Gaussian and Morlet wavelets, we can derive differential equations for the maxima lines and ridges.

4.1 Differential equations for the maxima lines

For a fixed scale a_0, local maxima of $|W^n f(a_0,b)|$ are obtained from

$$|W^n f(a_0,b)|_b = 0 \qquad \text{and} \qquad |W^n f(a_0,b)|_{bb} < 0. \tag{14}$$

Here, we used the abbreviation $\frac{\partial F}{\partial b} = F_b$. Connected lines of local maxima are called maxima lines. For the moment, let us restrict ourselves here to the Gaussian family of wavelets. Instead of calculating local maxima of $|W^n f(a_0,b)|$, we calculate those of $|W^n f(a_0, b)|^2$. Describing the maxima lines in a parametric form $\{a(s), b(s)\}$, we may approximate the change of $(W^n f)_b$ along the line by the first terms in a Taylor expansion, yielding

$$\frac{\mathrm{d}}{\mathrm{d}s}(W^n f)_b \approx (W^n f)_{ba}\frac{\mathrm{d}a}{\mathrm{d}s} + (W^n f)_{bb}\frac{\mathrm{d}b}{\mathrm{d}s} = 0. \tag{15}$$

The differential equations for maxima lines can thus be written in the form

$$\frac{\mathrm{d}a}{\mathrm{d}s} = -c\,(W^n f)_{ba}, \qquad \frac{\mathrm{d}b}{\mathrm{d}s} = c\,(W^n f)_{bb} \tag{16}$$

with an arbitrary constant c regulating the parameterization. An extension to complex wavelets is straightforward.

The skeleton of maxima lines is obtained by first calculating the values b_i of all maxima of $(W^n f)^2$ on the smallest scale a_{min}. Each pair (a_{min}, b_i) $(i = 1, \cdots, m)$ is then used as an initial condition for the differential equations (16), which can be integrated numerically.

As a first example, we evaluate the singularity spectrum of a fractal basin boundary, which is generated by a chaotically driven dynamical system

$$x_{n+1} = x_n{}^\gamma + b\,G(t_n) \qquad (\gamma > 1). \tag{17}$$

For weak forcing, the basin boundary can be evaluated explicitly as [11]

$$x(t) = \lim_{n\to\infty} x^n(t) \quad \text{where} \quad x^n(t) = 1 - b\sum_{k=1}^{n}\frac{G^k(t)}{\gamma^k} \tag{18}$$

For a tent map $G(t)$ (fig. 1b) and the parameters $\Delta = 0.7$, $\gamma = 1.3$, $b = 0.1$, the basin boundary is shown in fig. 1a. For all initial conditions (x_0, t) chosen in the prisoner set, $x_n(t)$ remains finite. In fig. 1c, the skeleton of maxima lines based on ψ_2 is shown. Application of the WTMM method [2,8] leads to the singularity spectrum $D(h)$ displayed in fig. 1d. The result is in good agreement with theoretical results obtained by means of the structure function method [11] (green left branch). In contrast to the WTMM method, the structure function method does not give access to the full spectrum.

The next example illustrates the advantage of using complex progressive wavelets. Fig. 2a shows the graph of the oscillating singularity given in eq. (6) for $h = 4/3$ and $\beta = 1$. In figs. 2b,c, the modulus of the CWT is plotted using the real Mexican hat wavelet ψ_2 and the complex progressive wavelet Ψ_2 respectively. The corresponding skeletons of maxima lines are shown in figs. 2d,e. Arnéodo et al. [7,6] have used a skeleton like the one shown in fig. 2d for an estimation of h and β. The advantage of using complex wavelets is obvious. The modulus of the CWT concentrates along two curves $a = |b - t_0|^{\varphi(t_0)}$ and scales like $|W^2 f| \sim a^{\alpha(t_0)}$ for $a \to 0$ (for details, see [10]). From the log-log plots shown in fig. 2f, the exponents $\alpha(t_0)$ and $\varphi(t_0)$ can be extracted. The Hölder and the oscillating exponents may finally be calculated from $\beta(t_0) = 1/\varphi(t_0) - 1$ and $h(t_0) = \alpha(t_0)/\varphi(t_0)$.

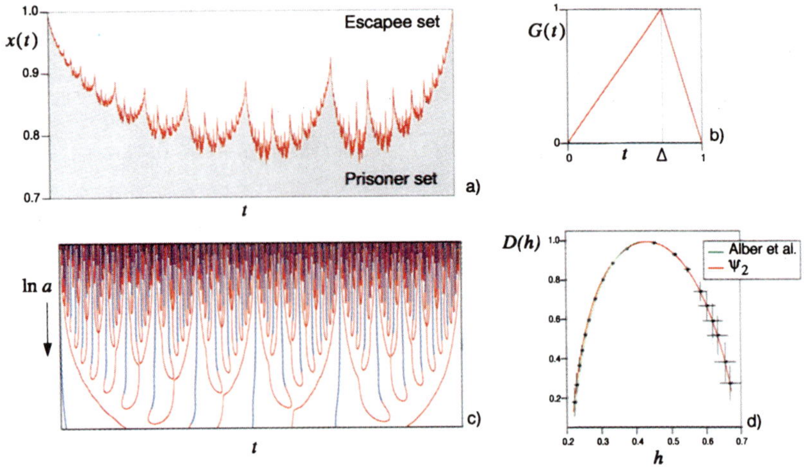

Figure 1: Dynamical system driven by chaotic forcing. (a) Fractal basin boundary, (b) tent map $G(t)$, (c) skeleton of maxima lines, (d) singularity spectrum $D(h)$.

4.2 *Differential equations for the ridges*

An arbitrary real signal $f(t)$ can always be written as $f(t) = \sin \tilde{A}(t) \cos \tilde{\Phi}(t)$. However, such a representation is not unique. To achieve uniqueness, it is common to introduce the analytic function $z(t) = f(t) + i\, Hf(t)$ with $f(t)$ as the real part and the Hilbert transform $Hf(t)$ as the complex part [3]. By definition, $z(t)$ is entirely characterized by its real part and its Fourier transform $\hat{z}(\omega)$ is zero for negative frequencies. The so-called *canonical representation* of $f(t)$

$$z(t) = A(t)e^{i\Phi(t)} \tag{19}$$

is then unique if we assume $A(t) \geq 0$ and $\Phi(t) \in [0, 2\pi)$. This allows us to introduce the instantaneous frequency $\omega(t) = \Phi'$, where the sign $'$ denotes the derivative with respect to time t. The physical significance of $\omega(t)$ might be doubtful in specific cases unless we restrict considerations to *asymptotic* signals with $A(t)$ and $\Phi'(t)$ slowly varying [4]. The envelope $A(t)$ and $\omega(t)$ then have a physical meaning. Using a wavelet of the Morlet family, the CWT can be approximated by the first term of an asymptotic expansion using the stationary phase argument [3]

$$Wz(a,b) = A(b)e^{i\Phi(b)}\overline{\hat{\Psi}(a\Phi'(b))} + \mathcal{O}\left(\frac{|A'|}{|A|}, \frac{|\Phi\Phi''|}{\Phi'^2}\right). \tag{20}$$

The modulus of the wavelet transform is concentrated in the neighbourhood of a curve, called *ridge*, satisfying the condition

$$a = a_r(b) = \frac{\omega^*}{\Phi'(b)} \tag{21}$$

where ω^* denotes the peak frequency of $\hat{\Psi}(\omega)$. By inserting this relation into (20), we see that, along the ridge, the wavelet transform is approximately proportional to the analytic signal $z(t)$ given in eq. (19) with a constant factor $C = \overline{\hat{\Psi}(a\,\Phi'(b))} = \overline{\hat{\Psi}(\omega^*)}$.

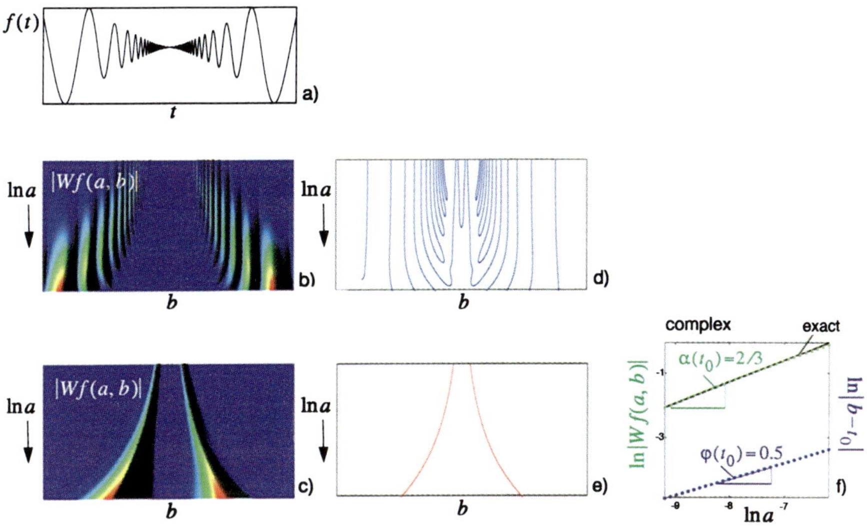

Figure 2: Oscillating singularity. (a) Graph of $f(t)$, (b) modulus of CWT using real ψ_2, (c) modulus of CWT using complex Ψ_2, (d),(e) corresponding skeletons of maxima lines, (f) extraction of Hölder and oscillating exponents from (e).

The transient vibration behaviour of structures can be described by a function $f(t)$ given as a superposition of asymptotic components with slowly varying amplitudes and phase variations. If we use wavelets $\Psi_n(t)$ of the Morlet family, the modulus of the CWT shows high concentrations along a series of curves denoted as ridges and given by eq. (21). For the determination of the ridges, it is in general sufficient to determine local maxima of $|W^n f(a, b_0)|$ for a fixed time b_0, see[4]. Hence, the conditions for ridges can be written in the form

$$|W^n f(a, b_0)|_a = 0 \qquad \text{and} \qquad |W^n f(a, b_0)|_{aa} < 0. \tag{22}$$

In analogy to Section 4.1, we derive differential equations for the ridges, which are written in a parametric form $\{a(s), b(s)\}$. Since we use complex wavelets here, they have a slightly different form

$$\frac{\mathrm{d}a}{\mathrm{d}s} = -C\, G_b(a, b) \qquad \frac{\mathrm{d}b}{\mathrm{d}s} = C\, G_a(a, b) \tag{23}$$

where $G(a, b) = \Re\left[(W^n f)_a\, \overline{W^n f}\right]$ and C is an arbitrary constant. The ridges are then obtained by first calculating the local maxima a_i for a fixed time b_0 and then using the pairs (a_i, b_0) $(i = 1, \cdots, m)$ as initial conditions for the integration of the differential equations (23).

The CWT restricted to ridges can be used for detecting changes in the microstructure of materials [13,12]. A change in the free response of a structure to a short-term impulse excitation can be expressed in terms of a variation of damping characteristics and vibration frequencies.

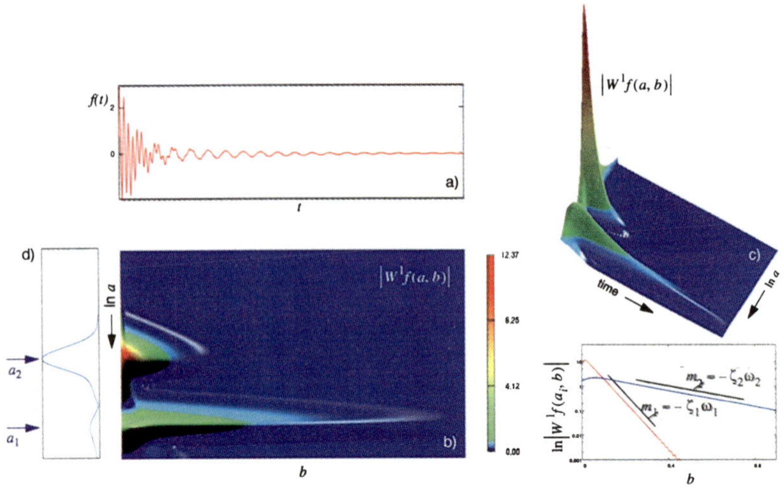

Figure 3: System identification. (a) Graph of vibration $f(t)$, (b), (c) modulus of CWT using complex Ψ_2, (d) extraction of scales and natural frequencies, (e) determination of damping coefficients.

Let us illustrate the method by means of a free vibration of a linear system with 2 degrees of freedom

$$f(t) = \sum_{j=1}^{2} A_j(t) \cos \Phi_j(t) \tag{24}$$

where $A_j(t) = C_j e^{-\zeta_j \omega_j t}$ and $\Phi_j(t) = \sqrt{1-\zeta_j^2}\, \omega_j t$; ω_j are the natural frequencies, ζ_j the damping ratios and a_j the amplitudes ($C_1 = 0.5$, $C_2 = 3.0$, $\zeta_1 = 0.03$, $\zeta_2 = 0.045$, $\omega_1 = 40\pi = 125.66$, $\omega_2 = 156\pi = 490.09$). In fig. 3a, the graph of the impulse response function is shown for $0 \leq t \leq 1$. In figs. 3b,c, two views of the modulus of the CWT are plotted using Ψ_1 with $\omega_0 = 5$. It can be seen that the CWT decouples the vibration modes automatically. However, for our method, it is not necessary to calculate the full CWT. We only have to evaluate the CWT for a fixed time b_0 (fig. 3d) and extract the scales a_1, a_2 of the maxima yielding $a_1 = 4.159 \cdot 10^{-2}$ and $a_2 = 1.046 \cdot 10^{-2}$. The corresponding frequencies are obtained from eq. (21) using the peak frequency $\omega^* = (\omega_0 + \sqrt{\omega_0^2 + 4})/2$ of Ψ_1 as $\tilde{\omega}_1 = 124.85$ and $\tilde{\omega}_2 = 496.42$. The damping parameters are found by plotting $\ln |W^1 f(a_j, b)|$ versus b along the ridges (see fig. 3e). From the slopes $m_1 = -21.34$, $m_2 = -3.76$, the damping coefficients are calculated as $\zeta_1 = 0.0430$ and $\zeta_2 = 0.0301$. As for monochromatic signals, eq. (19), $f(t)$ can be easily reconstructed from the CWT along the ridges as a superposition of two modes, see eq. (20,21). Thus, all information about the modal parameters and the vibration can be extracted from the ridges. The simplicity of the approach becomes even more pronounced, if we analyse nonlinear systems with time-dependent frequencies.

5 Application to musicology

With regard to musicology, a detailed analysis of recorded sounds is of high interest. Examples are the vibrato of singers, the exact intonation of choirs, the beat of sounds produced by two frequencies, which are $\Delta f < 20$ Hz apart (e.g. the ombak of the Indonesian Gamelan Gong Gede, the Nepalese sound shells or the 12-string western guitar with its sympathetic strings). The fine structure of transients, and especially initial transients, is very important for the quality and identification of instruments and instrument classes [14,15]. In western music, the steady state after the instrument has been plucked or stuck always produces a harmonic overtone structure. This structure consists of more or fewer overtones. Some instruments tend to an odd or even overtone structure (e.g. the clarinette tends to an odd one, because it is a tube with one end closed). A rough classification of instruments (bowed, plucked, struck or blown) is also possible during the steady state phase. However, a more precise identification of the instrument and also of the musical expression by means of semiology is found within the initial transient. Another important point is the duration of transients. For example, a good violin is said to react quickly on the bowing and reaches the steady state very fast [16].

Wavelet transform is a promising method for analysing these fine structures because, bearing the uncertainty principle in mind as well as the fact that music is constructed of mixtures of frequencies, it offers the opportunity of varying all parameters, such as window length, analysing wavelet, peak frequency and time resolution. So one can zoom in and out as necessary.

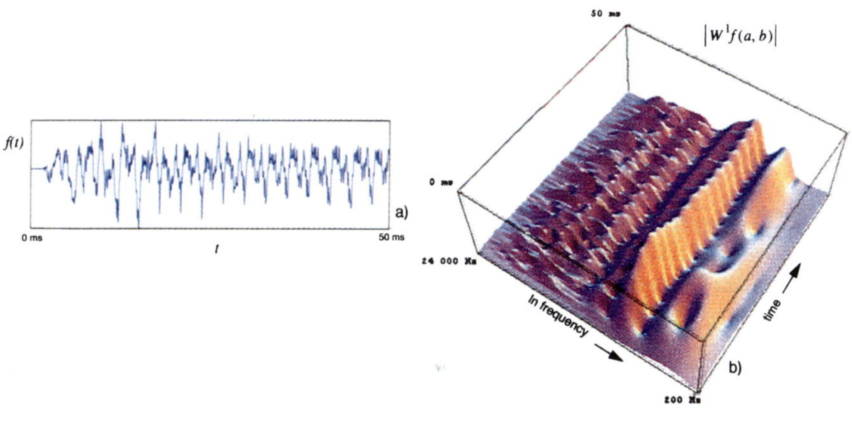

Figure 4: Guitar played apoyando on upper E-string struck loud. (a) Graph of vibration, (b) modulus of CWT using complex Ψ_1.

Let us take as an example the initial transient of a classical guitar tone, played apoyando on the upper E-string struck loud. There is a double peak of 215 Hz and 334 Hz (329.6 Hz is the theoretical value of the fundamental frequency of this tone e1). This double peak disappears in the steady state and is most likely caused by an eigenmode of the guitar, because the 215 Hz peak also appears with higher notes. The Fourier spectrum does not contain information about the fine structure of initial transients and thus is of no use for answering questions like, how the second peak interacts with the fundamental or how the higher overtones come in. In contrast, this can be analysed with the wavelet techniques.

6 Conclusions

We have derived differential equations for two families of wavelets, which allow us a direct tracing of maxima lines and ridges. The essential information on *local* scaling and instantaneous frequencies of a signal is concentrated in the skeleton of maxima lines and ridges. This can be seen as a fingerprint of the signal. We have shown how the Hölder and oscillating exponents can be easily extracted from the maxima lines for an oscillating singularity. In analogy to system identification in engineering problems, this approach can be used to identify the typical timbre of an instrument in musicology. Further investigations are needed to study the capability of the method. Among these are the study of modes with close frequencies [3,13] and the extension of the WTMM method to estimate the singularity spectrum of singular functions containing both cusp and oscillating singularities.

References

1. S. Mallat and W.L. Hwang, *IEEE Trans. Inf. Theory.* **38**, 617–643 (1992).
2. J.F. Muzy, E. Bacry and A. Arnéodo, *Int. J. Bif. Chaos* **4**, 245–302 (1994).
3. N. Delprat, B. Escudié, P. Guillemain, R. Kronland-Martinet, Ph. Tchamitchian, B. Torrésani, *IEEE Trans. Inf. Theory* **38** (1992) 644–664.
4. R. Carmona, W.L. Hwang, B. Torrésani, *Practical Time-Frequency Analysis*, Academic Press, San Diego (1998).
5. S. Jaffard, in *Fractals in Engineering*, (J. Lévy Véhel et al. eds.), Springer, London (1997).
6. A. Arnéodo, E. Bacry, S. Jaffard, J.F. Muzy, *J. Stat. Phys.* **87**, 179–209 (1997).
7. A. Arnéodo, E. Bacry, and J.F. Muzy, *Phys. Rev. Lett.* **74**, 4823–4826 (1995).
8. M. Haase, B. Lehle, in: *Fractals and Beyond*, (M. M. Novak ed.), World Scientific, Singapore 241–250 (1998).
9. M. Haase, in: *Paradigms of Complexity*, (M. M. Novak ed.), World Scientific, Singapore 287–288 (2000).
10. S. Jaffard, Y. Meyer, Memoirs of the AMS **587** (1996).
11. M. Alber, T. Galla, J. Peinke, preprint (1999).
12. M. Haase, J. Widjajakusuma, submitted to *Int.J.Eng.Science.*
13. W.J. Staszewski, *J. Sound and Vibration*, **214**, 639–658 (1998).
14. N. Fletcher, T. Rossing, *The Physics of Musical Instruments*, Springer, New York (1999).
15. K.D. Martin, E.K. Youngmoo, presented at 136th Meeting of the Acoust. Soc. Am. (1998).
16. C.M. Hutchins, V. Benade, *Research Papers in Violin Acoustics 1975-1993*, Acoust.Soc.Am., **2**, Woodbury, New York (1997).

RANDOM WAVELET SERIES: THEORY AND APPLICATIONS

JEAN-MARIE AUBRY

Centre de Mathématiques, Université Paris 12, 94010 Créteil Cedex, France
E-mail: aubry@univ-paris12.fr

STÉPHANE JAFFARD

Centre de Mathématiques, Université Paris 12, 94010 Créteil Cedex, France and Institut Universitaire de France
E-mail: jaffard@univ-paris12.fr

Random Wavelet Series were introduced in Aubry *et al.*[1] as a generalization of the Lacunary Wavelet Series of Jaffard.[2] They form a fairly broad class of random processes, with multifractal properties. We give three applications of this construction. First, we can synthesize random functions with a given spectrum of singularities, which is not necessarily concave. Secondly, we derive a multifractal formalism (a way of computing numerically the spectrum of singularities of a function) with a domain of validity not limited to concave spectra. Finally, we show that a particular case of our process satisfies a generalized selfsimilarity relation proposed in the theory of fully developed turbulence.

1 Introduction: Random Wavelet Series

Signals or images are often stored through their wavelet coefficients so that, in practice, the distributions of wavelet coefficients at each scale are often directly available; therefore it is natural to wonder which information on a function can be derived from them. In several situations these distributions have been numerically computed and statistical models have been fitted to the observed data. Let us mention a few examples:

- Several authors (Buccigrossi *et al.*,[3] Huang *et al.*,[4] Mallat,[5] Simoncelli,[6] Vidakovic[7]) have observed that the wavelet coefficients of natural images have highly non-Gaussian statistics. Exponential power distributions (of density $Ce^{-A|x|^{\alpha}}$) fit very well the distributions of wavelet coefficients for large classes of images.

- Cascade-type models for the evolution of the probability density function of the wavelet coefficients through the scales have been proposed to model the velocity in the context of fully developed turbulence (these models were initially proposed by Castaing *et al.*[8] for the increments of the velocity, and then fitted to the wavelet setting by Arnéodo *et al.*[9]). Random multiplicative models have also been considered in statistics, see Vidakovic.[10]

- In multifractal analysis, several formulas (Lévy-Vehel *et al.*,[11] Evertsz *et al.*,[12] Meneveau *et al.*,[13] Riedi[14]) were also proposed in order to derive spectra of singularities from distributions of increments of the function. These formulas are referred to as *large deviation multifractal formalisms*; they can be extended to a wavelet setting.

- Bayesian inference techniques based on a priori models for the distributions of wavelet coefficients at each scale have been widely studied (see for instance Abramovich *et al.*,[15] Johnstone,[16] Müller *et al.*[17]); the starting point of these statistical models is the following question: can the usual wavelet-based denoising algorithms (typically, soft thresholding on wavelet coefficients) be improved if the a priori information on the initial signal used is no more a "functional" information (for instance, f belongs to a given Besov space, or a given intersection of Besov spaces), but an information on the distributions of wavelet coefficients? Indeed, Bayesian approaches based on mixture models for wavelet coefficients lead to wavelet shrinkage algorithms which differ from those based on Besov-type a priori knowledge.

Our purpose in this paper is to study from a theoretical and numerical point of view the multifractal properties of some of these models.

Let us first recall the definition of Random Wavelet Series, and their main properties. We consider functions on $\mathbb{T} = \mathbb{R}/\mathbb{Z}$ (1-periodic functions). Let ψ be a mother wavelet such that the periodized wavelet family

$$\left\{\psi_{jk} : x \mapsto \sum_{l\in\mathbb{Z}} \psi(2^j(x-l)-k), j \in \mathbb{N}, 0 \le k < 2^j\right\}$$

form, together with the constant function $x \mapsto 1$, a basis of $L^2(\mathbb{T})$ (see Daubechies[18]). Note that this basis is not orthonormal, the L^∞ normalization being more convenient for our purpose.

1.1 *Random wavelet coefficients*

Definition 1. *f is a Random Wavelet Series (R.W.S.) if its wavelet coefficients C_{jk} in the basis above satisfy the following requirements:*

1. *for all j,k, C_{jk} is a random variable such that $-\frac{\log_2(|C_{jk}|)}{j}$ has density ρ_j;*
2. *these random variables are independents;*
3. *there exists $\gamma > 0$ such that*

$$\rho(\alpha) := \inf_{\varepsilon>0} \limsup_{j\to+\infty} \frac{\log_2\left(2^j \int_{\alpha-\varepsilon}^{\alpha+\varepsilon} \rho_j(t)dt\right)}{j} \tag{1}$$

is strictly negative for $\alpha < \gamma$.

The function ρ thus defined is called the *upper logarithmic density* of the process. It is upper semi-continuous, but not necessarily monotonous. Note that we do not make any assumption on the shape of ρ_j; it needs even not to be a function: it can be a probability measure on $\mathbb{R} \cup \{+\infty\}$, the measure $\rho_j(\{+\infty\})$ being the probability that $C_{jk} = 0$.

1.2 *Wavelet coefficients histograms*

Consider now an arbitrary function f, which can be (or not) a realization of a R.W.S. We form $N_j(\alpha) := \#\left\{k, |C_{jk}| \geq 2^{-\alpha j}\right\}$, and

$$\rho(\alpha) := \inf_{\varepsilon>0} \limsup_{j\to+\infty} \frac{\log_2(N_j(\alpha+\varepsilon) - N_j(\alpha-\varepsilon))}{j}. \tag{2}$$

In the case where f is a R.W.S., $\boldsymbol{\rho}$ is a deterministic function whereas ρ is random. The first important result of our previous paper[1] links these two functions.

Theorem 1. *For a R.W.S., almost surely, for all* α,

$$\rho(\alpha) = \begin{cases} \boldsymbol{\rho}(\alpha) \text{ if } \geq 0; \\ -\infty \text{ else.} \end{cases}$$

1.3 *Spectrum of singularities*

We recall that the spectrum of singularities of a function f is defined by

$$d(h) := \dim_H\left(\{x, f \text{ has Hölder exponent } h \text{ in } x\}\right),$$

with the usual convention that $\dim_H(\emptyset) = -\infty$.

Our second result gives the spectrum of singularities for a R.W.S. Let us first define

$$h_{\min} := \sup\left\{\alpha, \gamma < \alpha \Rightarrow \boldsymbol{\rho}(\gamma) < 0\right\}$$

and

$$h_{\max} := \left(\sup_{\alpha>0} \frac{\boldsymbol{\rho}(\alpha)}{\alpha}\right)^{-1}.$$

Theorem 2. *Let f be a R.W.S. Almost-surely,*

- *the almost-everywhere Hölder exponent is* $h_{\max}$;
- *for all* $h \in [h_{\min}, h_{\max}]$, $d(h) = h \sup_{\alpha\in(0,h]} \frac{\boldsymbol{\rho}(\alpha)}{\alpha}$;
- $d(h) = -\infty$ *else.*

See Figure 3 for an illustration. As a by-product of this theorem, we also proved for any uniform Hölder function ($f \in C^{\epsilon}(\mathbb{T})$, $\epsilon > 0$), the inequality

$$d(h) \leq h \sup_{\alpha\in(0,h]} \frac{\rho(\alpha)}{\alpha}. \tag{3}$$

2 Synthesis of multifractal processes

When a multifractal model is proposed, a natural question is to synthesize a function or a random process having a given spectrum of singularities $d(h)$. According to Theorem 2, the candidates for R.W.S. spectra are right-continuous functions $d \leq 1$ such that there exists $\gamma > 0$, $d(h) = -\infty$ for all $h < \gamma$, and such that the function $h \mapsto \frac{d(h)}{h}$ is increasing wherever $d(h) > -\infty$. This is also sufficient.

Proposition 2.1. *Let $d(h)$ satisfy the conditions above, and let $h_{\max}$ be the (only) solution to $d(h) = 1$. Take, for all $j \in \mathbb{N}$, for all $\alpha \in \mathbb{R}$,*

$$\boldsymbol{\rho}_j(\alpha) := \frac{j\ln(2)}{h_{\max}} 2^{j(d(\alpha)-1)}$$

and $\boldsymbol{\rho}_j(+\infty) := 1 - \int_0^{h_{\max}} \boldsymbol{\rho}_j(\alpha)d\alpha$. Then $\boldsymbol{\rho}_j$ is a probability density on $\mathbb{R} \cup \{+\infty\}$, and its upper logarithmic density is $\boldsymbol{\rho}(\alpha) = d(\alpha)$.

Proof. Note that we always have $d(\alpha) \leq \frac{\alpha}{h_{\max}}$. It follows that

$$\begin{aligned}\int_{\mathbb{R}} \boldsymbol{\rho}_j(\alpha)d\alpha &= \int_0^{h_{\max}} \boldsymbol{\rho}_j(\alpha)d\alpha \\ &\leq \frac{j\ln(2)}{h_{\max}} 2^{-j} \int_0^{h_{\max}} 2^{j\frac{\alpha}{h_{\max}}} d\alpha \\ &\leq 1,\end{aligned}$$

which proves, together with the definition of $\boldsymbol{\rho}_j(+\infty)$, that $\boldsymbol{\rho}_j$ is a probability density on $\mathbb{R} \cup \{+\infty\}$.

Let us now compute the upper logarithmic density. We have

$$\varepsilon\boldsymbol{\rho}_j(\alpha) \leq \int_{\alpha-\varepsilon}^{\alpha+\varepsilon} \boldsymbol{\rho}_j(t)dt \leq 2\varepsilon\boldsymbol{\rho}_j(\alpha+\varepsilon)$$

$$\frac{\log_2\left(\varepsilon\frac{j\ln(2)}{h_{\max}}\right)}{j} + d(\alpha) \leq \frac{\log_2\left(2^j\int_{\alpha-\varepsilon}^{\alpha+\varepsilon}\boldsymbol{\rho}_j(t)dt\right)}{j} \leq \frac{\log_2\left(2\varepsilon\frac{j\ln(2)}{h_{\max}}\right)}{j} + d(\alpha+\varepsilon)$$

$$d(\alpha) \leq \limsup_{j\to+\infty} \frac{\log_2\left(2^j\int_{\alpha-\varepsilon}^{\alpha+\varepsilon}\boldsymbol{\rho}_j(t)dt\right)}{j} \leq d(\alpha+\varepsilon)$$

hence, letting $\varepsilon \to 0$, $\boldsymbol{\rho}(\alpha) = d(\alpha)$. □

It is also clear that $\boldsymbol{\rho}(\alpha) = -\infty$ for $\alpha < \gamma$, and that $h\sup_{\alpha\in(0,h]} \frac{\boldsymbol{\rho}(\alpha)}{\alpha} = d(h)$ for all h. To synthesize the corresponding R.W.S., for all $j \in \mathbb{N}$ draw independently 2^j random variables α_{jk} with density $\boldsymbol{\rho}_j$ (using the rejection method for non-trivial densities), and let $C_{jk} = \chi_{jk}2^{-j\alpha_{jk}}$ (χ_{jk} is an arbitrary or random sign). Then

$$f(x) = \sum_{j\in\mathbb{N}} \sum_{k=0}^{2^j-1} C_{jk}\psi_{jk}(x) \tag{4}$$

is a realization of the process that has almost surely d as multifractal spectrum.

This allows in particular to synthesize processes with non-concave spectra. An example is given on Figure 1 with $d(h) = (h - \frac{1}{2})^2$ on $[\frac{1}{2}, \frac{3}{2}]$ and $-\infty$ elsewhere. The wavelet used is Daubechies 10 (extremal phase).

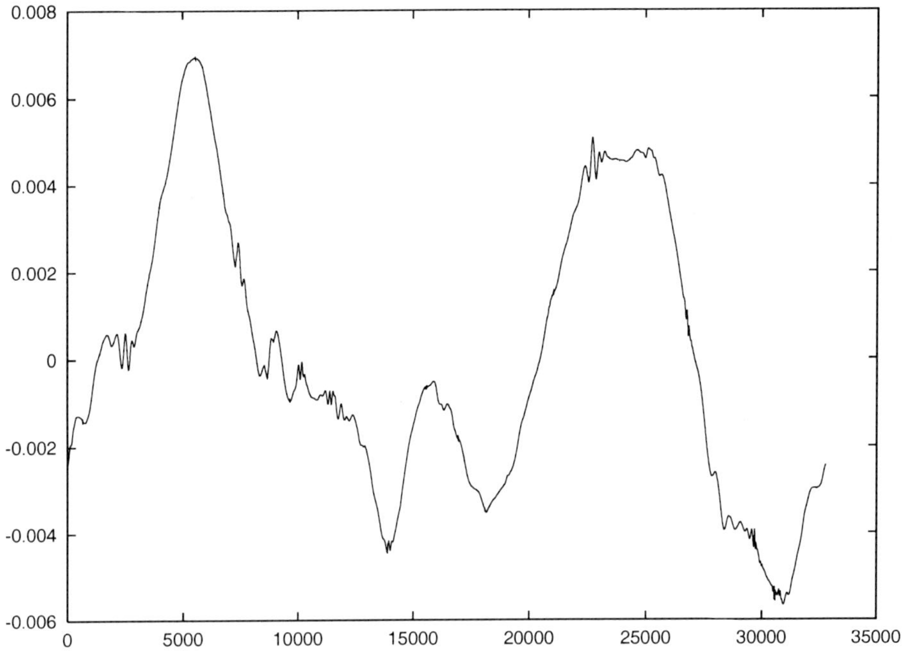

Figure 1. Multifractal process with non-concave spectrum

3 Multifractal formalisms

For a function given in a sampled form (signals, images), one often wishes to compute its singularity spectrum. Just applying the definition (identifying the sets of common Hölder exponent and computing their Hausdorff dimension) is clearly not feasible, so numerical procedures called *multifractal formalisms* have been proposed instead. In general, a multifractal formalism does not yield the correct spectrum, but an upper bound, and it is proved valid only on certain functions. For instance, the "classical" multifractal formalism, derived from the original ideas of Frisch *et al.*,[19] consists in computing the structure function

$$\tau(q) := \liminf_{j \to +\infty} \frac{\log_2 \left(\sum_k |C_{jk}|^q\right)}{-j},$$

and take its Legendre transform

$$d_1(h) := \inf_{q \geq q_c} hq - \tau(q) \tag{5}$$

as an estimation for $d(h)$. It was proved in Jaffard[20] that for any function f, if q_c is the (only) solution to $\tau(q_c) = 0$, then $d(h) \leq d_1(h)$; moreover, equality in (5) holds for selfsimilar functions (Jaffard[21]). Note, however, that this cannot be true for all functions; in particular the right-hand term of (5) is concave, whereas in general the spectrum is not.

One can also prove that, for any function (even for a tempered distribution), for all h,

$$h \sup_{\alpha \in (0,h]} \frac{\rho(\alpha)}{\alpha} \leq \inf_{q \geq q_c} hq - \tau(q); \tag{6}$$

this implies that (3) is sharper than (5). We can thus propose the formula

$$d_2(h) := h \sup_{\alpha \in (0,h]} \frac{\rho(\alpha)}{\alpha}$$

as a new multifractal formalism; we know from Theorem 2 that it is almost surely valid for R.W.S. Because of (6), this new multifractal formalism will be valid whenever the classical one is valid; moreover, since the spectrum thus obtained is not necessary concave, its domain of validity is strictly larger than the classical one.

Note that $\rho(\alpha)$ may not be easy to compute numerically, because (2) involves a double limit. But if we define

$$\lambda(\alpha) := \limsup_{j \to +\infty} \frac{\log_2(N_j(\alpha))}{j},$$

which is increasing, we can show that its upper closure $\bar{\lambda}$ (whose hypograph is the closure of the hypograph of λ) satisfies $\bar{\lambda}(\alpha) = \sup_{\alpha' \geq \alpha} \rho(\alpha)$. Hence $d_2(h) = \sup_{\alpha \in (0,h]} \frac{\lambda(\alpha)}{\alpha}$ as well, which is easier to compute.

We tested this algorithm on the process that we synthesized in § 2, with the same parameters as on Figure 1, except that it was computed with 2^{22} points to get a sufficient scale range. The analyzing wavelet is Daubechies 3 (extremal phase), different from the synthesizing wavelet in order to avoid "cheating". Implementation is straightforward; we used a simple linear regression on the 10 largest scales to compute $\lambda(\alpha)$, and then $d_2(h)$. Results are shown on Figure 2.

4 Generalized selfsimilarity

The model for fully developed turbulence proposed in Castaing *et al.*[8] asserts that the velocity field is a random process X, with increments at scale l following a law of density P_l, and that if $l < L$,

$$P_l(x) = \int g_{lL}(u) e^{-u} P_L(e^{-u}x) du, \tag{7}$$

where the *selfsimilarity kernel* satisfies $g_{lL} = g_{ll'} * g_{l'L}$ for $l < l' < L$. This is a generalization of the notion of selfsimilar process, because taking $g_{lL} := \delta_{H \ln \frac{l}{L}}$ in (7) yields

$$P_l(x) = \left(\frac{L}{l}\right)^H P_L\left(\left(\frac{L}{l}\right)^H x\right) \Longleftrightarrow X_l \overset{\mathcal{L}}{=} \left(\frac{l}{L}\right)^H X_L.$$

The construction of such a process for general g is still an open problem, but an approach can be done using wavelets. Assuming that for all k, $C_{jk} \overset{\mathcal{L}}{=} X_{2^{-j}}$, and

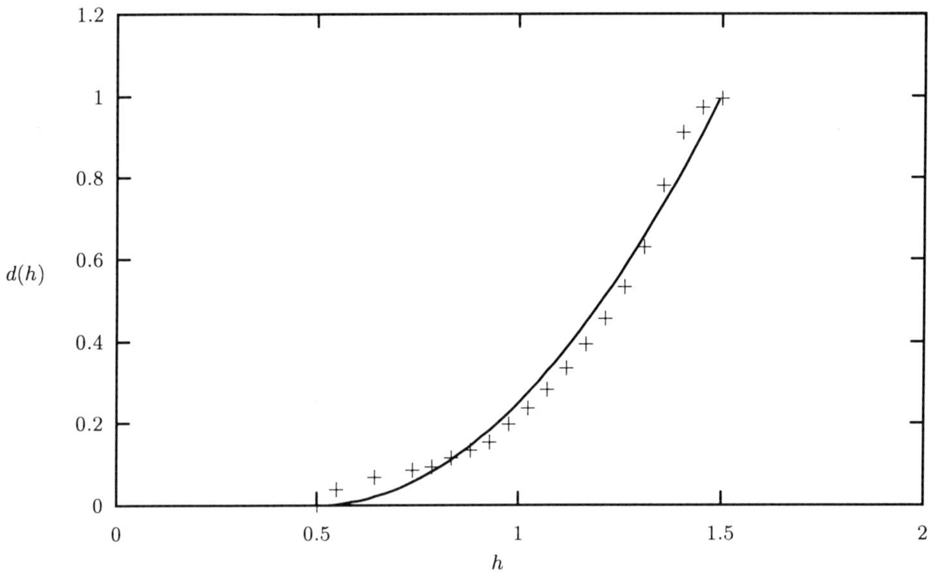

Figure 2. Computed spectrum of singularities: numerical (+) and theoretical (—) results.

that, at j fixed, the common probability density for $-\log_2(|C_{jk}|)$ is $\tilde{\boldsymbol{\rho}}_j$, (7) becomes for $j > J$

$$\tilde{\boldsymbol{\rho}}_j = G_{jJ} * \tilde{\boldsymbol{\rho}}_J, \tag{8}$$

where $G_{jJ}(u) := g_{2^{-j}2^{-J}}(-u)$ must satisfy $G_{jJ} = G_{jj'} * G_{j'J}$ for $j > j' > J$. One can furthermore assume that G_{jJ} depends only on $j - J$, in which case $G_{jJ} = G^{*(j-J)}$. Remark that, with the notation of Definition 1, $\boldsymbol{\rho}_j(\alpha) = j\tilde{\boldsymbol{\rho}}_j(j\alpha)$.

A process satisfying (8) is for instance the $\mathcal{W}$-cascade of Arnéodo *et al.*,[9] where the wavelet coefficient C_{jk} is obtained by multiplying its father in the wavelet tree by a random variable W_{jk} (the W_{jk} being i.i.d.); in that case G is simply the density of $-\log_2(|W_{jk}|)$. The scaling function $\tau(q)$ is equal to $\log_2(\hat{G}(iq)) - 1$; however, the spectrum of singularities of such a process can be random, which means that it will not satisfy a multifractal formalism.

We can construct a R.W.S. satisfying (8) by simply taking $\boldsymbol{\rho}_j(\alpha) = jG_{j0}(j\alpha)$. Then its spectrum of singularities can be computed almost surely and satisfies a multifractal formalism thanks to Theorem 2.

Example 1: $G = \delta_{\alpha_0} * \gamma_{\nu,\beta}$, with $\nu, \beta > 0$; $\alpha_0 > \alpha^*(\nu,\beta)$ which is the largest solution to $1 + \nu\log_2(-\alpha^*) + \beta\log_2(e)\alpha^* + \nu\log_2\left(\frac{\beta e}{\nu}\right) = 0$. For $\alpha > \alpha_0$,

$$\boldsymbol{\rho}(\alpha) = 1 + \nu\log_2(\alpha - \alpha_0) - \beta\log_2(e)(\alpha - \alpha_0) + \nu\log_2\left(\frac{\beta e}{\nu}\right).$$

The corresponding spectrum of singularities is shown on Figure 3.

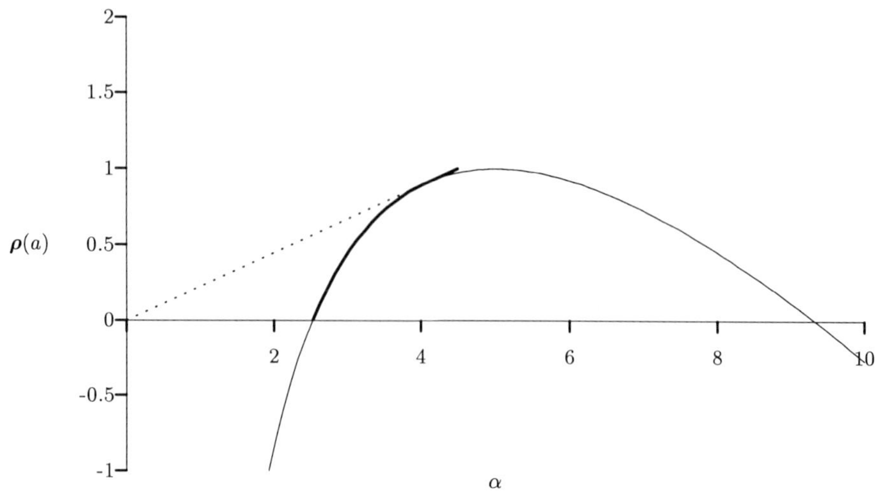

Figure 3. Upper logarithmic density $\rho(\alpha)$ for the γ kernel with $\alpha_0 = 1$, $\beta = \frac{1}{2}$, $\gamma = 2$. In bold the spectrum of singularities $d(\alpha)$.

Example 2: $G = N_{m,\sigma^2}$, with $m > \sigma\sqrt{\frac{2}{\log_2(e)}}$. For all α,

$$\rho(\alpha) = 1 - \log_2(e)\frac{(\alpha - m)^2}{2\sigma^2}.$$

The wavelet coefficients follow a log-normal law; the spectrum of singularities is a segment of a parabola followed by a segment of a line.

References

1. J.-M. Aubry and S. Jaffard. Random wavelet series. Preprint, 2000.
2. S. Jaffard. Lacunary wavelet series. *Ann. Appl. Probab.*, 10(1):313–329, 2000.
3. R. Buccigrossi and E. Simoncelli. Image compression via joint statistical characterization in the wavelet domain. Preprint, 1997.
4. J. Huang and D. Mumford. Statistics of natural images and models. Preprint.
5. S. Mallat. A theory for multiresolution signal decomposition: the wavelet representation. *IEEE Trans. on Pattern Anal. Machine Intell.*, 11:674–693, 1989.
6. E. Simoncelli. Bayesian denoising of visual images in the wavelet domain. *Lect. Notes Stat.*, 141:291–308, 1999.
7. B. Vidakovic. *Statistical Modeling by Wavelets*. John Wiley & Sons, 1999.
8. F. Chillà, J. Peinke, and B. Castaing. Multiplicative process in turbulent velocity statistics: a simplified analysis. *J. Phys. II France*, 6(4):455–460, 1996.

9. A. Arneodo, E. Bacry, and J.-F. Muzy. Random cascades on wavelet dyadic trees. *J. Math. Phys.*, 39(8):4142–4164, Aug. 1998.
10. B. Vidakovic. A note on random densities via wavelets. *Stat. Proba. Let.*, 26:315–321, 1996.
11. J. Lévy-Véhel and R. Riedi. Fractional brownian motion and data traffic modeling: The other end of the spectrum. In J. Lévy Véhel, E. Lutton, and C. Tricot, editors, *Fractals in Engineering*, pages 185–202. Springer-Verlag, 1997.
12. C. Evertsz and B. Mandelbrot. Multifractal measures. In H.-O. Peitgen, H. Jürgens, and D. Saupe, editors, *Chaos and Fractals, New frontiers of science*. Springer-Verlag, 1992.
13. C. Meneveau and K. Sreenivasan. Measurement of $f(\alpha)$ from scaling of histograms and applications to dynamical systems and fully developed turbulence. *Phys. Letters A*, 137:103–112, 1989.
14. R. Riedi. Multifractal processes. In Doukhan, Oppenheim, and Taqqu, editors, *Long range dependence : Theory and applications*. Birkhäuser, 2001.
15. F. Abramovich, T. Sapatinas, and B. W. Silverman. Wavelet thresholding via a Bayesian approach. *J. Roy. Statist. Soc. Ser. B*, 60(4):725–749, 1998.
16. I. Johnstone and B. Silverman. Empirical bayes approaches to mixture problems and wavelet regression. Preprint.
17. P. Müller and B. Vidakovic, editors. *Bayesian Inference in Wavelet Based Models*, volume 141 of *Lect. Notes Stat.* Springer-Verlag, 1999.
18. I. Daubechies. *Ten Lectures on Wavelets*, volume 61 of *CBMS-NSF regional conference series in applied mathematics*. SIAM, 1992.
19. U. Frisch. Fully developed turbulence and intermittency. In M. Ghil, editor, *Turbulence and Predictability in Geophysical Fluid Dynamics and Climate Dynamics*, volume 88, pages 71–88. International School of Physics Enrico Fermi, North-Holland, June 1983.
20. S. Jaffard. Multifractal formalism for functions Part I : results valid for all functions. *SIAM J. Math. Anal.*, 28(4):944–970, Jul. 1997.
21. S. Jaffard. Multifractal formalism for functions Part II : selfsimilar functions. *SIAM J. Math. Anal.*, 28(4):971–998, Jul. 1997.

SELF-AFFINE FRACTAL MEASUREMENTS ON FRACTURE SURFACES OF POLYMERS AND OPAL-GLASS

EDGAR REYES, CARLOS GUERRERO and MOISES HINOJOSA

DIMAT-FIME-UANL, A.P. 076 Suc. "F" Cd.Universitaria, San Nicolás de los Garza, N.L. México, C.P. 66450

E-mail: edgreyes@ccr.dsi.uanl.mx, cguerrer@ccr.dsi.uanl.mx, hinojosa@gama.fime.uanl.mx

In this work we report the self-affinity analysis of the fracture surfaces of an amorphous polymer and an opal-glass. In the case of the plastic material, samples of polystyrene were broken in bend test after being immersed in liquid nitrogen. In the case of the opal-glass, samples with different sizes of the opacifying particles, obtained by different thermal treatments, were broken in a punch test. Scanning Electron Microscopy images of fracture surfaces of both amorphous materials show the mirror and hackle zones.

The average roughness exponent, ζ, of height profiles generated by Atomic Force Microscopy images was estimated for both materials by applying the variable bandwidth method, covering a range of length scales spanning from a few nanometers up to ten micrometers. The roughness exponent obtained for both materials was close to 0.8. These results are in very good agreement with the claimed universal exponent of 0.8, reported in the literature for other materials.

1 Introduction

Quantitative descriptions of fracture surfaces using fractal geometry were first published in 1984 by Mandelbrot [1]. Nowadays it is accepted that fracture surfaces have self-affine character. A self-affine object has unequal scaling invariance in different directions. The irregularity of fracture surfaces is measured by the roughness exponent, ζ. When $\zeta=1$ surfaces are smooth while decreasing values reflect increasing degrees of surface roughness. A self-affine surface is fractal up to distances of the order of a characteristic length called correlation length; beyond this length the surface can be considered flat.
Studying fracture surfaces obtained in rapid kinetic conditions and analyzed, mainly by electron microscopy, Bouchaud proposed the idea of a *universal* roughness exponent, $\zeta \approx 0.78$, independent of the fracture mode, the microstructure and properties [2,3,4]. This universality was questioned by the finding of a self-affine regime with a low roughness exponent $\zeta \approx 0.5$ for fracture surfaces generated at low crack propagation speeds and/or analyzed at the nanometer scale using Scanning Probe Microscopy [5]. At present, the co-existence of the two self-affine regimes had been reported in several materials, both regimes separated at a cut-off length, which appears to be dependent on the kinetic conditions [3,6]. Attempts have been made to correlate the cut-off length with microstructural parameters [8].
The aim of this work was to determine and to compare the self-affine behavior of fracture surfaces of two glass-materials, polystyrene (PS) and opal-glass, using Scaning Electron Microscopy and Atomic Force Microscopy techniques. The roughness exponent was measured applying the variable bandwidth method.

2 Experiment

The materials whose fracture surfaces were analyzed in this work are representative of the plastic and ceramic materials. The plastic material selected for our analysis is an amorphous polystyrene, PS. Characterization of this material by Gel Permeation Chromatography showed that the number average molecular weight is 76,775, and the polydispersity index was 3.09. The glass transition temperature, T_g, was 86°C according to Differential Scanning Calorimetry measurements. The specimens were obtained by capillary extrusion at 180°C. Filaments (about 1 mm diameter and 20 mm long) were randomly selected and cooled at room temperature. After that, the specimens were immersed in liquid nitrogen for 15 minutes then broken in bending mode without control of the applied load. This implies a high enough crack propagation speed without preferential direction.

In the case of the glass materials, two samples of commercial opal glass were analyzed. As a result of different heat treatment during the manufacturing process, the two samples had different sizes of the opacifying precipitates. Flat specimens were broken using a punch tester.

For SEM analysis, some samples of both materials were gold sputtered. In opal-glass, the size of the opacifying particles was measured from SEM images as well as by AFM observations.

The fractometric study was carried out with the Atomic Force Microscope in contact mode for both materials. The maximum scan size was 10 micrometers and the best resolution of the 512 x 512 images was about 10 nm. Height profiles were extracted from these AFM images, at least 30 profiles were obtained for each specimen. The average roughness exponent was then calculated by applying the variable bandwidth method. In this method we computed $Z_{max}(r)$:

$$Z_{max}(r) = \left\langle \max\{z(r')\}_{x<r'<x+r} - \min\{z(r')\}_{x<r'<x+r} \right\rangle_x \ \alpha \ r^{\zeta}$$

Where "r" is the width of the window, $Z_{max}(r)$ is the difference between the maximum and the minimum heights "z" within this window, averaged over all possible origins "x" of the window.

3 Results

The fracture surfaces were analyzed using both SEM and AFM. In PS samples, Figure 1, we observed the mirror and hackle zones on the fracture surface. These zones are typical on fracture surfaces of soda-lime glass. Mirror zone corresponds to where the crack front has begun to propagate. In hackle zone, features are attributed to the onset of localized crack branching when a critical stress intensity or fracture energy is reached along the growing crack front [7]. Figure 2 is an AFM image showing the mirror and hackle zones.

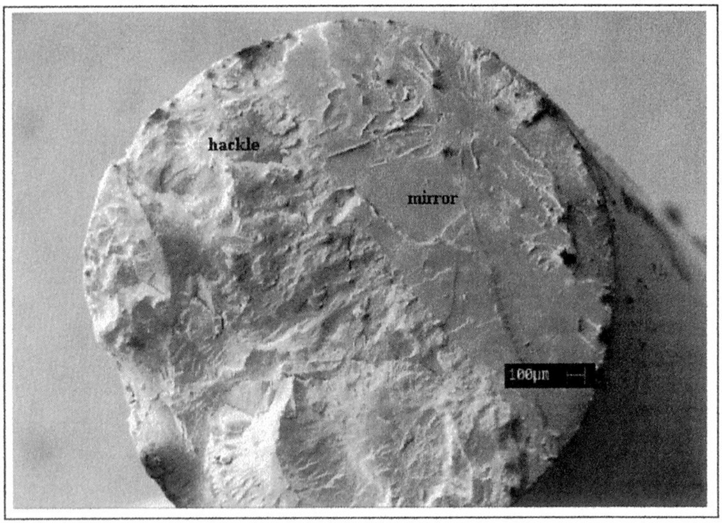

Figure 1.- SEM image of PS fracture surface.

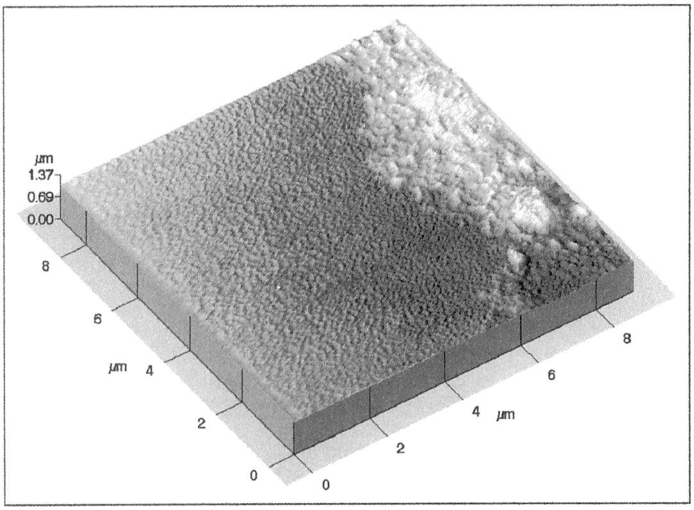

Figure 2.- 3-D images of fracture surface of PS on mirror-hackle transition zone by AMF.

Figure 3 and 4 show the results of the self-affinity analysis of the height profiles obtained from AFM images of both mirror and hackle zones. The slope of the straight line is the roughness exponent. For the mirror zone ζ=0.805 in the length scale from 0.002 to 0.1 µm, and for hackle zone ζ=0.810 in the length scale from 0.006 to 3.0 µm.

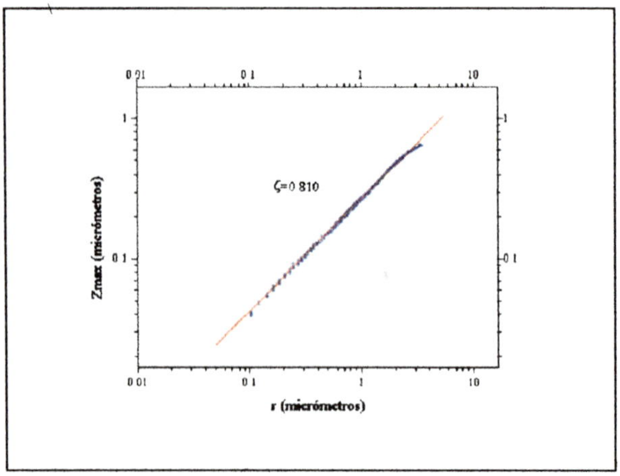

Figure 3.- Z_{max} –vs- r plot of PS fracture surface on hackle zone

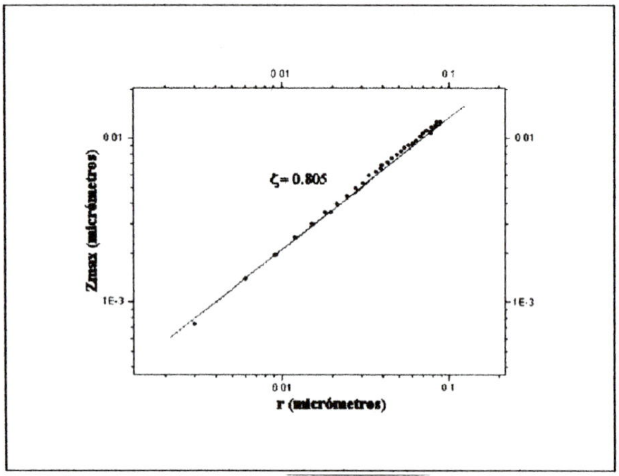

Figure 4.-. Z_{max} –vs- r plot of PS fracture surface on mirror zone

The mirror and hackle zones are also observed in the case of the opal glass for both the fine and coarse opacifying particles samples, Figure 5 is a SEM image of the sample with coarse opacifying particles. In the AFM images it was possible to measure the size of opacifying particles on the mirror zone. One of the sample had a homogeneous distribution of fine particles with an average size of 0.34 μm, the other had a distribution of a coarse particles of about 4 μm together with fine particles of 0.47 μm. Figure 6 is an AFM images, it shows clearly the distribution of fine opacifying particles.

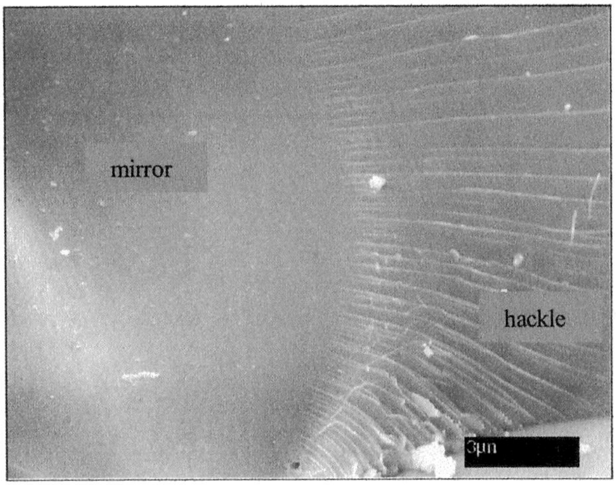

Figure 5.- SEM image of opal-glass fracture surface

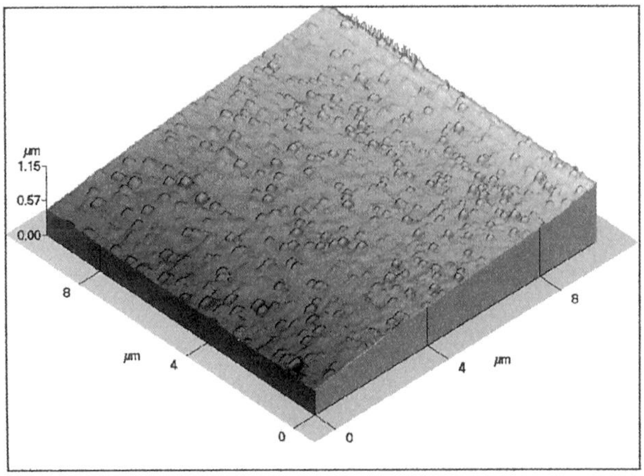

Figure 6.- 3-D images of opal-glass fracture surface with coarsening size particles on mirror zone by AFM.

For the two samples of opal-glass the self-affine analysis was concentrated on the mirror zone. On the figure 7 shows the results obtained.

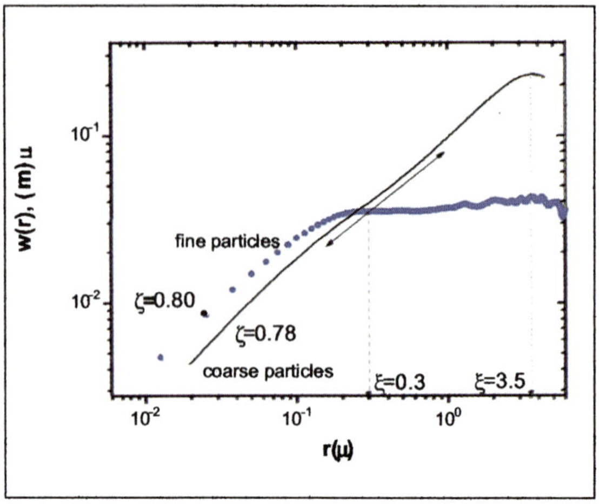

Figure 7.- Z_{max} –vs- r plots for the two samples of opal –glass fracture surfaces, both on mirror zone.

In the case of the fine-particle sample, the roughness exponent has a value of 0.8 and the correlation length is clearly seen at a value of around 0.3 μm, which agrees with the size of the opacifying particles. For the coarse-particle sample, the self-affine regime is apparently disturbed by the presence of the two types of fine and coarse particles, the roughness exponent has a value of 0.78 and the correlation length is about 3.5 μm, which again is of the order of the coarse opacifying particles. In figure 7 The arrow indicates the region where the self-affine regime corresponding to the coarse-particle specimen is disturbed presumably by the presence of the two distribution of the particles with different size.

4 Conclusions

The analyzed fracture surfaces of PS and opal glass exhibit self-affine behavior. Our results in opal-glass support the idea that the correlation length of self-affine fracture surfaces is determined by the largest heterogeneities in the microstructure. In all cases the roughness $\zeta \approx 0.8$ was obtained, in good agreement with previously reported results for similar kinetic conditions. These results represent an important advance in the quest of useful relationships among the microstructure, macroscopic properties, and self-affine or fractal parameters. Our results are representative of real ecological situations.

5 Acknowledgements

Financial assistance from the National Science and Technology Council (CONACyT) and from the Science and Technology Research program (PAICyT) is greatly appreciated.

References

1. Mandelbrot B.B.., Passoja D.E., and Paullay A.J.; Nature, **308**, p. 721-722 (1984).
2. Bouchaud E., Lapasset, and Planes J.; Europhys. Lett., 13, p. 73, (1990).
3. Daguier P., Henaux S., Bouchaud E., and Creuzet F.; Phys. Rev. E., **53**, p. 5637-5642 (1992).
4. Bouchaud E.; J. Phys.: Condens. Matter, **9**, p. 4319-4344 (1997).
5. Milman V.Y., Blumenfeld R., Stelmashemko N.A., and Ball R.C.; Phys. Rev. Lett., **71**, p. 204, (1993).
6. Arribart H., and Abriou D.; Ceramics-Silikaty, **44** (4), p. 121-128 (2000).
7. Mecholsky J.J., Freiman S.W., and Rice R.W.; In Fractography in Failure Analysis. ASTM publishing. p. 363-379 (1978).
8. Hinojosa M., Bouchaud E., and Nghiem B., *Long Distance Roughness Of Fracture Surfaces In Heterogeneous Materials*, Materials Research Society Symposium Proceedings Vol. 539 (1999), pp. 203-208.

SELF-AFFINE PROPERTIES ON FRACTURE SURFACES OF IONIC EXCHANGED GLASS

FRANCISCO J. GARZA
CIDEMAC-FCQ-UANL, A.P. 048 Suc. "F" Cd. Universitaria, San Nicolás de los Garza N.L. México, C.P. 66450
E-mail :fjgarza@gama.fime.uanl.mx

MOISÉS HINOJOSA and LEONARDO CHÁVEZ
DIMAT-FIME-UANL, A.P. 076 Suc. "F" Cd. Universitaria, San Nicolás de los Garza N.L. México, C.P. 66450
E-mail: hinojosa@gama.fime.uanl.mx

The effect of ionic exchange treatment on the self-affine properties of the fracture surfaces of soda-lime-silica glass is explored. The atomic force microscopy (AFM) was used to record the topometric data from the fracture surfaces. The roughness exponent (ζ) and the correlation length (ξ) were calculated by the variable bandwidth method. The analysis for both glasses (strengthened and non-treated) for the roughness exponent shows a value $\zeta \sim 0.8$, which agrees well with that reported in literature for high speed of crack propagation in different kind of materials. The correlation length shows different values for each condition. Our results suggest that the self-affine correlation length is influenced by the complex interactions of the stress field at the crack tip with that resulting from the collective behavior of the point defects introduced on glass by the ionic exchange treatment.

Key words: *fracture surfaces, roughness exponent, correlation length, glass fractography, ionic exchange.*

1 Introduction

1.1 Glass structure and mechanical behavior

At the present time glasses are used as structural materials in a variety of applications including transportation, aerospace, architecture, etc. These are amorphous materials with short range order and have not microstructure [1]. Due to their covalent bonding the theoretical strength of glasses is high (~7000 MPa), However the strength of these materials rarely exceeds 100 MPa [2]. The first attempt to explain the discrepancy between the theoretical and the real strength of these materials is due to Griffith [3]. Griffith suggested that there are microcracks at the free surface of glass, this defects acting as a stress amplifiers reducing the strength of these materials. The analytical estimation of Griffith is based on the early work of Inglis [4] who calculated the stress necessary to propagate a crack on a plate containing an elliptical hole. Equation (1) gives the stress necessary to propagate a crack for a plane stress condition, while equation (2) represent the plane strain condition (σ_c) [5].

$$\sigma_c = \sqrt{\frac{2E\gamma_S}{\pi a}} \quad (1)$$

$$\sigma_c = \sqrt{\frac{2E\gamma_S}{\pi a\left(1-\upsilon^2\right)}} \quad (2)$$

Here E is the Young modulus, γ_s is the surface energy due to crack growth, v is the Poisson ratio and a is the radius of the crack. On this basis, strengthening treatments rely on the elimination of the microcracks from the glass surface reducing the susceptibility to crack propagation and therefore enhancing the strength. Among these traditional methods are etching and flame polishing. Other methods employed consist in the formation of a compressive layer, a useful technique is the physical tempering of glass were it is rapidly cooled from a temperature above the glass transition. Due to the rapid cooling the surface acquires a rigid state faster than the bulk and therefore a compressive layer is formed. This method has been successfull employed in the automotive industry. Another technique is the chemical tempering of glass which consist in the substitution of small ions with ions of greater ionic radius. The collective behavior of these introduced point defects produces a stress field that interacts with the tensile stress field at the crack tips, thus improving the mechanical properties.

1.2 Fractography and self-affine surfaces

The term fractography was adopted by Carl A. Zapffe in 1944 to give a name to the method of fracture description [6], since his early work the traditional publications about fracture studies were often merely descriptive. On the other hand, due to their irregular morphology, fracture surfaces are ideal for fractal analysis. In 1984 it was proposed that there exist a direct correlation between the fractal dimension of fracture surfaces of metals and their toughness [7]. Nevertheless the accumulated evidence tells that there is not a direct correlation between the fractal dimension, or more properly, the self-affine parameters with the macroscopic mechanical properties [8]. At the present time it is accepted that fracture surfaces show statistical self-similar character, which can be quantified by the roughness exponent, ζ [9]. Different scientists [8, 9, 10, 11, 12] conclude that this exponent shows a "universal" behavior and the existence of a characteristic regime with ζ~0.8 is accepted for rapid crack propagation conditions. If there exist a universal value for ζ, then a new question arises: What is the role of the microstructure on self-affine parameters? In 1998 Hinojosa *et al* showed that the correlation length ξ has a direct relation with the largest heterogeneities present on the microstructure of metals [9], this behavior has been perfectly corroborated by J. Aldaco who analyses a very wide range of length scales [11]. This correlation length is the upper limit of the self-affine regime, that is to say, beyond this length the surface can be considered flat. Now, if glass fracture surfaces exhibit self-affine parameters, are these parameters affected by the ionic exchange? In this work we explore the possible effects of the collective behavior of the point defects introduced by the ionic exchange on the self-affine parameters of soda-lime-silica glass fracture surfaces, broken in bending.

2 Experimental procedure

2.1 Material and ionic exchange treatment

The material used in this work is soda-lime-silica glass in the form of flat plates, the dimensions of the specimens are: 25x75mm and a thickness of 1mm. The ionic exchange treatment consists in the substitution of sodium ion (Na^+) with potassium ion (K^+), the ionic radius of K^+ is 0.133 nm, while Na^+ has a radius f 0.095 nm. This difference produces a compressive stress layer that interacts with the tensile stress at the tip of the crack. Experimentally, the glass specimens are immersed on a bath of melted potassium salt (KNO_3) at 400 °C during 24 hr, after this treatment they are cooled at a rate of 17 °C/hr.

2.2 Microhardness indentation tests and the fractographic analysis

Sets of both treated and non-treated specimens were broken in rapid three-point bend tests. Microhardness indentations were used to estimate the effect of ionic exchange on the resistance to crack propagation. A load of 0.5 Kg was applied with a Vickers indenter. Measurements of the diagonals and crack lengths were made employing an optical microscopy assisted with a CCD camera and image analysis software. The classical fractographic analysis was performed employing scanning electron microscopy (SEM).

2.3 Topometric analysis

AFM scans in contact mode allowed us to obtain the topography of the fracture surfaces as well as the topometric profiles. A force of 9 nN with a scanning rate of 1 Hz was used. The statistical topometric analysis was performed using the variable bandwidth method [13] applied to the AFM profiles. In this method, a profile of length L is divided in bands of width Δ, inside of each band *Zmax* is computed, which is the difference between the lower and maximum heights. *Zmax* and w_2 follow a power law with the window size, r, as is expressed equation (3):

$$Z_{\max}(r) \propto r^{\zeta} \quad (3)$$

3 Results and discussion

Figure 1 shows a representative microhardness indentation obtained for the treated specimens. In the case of the non-treated specimens, Figs. 2 and 3, slowly moving microcracks departing from the indentation vertex were observed. This effect is not present in the treated specimens, Fig. 1. This behavior qualitatively illustrates the effect of ionic exchange in improving the resistance to both crack nucleation and propagation. The results from the microhardness tests indicated that the average Vickers Hardness Number is 727.5 and 579.5 for the treated and non-treated specimens respectively, thus the treatment caused a remarkable increase in "hardness", though the material is in reality tougher. It is pertinent to recall that microhardness measurements are commonly used to estimate the "brittleness" of glasses [14], however the presence of moving microcracks renders this estimation imprecise if time is not taken into account, considering that these cracks continue to advance even for times as long as 30 min.

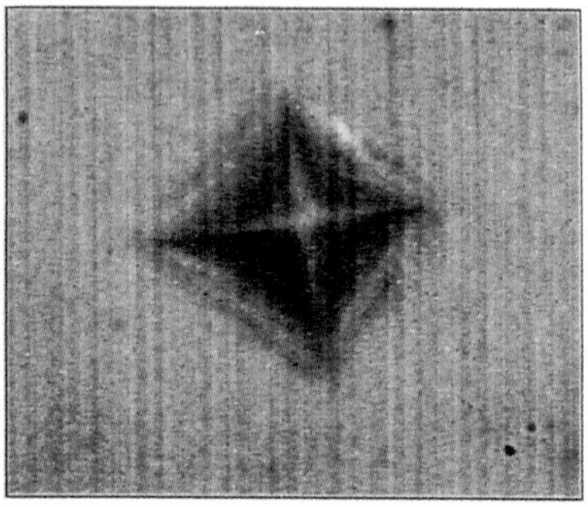

Figure 1.- Typical microhardness indentation on a treated specimen.

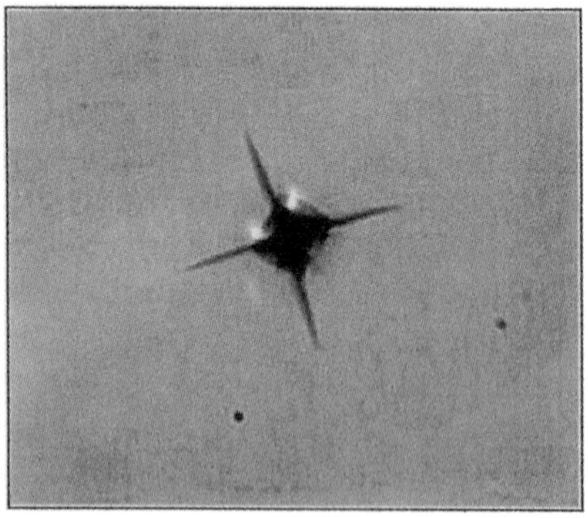

Figure 2.- A typical trace of the microhardness indentation on a specimen without treatment.

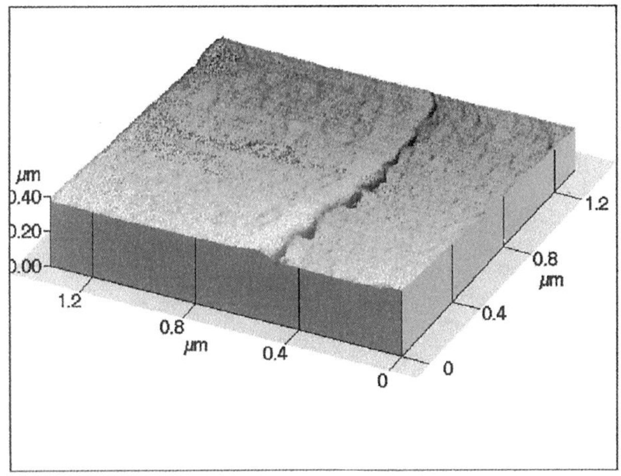

Figure 3.- A microcrack caused by the microhardness indentation in a non-treated specimen.

Figure 4 shows the typical fractographic analysis of glass, this image was obtained employing SEM in the mode of secondary electrons. In this picture the three familiar zones of fracture: mirror, mist and hackle, can be observed. The AFM observation shows that the mirror region is not flat, but shows a remarkable roughness, Fig. 5. The image on figure 5 shows the topography of the zone on the glass without ionic exchange, while figure 6 shows the same zone on glass with ionic exchange. Even if both images exhibit roughness, their morphologies are different. Figure 7 shows the topography from the hackle zone and, figure 8 show the mist zone. These results can be interpreted as a qualitative evidence of the self-affine character of the three fracture zones.

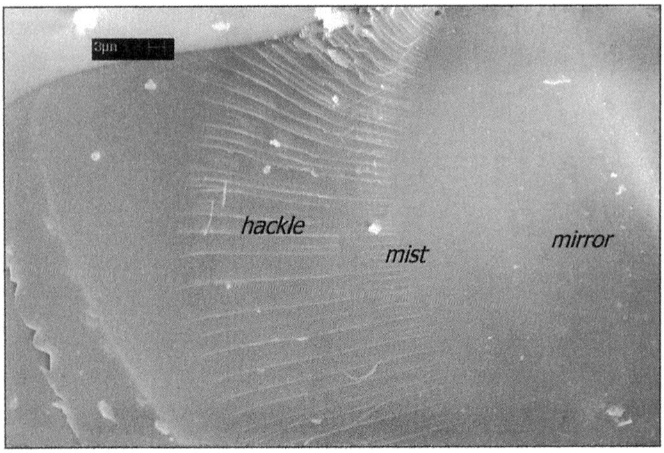

Figure 4.- SEM image showing the three fracture zones.

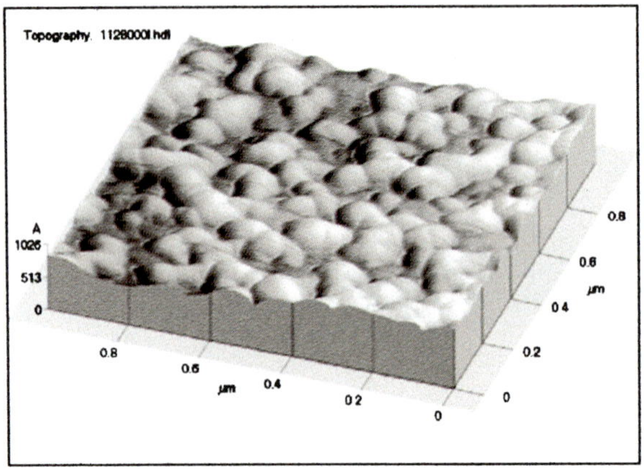

Figure 5.- AFM image of the mirror zone in a non-treated specimen.

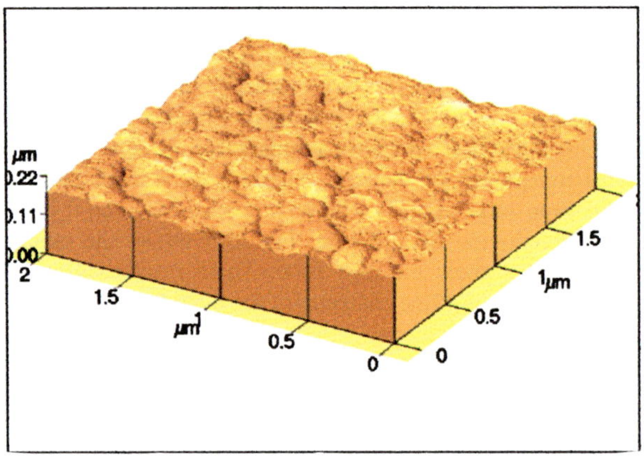

Figure 6.- AFM image of the mirror zone in treated specimen.

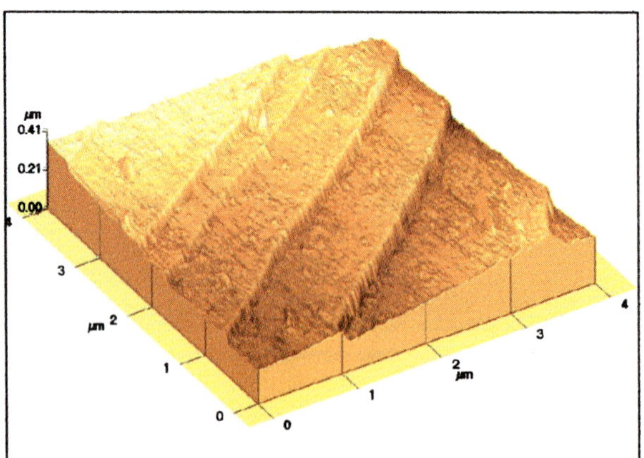

Figure 7.- The topography of the hackle zone as observed by AFM

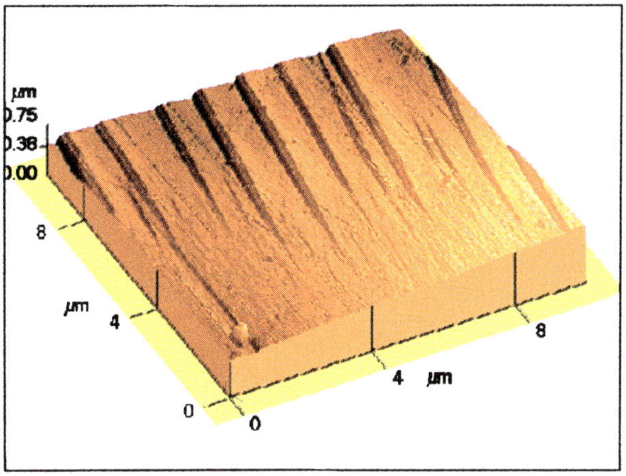

Figure 8.- AFM image showing the topography of the mist zone

The self-affine character of the mirror zone is revealed and quantified by the self-affine curves shown in Fig 9 and 10, which corresponds to the non-treated and treated specimens, respectively. In both cases the roughness exponent has values consistent with the claimed universal value $\zeta \sim 0.8$ [6, 7, 8, 9]. This implies that the ionic exchange treatment does not modify the exponent characterizing the self-affine regime. However, the curve corresponding to the treated samples suggest that the deviation from the power law occurs for lower values of r compared to the non-treated case. Confirmation of this result would imply that the correlation length is modified by the ionic exchange treatment as a result of the collective influence of the point defects introduced.

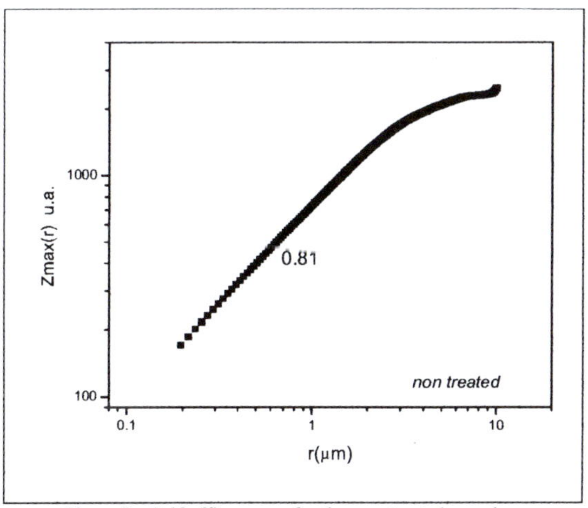

Figure 9.- Self-affine curve for the non-treated samples.

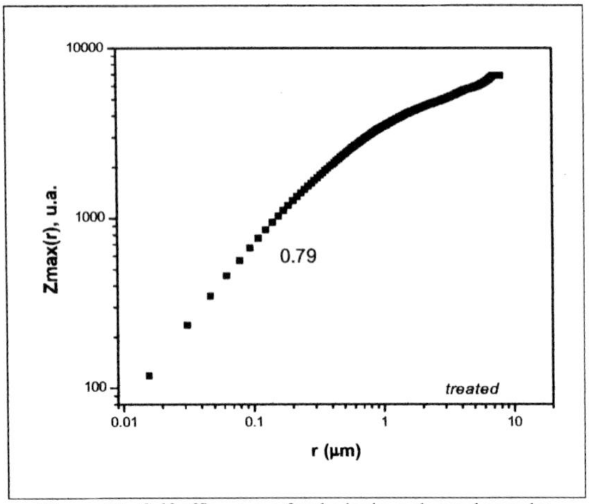

Figure 10.- Self-affine curve for the ionic exchanged samples.

4 Conclusions

The possible effects of ionic exchange treatment on the self-affine parameters associated with the fracture surfaces of glass were explored in this work. Microhardness measurements showed that the exchange treatment effectively improves both hardness and resistance to crack propagation. The self-affine character of the three characteristic fracture zones: mirror, mist and hackle, was corroborated and quantitatively characterized. The roughness exponent found for the mirror zone agrees well with the universal value $\zeta \sim 0.8$ for both treated and non-treated glass, thus the ionic exchange treatment does not modify this parameter, though the self-affine correlation length could be influenced by the collective behavior of the point defects introduced by the ionic exchange treatment.

5 Acknowledgements

Authors wish to express their gratitude to CONACYT (Grant red 600-1-1) and the Universidad Autónoma de Nuevo León for financial support through the PAICYT program. Thanks to Edgar Reyes for his incommensurable technical support.

References

1. Richard Zallen, *The Physics of Amorphous Solids, Wiley Classics Library Edition*, JWS publishing (1998) pp. 12.
2. I. W. Donald. *Methods for improving the mechanical properties of oxide glasses.* Journal of Materials Science 24 (1989), pp. 4177-4208.
3. George E. Dieter. *Mechanical Metallurgy*, Mc Graw Hill Series in Materials Science and Engineering, (1986)pp. 246.
4. C.E. Inglis. *Proc. Phys. Soc. Vol. 58*, London (1946), pp. 729.

5. Marc A. Meyers and Krishan K. Chawla. *Mechanical Metallurgy Principles and Applications*. Prentice Hall Intl. Inc. (1984) pp. 136
6. C. A. Zapffe and M. Clogg Jr., *Fractography-A New Tool for Metallurgical Research*, ASM (1944), Vol. 34, pp. 71-107.
7. B.B. Mandelbrot, D. E. Passoja and A.J. Paullay, *Fractal Character Of Fracture Surfaces Of Metals*", Nature 308 (1984) pp 721-722.
8. E. Bouchaud, *J. Phys: Condens Matter Vol. 9* (1997), pp. 4319-4344, and references therein.
9. Bouchaud E., Lapasset G., and Planès J., *Europhys. Lett. 13* (1990) pag. 73
10. Hinojosa M., Bouchaud E., and Nghiem B., *Long Distance Roughness Of Fracture Surfaces In Heterogeneous Materials*, Materials Research Society Symposium Proceedings Vol. 539 (1999), pp. 203-208.
11. J. Aldaco, F.J. Garza, M. Hinojosa, *Long Distance Surface Roughness On A Dendritic Aluminum Alloy*, Materials Research Society Symposium Proceedings, Vol. 578 (1999), pp. 351-356.
12. Edgar Reyes Melo, *Master in Science Thesis*, UANL, MEXICO, (1999), (in spanish).
13. Schmittbuhl, Vilote and Roux, *Phys. Rev. E., Vol 51, No 1*, (January 1995), pp. 131
14. J. Seghal, Y. Nakao, H. Takahashi, S. Ito, *Brittleness Of Glasses By Indentation"*, Journal of Material Science Letters, Vol. 14 (1995) pp 167-169.

CORRELATION DIMENSION OF DISSIPATIVE CONTINUOUS DYNAMICAL SYSTEMS STOCHASTICALLY EXCITED BY TEMPORAL INPUTS

KAZUTOSHI GOHARA AND JUN NISHIKAWA

Department of Applied Physics, Hokkaido University, Sapporo 060-8628, Japan
E-mail: gohara@nd-ap.eng.hokudai.ac.jp

We have presented a theory for dissipative continuous dynamical systems stochastically excited by external temporal inputs. The theory shows that the dynamics is characterized by a set $\Gamma(C)$ of trajectories in hyper-cylindrical phase space, where C is a set of initial states on the Poincaré section. Two sets, $\Gamma(C)$ and C, are attractive and invariant fractal sets. In this paper, using a nonlinear Duffing equation, we show numerically that the correlation dimension of the set C is approximately inversely proportional to the time length of inputs while the dimension is independent of the input amplitude. These obtained results might be universal characteristics of dissipative continuous dynamical systems stochastically excited by temporal inputs.

1 Introduction

Fractals generated by IFS (Iterated Function Systems) have been widely used for image processing [1,2]. From a dynamical systems view point, IFS is a set of discrete dynamical systems, i.e., a set of maps. One of authors has presented a set of continuous dynamical systems, i.e., a set of ODE (Ordinary Differential Equations), for modeling complex systems interacting with other systems through external temporal inputs [3]. We call the set of vector fields defined by temporal inputs as VFS (Vector Field Systems). VFS is reduced to IFS on the Poincaré section defined at the switching interval of the input. Thus, a fractal can be observed on the Poincaré section as an initial set of continuous trajectories that are solutions of ODE stochastically switched by external temporal inputs. Fractals generated by VFS have been observed in different domains such as hybrid dynamical systems [4,5], human behavior [6], a brain model of recurrent neural networks [7], a forced damped oscillator [8], and electronic circuit [9]. These examples indicate that fractals generated by the switching of ODE are ubiquitous in nature.

To quantitatively characterize these fractals generated by VFS, a correlation dimension is numerically investigated using a nonlinear Duffing equation stochastically driven by external temporal forces. In this paper, we show the dependence of the correlation dimension on two external parameters, i.e., time interval and amplitude, respectively.

2 Theory of Dynamical Systems Stochastically Excited by Temporal Inputs

In this section, we briefly summarize the theoretical framework for dissipative dynamical systems that interact with other systems through an input. More precise descriptions are given in reference [3]. We focus on dynamics expressed as dissipative non-autonomous dynamical systems defined by the following ordinary differential equations:

$$\dot{x} = f(x, I(t)), \tag{1}$$

$$x, I \in R^n,$$

where x, f, I, and t are state, vector field, input, and time, respectively. Equation (1) implies that a system is influenced by other systems through the input I.

2.1 Dynamics of Periodic Input

For the periodic input $I(t) = I(t+T)$ with period T, by introducing the angular variable $\theta \equiv \frac{2\pi}{T}t \bmod 2\pi$ and the new state variable $y \equiv (x, \theta)$ we can transform the non-autonomous system expressed by Eq. (1) into the following autonomous system:

$$\dot{y} = f_I(y), \tag{2}$$

$$y \in R^n \times S^1.$$

The vector field f_I is defined on a manifold $\mathcal{M} : R^n \times S^1$ that is a hyper-cylindrical space. In the space $\mathcal{M}$, we can globally define the Poincaré section, $\Sigma = \{(x, \theta) \in R^n \times S^1 | \theta = 2\pi\}$, where a trajectory starting from an initial state at $\theta = 0$ returns at $\theta = 2\pi$. On the section Σ, a mapping can be defined:

$$x_{\tau+1} = g_I(x_\tau), \tag{3}$$

$$x_\tau \in R^n,$$

where g_I is an iterated function that transforms a state x_τ to another state $x_{\tau+1}$ after interval T. We can summarize the dynamics with a periodic input as follows. The periodic input I defines two dynamical systems, one continuous and the other discrete, expressed by Eqs. (2) and (3), respectively. In the hyper-cylindrical phase space $\mathcal{M}$, a solution $\phi(t, x_0)$ starting from an initial state x_0 at $t = 0 (\theta = 0)$ converges to an attractor at $t \to \infty$. For periodic driven systems, there are several kinds of attractors, including point, limit cycle, torus, and chaos. We call the attractor corresponding to a periodic input the *excited attractor* in order to emphasize that the attractor is excited by the external input.

2.2 Dynamics of Switching Input

Next, we consider the dynamics when plural input patterns are stochastically fed into a system one after another. Let us suppose that *each input is one period of a periodic function*. For example, we can define a parameterized periodic function by the finite Fourier series with the amplitude vector $A \in R^m$ for Fourier coefficients and with the period $T \in R^1$. The set of these parameters defines the *input space*:

$$\mathcal{I} = \mathcal{I}(A, T). \tag{4}$$

The input space $\mathcal{I} : R^{m+1}$ is a functional space. Within this space, an arbitrary point represents an external temporal input. We consider the input as a set $\{I_l\}_{l=1}^{L}$

of time functions I_l sampled on the parameterized space $\mathcal{I}$. We sometimes abbreviate the subscripts and express individual sets as $\{\cdot\}$ for simplicity. In the same way as in the case of periodic input, we can define two sets of dynamical systems corresponding to the set $\{I_l\}$. One is the set of continuous dynamical systems that is defined by the set $\{f_l\}$ of vector fields on the hyper-cylindrical space $\mathcal{M}$. The other is the set of discrete dynamical systems that is defined by the set $\{g_l\}$ of iterated functions on the Poincaré section Σ. When the inputs I_l are stochastically fed into the system one after another, the vector fields f_l and the iterated functions g_l are also stochastically switched. To emphasize the relation among the set $\{I_l\}$, $\{f_l\}$ and, $\{g_l\}$, we use the following schematic expression:

$$\{I_l\} \to \{f_l\} \to \{g_l\}\,. \tag{5}$$

The set of iterated functions is designated as the *Iterated Function System* (IFS) by Barnsley [2]. Similarly, we call the set of vector fields the *Vector Field System* (VFS). Therefore, the two sets, $\{f_l\}$ and $\{g_l\}$, are examples of the VFS and the IFS, respectively. The discrete dynamics on the Poincaré section Σ correspond to the random iteration algorithm using the IFS with probabilities. When $Lip(g_l) < 1$, the map $g_l : x \to x$ is called the *contraction* or the *contraction map*, where $Lip(g_l)$ is the Lipschitz constant for g_l. If all the iterated functions are contractions, the state x_τ on the Poincaré section changes on the attractive and unique invariant set C after sufficient random iterations. The set C satisfies the following equation:

$$C = \bigcup_{l=1}^{L} g_l(C). \tag{6}$$

The property of set C of having a fractal-like structure affects the trajectory in the hyper-cylindrical space $\mathcal{M}$. In the space $\mathcal{M}$, the trajectory set $\Gamma(C)$ starting from the initial set C is obtained by the union of the trajectory set $\gamma_l(C)$ for each input I_l:

$$\Gamma(C) = \bigcup_{l=1}^{L} \gamma_l(C). \tag{7}$$

$\Gamma(C)$ is also the attractive and unique invariant set with a fractal-like structure. We have analytically proved Eqs. (6) and (7) when all the iterated functions $g_l(l = 1, 2, ..., L)$ are contractions. Even without this condition, we have shown numerically that these equations are valid [3]. C and $\Gamma(C)$ are also characterized by the hierarchy of a tree structure [10]. By introducing the interpolating system under some conditions, we can also show that the set $\Gamma(C)$ in the cylindrical phase space is enclosed by the tube structure whose initial set is the closure of the fractal set C on the Poincaré section [11].

The following expression of correlation dimension of the fractal set C has previously been derived for such a strictly self-similar set as the Sierpinski gasket constructed by a set of iterated functions that uniformly contract any two points on the Poincaré section [10]:

$$d = -\frac{\ln L}{\lambda T}, \tag{8}$$

where L, λ, and T are the number of the inputs, the real part of the eigenvalue of a matrix, and the input time length, respectively. Equation (8) holds to a self-affine set whose iterated functions are non-contractions that expand two points in some regions [9] . At this point, we have to notice that this equation does not include amplitude parameters of external inputs. This implies that the correlation dimension of the fractal set C is independent of details of the temporal structure of inputs, except for time interval. It is interesting that Eq. (8) does or does not hold to the nonlinear equation stochastically excited by temporal inputs. In the following section, we numerically investigate this issue.

3 Numerical Experiment

We use the following Duffing equation [12] as an example of nonlinear equations:

$$\ddot{x} + k\dot{x} + x + \varepsilon x^3 = A_l \cos \frac{2\pi}{T_l} t, \tag{9}$$

where $k(>0)$, ε, A_l, and $T_l(>0)$ are parameters for dissipation, nonlinearity, lth input amplitude, and lth input time length, respectively. k and ε are internal parameters while A_l and T_l are external ones. From a dynamical systems viewpoint, this second-order ordinary equation can be transformed into the following first-order equations:

$$\begin{cases} \dot{x}_1 = x_2 \\ \dot{x}_2 = -x_1 - \varepsilon x_1^3 - kx_2 + A_l \cos x_3 \\ \dot{x}_3 = \frac{2\pi}{T_l}. \end{cases} \tag{10}$$

This is a dynamical system, $D = (\mathcal{M}, f_l)$, where $\mathcal{M} = R^2 \times S^1 = (x_1, x_2, x_3)$ and $f_l = (x_2, -x_1 - \varepsilon x_1^3 - kx_2 + A_l \cos x_3, \frac{2\pi}{T_l})$. D is a dissipative dynamical system

$$div f_l = -k < 0. \tag{11}$$

Then, for periodic input, an initial state converges to an attractor after a sufficient period. Our main interest is the switching input dynamics. In this paper, for simplicity, we limit our description of numerical examples to equal switching probability using two inputs, i.e., I_1 and I_2, whose amplitudes differ from each other, i.e., $A_1 \neq A_2$, while time intervals are the same, i.e., $T_1 = T_2 = T$.

Equation (9) with $\varepsilon = 0$ results in a linear equation driven by external force. In this case, the real part of the eigenvalue of the linear equation is $-\frac{k}{2}$. Then, Equation (8) for two inputs, i.e., $L = 2$, becomes as follows:

$$d = 2\frac{\ln 2}{kT}. \tag{12}$$

This expression implies that the correlation dimension does not depend on the two parameters, i.e., ε and A_l. To verify this equation, we calculated the correlation dimension using the method in the reference [13] for fractal set C constructed numerically by switching inputs. In the following simulations, $k = 0.2$ is used.

3.1 Dependence on Input Time Length

Figure 1 (below) shows T dependence of correlation dimension d using three different values of ε, i.e., 0, 0.01, and 1.0, respectively. The amplitudes of two inputs were $A_1 = 1$ and $A_2 = -1$, respectively. Ten thousand iterations of switching inputs were plotted after discarding the initial 100 iterations. The solid line is calculated by Eq. (12). At the longer interval the dimension is nearly zero, while at the shorter interval it is nearly two. These tendencies on both sides are easy to intuitively understand as follows. At the limit $T \rightarrow \infty$, a trajectory could reach excited attractors that are points in this example. At the limit $T \rightarrow 0$, a trajectory might be disordered around two attractors due to the rapid transitions between them. On the other hand, we could not intuitively imagine what happens in the middle range.However, not only for $\varepsilon = 0$ but also for $\varepsilon = 0.01$ and $\varepsilon = 1.0$, Equation (12) shows close agreement, except for short time length where the points overlap each other.

For reference, examples of fractal set C on the Poincaré section are shown in the above. Three rows and four columns of panels correspond to three different values of ε and four different length of T, respectively. In each row of four panels, the points spread in every direction at $T = 1$ gradually cluster around to specific region as T becomes larger. At $T = 50$, two clusters around two excited attractors are observed at most. At both sides, i.e., $T = 1$ and $T = 50$, three panels in the same column share very similar features. However, in the middle range of $T = 5$ and $T = 10$, the fractal patterns in the same column show features quite different from each other. These observations indicate that the features of the fractal patterns are very sensitive to nonlinearity, while their correlation dimensions remain almost the same.

3.2 Dependence on Input Amplitude

Figure 2 shows A_1 dependence corresponding to Fig. 1. The time length and the amplitude of the second input are fixed at $T = 6$ and $A_2 = -1$, respectively. Below, $A_1 - d$ graph plotted for three different ε values. The solid line is $d = 1.16$ obtained by Eq. (12). Although a slight deviation can be observed for $\varepsilon = 1.0$ and for 0.01, we can say that Eq. (12) shows close agreement.

For reference, the example of fractal set C is shown in the above. Three rows and four columns of panels correspond to three different values of ε and four different amplitudes of A_1, respectively. Fractal patterns in each row for $\varepsilon = 1.0$ and for $\varepsilon = 0.01$ are quite different in feature while they are almost the same for $\varepsilon = 0$. In each column for the same amplitude A_1, they are completely different in feature. These observations imply that their fractal patterns are very sensitive to amplitude and nonlinearity while the correlation dimension depends negligibly on both parameters.

4 Summary and Discussion

In this paper, we demonstrated that Eq. (8) of the correlation dimension holds approximately to the support of the fractal set obtained by stochastically switching

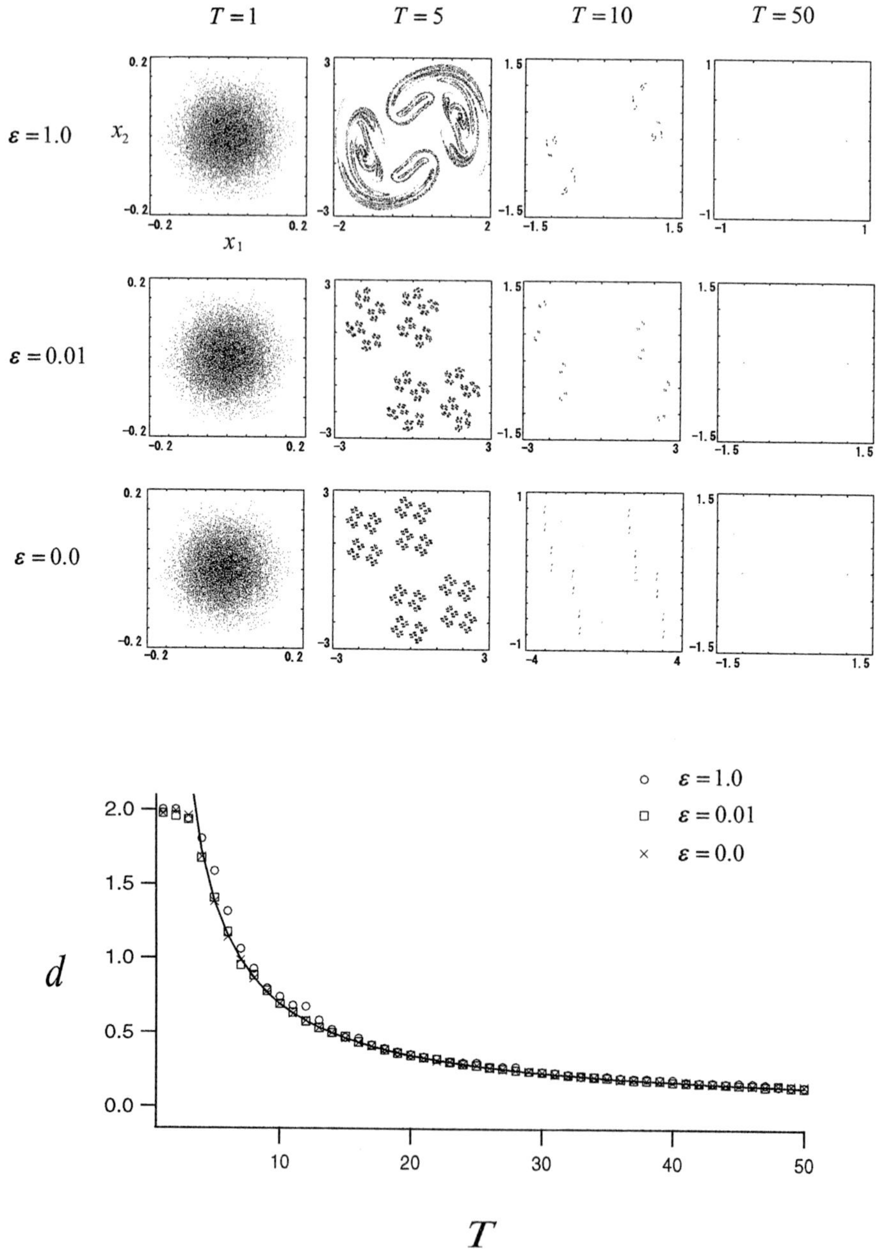

Figure 1. Dependence of fractal set (above) and correlation dimension d (below) on time length T and on nonlinearity ε.

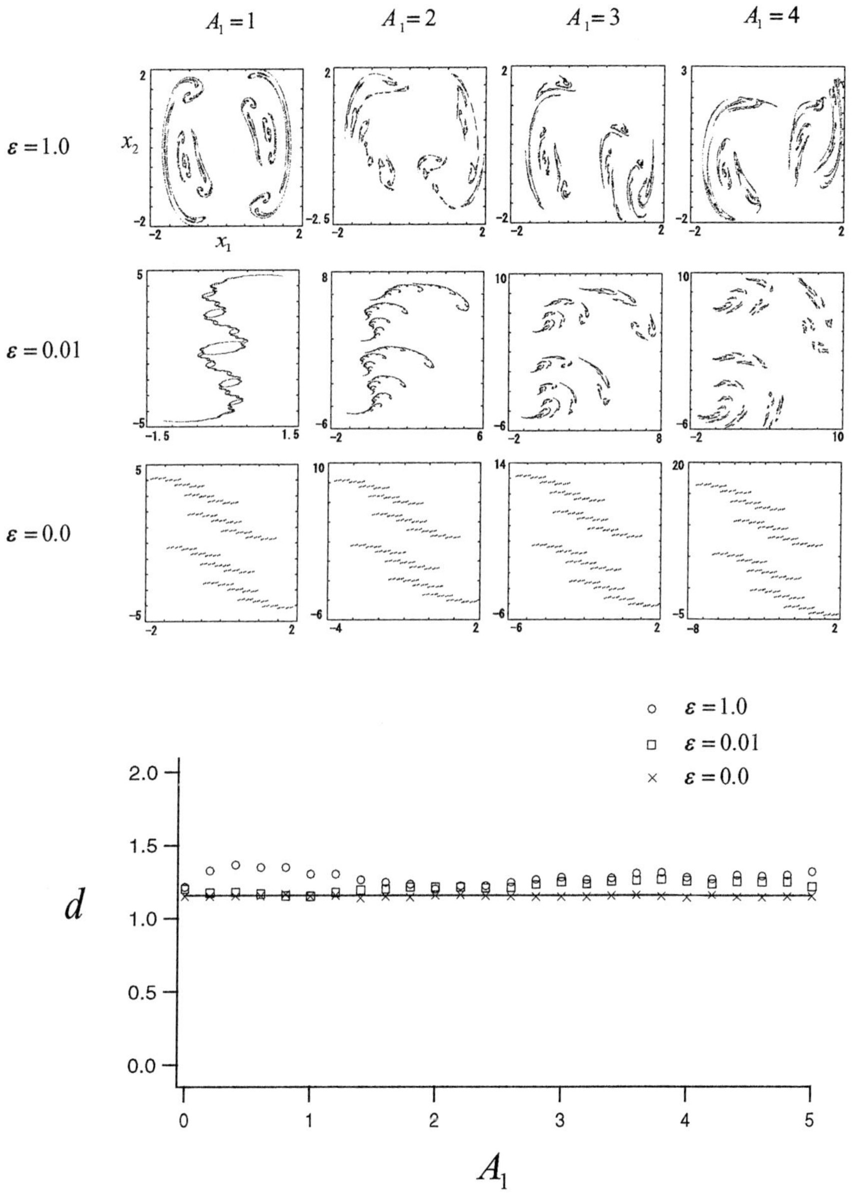

Figure 2. Dependence of fractal set (above) and correlation dimension d (below) on amplitude A_1 and on nonlinearity ε.

the Duffing equation through external temporal inputs. The experimental results clearly show that the correlation dimension is approximately inversely proportional to input time length while the dimension is independent of input amplitude.

We can classify parameters into two categories, i.e., internal ones and external ones. The example presented in this paper includes two internal parameters, i.e., ε for nonlinearity and k for dissipation, and two external parameters, i.e., A_l for amplitude and T_l for the time length of inputs, respectively. Equation (8) and the numerical experiment show that $k(= div f)$ and $T_l(= T)$ are essential for defining the correlation dimension. Although the correlation dimension is just one of infinite dimensions, our results suggest that the statistical properties of ODE stochastically switched by temporal inputs might be universally defined by the dissipation and switching interval independently of the details of the input temporal structure. We expect that our results could be extended to the multifractal formalism [14,15].

References

1. M.F.Barnsley and S.G.Demko, Proc. Roy. Soc. A**399**, 243 (1985).
2. M.F.Barnsley, Fractals Everywhere (2nd ed.) (Academic Press, Boston, 1993).
3. K.Gohara and A.Okuyama, Fractals **7**, 205 (1999).
4. M. S. Branicky, IEEE trans. Automat. Contr., **43**, 475 (1998).
5. R.Wada et al, Proceedings of the 4th International Conference automation of Mixed Process: Hybrid Dynamical Systems (ADPM2000), 93 (2000).
6. Y.Yamamoto and K.Gohara, Journal of Experimental Psychology **19**, 341 (2000).
7. S.Sato and K.Gohara, Int. J. of Bifurcation and Chaos **11**, 421, (2001).
8. K.Gohara, H.Sakurai, and S.Sato, Fractals **8**, 67 (2000).
9. J.Nishikawa and K.Gohara, Int. J. of Bifurcation and Chaos, in press.
10. K.Gohara and A.Okuyama, Fractals **7**, 313 (1999).
11. R.Wada and K.Gohara, Int. J. of Bifurcation and Chaos **11**, 755, (2001).
12. Y.Ueda, New Approaches to Nonlinear Problems in Dynamics edited by P.J.Holmes, 311, (SIAM, Philadelphia, 1980).
13. P.Grassberger and I.Procaccia, Characterization of Strange Attractors, Phys. Rev. Lett. **50**,346-349 (1983).
14. T.C. Halsey et al., Phys. Rev. A. **33**,1141 (1986).
15. K. J. Falconer, Technique in Fractal Geometry (John Wiley & sons (1997)).

ANOMALOUS DIFFUSION ON A ONE-DIMENSIONAL FRACTAL LORENTZ GAS WITH TRAPPING ATOMS*

V. V. UCHAIKIN

Ulyanovsk State University, Institute for Theoretical Physics
42 L. Tolstoy str., 432700 Ulyanovsk, Russia
e-mail: uchaikin@sv.uven.ru

The problem of diffusion of a particle on one-dimensional stochastic fractal is solved. This fractal is a set of points (atoms) correlated on the axis along which the particle is walking. The general solution of the problem is found. It is shown that the exponent of growth of diffusion packet is twice less then in the case of a fractional diffusion. This is an effect of correlations of consecutive free paths.

1 Introduction

The term "anomalous diffusion" relates to the case where the size $\Delta(t)$ of a diffusion packet grows with time slower or faster then in the normal (Gaussian) case where $\Delta(t) \propto t^{1/2}$. Numerous anomalous phenomena have been investigated for a few decades: charge carrier transport in amorphous semiconductors, transport in percolative, polymeric and porous systems, diffusion in convection rolls and rotating flows, Richardson turbulent diffusion and diffusion in turbulent plasma, processes in genetics, physiology and so on. The list of the phenomena with corresponding references can be found in reviews [1–5] and others.

It is well known that in spite of different specific mechanisms generating normal diffusion process in different physical phenomena, the main features of diffusion can be obtained from the *continuous time random walk* (CTRW) scheme as time $t \to \infty$. This raised the hopes that the CTRW model could also describe a large number of different anomalous processes without regard for their specific mechanisms. Many works develop and improve on this model [6–32].

The basic idea of the simplest (one-dimensional decoupled) version of CTRW model is that different jump lengths R_j, as well as waiting times T_j between two successive jumps are independent random variables. One supposes that jump lengths have common distribution

$$\mathsf{P}\{R > r\} \propto r^{-\alpha}, \quad \alpha > 0, \quad r \to \infty$$

and the waiting time's distribution obeys the condition

$$\mathsf{P}\{T > t\} \propto t^{-\beta}, \quad \beta > 0, \quad r \to \infty$$

and both directions for a test particle leaving a trap are equal in probabilities. If $\alpha > 2$ and $\beta > 1$ we observe normal diffusion, all other values of α and β lead to anomalous diffusion with characteristic exponents α and β. The asymptotic term of the probability density function $p(x, t)$ (propagator) obeys the fractional diffusion

*THIS RESEARCH WAS PARTIALLY SUPPORTED BY RFBR GRANTS 0001–00284 AND 0002–17507

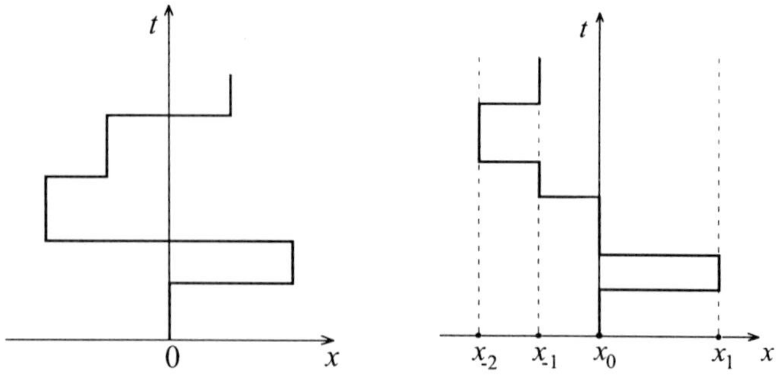

Figure 1. Fractional diffusion (the left panel) and diffusion on fractal (the right panel).

equation

$$\frac{\partial^\beta p(x,t)}{\partial t^\beta} = -D(-\Delta)^{\alpha/2} p(x,t) + \frac{t^{-\beta}}{\Gamma(1-\beta)}\delta(x), \quad 0 < \alpha \le 2, \quad 0 < \beta \le 1.$$

Here

$$\frac{\partial^\beta p(x,t)}{\partial t^\beta} = \frac{1}{\Gamma(1-\beta)} \frac{\partial}{\partial t} \int_0^t \frac{p(x,\tau)d\tau}{(t-\tau)^\beta}$$

is the Riemann–Liouville fractional derivative and $(-\Delta)^{\alpha/2} \equiv -\left(\frac{\partial^2}{\partial x^2}\right)^{\alpha/2}$ is a fractional power of the Laplace operator (see for details [33]). This equation has a solution in self-similar form

$$p(x,t) = \left(Dt^\beta\right)^{-1/\alpha} \psi^{(\alpha,\beta)}\left(|x|(Dt^\beta)^{-1/\alpha}\right),$$

where the function

$$\psi^{(\alpha,\beta)}(r) = \int_0^\infty g^{(\alpha,0)}(r\tau^{\beta/\alpha}) g^{(\beta,1)}(\tau)\tau^{\beta/\alpha} d\tau$$

is expressed through the one-sided stable density $g^{(\beta,1)}(\tau)$ and symmetrical isotropic stable density $g^{(\alpha,0)}(r)$. Note that the stable densities are defined by their characteristic functions. In the one-dimensional case considered below

$$\tilde{g}^{(\alpha,\theta)}(k) = \int_{-\infty}^\infty e^{-ikx} g^{(\alpha,\theta)}(x)dx = \exp\left\{-|k|^\alpha \exp[-i\frac{\pi}{2}\theta\alpha \text{ sign } k]\right\}.$$

It is easy to see from above that the width of the anomalous diffusion packet $\Delta(t) \propto t^{\beta/\alpha}$. When $\beta/\alpha < 1/2$ we have a subdiffusion regime, when $\beta/\alpha > 1/2$ we observe a superdiffusion regime.

We will call the process *fractional diffusion* (FD) in order to distinguish it from another process—*diffusion on fractals* (DF) which will be considered below. Whereas in the first case two consecutive jump lengths are independent, in the second case they are correlated. The sketch in Fig. 1 shows why the difference occurs: when the walker goes back, he meets with the same atoms which were visited when he went forward. The back jump has the same length as the forward jump, this is a cause of the correlations between consecutive free paths.

The problem is to find FD-propagator and to compare it with DF-propagator.

2 Stochastic Lorentz gas

The first stage of solution of the problem is choosing an appropriate stochastic ensemble for representation of the random medium. We consider one-dimensional case where the substance is concentrated at points $\{X_j\} = \ldots, X_{-2}, X_{-1}, X_0 = 0, X_1, X_2, \ldots$ (atoms), randomly placed on x-axes. In order to construct the statistical ensemble we suppose that $X_j - X_{j-1} = Y_j$ are independent identically distributed random variables with common distribution function

$$F(x) = \mathsf{P}\{Y < x\} = \int_0^x f(x)dx.$$

With this assumption, the random values X_j form a correlated random sequence, or one can say, two independent sequences $X_1, X_2, \ldots$ and $X_{-1}, X_{-2}, \ldots$ with a common initial point $X_0 = 0$. This model is called *one-dimensional Lorentz gas* [26].

Let $F_n(x)$ be the n-fold convolution of the distribution function $F(x)$:

$$F_{n+1}(x) = \int_0^x F_n(x-y)dF(y), \qquad F_1(x) = F(x)$$

and let $N_+(x)$ be the random number of atoms belonging to the interval $(0, x]$. Then

$$W(n, x) \equiv \mathsf{P}\{N_+(x) = n\} = F_n(x) - F_{n+1}(x).$$

The similar relation takes place for the number $N_-(x)$ of atoms in the interval $[-x, 0)$. Note that the total number of atoms in the interval $[-x, x]$ is the sum

$$N(x) = N_+(x) + N_-(x) + 1.$$

Thus, choosing different functions $F(x)$ we obtain random sets of various kinds. Let us consider a few examples.

Example 1 *Choosing the distribution function in the form of the Heaviside function*

$$F(x) = H(x-a) \equiv \begin{cases} 0, \ x < a, \\ 1, \ x \geq a \end{cases}$$

we obtain a one-dimensional lattice with parameter a

$$W(n, x) = H(x - an) - H(x - an - a) = \begin{cases} 1, \ an \leq x < an + a, \\ 0, \ x < an \text{ or } x \geq an + a. \end{cases}$$

Example 2 *Taking exponential distribution for $F(x)$*

$$F(x) = 1 - \exp\{-\mu x\}$$

we have got

$$W(n,x) = \int_0^x \mu \frac{(\mu y)^n}{n!} e^{-\mu y} dy - \int_0^x \mu \frac{(\mu y)^{n+1}}{(n+1)!} e^{-\mu y} dy = \frac{(\mu x)^n}{n!} e^{-\mu x}.$$

This is nothing but a homogeneous Poisson process with the mean value

$$\langle N(x) \rangle = \mu x.$$

The relative deviation of the random variable $N(x)$

$$\frac{\Delta(x)}{\langle N(x) \rangle} \equiv \sqrt{\langle N^2(x) \rangle - \langle N(x) \rangle^2} \Big/ \langle N(x) \rangle = \langle N(x) \rangle^{-1/2} = (\mu x)^{-1/2}$$

disappears as $x \to \infty$. *Note that in this case the numbers of atoms belonging to different disjoint domains are independent.*

Example 3 *Now we do not use any concrete expression for* $F(x)$. *We suppose only that its variance* σ^2 *is finite. Naturally, with this assumption only asymptotical results are available. The larger* x, *the larger values of* n *play the leading role in consideration. Application of the central limit theorem yields*

$$F_n(x) \sim \Phi(z_n),$$

where

$$\Phi(z) = \frac{1}{\sqrt{2\pi}} \int_{-\infty}^{z} \exp\left(-\frac{x^2}{2}\right) dx$$

is the normal distribution function, $z_n = \dfrac{x - n/\mu}{\sigma\sqrt{n}}$ *and* $\mu = \langle Y \rangle^{-1}$. *Asymptotically*

$$F_n(x) - F_{n+1}(x) = \Phi(z_n) - \Phi(z_{n+1}) \sim \Phi'(z_n)(z_n - z_{n+1}),$$

and we arrive at a normal distribution with mean value $\langle N(x) \rangle = \mu x$ *and a relative deviation*

$$\frac{\Delta(x)}{\langle N(x) \rangle} \sim \frac{\mu\sigma}{\sqrt{\mu x}}, \qquad x \to \infty.$$

Example 4 *Let the distribution function* $F(x)$ *obey the asymptotic relation*

$$1 - F(x) \sim \frac{A}{\Gamma(1-\alpha)} x^{-\alpha}, \quad \Delta > 0, \quad x \to \infty.$$

For $\alpha > 2$ *the variance of* Y *is finite and we arrive at the previous example. For* $\alpha < 2$ *the variance is infinite and we have a qualitatively another kind of medium called Lévy–Lorentz gas* [26]. *Two cases arise:* $1 < \alpha < 2$, *where mathematical expectation exists, and* $0 < \alpha < 1$, *where mathematical expectation does not exist.*

In the first case, according to the generalized limit theorem

$$F_n(x) \sim G^{(\alpha,\theta_\alpha)}(z_n), \qquad z_n = \frac{x - n/\mu}{c(An)^{1/\alpha}}, \quad n \to \infty$$

where $G^{(\alpha,\theta_\alpha)}(z_n)$ is a totally skewed stable distribution function with characteristic exponent $\alpha \in (1,2)$, and c is some positive constant. One can show that in this example $\langle N(x)\rangle \sim \mu x$, but

$$\frac{\Delta(x)}{\langle N(x)\rangle} \propto (\mu x)^{-1/\alpha}, \quad x \to \infty.$$

The latter means that the relative deviation disappears at large distances although slower then in the previous examples. Nevertheless, if we represent some characteristic of the medium as a smooth function $f(N(x), x)$ of the random variable $N(x)$ then we get

$$f(N(x), x) \to f(\langle N(x)\rangle, x), \quad x \to \infty.$$

This means that the depth growth provides for averaging over whole statistical ensemble on the basis of a unique sample only. The property called self-averaging joins all above cases to a class of asymptotically regular media.

In the second case the generalized limit theorem yields

$$F_n(x) \sim G^{(\alpha,1)}(z_n), \quad z_n = \frac{x}{(An)^{1/\alpha}}$$

and therefore

$$W(n,x) \sim g^{(\alpha,1)}\left(\frac{x}{(An)^{1/\alpha}}\right)\frac{x}{\alpha n(An)^{1/\alpha}}, \quad x \to \infty.$$

For $\alpha < 1$, the density $g^{(\alpha,1)}(x)$ is one-sided, i.e., is distributed on the positive semiaxes only and its Laplace image has a form of a stretched exponent

$$\int_0^\infty e^{-\lambda x} g^{(\alpha,1)}(x)dx = e^{-\lambda^\alpha}, \qquad \lambda > 0.$$

All its moments of orders $\nu \geq \alpha$ diverge, all moments of orders $\nu < \alpha$ (including negative orders) exist and are expressed through the gamma-function:

$$\int_0^\infty x^\nu g^{(\alpha,1)}(x)dx = \frac{\Gamma(1-\nu/\alpha)}{\Gamma(1-\nu)}.$$

Note that if $\alpha \to 1$ then $g^{(\alpha,1)}(x) \to \delta(x-1)$ and we arrive at the lattice with the parameter $a = 1$.

As to moments of the random number of atoms on $(0, x]$, all of them exist and are of the form

$$\langle N^\nu(x)\rangle = \frac{\Gamma(1+\nu)}{\Gamma(1+\alpha\nu)}\frac{x^{\alpha\nu}}{A^\nu}.$$

One can see from here that asymptotically

$$\langle N(x)\rangle = \frac{1}{\Gamma(1+\alpha)}\frac{x^\alpha}{A} \equiv N_1 x^\alpha, \quad N_1 = \langle N(1)\rangle$$

and

$$\frac{\Delta(x)}{\langle N(x)\rangle} = \text{const} \neq 0.$$

Moreover, the distribution itself takes a scaling form

$$W(n,x)\,dn = w(z,\alpha)dz, \qquad z = n/\langle N(x)\rangle,$$

$$w(z,\alpha) = \frac{z^{-1-1/\alpha}}{\alpha\Gamma(1+\alpha)}\, g^{(\alpha,1)}\left(\frac{z^{-1/\alpha}}{\Gamma(1+\alpha)}\right).$$

Thus we obtain a random point structure with the following properties.

1. *All atoms of the set are equal in rights by construction so that all processes* $(X_j, X_j + x)$ *are statistically equivalent:* $N(X_j, X_j + x) \stackrel{d}{=} N(x)$ ($\stackrel{d}{=}$ *means equality in distribution).*
2. *The mean number of atoms grows with depth* x *according to inverse power low*
$$\langle N(x)\rangle \propto x^\alpha, \quad 0 < \alpha < 1.$$
3. *The relative deviations, i.e., statistical fluctuations are the same at all depths and this structure is random at all scales.*

This is a case of a stochastic fractal Lorentz gas with fractal dimension α *embedded in a one-dimensional space. On the contrary to regular media stochastic fractals do not possess the property of self-averaging: while for a regular medium we have*

$$\langle f(N(x),x)\rangle \to f(\langle N(x)\rangle, x), \quad x \to \infty,$$

for a fractal Lorentz gas we obtain

$$\langle f(N(x),x)\rangle \to \int_0^\infty f(N_1 x^\alpha z, x) w(z,\alpha) dz, \quad x \to \infty.$$

3 Diffusion on a single sample

The second stage of the work consists in consideration of a test particle walking on a fixed set of atoms $\{x_j\}$ arranged irregularly on a line. The test particle appears at time $t = 0$ at the origin $x_0 = 0$ and stays there up to time $T_1 > 0$. At time $t = T_1$, it performs an instantaneous jump with equal probability to one of the neighbouring atoms x_{-1} or x_1. It waits there for the time T_2 and at the moment $T_1 + T_2$ jumps to one of neighbours again and so on. Waiting times $T_1, T_2, \ldots$ are independent identically distributed random variables with a common distribution function $Q(t) = \mathsf{P}\{T_j < t\}$.

This algorithm generates an ensemble the of particle trajectories $\left\{X(t;\{x_j\})\right\}$ on the set of fixed atoms $\{x_i\}$. Averaging over this ensemble will be denoted by the overbar:

$$F\Big(x,t|\{x_j\}\Big) \equiv \mathsf{P}\Big\{X(t;\{x_j\}) < x\Big\} = \overline{H\Big(x - X(t;\{x_j\})\Big)}.$$

Passing to a new random variable $J(t)$ via relation

$$X\Big(t;\{x_j\}\Big) = x_{J(t)}$$

and taking into account monotonicity of the function x_j $(j = \dots, -1, 0, 1, \dots)$ we obtain

$$F\Big(x, t|\{x_j\}\Big) = \overline{H\Big(n(x) - J(t)\Big)}$$

where $n(x)$ obeys the equation

$$x_n \leq x \leq x_{n+1}, \quad n = \dots, -1, 0, 1, \dots$$

Further,

$$J(t) = \sum_{j=1}^{K(t)} U_j$$

where the independent random variables $U_j = \pm 1$ with equal probabilities and $K(t)$ is the random number of jumps up to time t (U_j and K are independent of each other). According to the central limit theorem

$$\mathsf{P}\Big\{J < n|K = k\Big\} \sim G^{(2,0)}(n/\sqrt{k}), \qquad k \to \infty.$$

Then

$$\mathsf{P}\Big\{J < n\Big\} = \overline{G^{(2,0)}(n/\sqrt{K})}$$

and as a result

$$F\Big(x, t|\{x_j\}\Big) \sim \sum_{k=1}^{\infty} G^{(2,0)}\Big(n(x)/\sqrt{k}\Big) W(k, t), \qquad t \to \infty$$

where

$$W(k, t) = \mathsf{P}\Big\{K(t) = k\Big\}.$$

The random process $K(t)$ can be defined in the same manner as the $N_+(x)$. Assuming in particular

$$1 - Q(t) \sim \frac{B}{\Gamma(1-\beta)} t^{-\beta}, \qquad t \to \infty, \qquad 0 < \beta < 1,$$

we obtain

$$W(k, t)dk \sim w(z, \beta)dz$$

where

$$z = k/\langle K(t)\rangle, \qquad \langle K(t)\rangle = \frac{1}{\Gamma(1+\beta)} \frac{t^\beta}{B} \equiv K_1 t^\beta.$$

Thus the times of jumps $T_1, T_1 + T_2, T_1 + T_2 + T_3, \dots$ form a fractal set on the time axes with fractal dimensionality β. We meet here with subdiffusion behavior witch is described by the distribution function

$$\begin{aligned} F\Big(x, t; \{x_j\}\Big) &\sim \int G^{(2,0)}(n(x)) \left(zK_1 t^\beta\right)^{-1/2} w(z, \beta)\, dz \\ &= \Psi^{(2,\beta)}\Big(n(x)(Dt^\beta)^{-1/2}\Big), \qquad x \to \infty \end{aligned}$$

where

$$\Psi^{(2,\beta)}(x) = \int_{-\infty}^{x} \psi^{(2,\beta)}(x)dx$$

and $D = \dfrac{1}{2B}$ is the subdiffusion coefficient.

4 Diffusion on a stochastic Lorentz gas

The third, final stage of the problem solution is averaging the conditional distribution function $F\Big(x,t|\{x_j\}\Big)$ over all possible arrangements of atoms. We will denote this operation by angle brackets as above:

$$F(x,t) = \Big\langle F(x,t)|\{X_j\}\Big\rangle.$$

For an asymptotically regular medium we obtain

$$\begin{aligned} F(x,t) &\sim \Big\langle \Psi^{(2,\beta)}(N(x)(Dt^\beta)^{-1/2})\Big\rangle \\ &\sim \Psi^{(2,\beta)}\Big(\langle N(x)\rangle (Dt^\beta)^{-1/2}\Big) \\ &= \Psi^{(2,\beta)}\Big(N_1(Dt^\beta)^{-1/2}x\Big). \end{aligned}$$

This result coincides with the corresponding FD distribution (formulas (35), (36) in [29]).

For a fractal Lévy-Lorentz gas we have got for $x \geq 0$ (the distribution is symmetrical with respect to $x = 0$)

$$F(x,t) \sim \Big\langle \Psi^{(2,\beta)}\Big(N(x)(Dt^\beta)^{-1/2}\Big)\Big\rangle = \int_0^\infty \Psi^{(2,\beta)}\Big(zN_1(Dt^\beta)^{-1/2}x^\alpha\Big)w(z,\alpha)dz.$$

Passing to the variate

$$y = \frac{z^{-1/\alpha}}{\Gamma(1+\alpha)}$$

and taking into account that

$$w(z,\alpha)dz = g^{(\alpha,1)}(y)dy$$

we rewrite the above result in the form

$$F(x,t) \sim \Xi^{(\alpha,\beta)}\Big((D't)^{-\beta/2\alpha}x\Big)$$

where

$$\Xi^{(\alpha,\beta)}(x) = \int_0^\infty \Psi^{(2,\beta)}\left((x/y^\alpha)\right)g^{(\alpha,1)}(y)dy$$

is the new distribution function describing the form of diffusion packet of the particle walking on the fractal with fractal dimensionality α, $D' = \text{const} > 0$.

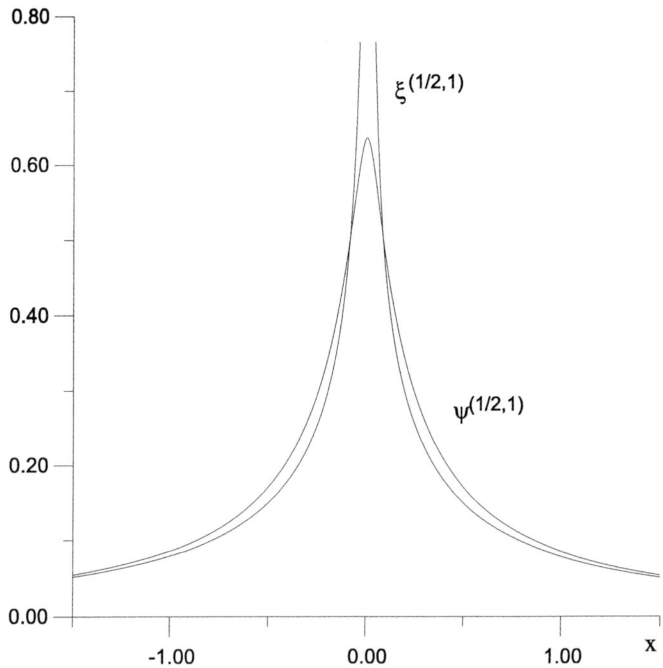

Figure 2. DF- and FD- distribution densities with parameters $\alpha = 1/2$ and $\beta = 1$.

5 A particular case

Consider the case $\alpha = 1/2$ and $\beta = 1$. In this case

$$\Xi^{(1/2,1)}(x) = \int_0^\infty \Psi^{(2,1)}(\sqrt{x/y})g^{(1/2,1)}(y)dy$$

and the density function

$$\xi^{(1/2,1)}(x) = d\Xi^{(1/2,1)}(x)\Big/dx = (\alpha/x^{1-\alpha})\int_0^\infty \psi^{(2,1)}(\sqrt{x/y})y^{-1/2}g^{(1/2,1)}(y)dy.$$

Substituting here

$$\psi^{(2,1)}(x) = \frac{1}{2\sqrt{\pi}}\exp(-x^2/4)$$

and

$$g^{(1/2,1)}(y) = \frac{1}{2\sqrt{\pi}}y^{-3/2}\exp(-1/4y),$$

we easily obtain

$$\xi^{(1/2,1)}(x) = \frac{1}{2\pi\sqrt{x}(x+1)}, \quad x > 0.$$

The DF distribution density $\xi^{(1/2,1)}(x)$ with the corresponding DF distribution $\psi^{(1/2,1)}(x)$ are shown in Fig. 2.

6 Concluding remarks

Three important conclusions can be extracted from obtained results.

1. The fractal media does not possess the self-averaging property:
$$p\Big(x,t|\{x_j\}\Big) \neq \Big\langle p(x,t|\{X_j\})\Big\rangle$$
for $t \to \infty$.

2. The DF-packet grows in width as $t^{\beta/2\alpha}$, i.e., much slower than the corresponding FD-packet whose width $\sim t^{\beta/\alpha}$. This is the effect of neighbouring atoms playing the role of some kind of traps (see Fig. 1).

3. The DF- and FD- packet forms essentially differ from each other (see Fig. 2) but both of them are expressed through the stable distribution densities. The explicit expressions are brought above.

References

1. J.-P. Bouchaud, A. Georges. *Phys. Rep.* **195**, 127 (1990).
2. M.B. Isichenko. *Rev. Mod. Phys.* **64**, 961 (1992).
3. B.J. West, W. Deering. *Phys. Rep.* **246**, 1 (1994.)
4. V.Uchaikin, V. Zolotarev. *Chance and Stability.* (VSP, Utrecht, The Netherlands, 1999).
5. R.Metzler, J.Klafter. *Phys. Rep.*, **339**, 1 (2000).
6. E.W. Montroll, G.H. Weiss. *J. Math. Phys.* **6**, 167 (1965).
7. G.H. Weiss, R.J. Rubin. *J. Stat. Phys.* **14**, 333 (1976).
8. M.F. Shlesinger, J.Klafter, Y.M. Wong. *J. Stat. Phys.* **27**, 499 (1982).
9. G.H. Weiss, R.J. Rubin. *Adv. Chem. Phys.* **52**, 363 (1983).
10. R. Nigmatullin. *Phys. Status Solidi* B **124**, 389 (1984).
11. W. Wyss. *J. Math. Phys.* **27**, 2782 (1986).
12. J. Klafter, A. Blumen, M.F. Shlesinger. *Phys. Rev. A* **35**, 3081 (1987).
13. M.F. Shlesinger, B. J. West, J. Klafter. *Phys. Rev. Lett.* **58**, 1100 (1987).
14. V.V. Afanasiev, R.Z. Sagdeev, G.M. Zaslavsky. *Chaos* **1**(2), 143 (1991).
15. M. Kotulski. *J. Stat. Phys.* **81**, 777 (1995).
16. A. Compte. *Phys. Rev. E* **53**, 4191 (1996).
17. A. Compte, D. Jou, Y. Katayama. *J. Phys. A: Math. Gen.* **30**, 1023 (1997).
18. B.J. West, P. Grigolini, R. Metzler. T.F. Nonnenmacher. *Phys. Rev. E* **55**, 99 (1997).
19. A.I. Saichev, G.M. Zaslavsky. *Chaos* **7**(4), 753 (1997).
20. R. Gorenflo, F. Mainardi. *Frac. Calc. Appl. Analys.* **1**, 167 (1998).
21. R. Gorenflo, F. Mainardi. *J. Analys. Appl.* **18**, 231 (1999).
22. D. Kusnezov, A. Bulgas, G.D. Dang. *Phys. Rev. Lett.* **82**, 1136 (1999).
23. W.A. Curtin. *J. Phys. Chem. B* **104**, 3937 (2000).
24. I.M. Sokolov. *Phys. Rev. E* **63**, 011104-1 (2000).
25. G. Margolin, B. Berkowitz. *J. Phys. Chem. B* **104**, 3942 (2000).
26. E. Barkai, V. Fleurov, J. Klafter. *Phys. Rev. E* **61**, 1164 (2000).

27. F.Mainardi, Yu.Luchko, G.Pagnini. *Frac. Calc. Appl. Analys.* **4**, 153 (2001).
28. I.M. Sokolov. *Phys. Rev. E* **63**, 056111-1 (2001).
29. V.V. Uchaikin. *J. Exper. Theor. Phys.* **88**, 1155 (1999).
30. V.M. Zolotorev, V.V. Uchaikin, V.V. Saenko. *J. Exper. Theor. Phys.* **88**, 780 (1999).
31. V.V. Uchaikin. *Intern. Journ. Theor. Phys.* **39**, 2087 (2000).
32. V.V. Uchaikin. In *Paradigms of Complexity: Fractals and Structures in the Sciences (M. M. Novak, Ed.)* (World Scientific, Singapore, 2000), p.41.
33. S.G. Samko, A.A. Kilbas, O.I. Marichev. *Fractional Integrals and Derivatives – Theory and Applications* (Gordon and Breach, New York, 1993).

IS FRACTAL ESTIMATION OF A GEOMETRY WORTH FOR ACOUSTICS ?

PHILIPPE WOLOSZYN
Cerma UMR CNRS 1563, EAN, BP 81931 F-44319 Nantes Cedex 3, France
E-mail : woloszyn@cerma.archi.fr

Through the presence of stonework, windows and balconies, the frontage of a typical urban building presents irregularity depths comparable to the sound wavelengths. Thus, the major type of reflexions on the buildings is produced through an anomalous back-scattered region, creating an acoustic interference field in its neighborhood. In order to be able to detect that scattered energy's minima and maxima, we have to take into account both the acoustic sources incidence angles and the fractal characterization of urban surfaces through their multiscale dilation , involving the *Minkowski sausage* technique.

For a given regular plane division, repeating n times a spatial unit of width Λ, the propagation directions are defined as the *characteristic directions of scattering* associated with the localization length Λ:

$$\sin\alpha_d = \frac{n\Lambda}{p\lambda} - \sin\alpha \text{ , } n = 0, +\text{-}1, \ldots \tag{1}$$

where α_d is the grazing diffraction angle and p the diffraction order. For $\Lambda = 0$, equality between incident and reflected angle remaining true (specular conditions).

The two-dimensional polar response of a given indented surface can be expressed through the diffraction orders (p, q) [1]:

$$\sqrt{p^2 + q^2} = \Lambda\frac{\sin\alpha_d + \sin\alpha}{\lambda}. \tag{2}$$

The characteristic directions are those along which the wavelets emanating from the individual localization lengths Λ_n (periods of the surface) are exactly in phase. This constructive interference condition is conditioned with the a-dimensional modulus Λ_n/λ, which quantifies the energy of non-evanescent scattering losses, represented by the area of scattering intensity pattern lobes. Taking the multiscale evolution of a D-dimensioned prefractal structure into account, localization length Λ follows the following expression, related to the Hurst coefficient:

$$\Lambda_n \to \Lambda^{(d-D)} \propto \frac{S_x(d-D)\Lambda}{P\upsilon} = (d-D)^2\Lambda = H^2\Lambda \tag{3}$$

This leads us to reconsider the evolution of the diffraction orders with a multiscale reformulation of the polar response of a D-dimensioned structure as:

$$\sqrt{\sum_0^{\Lambda_{max}} k^2} = H^2\Lambda\frac{\sum_0^{\Lambda_{max}} \sin\alpha_d + \sin\alpha}{\lambda} \tag{4}$$

with k, the cumulative number of the diffraction orders, and $\Lambda^H/\lambda \propto H^2\Lambda/\lambda$, the adimensional modulus of the local constructive interference conditions. Calculation of the mean square diffracted sound pressure P_d by the whole surface of area S is given by [3]:

$$P_d^2 = \sigma\iint_S \frac{\cos\alpha\cos\alpha_d}{\pi r_0^2} P_i^2\, dS \tag{5}$$

where r_0 represents the distance from the receiver to the structure, and P_i the incident waves pressure. The angular repartition of the sound energy is computed through the prefactor σ, which represents the *scattering pressure function* of the surface, involving the density distribution of the sonic particles *P(r,t)*, with the following relationship [4]:

$$\sigma(k_d, i\omega) = \int dr \exp(ik_d r) \int_0^\infty dt \exp(-i\omega\, t) P(r,t) \tag{6}$$

using the imaginary part of the wave number ω, the fractional diffusion order k_d, and the density distribution of the sonic particles *P(r,t)*, which is the probability for a sonic particle to walk to location *r* during the time *t* into a *fractional Brownian walk* in a *D*-dimensional space [2]:

$$P(r,t) \to \frac{1}{(4\pi\delta t)^{D/2}} \exp\frac{r^2}{4\delta t}, N \to \infty \tag{7}$$

where δ is the diffusion coefficient of the *D*-dimensional space. This function can be evaluated using a Laplace transform [4]. After integrating over the angles, density distribution becomes:

$$\sigma(k_d, i\omega) = \frac{2\pi G}{i\omega^{1-D/d_w}} \int dr\, r^{D-1} \frac{\sin(k_d r)}{k_d r} * \frac{1}{(i\omega\, r^{1/d_w})^{D-1/2}} \exp(-i\omega\, r^{1/d_w} / G) \tag{8}$$

where *G* is the *angular distribution factor* along the characteristic directions of scattering and *D* the fractal dimension. Fractional diffusion order k_d can be locally expressed with the corresponding *k* diffraction orders, involving the polar response of the structure as:

$$k_d \propto \sqrt{\sum_0^{\Lambda_{max}} k^2} = V_D \sum_0^{\Lambda_{max}} \sin\alpha_d + \sin\alpha \tag{9}$$

Through the expression of the *diffusion volume* V_X of the structure, involving the global neighborhood volume $V_{d(r_0)}$, the local measure $V_{d(\Lambda max)}$ and the generalized Euclidean pre-factor F_X, [5]:

$$V_x \prec \frac{V_d(\Lambda_{max})}{V_d(r_0)} = \left(\frac{\Lambda_{max}}{r_0}\right)^{(d-D)} = x^H = \frac{H^2 \Lambda_{max}}{F_x \lambda}. \tag{10}$$

At each step of the measure of the structure, the Minkowski scrutation technique provides a quantification of the scattering pressure function at each incidence angle through the polar response involving successive diffusion sections S_x, which traduces the behavior of the scattering pressure function $\sigma(k_d)$ for a given wavenumber ω of the entire structure:

$$\sigma(k_d) \propto \frac{F_x}{V_x} r_0 \sum_0^{\Lambda_{max}} \sin\alpha_d + \sin\alpha \tag{11}$$

This research work defines the diffusion process as a geometry-dependant phenomenon, confirming the multiscale influence of the built structure on acoustical early propagation.

References

1. D'Antonio P., Konnert J., " The QRD Diffractal: A New One- or Two-Dimensional Fractal Sound Diffusor " J. Audio Eng. Soc., Vol. **40**, No. **3**, March 1992.
2. Gouyet J.F. : " Physique et structures fractales ", Paris, *Masson*, 1992, pp. 47-48.
3. Makarewicz R & Kokowski P. : " Reflection of Noise from a Building's Facade", Applied Acoustics **43**, Elsevier, London, pp. 149-157 (1994)
4. Roman, 1997, " Diffusion on self-similar structures ", Fractals, World Scientific, Vol. **5**, No.**3**, September 1997 pp. 379-393.
5. Woloszyn P. : " Squaring the circle: diffusion volume and acoustic behavior of a fractal structure ", Fractal 2000, World Scientific, Singapour, 2000, pp. 299-300.

WIND VELOCITY TIME SERIES ANALYSIS

ANA M. TARQUIS

Dpto. de Matemática Aplicada a la Ingeniería Agronómica. E.T.S. de Ingenieros Agrónomos, UPM. Ciudad Universitaria sn, 28040 Madrid, Spain.
E-mail: atarquis@mat.etsia.upm.es

ROSA M. BENAVENTE, ANTONIO ROMERO, AND JOSÉ L. GARCÍA

Dpto. de Ingeniería Rural. E.T.S. de Ingenieros Agrónomos, UPM. Ciudad Universitaria sn, 28040 Madrid, Spain.
E-mail: rbenaven@iru.etsia.upm.es jlgarcia@iru.etsia.upm.es

PHILIPPE BAVEYE

Dept. of Agricultural and Biological Engineering, Cornell University. Ithaca, NY 14853 - USA
E-mail: pcb2@cornell.edu

The study of wind velocity (v) is important for greenhouse control (heating and ventilation), since v influences both. Wind increases heat losses during winter nights. On the other hand, the opening of the windows must be reduced with high values of v [1]. Therefore, it is necessary to include in the control greenhouse algorithm a decision rule as function of v. The aims of our work are: a) to explore the use of multifractal analysis for defining a statistically equivalent situation and b) to simulate it easily improving the traditional smooth function approach of v data. The study of the multifractal nature of this type of data has been already done in several works, as for example in Tuck and Hovde [2]. However, fewer works have been published applying deterministic fractal interpolating functions or fractal-multifractal representation (FM) introduced by Puente [3]. The basic idea of the FM approach is to think of intricate patterns as projections of fractal functions, which go through simple multifractal measures, and consequently this methodology has a deterministic nature [3, 4].

The fractal interpolation functions are continuous functions f that interpolate a given set of N+1 points in the plane { (x_0,y_0), (x_1,y_1), (x_2,y_2), ..., (x_N,y_N), where $x_0 \le x_1 \le ... \le x_n$ }. These points are obtained iterating N contractile affine mappings w_n. The w_n used are calculated as follow [3]:

$$w_n \begin{pmatrix} x \\ y \end{pmatrix} = \begin{pmatrix} a_n & 0 \\ c_n & d_n \end{pmatrix} \begin{pmatrix} x \\ y \end{pmatrix} + \begin{pmatrix} e_n \\ f_n \end{pmatrix} \qquad (1)$$

and the conditions that guarantee the existence of a unique set { (x, f(x)) / x $\in [x_0, x_n]$} are

$$w_n\begin{pmatrix}x_0\\y_0\end{pmatrix}=\begin{pmatrix}x_{n-1}\\y_{n-1}\end{pmatrix},\quad w_n\begin{pmatrix}x_N\\y_N\end{pmatrix}=\begin{pmatrix}x_n\\y_n\end{pmatrix}\qquad(2)$$

and $0 \le |d_n| < 1$, for n = 1, 2, ... , N. The equations (1) and (2) allow solving the parameters a_n, b_n, c_n, e_n and f_n in terms of the interpolating points' coordinates and the scaling parameters d_n [3]. Once those f functions are calculated the derived distributions dx and dy are determinate. The measure dy can be interpreted as a weighted projection of the function f, with the weights given by dx (for further details see Puente, 1992). A great variety of derived measures dy are obtained by varying f and dx [3]. An advantage of this approach is that doesn't need assumptions of stationarity, ergodicity or a minimal length of the data set. Data used in this study was acquired from the climatic station of the Dpto. de Producción Vegetal, placed in the experimental fields of the Agricultural School of Madrid. The station recorded hourly mean values of v (m/s) and from this data a daily average is calculated. Data from August of 1994 till July of 1995 was used, having a total of 8760 data points. To apply the FM approach a three-dimensional case is considered as described by Puente [4]. The results show that multifractal analysis provides rich information for the statistical characteristics of v. The projections of fractal interpolating functions weighted by multifractal measures can simulate high-resolution v time series, as it has been reported for rainfall time series [4]. Future work should be devoted to relate the v simulation FM model with the control greenhouse algorithm.

References

1. García, J.L.; de la Plaza, S.; Navas, L.; Benavente, R.M.; Luna, L.. Evaluation of the Feasibility of Alternative Energy Sources for Greenhouse Heating. *J. agric. Engng. Res.*, 69 (1998) pp. 107-114.
2. Tuck, A.F.; Hovde, S.J.. Fractal behavior of ozone, wind and temperature in the lower stratosphere.. *Geophysical-Research-Letters*, 26 (1999) pp. 1271-1274.
3. Puente, C.E. Multinomial Multifractals, fractal interpolators, and the Gaussian distribution. *Phisycs Letters A*, 161 (1992) pp. 441-447.
4. Puente, C.E. and Obregón, N. A. deterministic geometric representation of temporal rainfall: Results for a storm in Boston. *Water Resource Research*, 9 (1996) pp. 2825-2839.

SPATIAL LEAF AREA DISTRIBUTION OF A FABA BEAN CANOPY

ANA M. TARQUIS AND VALERIANO MÉNDEZ

Dpto. de Matemática Aplicada a la Ingeniería Agronómica. E.T.S. de Ingenieros Agrónomos, UPM. Ciudad Universitaria sn, 28040 Madrid, Spain.

E-mail: atarquis@mat.etsia.upm.es v.mendez@grupobbva.com

CARLOS H. DÍAZ-AMBRONA, MARGARITA RUÍZ-RAMOS AND INÉS MÍNGUEZ

Dpto. de Producción Vegetal: Fitotecnia. E.T.S. de Ingenieros Agrónomos, UPM. Ciudad Universitaria sn, 28040 Madrid, Spain.

E-mail: chernandez@pvf.etsia.upm.es clima@pvf.etsia.upm.es iminguez@pvf.etsia.upm.es

Plant canopy structure is the spatial arrangement of the aboveground organs of plants in a plant community. It affects the radiative and convective exchange of the plant community being the leaves one of the main elements. The foliage is often assumed to be randomly distributed throughout the canopy space [1], but these assumptions are rarely found in the field. Lately Lindenmayer systems (L-system) models based on the growth of a *Faba bean* plant (HABA) have been designed to estimate by simulation the leaf area at the different stages during plant growth [2]. Díaz-Ambrona et al. work shows that it is possible to reproduce the canopy structure closest to real field situations using the HABA model [2]. On the other hand, many complex processes occurring in nature are believed to be organized in a selfsimilar (selfaffinity) way arising to statistically selfsimilar (selfaffinity) measures [3]. The aim of our research was to apply a multifractal analysis to the leaf area distribution (LAD) defined by HABA model to find the structure of such distribution.

The canopy model is based on physiological time named thermal time (°Cd), commonly used in agronomy. Thermal time is defined as the summation of the products of each temperature above 0 centigrade (°C) by the corresponding time, e.g. number of days (d). The simulated canopy was built with nine plants with three stems at 1700°Cd growth stage. More details can be found in Díaz-Ambrona *et al* [2]. The image was generated by the leaves projection on the field in gray tones and with a resolution of 500x500 pixels. It is obvious that the gray level in each pixel (from 0 to 255) correlates with the amount of leaf area in that pixel (see Figure 1A). For this reason, the gray level distribution is analyzed.

In this work the Chhabra and Jensen method [4], based on multipliers, has been applied to obtain the f(α) function. The up-scaling process selected was the gliding box method [5], that constructs samples by gliding a box over the grid map in all possible ways providing that the box is completely bounded by the grid map.

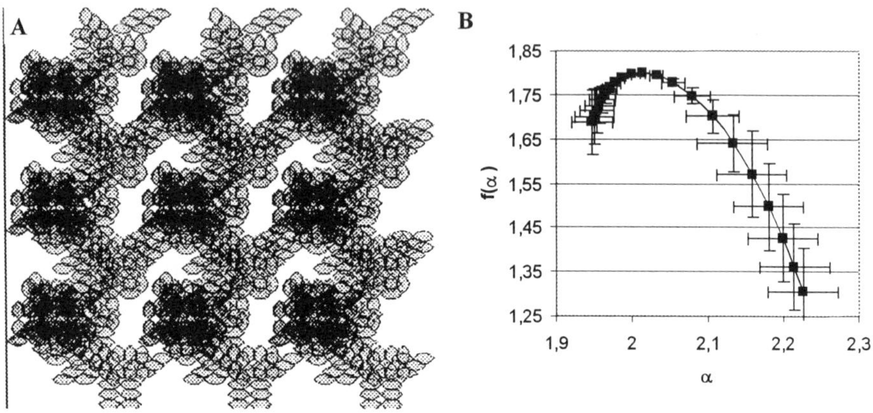

Figure 1. Simulated canopy: A) field projection of LAD, B) multifractal spectrum (bars represent s.e.).

The above figure shows a clear spatial scaling structure of the LAD at the scales considered in this work (see Figure 1B). It also suggests that such structure may arise from the L-system model applied to simulate each plant geometry.

References

1. Chen, S.G.; Ceulemans, R. and Impens, I.. A fractal-based *Populus* canopy structure model for the calculation of light interception. *Forest Ecology and Managment*, 69 (1994) pp. 97-110.
2. Díaz-Ambrona, Carlos H., Tarquis Ana M. and Mínguez M. Inés. Faba bean canopy modeling with a parametric open L-system: a comparison with the Monsi and Saeki model. *Field Crops Research* 58 (1998) pp. 1-13.
3. Solé, R.V. and Manrubia, S.C.. Are Rainforests Self-organized in a Critical State?. *J. of Theoretical Biology* 173 (1995) pp. 31-40.
4. Chhabra, A.; Jensen, R. V.. Direct determination of the f(α) singurlarity spectrum. *Phy. Rev. Lett.*, *62* (1989) pp. 1327-1330.
5. Cheng, Q.. The gliding box method for multifractal modeling. *Computers&Geosciences*, 25 (1999) pp. 1073-1079.

POPULATION CHANGE OF ARTIFICIAL LIFE CONFORMING TO A PROPAGATING RULE —GENERATION OVERLAPPING AND FRACTAL STRUCTURE CHANGE—

KENICHI KAMIJO

Department of Life Sciences, Toyo University, Itakura, Gunma, 374-0193, Japan
E-mail: kamijo@itakura.toyo.ac.jp

MASAHIDE YONEYAMA

Department of Information and Computer Sciences, Toyo University, Kawagoe, 350-8585, Japan
E-mail: yoneyama@cs.toyo.ac.jp

In order to study the intrinsic behavior of actual organisms, idealized Artificial Life (AL) ecological system has been considered on a computer. The purpose of this paper is to investigate the fractal dimension in the population change for AL entities conforming to the so-called logistic reproduction rule. Especially chaotic population change possessing the generation overlapping has been simulated by a computer. The second purpose is to find a law underlying the said chaotic population change and the population change converted by the generation overlapping. The chaotic phenomena in the population change of AL entities are noted as a major objective of this simulation. In connection with such measure, supposition is simplified as far as possible so that the generality of the measure will not be lost in this AL ecological system. The measure in the population change is assumed as a discrete change of AL entities conforming to the so-called logistic reproduction rule. In other words, the said change is a discrete process that is to be determined exclusively by the population in the previous generation. Accordingly, discussion is to be made about the chaotic problems especially lurking in the population change from the viewpoint of the complex system science.

In this paper, the macro population fluctuation of AL entities will be discussed on the supposition of the logistic reproduction rule as shown below.

$$x_n = a\, x_n\,(1 - x_n), \tag{1}$$

where x_n is population ratio for the maximum population of AL, which can inhabit in the restricted environment ($0 \leqq x_n \leqq 1$), and 'a' is the environment parameter ($0 < a \leqq 4$).

Except near the values of 3.63, 3.74, and 3.83 to 3.86 in the environment parameter 'a', the fractal dimension in both the original population change and the population change converted by the moving total can be well denoted by the polynomial of degree 6 with the environment parameter with high contribution ratio, and the polynomials of degree 6 are fitted best among the elementary functions. Then it may be possible to inverse-estimate the environment parameter using the observed fractal dimension and the polynomials of degree 6.

In case of k=1,

$$D = -6649.8a^6 + 152477a^5 - 10^6a^4 + 7*10^6a^3 - 2*10^7a^2 + 3*10^7a - 2*10^7$$
$$(R^2 = 0.9222), \tag{2}$$

and in case of k=10,

$$D = -6451.9a^6 + 146316a^5 - 10^6a^4 + 7*10^6a^3 - 2*10^7a^2 + 3*10^7a - 2*10^7$$
$$(R^2 = 0.7794), \quad (3)$$

where 'a' is the environment parameter, D is the fractal dimension and R^2 is the contribution ratio of the regressive fitting

When the overlapping conversion is carried out, the relationship between the overlap number, which means the number of the consisting populations in every generation, and the fractal dimension was investigated. It has been shown that the fractal dimension will decrease in proportion to the increase of the overlap number. Also the oscillation of fractal dimension can be seen for discrete change of the overlap number in case of a=3.6 and 3.7. Then the fractal dimension in some cases of the environment parameter can be well denoted by the polynomial of degree 6 with the overlap number with high contribution ratio, and the polynomials of degree 6 are fitted best among the elementary functions.
That is, for example,
in case of a=3.8,

$$D = 10^{-5}k^6 - 0.0004k^5 + 0.0058k^4 - 0.0426k^3 + 0.1729k^2 - 0.4164k + 1.4148$$
$$(R^2 = 0.9287), \quad (4)$$

in case of a=3.9,

$$D = -5*10^{-5}k^6 + 0.0018k^5 - 0.0218k^4 + 0.1274k^3 - 0.339k^2 + 0.2378k + 1.1649$$
$$(R^2 = 0.8074), \quad (5)$$

and in case of a=4,

$$D = -6*10^{-5}k^6 + 0.0021k^5 - 0.0254k^4 + 0.1538k^3 - 0.4509k^2 + 0.4461k + 0.9216$$
$$(R^2 = 0.9971), \quad (6)$$

where 'k' is the overlap number in the overlapping conversion.

The analytical results for AL entities in this computer simulation can be summarized as shown below.

1) The generation overlapping conversion has been applied to the chaotic population change of the AL entities in the AL ecological system on a computer. In order to investigate the fractal structure change in the chaotic population change, the Change Integration Method and the 6-Point Evaluation Method have been adopted as a fractal dimension measuring method.
2) The relationship between the generation overlap number and the fractal dimension has been investigated in the overlapping conversion. The overlap number means the number of the consisting populations in every generation. It has been shown that the fractal dimension will decrease in proportion to the increase of the overlap number. Then the fractal dimension in some cases of the environment parameter can be well denoted by the polynomial of degree 6 with the overlap number with high contribution ratio, and the polynomials of degree 6 are fitted best among the elementary functions.
3) Except near the values of 3.63, 3.74, and 3.83 to 3.86 in the environment parameter, the fractal dimension in both the original population change and the population change converted by the generation overlapping can be well denoted by the polynomial of degree 6 with the environment parameter with high contribution ratio, and the polynomials of degree 6 are fitted best among the elementary functions. Then it may be possible to inverse-estimate the environment parameter using the observed fractal dimension and the polynomials of degree 6 obtained in this paper.

DO THE MESOAMERICAN ARTISTIC AND ARCHITECTURAL WORKS HAVE FRACTAL DIMENSION?

GERARDO BURKLE-ELIZONDO

Universidad Autónoma de Zacatecas. Unidad de Postgrado II. Doctorado en Historia. Ave. Preparatoria s/n, Col. Hidráulica. CP 98060, Zacatecas, Zac. México E-mail: burklecaos@hotmail.com

RICARDO DAVID VALDEZ-CEPEDA

Universidad Autónoma Chapingo. Centro Regional Universitario Centro Norte. Apdo. Postal 196, CP 98001, Zacatecas, Zac. México E-mail: vacrida@hotmail.com

There is widely known Roman and Greek architects liked circles and golden rectangles. Also, Egyptians used the golden rectangle in art, architecture and hieroglyphics [4]. Mesoamerican art, sculptures, codex, and pyramids and urban architectural designs have a strong influence of golden measures on them [2], and Olmeca monumental heads were made under the basis of golden rectangles as harmonic units [1]. However, there is no form to know the sequence in which the lines or pictures were originally traced or drawn in all these works. This means there are no equations or temporal information useful to characterize ancient artistic and architectural works when treated as complex systems. Thus geometric analysis and mathematics used in art composition and design of buildings are not yet clearly elucidated. By this way, the Mesoamerican artistic and architectural works can be considered as static objects, and so they may be having an inherent dimension. Therefore, the fractal dimension is an experimentally accessible quantity that might be related to the aesthetic of the pattern(s) of these works. Then it would be interesting to know if the artists and architects preferences were different for groups or types of work in the ancient Mesoamerican culture. In this paper, we present a fractal analysis of several Mesoamerican artistic and architectural works, and a comparison among them taking into account different groups or types of work.

We collected 90 images of Mesoamerican artistic and architectural work by reviewing literature on archeology. From the 90 figures analyzed, 61 correspond to the Maya culture (MC) during late preclasic (300 BC. to 250 AC.), and early and late clasic (250 to 700 BC.) periods, developed at Mexican Chiapas and Yucatán states, and Guatemala and Honduras; 26 to the Aztec or Mexica culture (AC) during clasic and epiclasic periods (300 to 1100 AC.), developed at Mexican Central Highplains; two to the ancient Olmec culture (OC) developed from 1350 to 900 BC., at Mexican Veracruz, state); and one to the Toltec culture (TC), developed from 700 to 1100 AC. corresponding to the first step of Nahua civilization, at Mexican Hidalgo State.

All these 90 images were scanned using a Printer-Copier-Scanner (Hewlett Packard, Model LaserJet 1100A) and saved as bitmap (*.bmp) files in a Personal Computer (Hewlett Packard, Model Pavilion 6651). Thereafter, these images were analyzed with the program Benoit, version 1.3 [3] in order to calculate Box (D_b), Information (D_i), and Mass dimensions (D_M), and their respective standard errors and intercepts on log-log plots. It was taken under consideration that the information dimension differs from the box dimension in that it weigths more heavily boxes containing more points. In all the 90 cases a straight line was evidenced, so the three different approaches to estimate the fractal dimension works well.

For all the 90 cases the fractal dimension values were high from a $D_b = 1.803 \pm 0.023$ for the left and superior side of the 'Vase of seven gods' (MC), to a $D_M = 2.492 \pm 0.195$ for the left side of the 'Door to underworld of the Temple 11, platform' at Copán (MC).

This late case could be related to the Mayan vases are less complex than the other figures and groups from other cultures because they contain wider empty but painted rectangular or squared spaces. Certainly, there is unknown the sequence in which the lines were traced in those works having high D_M values as the left and the superior side of the 'Door to underworld of the Temple 11, platform' at Copán, which contains a lot of human like figures representing gods and ancestors but they are not concentrically distributed in a trapezoidal plane explaining its high D_M value surpassing the dimension of the plane. In this figure the traces are in fact irregularly distributed which makes really a complex composition able to fill the trapezoidal plane, and this characteristic is common to other works from the same civilization and Aztec culture such as the whole and parts of the 'Temple of foliated cross tablet' (MC); the whole center east of the 'Ball Game Tablet' at Chichen-Itzá (MC); 'Mural of the 4 Ages' at Toniná (MC); the whole 'Tablet of 96 Hieroglyphs' at Palenque (MC); 'Temple of the Cross, Door panel, Glyphs 2 and 14' (MC); 'Temple of the Sun, superior view' (AC); 'Temple of the Sun' at Palenque (MC); 'Pyramid of the Wizard' at Uxmal (MC); 'Pyramid Temple' at Tulún (MC); 'Palace of Hochob' at Tabasqueña (MC); 'Dresden Codex, page 13b' (MC); 'Borgia Codex, ritual 2, page 34' (AC); 'Aztlán Annals', page 3 (AC); 'Stela F' at Quirigua (MC); 'Stela A' at Copán (MC); 'Humboldt Disc' (AC); 'Huaquechula Disc' at Puebla (AC); 'Jaguar, portico 10, jaguars joint, zone 2' at Teotihuacan (AC); 'The Inferior Face of West Side of Chamber 1 of Murals' at Bonampak (MC); 'Mural of the battle' at Chichen-Itzá (MC); 'mayan vase with drawing of moon god with snake roll up' (MC); 'mayan vase' of Naranjo (MC); and 'disc of the Cenote sagrado' at Chichen-Itzá (MC).

A few of the circular astronomic and calendar great stones from Aztec culture, which really contain a lot of information radially distributed are well characterized by D_M values, that is, these values are similar to D_b and D_i values. Clearly, this occurs for 'Aztec Calendar' or 'Sun Stone' ($D_b = 1.92 \pm 0.005$, $D_i = 1.9 \pm 0.005$, D_M 1.901 ± 0.008); 'Tizoc Disc' ($D_b = 1.906 \pm 0.008$, $D_i = 1.882 \pm 0.004$, D_M 1.866 ± 0.008); and 'Chalco Disc' ($D_b = 1.885 \pm 0.006$, $D_i = 1.858 \pm 0.002$, D_M 1.842 ± 0.01).

The most probably correct answer to the question written in the title of this paper is yes, many of the Mesoamerican art and architectural works have fractal dimension. Meaningfully, Mesoasoamerican artistic and architectural works are characterized by a box fractal dimension $D_b = 1.912 \pm 0.009$, and/or by an information fractal dimension $D_i = 1.916 \pm 0.002$.

References

1. de la Fuente, B., Los Hombres de Piedra. Escultura Olmeca. (2nd Edition, Universidad Nacional Autónoma de México, Dirección General de Publicaciones. México, D.F. 1984). 390 p.
2. Martínez del Sobral, M., Geometría Mesoaméricana. (1st Edition, Fondo de Cultura Económica, México, D.F. 2000). 287 p.
3. TruSoft Int'l Inc. Benoit, version 1.3: Fractal Analysis System. (20437th Ave. No. 133, St. Petersburg, FL 33704, USA).
4. www.geocities.com/CapeCanaveral/Station/8228/arch.htm.

DOES RANDOMNESS IN MULTINOMIAL MEASURES IMPLY NEGATIVE DIMENSIONS?

Wei-Xing Zhou
Institute of Geophysics and Planetary Physics,
University of California, Los Angeles, CA 90095-1567, USA
E-mail:wxzhou@moho.ess.ucla.edu

Zun-Hong Yu
Institute of Clean Coal Technology, East China University of Science and Technology,
Box 272, Shanghai, 200237, PR China

Multifractal analysis has been adopted as a very useful technique in the analysis of singular measures in many fields, say fully developed turbulence, DLA and chaotic dynamical systems. Most of the physical processes are random, which has led to the sample-to-sample fluctuations of the multifractal function $f(x)$ by an amount greater than the error bars on any one sample would indicate and, in general, the negative dimensions arise in random multifractals. The randomness of multifractals at least arises in two situations where negative dimensions may emerge[1]. First, one may obtain a multifractal constructed by a multiplicative cascade that is inherently probabilistic and the negative dimensions, if exist, describe the rarely occurring events. Second, one may have to investigate the experiment in a probabilistic view. For instance, one-dimensional cuts of a deterministic measure carried by a deterministic Sierpinski sponge will inevitably introduce randomness and may be regarded as random samples of a population. Also, when one performs measurements upon turbulent flows with dual PDA or hot-wire anemometry, one-dimensional cuts are obtained, while two-dimensional cuts are obtained to describe the scalar fields when CCD and LIF technique are utilized[2].

The physical implication of negative dimensions was discussed foremostly by Cates and Witten. However, it was Mandelbrot who recognized the essential importance of negative dimensions for both physical phenomena and mathematical realizations and then studied them systematically based on turbulence and multinomial measures. Then, Evertsz and Mandelbrot applied Chernoff's theorem on large deviations to compute the tail distribution of the coarse Hölder exponent of a randomly picked interval of the multinomial measure of finite discrete multipliers and pointed out that more general Cramèr type large deviation theorems provide a full justification of the so-called thermodynamic formalism of multifractals based on the Legendre transforms. Also, with a simple but cogitative analytical example, Chhabra and Sreenivasan tried to propose a theory of negative dimensions. It is natural that different methods have been developed to extract $f(\alpha)$, such as the method of moments, histograms method, method of supersampling, canonical method, multiplier method, and gliding box method.

However, these points are intuitive heuristic interpretations rather than mathematical descriptions. It is well known that multifractal formalism involves decomposition of fractal measure into interwoven fractal sets each of which is characterized by its singularity strength α. When concerning with statistically self-similar measures, Falconer[3] proved that there is an open interval $A = (\alpha_{\min}, \alpha_{\max})$ such that, for all $\alpha \in A$, $\dim F_\alpha = f(\alpha)$ with probability one if $f(\alpha) \geq 0$. Although he said nothing in his theorem about the situation when $f(\alpha) < 0$, he gave an example where negative dimensions do appear. In a more recent work, Barral[4] improved the previous result and claimed that, with probability one, for all $\alpha \in A$, $\dim F_\alpha = f(\alpha)$. Moreover, Olsen[5] has made rigorous some aspects of Mandelbrot's intuition that negative dimensions may be explained geometrically by considering cuts of higher dimensional multifractals. The multifractal slice theorem presents a mathematical interpretation of negative dimensions and is the basis of experimental measurement of lower dimensional cuts when the ensemble measurement is difficult to be carried out. These works are the mathematical basis of discussing negative dimensions in physics.

It is an interesting remark to physicists that negative dimensions might be absent in random multinomial measures, that are employed extensively as cascade models in physics, say the well-known p-model in turbulence and random hierarchical resistor network. Define that $\overline{M} = \{i \in I : \ln m_i / \ln r_i = \alpha_{\max}\}$ and $\underline{M} = \{i \in I : \ln m_i / \ln r_i = \alpha_{\min}\}$, where m_i and r_i are multipliers and scale ratios respectively, and I is the index set. The partition equation is $\sum_{i=1}^{N} p_i w_i(q) = 1$, where $w_i(q) = m_i^q / r_i^{\tau(q)}$ and p_i is the probability of certain construction rule. There are three types of asymptotic behaviors at the endpoints of the multifractal function $f(\alpha)$. If $\sum_{i\in\overline{M}} p_i > 1$, then $0 < w_i(-\infty) < 1$ for all $i \in \overline{M}$, and hence $f(\alpha_{\max}) > 0$. If $\sum_{i\in\overline{M}} p_i = 1$, then $w_i(-\infty) = 1$ for all $i \in \overline{M}$, and hence $f(\alpha_{\max}) = 0$. If $\sum_{i\in\overline{M}} p_i < 1$, then $w_i(-\infty) > 1$ for all $i \in \overline{M}$, and hence $f(\alpha_{\max}) < 0$. Similar situation appears when $q \to \infty$. If $\sum_{i\in\underline{M}} p_i > 1$, then $0 < w_i(\infty) < 1$ for all $i \in \underline{M}$, and thus $f(\alpha_{\min}) > 0$. If $\sum_{i\in\underline{M}} p_i = 1$, then $w_i(\infty) = 1$ for all $i \in \underline{M}$, and thus $f(\alpha_{\min}) = 0$. If $\sum_{i\in\underline{M}} p_i < 1$, then $w_i(\infty) > 1$ for all $i \in \underline{M}$, and thus $f(\alpha_{\min}) < 0$. Since $f''(\alpha) < 0$ for nontrivial random multifractals, that is, the $f(\alpha)$ curve is somewhat $\cap$-shaped, negative dimensions exist if and only if $\sum_{i\in\overline{M}} p_i < 1$ or $\sum_{i\in\underline{M}} p_i < 1$, which is the so-called *latent dimensions condition*. In contrast, negative dimensions emerge automatically in continuous multifractals[2].

We thank L. Olsen and J. Barral for providing a variety of reprints and fruitful comments and suggestions. This work was supported by National Development Programming of Key and Fundamental Researches of China (No. G1999022103).

1. A.B. Chhabra and K.R. Sreenivasan, *Phys. Rev.* A **43**, 1114 (1991); *Phys. Rev. Lett.* **68**, 2762 (1992).
2. W.X. Zhou and Z.H. Yu, *Phys. Rev.* E **63**, 016202 (2001); *Physica* A **294**, 273 (2001).
3. K.J. Falconer, *J. Theor. Prob.* **7**, 681 (1994).
4. J. Barral, *J. Theor. Prob.* **13**, 1027 (2000).
5. L. Olsen, *Periodica Mathematica Hungaria* **37**, 81 (1998); *Hiroshima Math. J.* **29**, 435 (1999); *Progress in Probability* **46**, 3 (2000).

SHAPE PREDICTABLE IFS REPRESENTATIONS

LJUBIŠA M. KOCIĆ

Fac. Electronic Engineering, University of Niš
P.O.Box 73, 18000 Niš, F. R. Yugoslavia E-mail: kocic@elfak.ni.ac.yu

ALBA C. SIMONCELLI

Dip. Matematica ed Applicazioni, Università Federico II
Via Cinzia, M.te S. Angelo, 80126 Napoli, Italy E-mail: simoncel@unina.it

AMS : 28A80, 65D17
Key words: IFS, affine invariance, multidimensional fractal modeling

The idea of merging IFS techniques with classical tools of Computer Aided Geometric Design, such as *corner cutting* algorithms and *control points*, has been around for some time, yet. Methods have been proposed for the representation of parametric curves and tensor product surfaces as IFS attractors (see [1], [4], references in [2]) but, to our knowledge, not for arbitrary multidimensional attractors.

Here we deal with a general m-dimensional model, introduced formerly by the authors in [2], to treat the problem of *fractal form primitives* and of *control* over the shape of the attractor, in a linear algebra context. The AIFS model is an IFS containing only linear maps. We give results concerning the *linearization* of an arbitrary affine IFS, and state conditions on the AIFS under which the attractor possesses properties that are the basic ones in CAGD. Having to be a very short, we refer the reader to [3] for all the proofs and notation details, and for more results.

We work in $(\mathbb{R}^m, d)$ $(m \geq 2)$ with the orthonormal basis $\{\mathbf{e}_i\}_{i=1}^m$, and the subspaces $\mathbf{V} = \mathrm{aff}(\{\mathbf{e}_i\}_{i=1}^m)$ and $\mathbb{R}_\perp^{m-1}$ $(x_m = 1)$. Basic tools are the *canonical simplex* $\mathbf{E} = \mathrm{conv}(\{\mathbf{e}_i\}_{i=1}^m) \subset \mathbf{V}$ and the *standard simplex* $\mathbf{E}_0 = \mathrm{proj}_\perp \mathbf{E} = \mathrm{conv}(\{\mathrm{proj}_\perp \mathbf{e}_1, \ldots, \mathrm{proj}_\perp \mathbf{e}_{m-1}, \mathbf{e}_m\}) \subset \mathbb{R}_\perp^{m-1}$. We introduce the linear maps $\mathcal{L}_p(\mathbf{x}) = S_p^T \mathbf{x}$ (projection) and $\mathcal{L}_l = \mathcal{L}_p^{-1}$ (lifting), where S_p and $S_l = S_p^{-1}$ are feasible upper block triangular matrices, to map $\mathbf{V} \subset \mathbb{R}^m$ into $\mathbb{R}_\perp^{m-1} \subset \mathbb{R}^m$ and vice-versa, so to have $\mathbf{E}_0 = \mathcal{L}_p(\mathbf{E})$, $\mathbf{E} = \mathcal{L}_l(\mathbf{E}_0)$. By these tools we deal with the AIFS model [2](affine invariant IFS), essentially a linear IFS "working" in $(\mathbf{V}, d)$.

Definition 1 (AIFS) *With $m, n \geq 2$, let the vertices $\mathbf{T} = [\mathbf{T}_1, \ldots, \mathbf{T}_m]^T$ define a non-degenerate closed simplex in $\mathbb{R}^{m-1}$, and let $\{S_i\}_{i=1}^n$ be real square nonsingular row-stochastic matrices of order m. If the linear maps $\mathcal{L}_i(\mathbf{x}) = S_i^T \mathbf{x}$ are contractions in $(\mathbb{R}^m, d)$, the system $\Omega_\mathbf{T} = \{\mathbf{T}; \{S_i\}_{i=1}^n\}$ is a hyperbolic AIFS in $\mathbb{R}^m$.*

In this setting we can establish the following basic result.

Theorem 1 *Given any affine map of $\mathbb{R}^{m-1}$, say $w : \mathbf{x} \mapsto A\mathbf{x} + \mathbf{b}$, there is one and only one linear map of $\mathbb{R}^m$, call it $\mathcal{L} = S^T \mathbf{x}$, such that w is the orthogonal projection of $\mathcal{L}$, and S is a row stochastic matrix. More precisely, if S_w is the m-dimensional, upper triangular, block matrix having block rows $[\,A\ \mathbf{b}\,]$ and $[\,\mathbf{0}^T\ 1\,]$, the relation $S = S_p S_w^T S_l = S_p S_w^T (S_p)^{-1}$ ties up S and S_w (namely S and A, $\mathbf{b}$), and the spectra of the matrices S, A satisfy the relation $\sigma(S) = \sigma(A) \cup \{1\}$.*

At this point we can introduce the *fractal form primitives*, namely canonical attractor and standard attractor for a given set of linear maps, and their projection.

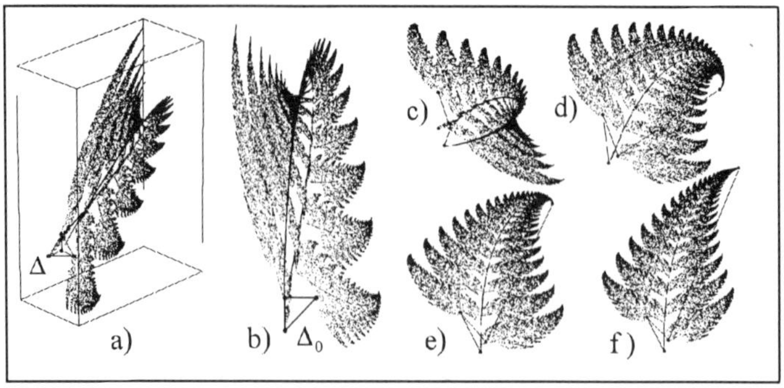

Figure 1. *a) Canonical Barnsley fern; b) Standard Barnsley fern; c) - f) Simplex controlled fern.*

Definition 2 (Canonical and standard AIFS) *Given the row stochastic matrices* $\{S_i\}_{i=1}^n$, *the IFS* $\Omega = \{\mathbf{E};\ ,\{S_i\}_{i=1}^n\} \equiv \{\mathbf{V};\ \{\mathcal{L}_i\}_{i=1}^n\}$ *is the corresponding canonical AIFS. Its attractor,* $\mathrm{att}(\Omega) \subset \mathbf{V}$, *is the canonical attractor. The corresponding standard AIFS is* $\Omega_0 = \{\mathbf{E}_0;\ \{S_i\}_{i=1}^n\} \equiv \{\mathbb{R}_\perp^{m-1};\ \{\mathcal{L}_p\mathcal{L}_i\mathcal{L}_l\}_{i=1}^n\}$, *and* $\mathrm{att}(\Omega_0) \subset \mathbb{R}_\perp^{m-1}$ *is the standard attractor. Furthermore, if* w_i *is the projection of* $\mathcal{L}_i$ *on* $\mathbb{R}^{m-1}$, $\forall i$, *then the IFS* $\Sigma = \{\mathbb{R}^{m-1};\ \{w_i\}_{i=1}^n\}$ *is called the projection of* Ω.

Lemma 1 *Given the IFS* $\Sigma = \{\mathbb{R}^{m-1};\ \{w_i\}_{i=1}^n\}$, *there is one and only one canonical AIFS* $\Omega = \{\mathbf{E}; \{S_i\}_{i=1}^n\}$ *in* $\mathbb{R}^m$ *such that* Σ *is the projection of* Ω.

Theorem 2 *Let the canonical AIFS* Ω *be given and let the IFS* Σ *be its projection. Then, if one of them is hyperbolic, the other one is hyperbolic too.*

Corollary 1 (Affine invariance) *If, in* $\mathbb{R}_\perp^{m-1}$, $\mathbf{T}$ *is a non degenerate simplex and* ϕ *is an invertible affine map such that* $\phi:\ \mathbf{E}_0 \to \mathbf{T}$, *then* $\mathrm{att}(\Omega_\mathbf{T}) = \phi(\mathrm{att}(\Omega_0))$.

Corollary 2 (Fractal dimension) *Let* $\dim(X)$ *denote the Hausdorff dimension of the set* X. *Then,* $\dim\big(\mathrm{att}(\Omega)\big) = \dim\big(\mathrm{att}(\Omega_\mathbf{T})\big)$.

Corollaries 1 and 2 guarantee controllability of the attractor (by the symplex $\mathbf{T}$) and consistency of the model with respect to fractal dimension. The tridimensional *controlled Barnsley fern* in Figure 1 illustrates the role of the simplex. For more examples, as well as for rendering algorithms and for CAGD relevant properties such as convex hull property, we must refer the reader to [2] and [3].

References

1. Kocić, Lj. M., Fractals and Bernstein polynomials. Periodica Mathematica Hungarica **33** (1996), 185–195.
2. Kocić, Lj. M., and Simoncelli, A. C., Towards Free-Form Fractal Modelling. In: Mathematical Methods for Curves and Surfaces II, M. Dæhlen, T.Lyche, and L. L. Schumaker (Ed.), pp. 287–294, Vanderbilt U.P., Nashville, 1998.
3. Kocic, Lj. M. and A. C. Simoncelli, Shape predictable IFS representations: the AIFS model and first results, Pubbl. Dip. Mat. Appl. "R. Caccioppoli" Univ. Federico II, Napoli, n.28 (2001) [Elsevier Math. Preprints Server, n.20010923]
4. Zair, C. E. , and Tosan, E., Fractal modeling using free form techniques, Computer Graphics forum **15**(1996), 3 (conference issue), 269–278

USE OF FRACTALS TO CAPTURE AND ANALYSE BIODIVERSITY IN PLANT MORPHOLOGY

A. BARI[1], A. MARTIN[2], D. BARRANCO[3]
J. L. GONZALEZ-ANDUJAR[2], G. AYAD[1] AND S. PADULOSI[1]
1. IPGRI,Via dei Tre Denari, 472/a, 00057 Maccarese, ITALY
2. Institute of Sustainable Agriculture, PO Box 4084, 14080 Cordoba, SPAIN
3. University of Cordoba, P.O. Box 3048, 14080 Cordoba, SPAIN
E-mail: a.bari@cgiar.org and E-mail: ge1mamua@uco.es

1 Introduction

Species exhibit geographical and temporal patterns of variation within boundaries set up by ecological and evolutionary processes [3]. The study of these patterns involves scoring either morphological characters or molecular markers or both. However, studies of scoring systems show that the interpretation of plant character state varies from person to person. A study conducted by Gift and Stevens [2] on morphological traits show that the researcher influences character state delimitation. They also reported that expert knowledge appeared to be of questionable value in delimiting character states. A large number of character states used in phylogenetic studies were found to represent overlapping quantitative variation. Genetic variation has also been found to exhibit chaotic, irregular patterns similar to other natural phenomena, such as population fluctuation [4]. This may make it difficult to score for traits that are continuous and fuzzy in nature, such as the surface of the stone of the olive fruit. The International Olive Oil Council (IOOC) uses stone features in the description and identification of cultivars worldwide [1]. Theses features are scored qualitatively and in this study we investigated the use of fractals to score these features and assign them quantitative values. The fractal dimension noted by D can then be used as quantitative trait to carry out further in depth analysis such as plant diversity analysis and in combination with other traits to locate maximum biodiversity areas.

2 Methods

Olives were collected from 28 cultivars. Images were taken of both the fruits and their stones. The colour images were treated in the same way and converted to black and white images to eliminate the background noises. The fractal dimension D_b was calculated using box counting in which pixel size is used to represent the lower limit of the range of scale invariance. Shannon diversity index (H'), a heterogeneity measure that includes both the richness and evenness of plant diversity, was used to investigate whether fractals D_b captured more variation in comparison to other variables. To measure its effect on the total variation general linear modeling was also performed. A program was developed to convert images to numerical values that will be used in the estimation of the fractal dimension D_v through the variogram of gray-scale variation in a follow up to this study. D_v estimator will be carried out on different resolutions to see whether it will improve discrimination between cultivars.

3 Analysis and Results

The trends of the curves of the log-log plot of the number of grid segments versus the size of the grid clearly show the fractal nature of olive stone characteristics. General linear modeling showed that D_b is able to discriminate between cultivars. Principal component analysis of the four factors that contribute most to total variation on the basis of their eigenvalues (greater than 1.0) showed that D_b accounted for 26% of total variation while and the other three variables together accounted for 67%. Having verified D_b as an accurate indicator of genetic variation, further analysis was performed to classify the total variation identified by dividing D_b into categories or groups. The total variation was divided into six groups using the mean and the variance. Then the diversity index H' was measured. Interestingly the diversity indices for D_b are higher than those of the other variables. D_b has detected more variation than other variables.

4 Discussion and future lines

Description of the surface of olive stones using fractals can complete the visual description of the stones and thus improve the characterization of genetic diversity. The deployment of fractal dimension as character can allow the capture of a large amount of variation. In this study the intercept of the log-log plot was shown to be more rigorous in identifying differences between cultivars than D_b, which is the slope, but D_b detected more variation. Both D_b and the intercept can be used to capture and detect genetic variation and thus both can be used as new characters in studies of genetic variation in olives. In a follow up to this study, another estimator D_v will be measured by estimating the best fitting line of the log transformed semi-variance function and neural networks will be used for identification purposes and pattern detection. This will be conducted by checking also on the image conversion and compression from 3D to 2D to tune data capture and analysis procedure of diversity.

5 Conclusion

Fractals captured the complexity of the surface of olive stones. Using the fractal dimension as new descriptor detected more variation than other characters. Fractals can play a major role in capturing genetic variation or expression of characters that exhibit continuous, chaotic or fuzzy patterns, which are otherwise difficult to score. This will enable better identification of patterns of genetic variation and will help in locating areas of maximum biodiversity. They can also assist in the identification of thousands of cultivars grown worldwide where the differences between them can be very delicate.

References

1. Barranco D., Cimato A., Fiorino P., Rallo L., Touzani A., Castañeda C., Serafini F., Trujillo I., World Catalog of Olive Varieties (IOOC, Madrid, 2000).
2. Gift N., Stevens P. F., Vagaries in the delimitation of character states in quantitative variation- an experimental study, *Systematic Biology* (1997) **46** pp. 112-125.
3. Maurer Brian A., Geographical Population Analysis: Tools for the Analysis of Biodiversity. (Blackwell Scientific Publications, Alden Press, Oxford, 1994)
4. May R. M., How many species, The fragile environnent (Cambridge U. Press, 1989).

THE NATURE OF THE GRAY TONES DISTRIBUTION IN SOIL IMAGES

ANA M. TARQUIS AND ANTONIO SAA

Dpto. de Matemática Aplicada a la Ingeniería Agronómica and Dpto. de Edafología. E.T.S. de Ingenieros Agrónomos, UPM. Ciudad Universitaria sn, 28040 Madrid, Spain.
E-mail: atarquis@mat.etsia.upm.es asaa@eda.etsia.upm.es

DANIEL GIMÉNEZ

Dept. of Environmental Sciences, Rutgers, The State University of New Jersey, 14 College Farm Road, New Brunswick, NJ 08901-8551, USA.
E-mail: gimenez@envsci.rutgers.edu

RICHARD PROTZ

Dept. of Land Resource Sci., Univ of Guelph, Guelph, ON, Canada N1G 2W1.
E-mail: rprotz@lrs.uoguelph.ca

M.C. DÍAZ, CHINQUINQUIRA HONTORIA AND J.M. GASCÓ

Dpto. de Edafología. E.T.S. de Ingenieros Agrónomos, UPM. Ciudad Universitaria sn, 28040 Madrid, Spain.
E-mail: mcd@eda.etsia.upm.es kirah@eda.etsia.upm.es gasco@eda.etsia.upm.es

With the recent rapid advancement in digital cameras, computer processing, physical memory, and software, complete image analysis systems can be readily built for the quantitative image analysis of soil morphology [1]. Black and white pictures of thin soil sections and preferential flow have been analyzed to get a detailed description of the geometry based on fractal dimensions [2, 3]. A very important step in determining fractal dimensions from images of soil sections is to decide on the threshold value that converts an image to a binary combination of solid and pores. Our work is focused to study the influence of the threshold value selected and the gray distribution in images of soil sections.

Three images of soil samples were selected to represent different soil void patterns (named G15-1, G16-9 and GD-21). Images had a 520x480 pixel lattice with a 256-level gray scale. Gray images were inverted (pores appearing black) and then a multifractal analysis was performed on the distribution of gray levels applying a direct determination of the $f(\alpha)$ [4]. The configuration of gray levels was related to the configuration of the pore distribution. A higher complexity in the distribution is reflected in a wider spectrum (Figure 1).

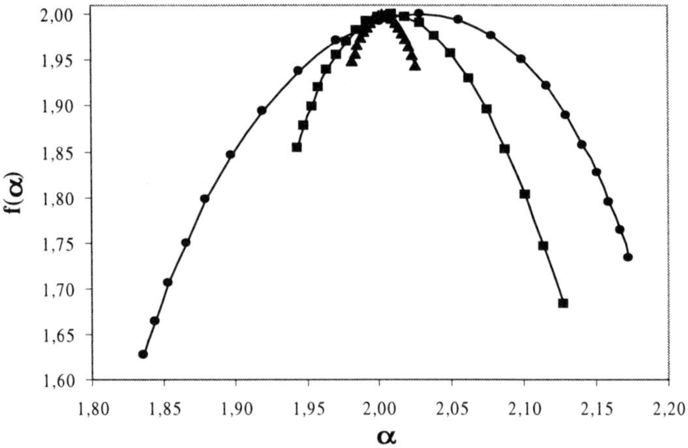

Figure 1. Multifractal spectrum for each image: G15-1 (●), G16-9 (■), GD-21 (▲).

Then the images were bi-partitioned with calculated threshold values: a) by the median of the histogram, b) by visual comparison between a derived binary image and the original gray image. With these two sets of images the configuration entropy (E(L)) and the characteristic length (L) were calculated [5]. A relation between the variation of entropy and characteristic length with the selected threshold values and the complexity observed in Figure 1 is discussed.

References

1. Protz , R. and VandenBygaart, A.J.. Towards systematic image analysis in the study of soil micromorphology. *Science Soils* 3:4. (1998) (available online at http://link.springer.de/link/service/journals/).
2. Ogawa, S.; Babeye, P. ; Boast, C.W.; Parlange, J.Y. and Steenhuis, T. Surface fractal characteristics of preferential flow patterns in field soils: evaluation and effect of image processing. *Geoderma*, 88 (1999) pp. 109-136.
3. Giménez, D.; Allmaras, R.R.; Huggins, D.R.; and Nater, E.A.. Mass, surface and fragmentation fractal dimensions of soil fragments produced by tillage. *Geoderma*, 86 (1998) pp. 261-278.
4. Chhabra, A.; Jensen, R. V.. Direct determination of the f(α) singurlarity spectrum. *Phy. Rev. Lett.*, *62* (1989) pp. 1327-1330.
5. Andraud, C.; Beghdadi, A.; Haslund, E.; Hilfer, R.; Lafait, J.; Virgin, B. Local entropy characterization of correlated random microstructures. *Physica A*, 235 (1997) pp. 307-318.

ON THE FRACTALITY OF MONTHLY MINIMUM TEMPERATURE

RICARDO DAVID VALDEZ-CEPEDA,[a] DANIEL HERNÁNDEZ-RAMÍREZ
Universidad Autónoma Chapingo. Centro Regional Universitario Centro Norte. Apdo. Postal 196. CP 98001 Zacatecas, Zac., México
[a]*E-mail: vacrida@hotmail.com*

BLANCA E. MENDOZA-ORTEGA,[b] JOSÉ VALDÉS-GALICIA, DOLORES MARAVILLA
Universidad Nacional Autónoma de México. Instituto de Geofísica. Departamento de Geomagnetismo y Exploración. CP 04510 Ciudad Universitaria, Coyoacán, México, D. F., México
[b]*E-mail: blanca@tonatiuh.igeofcu.unam.mx*

Interest in climate change has increased over the end of past century due largely to the global predictions associated with the greenhouse effect, which appear to lead to a substantial increase in planetary temperature. The continued buildup of greenhouse gases in the atmosphere will lead to a substantial increase in temperature, a rise in sea level, melting of ice caps and glaciers, and droughts in the continental interiors.

Implications of such results have led many scientists to examine the climatic records from different regions of the world in order to understand the temperature behavior. A great number of these studies have been realized by using records of over two centuries from European stations. Unfortunately, long-term monthly temperature records of over one century, or/and one and half century do not exist for many stations throughout America, particularly in Latin America.

A few studies [2, 3] have concluded that the temperature rise for mid-to-high latitude European land areas is predicted to be substantially greater than the increase expected for the planet as a whole. In these studies, investigators have considered simulation models taking into account adjusted monthly observations to a grid-box data (e. g. the Jones temperature record) or other artifact. In such a way, several procedures include quality problems as the presence of extreme values and large changes in the mean and variance [1]. However, many research workers have noted that changes in temperature variability are also important in determining the future temperature distributions.

Temporal variation of natural phenomena has been difficult of characterize and quantify. To solve these problems, fractal analysis was introduced by Mandelbrot. Time series can be characterized by a non-integer dimension (fractal dimension) when treated as random walks or self-affine profiles. Self-affine systems are often characterized by the roughness, which is defined as the fluctuation of the height over a length scale. For self-affine profiles, the roughness scales with the linear size of the surface by an exponent called the roughness or Hurst exponent. However, this exponent gives limited information about the underlying distribution of height differences [4]. There is the fact that the Hurst exponent, as well as the fractal dimension, measures how far a fractal curve is from any smooth function which one uses to approximate it. Thus, it may be instructive to know the complexity of temperature data sets when ordered as a sequence in time as well as to determine the existence of chaos.

In this context, we used the concept of fractal time series, or biased random walks, with the main aim of characterizing the long-term record of monthly extreme minimum temperature registered at the Guanajuato, México Station (2 037 masl, 21° 01' North latitude, and 101° 15' West longitude) through variography and power spectrum approaches to estimate the fractal dimension. Data was kindly provided by the National Meteorological Service of Mexico.

Data set was treated as a fractal profile to estimate the fractal dimension through variography (D_v) and power-spectral (D_s) approaches under two situations:

- Complete series, from January 1895 to December 1997 with 312 missing observations, and
- Partial series, from January, 1921 to April, 1963 without missing values.

A decreasing linear trend of 0.119°C decade^{-1} was identified, thus it was removed for the estimation of fractal dimension through the two approaches. Moreover, in the power-spectral technique we used the 'running sum' transformation to shift, by a factor of +2, the slope, and thereby the Hurst exponent and the D_S, because data trace had a slope between −1 and 1 on the log-log plot.

In both series we obtained similar values for the two types of fractal dimension meaning there are not significant effect of missing values. The estimated fractal dimensions for the partial series (508 observations) are near 1.5 ($D_v = 1.498 \pm 0.087$, $D_s = 1.486 \pm 0.155$), which means monthly extreme minimum temperature is almost equally characterized by both short-range and long-range variations.

What our finding implies is that prediction processes involving interpolated records as the adjusted monthly observations to a grid-box data (e. g. the Jones temperature record) can generate wrong results when predicting near future monthly extreme minimum temperature values because they do not take into account the possibility that the long-term records of temperature for at least several stations have a power-law spectra, which is the case for the series analyzed in this paper. Almost ten years ago, Turcotte [5] pointed out that an important application of power-law spectra is in interpolating measured data sets. Interpolation can make use of the fact that monthly extreme minimum temperature has a power-law spectrum. Interpolated data generated by this way may develop greater confidence in their capability and reliability of eventually forecasts of near future climate.

Aknowledgements

The authors are grateful for partial financial support by CONACyT grant No. 33057–T.

References

1. Balling, R.C. Jr., Vose, R.S. and Weber, G.-R., Analysis of long-term European temperature records: 1751-1995, *Clim. Res.* **10** (1998), pp. 193-200.
2. Cusbasch, U., von Storch, H., Waszkewitz, J. and Zorita, E., Estimation of climate change in Southern Europe derived from dynamical climate model output, *Clim. Res.* 7 (1996), pp. 129-149.
3. Dequé, M., Marque, P. and Jones, R.G., Simulation of climate change over Europe using a global variable resolution general circulation model, *Clim. Dyn.* **14** (1998), pp. 173-189.
4. Evertsz, C.J.G. and Berkner, K., Large deviation and self-similarity analysis of curves: DAX stock prices, *Chaos, Solitons & Fractals* **6** (1995), pp. 121-130.
5. Turcotte, D.L., *Fractals and Chaos in Geology and Geophysics* (Cambridge University Press, Cambridge, 1992).

Author Index